Novel Drug Delivery Systems

Scrivener Publishing
100 Cummings Center, Suite 541J
Beverly, MA 01915-6106

Publishers at Scrivener
Martin Scrivener (martin@scrivenerpublishing.com)
Phillip Carmical (pcarmical@scrivenerpublishing.com)

Novel Drug Delivery Systems

Fundamentals and Applications

Edited by

Mehrdad Hamidi

Pharmaceutical Nanotechnology Research Center, Zanjan University of Medical Sciences, Zanjan, Iran

and

Mahsa Sayed Tabatabaei

HIDA Pharmaceutical Technologies Development Center, HIDA Pharmaceuticals, Tehran, Iran

Scrivener Publishing

WILEY

This edition first published 2025 by John Wiley & Sons, Inc., 111 River Street, Hoboken, NJ 07030, USA and Scrivener Publishing LLC, 100 Cummings Center, Suite 541J, Beverly, MA 01915, USA

For more information about Scrivener publications please visit www.scrivenerpublishing.com.

Wiley Global Headquarters
111 River Street, Hoboken, NJ 07030, USA

For details of our global editorial offices, customer services, and more information about Wiley products visit us at www.wiley.com.

Library of Congress Cataloging-in-Publication Data

ISBN 978-1-119-85228-5

Cover image: Adobe Firefly
Cover design by Russell Richardson

Set in size of 11pt and Minion Pro by Manila Typesetting Company, Makati, Philippines

Printed in the USA

10 9 8 7 6 5 4 3 2 1

Contents

Preface

The ever-changing nature of diseases and the emergence of a series of new health complications, whether or not they are classified as diseases, have caused a big paradigm shift in drug therapy. Furthermore, vast changes in human lifestyle and economic situations have forced the pharmaceutical world to explore the development of novel, efficient and affordable drug products. Alternatively, the emergence of promising novel technologies that should improve the therapeutic performance of drugs are also a driving force in the development of new pharmaceuticals with enhanced pharmacokinetic/pharmacodynamics characteristics.

Novel Drug Delivery Systems (NDDSs) are defined as a series of attempts, approaches, scenarios, methods, and systems (e.g., drug carriers) that can deliver active drug molecules intentionally to a desired location with a desired time regimen, i.e., via spatially- and/or temporally-controlled release. In comparison to the discovery and development of new active molecules, new drug delivery systems—categorized in the pharma industry/regulatory as new dosage forms—are a remarkably inexpensive and timesaving process. There is no doubt that in the coming years, NDDSs will become increasingly prominent in different therapy categories, making the field more and more attractive worldwide.

This book presents the most recent updates in the field of NDDSs to a variety of audiences including university professors, students, researchers, policymakers, managers, industrial pharmacists, and even drug prescribers. It provides a cooperative space for more serious consideration of this relatively new class of pharmaceuticals and explores how these systems offer the safer, more efficacious, and more feasible management of emerging diseases.

The first section of the book is devoted to the general principles underlying the NDDSs, including the scientific and therapeutic basis of these systems, as well as epidemiologic and economic trends behind the development of NDDSs. It then discusses the market-oriented aspects of NDDSs including the major trends, events, and projections of NDDSs in the global market. In chapters 3 to 12, a taxonomy based on the route of entry of NDDSs is presented, which is more attractive and practical for the clinic, pharmaceutical industry, and pharmaceutical market. Finally, the book concludes with a forward-looking report that outlines the probable future directions of NDDSs.

The real futuristic scenarios for the development of NDDSs will depend on different determinants both from the market and technology sides. Nonetheless, the future of NDDSs is very promising and attractive to pharmaceutical companies that will drive the expansion of the current boundaries of the territory of NDDSs.

We hope that you find the resource informational, engaging, and enlightening. Our deepest thanks go out to Martin Scrivener and Scrivener Publishing for their assistance and the publication of this book.

Our deepest thanks go out to Martin Scrivener and Scrivener Publishing for their assistance and the publication of this book.

Mehrdad Hamidi and Mahsa Sayed Tabatabaei
October 2024

1

Introduction to Novel Drug Delivery Systems

Mehrdad Hamidi[1,2*], Mahsa Sayed Tabatabaei[2], Mohammad Moslehi[3] and Maedeh Barati[3]

[1]Pharmaceutical Nanotechnology Research Center (ZPNRC), Zanjan University of Medical Sciences, Zanjan, Iran
[2]HIDA Pharmaceutical Technologies Development Center, HIDA Pharmaceuticals, Tehran, Iran
[3]Department of Pharmaceutics and Pharmaceutical Nanotechnology, School of Pharmacy, Shahid Beheshti University of Medical Sciences, Tehran, Iran

Abstract
Conventional drug delivery systems are referred to as classical drug delivery approaches, mainly aimed at immediate drug release for rapid absorption, without making a deliberate attempt to solve the challenges ahead (e.g., repeated dosing, fluctuations, and systemic side effects) or create an added value (e.g., controlling the release rate). Such restrictions opened the arena for Novel Drug Delivery Systems (NDDSs), with major competitive advantages, to be raised. NDDSs is a broad term to describe all the *formulations, dosage forms, devices, techniques, technologies,* and *approaches* that are used intentionally to exert temporal and/or spatial control over drug delivery in the host body. It has been decades since the emergence of the NDDS concept and since then, numerous types of NDDSs have been introduced and developed. In this chapter, after a brief review of their historical background, an attempt has been made to present a comprehensive classification of what is known today as NDDSs (mainly including nanocarriers, microcarriers, cell-based carriers, and drug delivery devices). Then, following a Why-How-What Framework, the "Why" of developing NDDSs will first be discussed, looking at the positive impact they can have on each phase of pharmacokinetic phases (LADME: Liberation, Absorption, Distribution, Metabolism, and Excretion). Then, "How" to overcome the biological barriers facing NDDSs to achieve an optimal drug release at the right time in the right place, and the formulation considerations that

**Corresponding author*: hamidim@zums.ac.ir; hamidi@hidapharma.com

Mehrdad Hamidi and Mahsa Sayed Tabatabaei (eds.) Novel Drug Delivery Systems: Fundamentals and Applications, (1–34) © 2025 Scrivener Publishing LLC

should be taken into account to achieve this goal will be explained. Finally, "What" therapeutics can be potentially delivered by NDDSs, to combat the diseases, will be addressed.

***Keywords*:** Novel drug delivery systems, controlled release, nanocarriers, pharmacokinetics, drug delivery barriers

1.1 Historical Background

Conventional drug delivery encompasses classical approaches of drug administration that immediately release the drug and increase its plasma concentration, but are incapable of maintaining it in the therapeutic window, therefore, during an exponential decrease, the drug level rapidly enters the sub-therapeutic region [1]. Accordingly, conventional drug delivery faces two major challenges. First, in many cases, such a timeframe is not long enough to create the desired therapeutic effects; so, repeated dosing is inevitable. Second, they generally exhibit fluctuations; that is, they either step beyond the toxic level or fall below the therapeutic limit [2]. Moreover, conventional drug delivery suffers from a lack of target specificity, premature metabolism/excretion, and inadequate bioavailability [1]. These limitations have led to an upsurge of research in the last decades on novel drug delivery systems (generally referred to as NDDSs) to improve the therapeutic efficacy and safety of drugs.

The emergence of NDDSs dates back to the mid-1960s when the idea of "an implanted, zero-order, reservoir drug delivery device" was sparked by Folkman and Long [3]. From then until the late 1970s, the basics of controlled drug delivery namely drug release mechanisms were established. This led to the development of the first generation of NDDSs including once/twice-a-day oral delivery systems and once-a-day/week transdermal patches [4].

In the second generation of NDDSs (from around the 1980s to the 2010s), concepts such as zero-order drug release, smart polymers, and peptide/protein delivery using biodegradable depots attracted more and more attention [4]. Consequently, different zero-order–controlled drug delivery devices (such as ophthalmic inserts, transdermal patches, and osmotic pumps), swelling/gelling hydrogel systems, and polymer matrices for gradual delivery were developed. This period during which macroscopic drug delivery systems were developed is known as the "macro era". In parallel with these advancements, thanks to the development of biodegradable polymers, macroscopic zero-order systems evolved into microscopic sustained-release ones (the so-called "micro era") [5].

In the mid to late 1970s, polymer-drug conjugates arose based on three fundamental concepts of PEGylation, active targeting, and passive targeting. Later, the liposomes and polymeric micelles (previously introduced in the 1960s and 1970s, respectively [6]), were PEGylated, thus taking an important step in the field of nanoscale carriers. In the last decade of the second generation, the expansion of nanosystems ushered in a modern age in drug delivery, which was the "nano era" of targeted drug delivery systems [5]. The nanoscopic era brings to life the concept coined by Ehrlich a century ago, called "Magic Bullet" (selective targeting of a pathogen without harming the host organism) [7].

Starting around the 2010s and still ongoing, the third generation of NDDSs covers a wide range of recent developments, including modulated, long-term (6–12 months), self-regulated, and targeted drug delivery systems [5, 7]. Also, today, significant attention has been drawn to biologically precise and controlled delivery systems with more biological and less material-oriented properties—the so-called personalized medicine [8]. The drive to achieve better NDDSs and commercialize them has continued unabated to this day and will continue to do so.

1.2 Definition, Terminology, and Taxonomy

The general term "novel drug delivery systems" is regularly referred to as the *approaches, methods, formulations, dosage forms, devices, techniques,* and *technologies* that are, collectively, known as the *systems* intended to exert some kind of *time* and/or *spatial control on the drug delivery throughout the host body.*

Based on the mode of control exerted on the drug release, the technology/approach used, and the route of drug entry to the host body, a series of terms have been used interchangeably to describe the different NDDSs. Although there is no distinct separation between the application of the popular terms used to indicate the different types of NDDSs, some of them are used more frequently for per-oral NDDSs, including *sustained-release, extended-release, controlled-release, modified-release, delayed-release,* and *retard.* On the other side, a variety of terms, such as *long-acting, prolonged-release,* and *long-action,* have been used in recent years for injectable NDDSs. The important fact behind the application of a variety of attributes to assign to the NDDSs upon being launched to the pharma market is using the opportunity of having a new "surprising" name or, otherwise, attaching a new product to an existing successful product platform already in the market.

Different aspects of NNDSs have been used in an attempt to classify these products, the most popular of which is the route of administration. In Figure 1.1, a summary of different classification systems found in the literature for NDDSs is shown as an infographic. Figure 1.2 illustrates different types of NDDSs.

1.3 Novel Drug Delivery Systems: Why, How, and What to Deliver?

NDDSs undertake an intermediary role between the drug and the patient to not only mitigate the side effects of the pharmaceutical agents but also elicit efficacy, safety, and patient compliance to treatment [9, 10]. NDDSs can be considered from three perspectives. First, "why delivery?" the answer to which, in a word, is to improve the pharmacokinetic and pharmacodynamic profiles of drugs. Second, "how to deliver?" that is, how to overcome biological barriers and, then, attain a desirable drug release at the right timeframe in the right place *via* optimizing the design of drug carriers; and third, "what to deliver?" which includes the therapeutics that are intended to reach the site of action within the host body to combat the disease state(s) [9]. In the following, an attempt will be made to briefly address these issues.

1.3.1 Why Delivery?

An ideal drug delivery system can not only protect the drug from physical and chemical destabilizing agents but also deliver it to the right place within the right period [9]. The major parameters determining the appropriate dose and administration route of a drug to achieve the desired therapeutic response are a sequence of processes collectively known as the "LADMER" system [11]. It is a six-letter acronym that stands for liberation, absorption, distribution, metabolism, excretion, and response (see Figure 1.3). Obviously, the outcome of a drug undergoing the LADMER system depends on a complex interrelationship network between the drug, drug product, and the host body [12]. Following, it is discussed how NDDSs affect each LADMER process.

1.3.1.1 Liberation

Liberation is the process by which the drug is released from its administered form. Except for intravascular and peroral true solutions, liberation

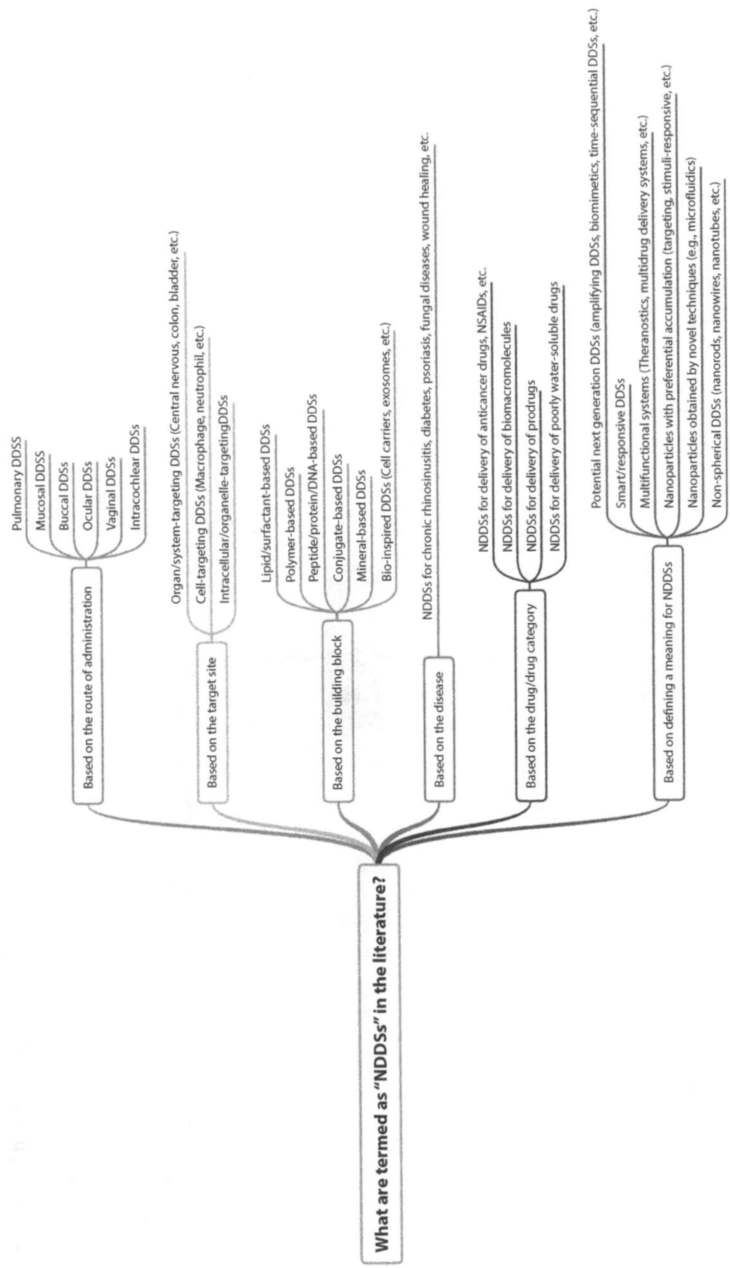

Figure 1.1 A summary of different classification systems for NDDSs.

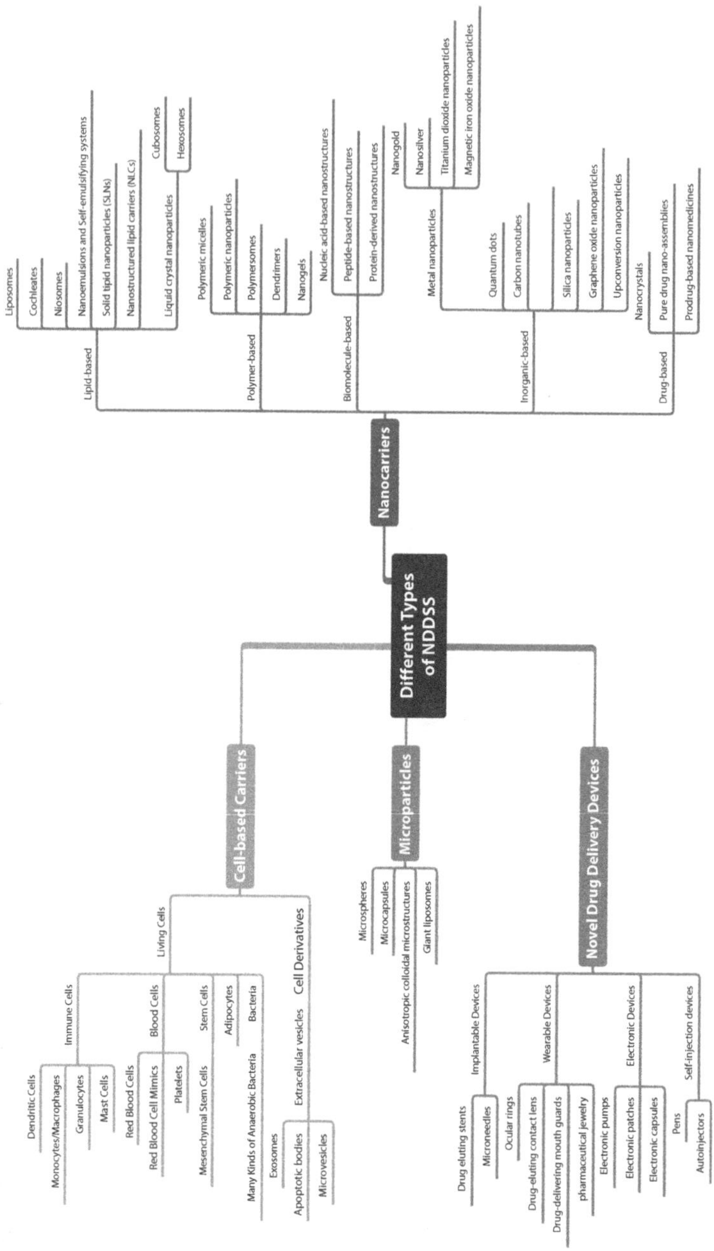

Figure 1.2 Some types of NDDSs.

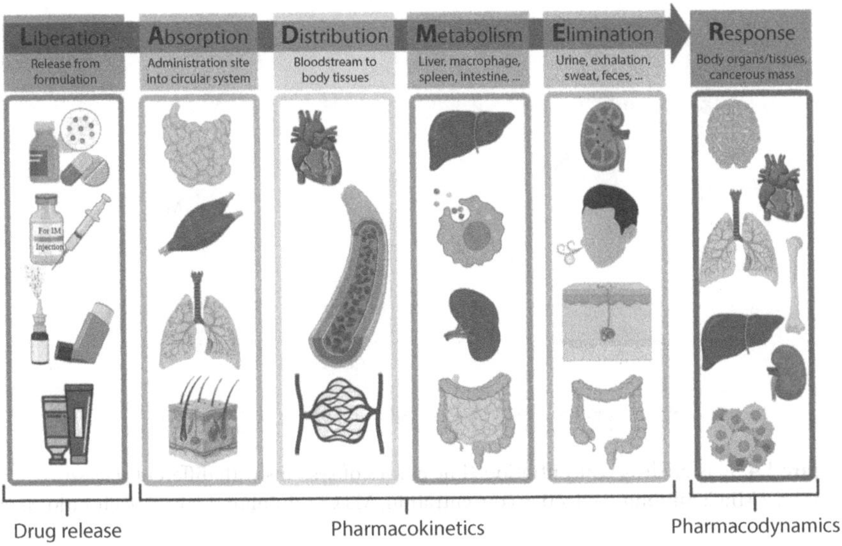

Figure 1.3 The steps of drug disposition in the body known by the acronym "LADMER" (Created with BioRender.com).

is considered the first step for all drug products with any administration route. Liberation potentially affects the onset of action, absorption rate, and bioavailability and is controlled by the features of the designed delivery system [12]. Conventional drug products follow an immediate-release profile, which generally leads to a quite rapid drug absorption and subsequent rapid onset and termination of the associated pharmacodynamic effects as well as concentration fluctuations. On the contrary, NDDSs aim to deliberately take control of the release to achieve a better therapeutic outcome and/or patient compliance with minimized fluctuations (Figure 1.4). These systems are generally known as the modified-release (or synonymously, controlled-release, sustained-release, extended-release, prolonged-release, and long-acting) products that endeavor to alter the timing of drug release [13]. It is also possible to improve the solubility and dissolution rate of the drug using solubility enhancement methods (e.g., solid dispersions [14], nanocrystals [15], etc.).

1.3.1.2 Absorption

To attain an adequate blood concentration and eventually the intended pharmacodynamic effect(s), a drug should be absorbed at an optimal rate via the desired site within the host body [11]. NDDSs can improve the drug absorption using two approaches [9, 16]:

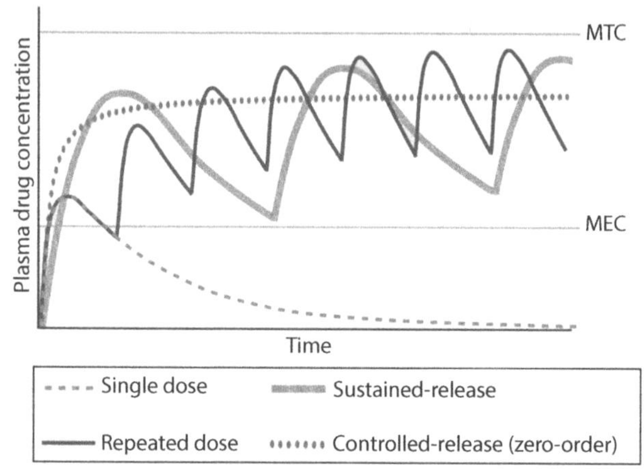

Figure 1.4 Plasma drug concentration-time curves of systems with different release profiles. (MEC: minimum effective concentration, MTC: minimum toxic concentration).

(i) Improvement of bioavailability by changing the solubility, dissolution rate, permeability, or residence time of a drug molecule by adopting different physical or chemical methods. For instance, the down-sizing of a drug crystal (e.g., through micronization and nanomization) enhances the solubility. Nanomization, in particular, can also improve the dissolution rate of drugs [17]. There are numerous physical [18] and chemical [19] approaches (like using microneedles and efflux inhibition, respectively) to enhance the permeability of drugs through different physiological barriers. Mucoadhesive formulations can prolong the residence time and thereby, provide sufficient opportunity for drug absorption [20]. Prodrugs can enhance the bioavailability of drugs by improvement of passive and or transport-mediated absorption [21, 22]. Also, they may circumvent efflux transporters [23].

(ii) Suggesting a new administration route to bypass the absorption barriers. For instance, some types of NDDSs improve the bioavailability of drugs by providing the possibility of nose-to-brain delivery as a potential alternative approach to accessing the brain without the hindrance of the blood-brain barrier (BBB) [24, 25]. Lymphatic targeted drug delivery is another example of the ability of

NDDSs to change the route of drug administration and to improve drug absorption, thereby the effectiveness of treatment [26].

1.3.1.3 Distribution

Drug distribution refers to the reversible movement of drugs from the bloodstream to the tissues [27]. This process is affected by the characteristics of the drug molecule/delivery system as well as the host body [28]. NDDSs can potentially modify the distribution pattern of drugs via two major strategies:

(i) Structural modifications of the drug molecule. For instance, prodrugs may demonstrate a significantly different distribution pattern than the parent drug due to their different physicochemical properties [29].
(ii) Adjustment of some drug/carrier properties, other than the molecular structure of the drug, based on a strategy intended for passive or active drug targeting. Passive targeting relies on some basic physical properties of the carrier, mainly the size, which favors the drug distribution to a particular location within the host body. Enhanced permeation and retention (EPR) is a typical example of this strategy which is based on the passing of a narrow-sized carrier population through the 'leaky vessels' of the defected (cancer and inflamed) tissues, thereby drug molecules accumulate in target tissue to a higher extent than they do in healthy tissues [30, 31]. EPR is applicable both for macromolecular therapeutics and small molecule drugs via encapsulation within the nanocarriers. Active targeting, on the other hand, is based on the differentiation between healthy and affected tissues by the drug carrier via some kind of "intelligence," thus leading to selective accumulation in the target site(s). Active targeting is generally realized by the surface decoration of nanocarriers with special ligands to bind to the receptors on the target tissue [30]. Also, it is possible to conjugate tumor-specific ligands to cytotoxic drugs to provide prodrugs with high selectivity [32].

1.3.1.4 Metabolism

NDDSs can change the pattern of drug metabolism and clearance in various ways, including changing drug transport pathways in the body, decreasing the rate of drug-metabolizing enzyme reactions, bypassing efflux transporters, and delaying drug release (hence, reducing the drug presentation rate to the hepatic or other kinds of enzymes) [9]. By the simplest definition, metabolism or biotransformation is the enzymatic modification(s) of different chemicals *in vivo* [33]. Biotransformation often leads to inactivation or detoxification. However, it may sometimes provide active metabolites, like what happens with prodrugs [34]. The logic of prodrug design is to take advantage of an intentional metabolism, which may occur in various organs/tissues depending on the design model [35]. If the prodrug has optimal metabolic stability, it is treated by the body as a new chemical until it enters the metabolism phase.

Encapsulation within the nanocarriers also influences the drug metabolism pathway. Immediately after entering the blood circulation, nanocarriers are sequestered by the mononuclear phagocytic system (MPS)—a part of the immune system that consists of phagocytic cells, in particular resident macrophages in the liver, spleen, and lymph nodes [36, 37]. Macrophages engulf nanoparticles with adsorbed opsonins (serum proteins) on their surface [38]. Ideally, the nanocarriers should be physicochemically stable in order to avoid the rapid disintegration upon contact with blood components, or premature release of their payload. Also, they must be able to escape the reticuloendothelial system (generally known as RES), i.e., the natural defense system of the host body against the foreign particulate "newcomers" [39]. How nanocarriers interact with the MPS provides a major indicator of their blood circulation half-lives and their clearance rate. Inadequate circulation half-life usually leads to reduced uptake and efficiency of the therapeutics [38].

MPS rapidly recognizes too hydrophobic and charged particles. So, the surface of nanocarriers is preferred to be hydrophilic with a neutral/slight anionic charge. To achieve this, water-soluble polymers (e.g., PEGs) have been widely studied to be conjugated on the surface of nanocarriers [30]. Cloaking of nanocarriers with cell membranes (biomimetic coating) is an alternative approach that not only enhances the longevity of carriers in the bloodstream but can also improve their targetability [40, 41]. Another recently noticed strategy to improve blood circulation is red blood cell (RBC)-hitchhiking, where nanocarriers are adsorbed (non-covalently) on the red blood cells (RBCs) surfaces; then, the associate is injected to reach the downstream organs [42–44]. Alternatively, plasma proteins can undertake the role of carriers. Recently, albumin-based prodrugs or albumin-bound nanocarriers have been developed to prolong the circulation half-lives in a way to improve the antitumor activity of anticancer drugs [45, 46]. In this regard, albumin could serve as a long-circulating carrier to enhance their accumulation through the EPR effect [47]. Engineering the mechanical and morphological characteristics of the nanocarriers is a straightforward strategy to keep them in circulation for an adequately long time [48].

Instead of regulating the drug-body interactions, it is also possible to control drug clearance using sustained-release systems. For instance, in the case of susceptible drugs that rapidly undergo metabolism or clearance, implantable depot systems provide an excellent platform to continuously release the drug for a long time [9].

1.3.1.5 Excretion

Drug excretion refers to several pathways that collectively remove the drug and/or its metabolite(s) from the body [49]. The time it takes for a drug to be excreted should be carefully considered; not so short that there is no opportunity for efficacy, and not so long that it jeopardizes safety [39, 50]. There are various routes of drug excretion including urine (kidneys), feces (liver), saliva, sweat, respiration, breast milk, and so on [49]. In the case of nanocarriers, the liver and spleen undertake a more prominent role. The route of administration has a key role in determining which organs contribute the most to accumulation and elimination [51]. Nanocarrier-dependent factors are also of crucial importance. Zhang *et al.* [52] have proposed a flowchart demonstrating what excretory path a nanoparticle takes based on its properties. Accordingly, how nanoparticles are disposed of depends on three main factors including (i) size, (ii) degradability, and (iii) their interaction with the liver [51].

Prodrugs may be eliminated by renal or biliary excretion mechanisms. Briefly, the bioavailable fraction of the prodrug is metabolized and activated as the parent drug. The formation of metabolite(s) other than the drug is also possible, which will subsequently be excreted. The drug may be further metabolized to a daughter metabolite or directly excreted in the urine. Depending on the circumstances, a prodrug may show a longer or shorter elimination half-life than its parent molecule [53]. Generally speaking, drug delivery systems can be engineered to be excreted in an expected pattern. Figure 1.5 represents some novel systems/strategies employed to improve the performance of the drug in each of the pharmacokinetic phases.

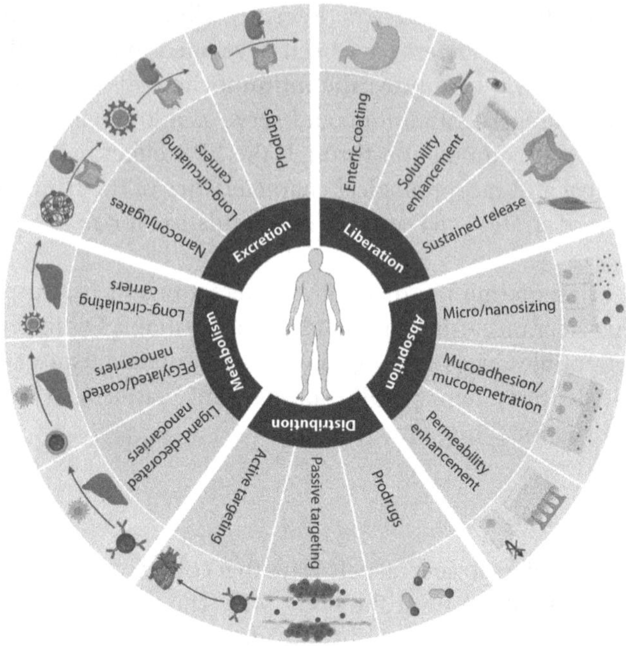

Figure 1.5 By affecting which phase of LADME, NDDSs improve pharmacokinetics? (Created with BioRender.com).

1.3.2 How to Deliver?

1.3.2.1 Drug Delivery Barriers

Being constantly exposed to various types of xenobiotics from different origins and via different routes (e.g., skin contact, ingestion, inhalation, etc.), the body is equipped with highly efficient barriers at the cell, tissue, organ, and system levels [54]. As an external substance, the drug is no exception to this rule. Therefore, drugs have to travel a long and winding road, as shown in Figure 1.6. Which barriers and to what extent they play a role varies depending on the route of drug administration, characteristics of the drug molecule/delivery system, and the disease state.

1.3.2.2 Overcoming Strategies

Three main scenarios assumed for NDDSs to overcome the drug delivery barriers, are discussed below.

Disruption of Barriers

In both epithelium and endothelium, tight junctions form a tremendous barrier to paracellular drug absorption [55]. Tight junction modulation

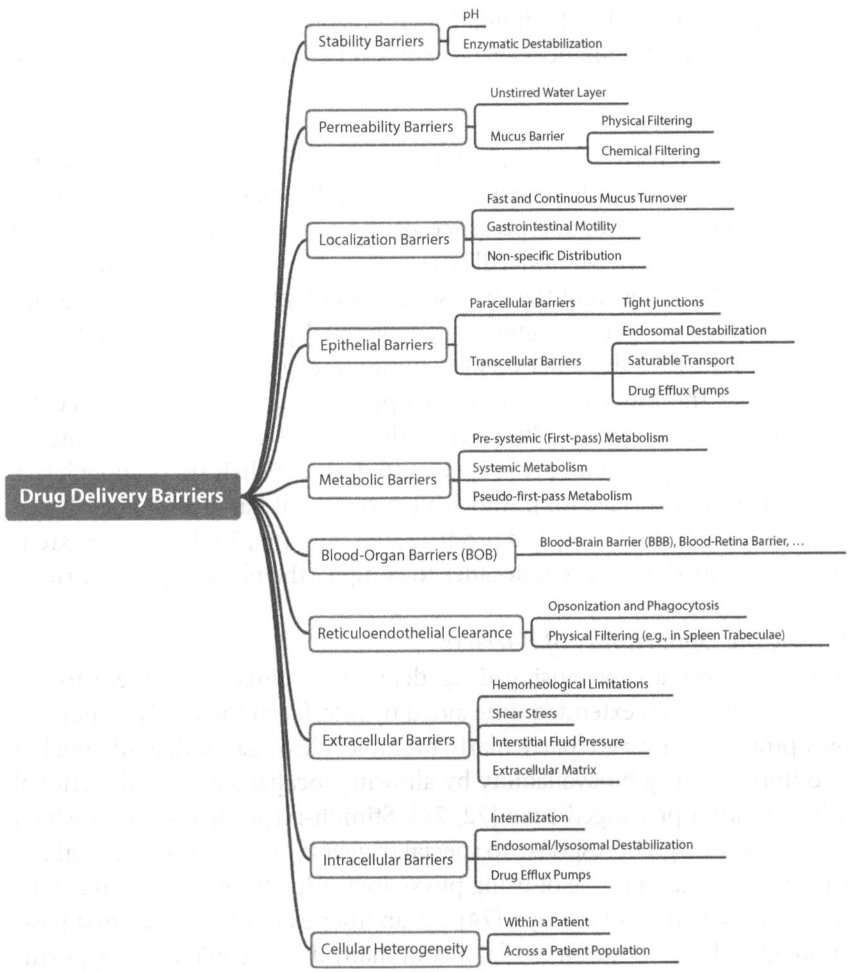

Figure 1.6 Drug delivery barriers [9, 34, 52, 75].

(using chelating agents, cationic polymers, toxins, etc.) is an attempt to address this issue [56]. However, despite their efficiency, the commercialization of tight junction modulators is hampered by safety concerns [55]. In the case of oral drug delivery, one of the major factors that lead to inadequate absorption is drug efflux. Therefore, suppressing the drug efflux pumps (using different excipients and inhibitors) or reducing their function (silencing transporters, for instance) is of great interest to improve the oral bioavailability of the drug [57]. Another example of impairing barrier performance is temporal disruption of the stratum corneum to improve percutaneous drug delivery. In this regard, different chemical (e.g., surfactants, solvents, lipids [58]) and physical (e.g., iontophoresis, sonophoresis [59]) enhancers have been studied with more or less success.

Bypassing Barriers

Delayed drug release that is commonly realized through enteric-coating is the best-known example of this category that bypasses the harsh stomach environment thanks to the presence of the gastro-resistant pH-activated polymeric layer [60]. Microneedles are other drug delivery systems whose rationale is to circumvent the successive physiological barriers [61]. Drugs administered *via* special routes (e.g., sublingual [62], lymphatic [63], and nose-to-brain [64] drug delivery) naturally escape first-pass metabolism. To conquer the mucus barrier, mucus-penetrating carriers have recently emerged that could freely diffuse across the mucus (e.g., in the gastrointestinal tract [65], vagina [66], eye [67], lung [68]) due to their small enough size and muco-inert surface [67]. Above all, a variety of nanotechnology-based drug delivery systems [54] and prodrug strategies [69, 70] have been extensively developed over recent decades to conquer the physiological barriers.

Seizing the Opportunity of Barriers

Mucoadhesives are promising drug delivery systems that adhere to the mucus layer for an extended time aided by interfacial forces [71]. They not only protect their cargo (particularly, biomolecules) against degradation but also improve drug bioavailability by allowing localized and unidirectional diffusion for a prolonged time [72, 73]. Stimuli-responsive systems which release their cargo in response to special triggers (pH, enzymes, microflora, etc.) are a prime example of using physiological/pathophysiological conditions for the benefit of therapy [74]. As another example, hepatic first pass, although a threat to the bioavailability of many drugs, could be an opportunity for liver-targeted prodrugs [75] and nanoparticles [76].

1.3.2.3 Designing Criteria of NDDSs

Thanks to the advances in different scientific fields such as material sciences (e.g., biomaterials), manufacturing techniques (e.g., 3D printing), and computational tools (e.g., artificial intelligence) along with learning from nature (e.g., bio-inspired design), it has been possible for pharmaceutical researchers to precisely design and manipulate the characteristics of drug delivery systems. The first step to achieving such systems is the logical selection of a drug molecule and its appropriate delivery system. Here, we briefly deal with the drug- and delivery system-related considerations that should be considered.

Drug-related considerations

Figure 1.7 highlights the most important considerations when selecting a drug molecule to be delivered by controlled-release systems. Also, the inclusion and exclusion criteria for choosing optimal molecules are listed. Accordingly, drug molecules with high molecular weights (e.g., proteins and peptides), inadequate solubility (e.g., BCS classes III and IV), excessive lipophilicity/hydrophilicity, and those unstable in the biological environments (e.g., biologicals) are poorly favorable candidates for controlled-release systems. On the contrary, small molecules, non-ionized drugs, and those whose absorption mechanism is not mediated to carriers are generally considered optimal candidates to be formulated as controlled-release systems [77]. More details are provided in Figure 1.7.

Many drugs in themselves lack the inclusion criteria listed in Figure 1.7. A drug with poor absorption, rapid metabolism, and/or excretion will require a higher dose to attain its therapeutic efficiency; which either leads to side effects and toxicity or is not practical. In these cases, the pharmaceutical attributes of the drug molecule should be modified through chemical modification or formulation (e.g., encapsulation) [78].

Structural modification of a drug molecule to reach a prodrug is considered an efficient tool to improve the physicochemical, pharmacokinetic, and pharmacodynamic properties of the parent drug. This goal is usually achieved by covalent conjugation of hydrophilic moieties (to improve solubility) or lipophilic functionalities (to enhance permeability) to the parent drug. In both cases, absorption and, thereby, bioavailability are positively affected [79]. For BCS classes II and IV, the prodrug increases the bioavailability through the enhancement of water solubility. In the case of BCS class III drugs, the prodrug strategy improves their permeability and may alter their position in BCS class II [80]. Additionally, prodrugs can circumvent efflux transporters [23]. The classical prodrug approach is beneficial for improving permeability through passive diffusion. More recently, however, thanks to the expansion of research on transporters and enzymes, a novel approach called "targeted-prodrug" has emerged that focuses on increasing absorption through transporters [79]. Unlike the BCS I molecules, whose permeation is mainly through passive diffusion, intestinal transporters possess a considerable contribution to the transportation of BCS II-IV drugs [81]. Consequently, targeted-prodrugs are prone to significantly improve the oral bioavailability of drugs via the enhancement of carrier-mediated transportation.

The AUC of a prodrug may differ significantly from that of the drug [29]. This is simply justified given that the volume of distribution (V_d) is strongly dependent on the physicochemical properties of the molecule.

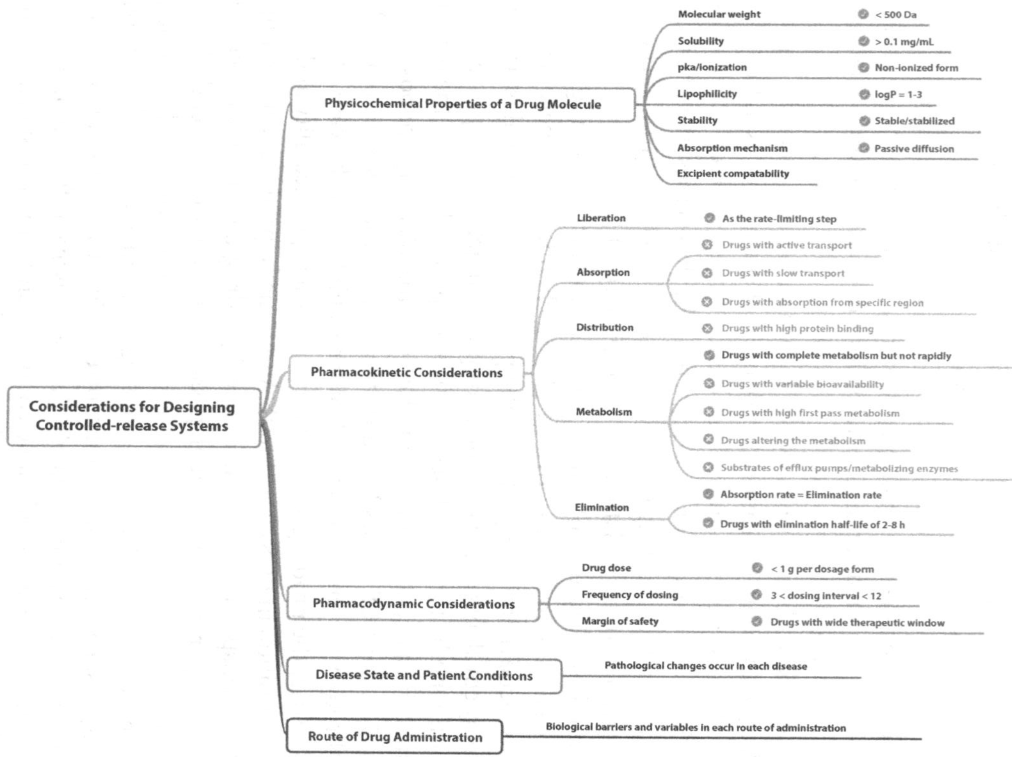

Figure 1.7 Considerations to be observed for designing controlled-release drug delivery systems [75, 97–99] (the inclusion and exclusion criteria of drug molecules are marked with a green tick and an orange cross, respectively).

For instance, basic drugs often have a V_d of more than 100 L, while this value is approximately 10~20 L for their acidic metabolites [53]. Additionally, targeted prodrugs may alter the tissue distribution pattern and decrease the toxicity of the drug due to their increased selectivity. Besides, prodrugs can be designed to avoid pre-systemic metabolism (whether in the intestinal membrane or the liver). So, they can pass the metabolic barriers intact until they are hydrolyzed in the blood circulation [82]. Additionally, prodrug is an intelligent approach for protecting drugs against enzymatic degradation [77]. They may also enable a new administration route. Such properties make structural modification a powerful strategy to overcome barriers posed by the body. How prodrugs improve the therapeutic efficiency of drugs is shown in Figure 1.8.

Formulation-related considerations

As mentioned in previous sections, a common strategy to achieve the desired safety and efficacy using drugs with inappropriate physicochemical and pharmacokinetic properties is to formulate them as a dosage form representing the intended attributes. Considerations related to the formulations used via each route of administration will be discussed in their respective chapters. Here, considering the wide coverage and importance of nanocarriers, considerations relevant to nanotechnology-based delivery systems will be reviewed with special attention to their pharmacokinetics.

During the tortuous journey from the site of administration to the target site, only a small fraction of nanocarriers is fortunate enough to reach their destination. Today, grasping the physiological environment and how it interacts with the carriers, the path has been paved to achieve more efficient nano-based drug delivery systems. As shown in Figure 1.9, the main parameters to be optimized to achieve ideal nanoformulations are their type (material composition), physical/mechanical properties, and surface characteristics. Of those, size, morphology, rigidity, charge, and surface chemistry are among the most important parameters affecting the biofate of nanocarriers. In the following, how these parameters impact each pharmacokinetic process is discussed.

Effect of Composition

Ideally, the nanoparticles should be physicochemically stable so that they do not disintegrate rapidly upon contact with blood components, or release their payload prematurely. Also, they must be able to escape the MPS. Eventually, they should be removed from the body after a while [39].

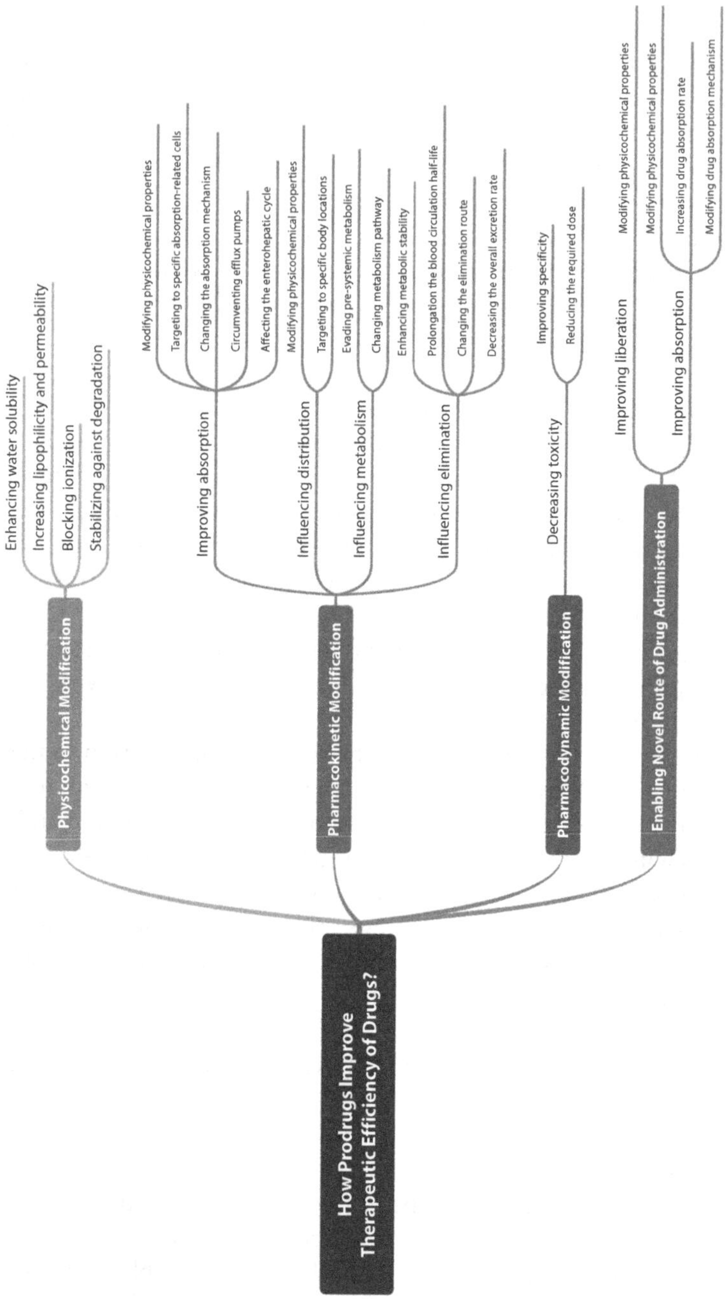

Figure 1.8 How prodrugs improve therapeutic efficiency of drugs.

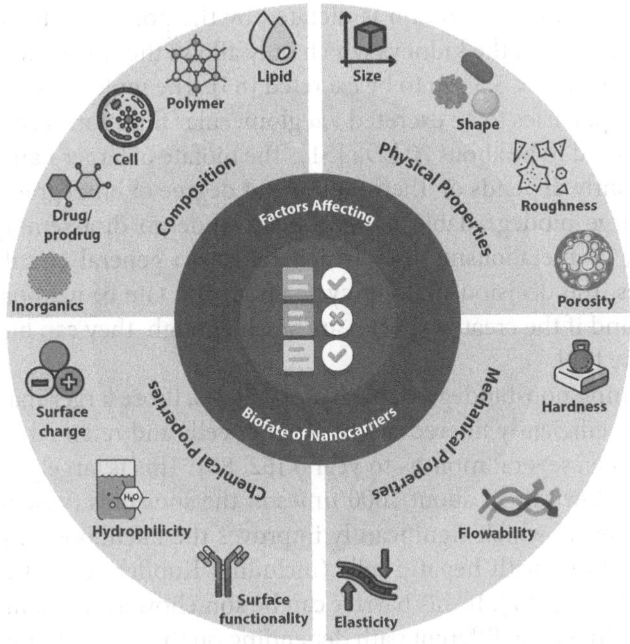

Figure 1.9 Considerations to be observed in design of nanocarriers.

Nanoparticles whose excretion is carried out within a certain period are favored by the FDA for biomedical applications [83].

The materials from which nanoparticles are made significantly affect their *in vivo* retention. In particular, whether nanocarriers are degradable and decomposable or not, determines their elimination pathway. It has been found that nanocarriers constructed from biodegradable materials or those with renal clearable characteristics are more fortunate for clinical translation. Contrarily, non-degradable nanocarriers accumulate in the body, so they are more likely to show toxicity. Most inorganic nanoparticles are neither biodegradable nor easily degraded under bio-relevant conditions. Therefore, their use, especially for the long term, raises concerns about their toxicity. Generally, organic-based nanocarriers are preferable due to being easily clearable from the host body and not creating cytotoxic degradation products. Recently, biodegradable organic-inorganic hybrid nanoparticles have been raised to combine the properties of inorganic moiety and biodegradation characteristics of the organic portion [84].

There are two main mechanisms for nanoparticles to be cleared from the blood and eliminated from the body including (i) renal and (ii) hepatobiliary elimination pathways [52]. A study on quantum dots has shown that efficient urinary excretion is dictated by the pore size cut-off of glomerular filtration in the kidneys which only allows the small nanoparticles with a diameter of < 5.5 nm to be excreted [85]. The molecular weight cut-off for nanoparticles to be excreted *via* glomerular filtration is conventionally considered to be about 70 kDa [51]. The biofate of larger nanoparticles predominantly depends on their nature and degree of biodegradability. In general, large biodegradable nanoparticles undergo disassembly, breaking down, or metabolism; after which return to general circulation. In these cases, the decision-making cycle about the fate of nanoparticles is repeated; and if the created species are small enough, they can be excreted in the urine [52].

Concerning non-biodegradable nanoparticles, those larger than around 200 nm are efficiently filtered out by Kupffer cells and retained there for a long time (i.e., several months to years) [52, 86]. This is largely due to the slow blood flow rate of about 1000 times in the sinusoids (vascular channels of the liver) which significantly improves the likelihood of nanoparticle interactions with hepatic cells (including Kupffer cells, B cells, and endothelial cells) [87]. If this barrier can be somehow avoided, nanoparticles will again have a different path depending on their size. Particles larger than liver sinusoidal endothelial cell fenestrae having a diameter of about 100 to 200 nm, will not be able to reach the hepatocytes; so, they inevitably re-enter the systemic circulation [52] and RES (e.g., spleen and lymph nodes) is responsible for their elimination [51]. Smaller nanoparticles extravasate through the fenestrae to enter the space of Disse—a location between a sinusoid and a hepatocyte. Thereafter, they are taken up by the hepatocytes; afterward, enter the bile duct and are excreted in the feces [52].

Effect of Size

There is an inverse relationship between the size of nanoparticles and their cellular uptake. In addition, size plays an important role in determining whether nanoparticles are specifically absorbed by the hepatic portal system or the lymphatic pathway; the larger the particle size, the greater the preference for lymph [88]. The average optimal size range of nanoparticles to be absorbed from the gastrointestinal tract is about 20 to 100 nm from enterocytes and 100 to 500 nm from M cells [54]. Larger nanoparticles could be taken up by M cells, but they will be entrapped within the Peyer's patches [88]. Besides, particle size specifies by which clearance system and how fast the elimination takes place. As a general rule, nanoparticles with a size of 5 to 200 nm are more fortunate to have extended blood circulation (leading to an improved propensity for extravasation). This is because those smaller than 5 nm are quickly excreted by the renal system. On the other hand, resident macrophages of the MPS rapidly clear the blood from particles larger than 200 nm [83]. Consequently, nanoparticles are non-specifically distributed in healthy organs; in particular, they highly accumulate in the liver (fenestrations size range of 50–100 nm) and spleen (with fenestrations of 200–500 nm). Lung capillaries are capable of accumulating micrometric particles with a range of 2 to 5 μm [36]. Except for the listed clearance organs, diffusion through the other healthy blood vessels is only allowed for small molecules. This rule, however, is violated in the inflammatory tissues and tumor vasculature (with fenestrations measuring 380–780 nm), where even particles as large as 200 nm are capable of extravasation and being trapped through the EPR effect—an exceptional opportunity for macromolecules and nanoparticles [31, 36, 83, 89]. As previously mentioned, the EPR effect is a passive targeting mechanism that makes accumulation possible by offering increased vascular permeability (arising from fenestrated leaky vessels) and prolonged retention (due to impaired lymphatic drainage) [31, 90].

Effect of Shape

Rod-shaped nanoparticles generally outperform the spherical ones; since they show higher cellular uptake and transcytosis, more efficient internalization, longer retention time in the gastrointestinal tract, better penetration into the interstitial space of villi, and greater lymphatic transport compared with the sphere nanoparticles [54, 91, 92]. This is mainly attributed to the higher chance of rod-shaped particles to contact the cell membrane [93]. Generally, the internalization rate of non-spherical particles with a high aspect ratio is faster than spherical ones [86]. Furthermore, it is worth noting that the aspect ratio of nanoparticles affects their internalization pathway. It has been shown that nanoparticles with high aspect ratios tend to internalize through macropinocytosis into the endothelial and epithelial cells, while sphere-shaped nanoparticles prefer clathrin-mediated endocytosis to enter the endothelial cells [94]. Nanoparticles with different shapes demonstrate special flowability; hence, they may have different circulating half-lives, cell membrane interactions, and MPS uptake. In general, spherical-shaped small nanoparticles migrate at a considerable distance from the capillary walls. Migrating in such a cell-free layer significantly impedes particle-cell contact points, and thereby, the targeting efficiency (whether active or passive accumulation). Accordingly, non-spherical nanoparticles (e.g., discoidal-shaped ones with unique tumbling characteristics) are more exposed to contact cells and to extravasate. Additionally, non-spherical particles typically exhibit longer circulating half-life, partly due to their propensity to align with the blood flow. Concerning macrophage internalization, morphological parameters like aspect ratio and curvature highly affect phagocytosis. Spherical particles with lower normalized curvature are picked up faster by macrophages than the non-spherical ones. Altogether, considerable attention has recently been paid to non-spherical morphologies (e.g., cylindrical, ellipsoidal, discoidal, and worm-like geometries) to achieve improved accumulation within the tumors [36].

Effect of Rigidity

Rigidity has a critical effect on the bioavailability of nanoparticles, and hence, modulating rigidity could help the nanoparticles to effectively conquer the barriers ahead like mucosal and tumor delivery. There is an optimal range for nanoparticle stiffness. Softer nanoparticles show weak mucosal penetration as well as cellular internalization. On the other hand, hard nanoparticles exhibit the appropriate cellular uptake, while they are not able to efficiently penetrate the mucus. Consequently, to overcome the biological multi-barriers of organs and tissues, rigidity modification is promising [95]. The degree of particle stiffness also impacts the biodistribution pattern, half-lives, and efficiency of nanoparticles (by affecting macrophage uptake as well as receptor-mediated cancer cell uptake). Rigid particles larger than the cut-off of splenic interendothelial slits are simply removed while passing through [36, 83, 96]. On the other hand, soft nanoparticles show a low internalization level into the MPS, thus reducing the endocytosis rate and extending blood circulation [96–98]. Consequently, softer particles with higher flexibility are expected to be more favored due to their extended circulation half-life and decreased non-specific accumulation in the spleen [36, 83].

Effect of Surface Properties

Surface charge (which is usually measured by zeta potential; ξ) and hydrophilicity/hydrophobicity are among the most important variables to achieve optimal particle features [99]. Surface properties of nanoparticles impact their overall behavior, both *in vitro* (e.g., aqueous dispersibility and encapsulation efficiency) and *in vivo* (e.g., aggregation in the gastrointestinal tract, interactions with biological membranes, biodistribution, biocompatibility, and clearance) [99]. Upon contact with the gut luminal fluids, the charge density of particles may be altered. Depending on the surface charge and interactions that take place between nanoparticles and mucus, the residence time of nanoparticles may be beneficially extended or, disadvantageously, the absorption of nanoparticles is hindered by mucus.

Typically, cationic nanoparticles exhibit improved uptake, enhanced internalization, transport (till lamina propria), and considerably increased bioavailability [91]. M cells tend to absorb non-ionic nanoparticles with hydrophobic properties compared with charged hydrophilic ones [88]. The surface charge is of crucial importance in determining circulation lifetime and accumulation of nanoparticles in the target sites. It has been shown that negative and neutral charged particles demonstrate lower adsorption to serum proteins leading to a prolonged circulation half-life. However, highly negative charge particles ($\xi < -10$ mV) demonstrate a considerable RES uptake. The least RES interaction belongs to around-neutral nanoparticles (-10 mV $< \xi < 10$ mV) [36, 100]. On the other hand, positively charged particles induce serum protein aggregation and show a high non-specific uptake, preferential internalization within the tumor cells and inflammation sites, and facilitated endosomal release. With all these in mind, an optimal nanoparticle to be accumulated in the tumor is expected to have a neutral or slightly negative surface charge in the blood circulation and a switched positive charge while arriving at the tumor [36].

Effect of Surface Modification and Functionalization

Nanoparticles must be able to protect themselves and the medicine; otherwise, the nanoparticles may form aggregates, drug molecules undergo precipitation, and an early clearance occurs [99]. To avoid this, surface modification of nanoparticles, using PEG or other polymer coatings, is usually carried out which provides hydrophilic surface chemistry. In oral drug delivery, for instance, this can minimize the mucin trapping thereby improving mucosal permeability. Additionally, in the case of lipid-based nanoparticles, the lipolysis is significantly suppressed. These performances could effectively improve the bioavailability of the drug to be delivered orally [91, 101].

Besides, surface modification is an efficient method to improve both passive and active targeting [91]. Concerning passive targeting, for instance, it has been shown that bioadhesive coating not only enhances nanoparticle absorption but also positively affects organ distribution patterns [102].

As mentioned previously, surface functionalization of nanoparticles with appropriate targeting moieties is a suitable strategy to improve the oral absorption of intact nanoparticles. If the intended receptor is also expressed in other organs rather than the intestine, then there is a potential for distribution in the targeted organ following the oral absorption, and thereby, improved accumulation and efficiency. However, targeting may be a double-edged sword, when there is an interaction between nanoparticle and off-target tissue receptors thus leading to a faster clearance. Although it may seem unpleasant in the first place, RES uptake can be advantageously utilized to generate an immune response upon oral vaccination [91].

To sum up, as a general trend, it has been proven that (i) larger and non-spherical nanoparticles have a higher probability of margination; (ii) positively charged and uncoated nanoparticles are more quickly cleared by MPS; (iii) rod-like, neutral and targeted nanoparticles more easily penetrate to tumor cells [54]; and (iv) non-spherical and flexible nanoparticles have a more prolonged circulation lifetime [86].

After the absorption phase, there is another obstacle ahead, namely local distribution which requires overcoming microenvironmental barriers. These barriers encompass chemical and physical obstacles restricting the penetration of nanoparticles into the target site. The condition of microenvironments is often fundamentally different from those of the circulation; hence, they may greatly affect the physical features and stability of nanoparticles. Microenvironmental characteristics are highly dependent on the state of the disease and often refer to factors like pH variations, hyperthermia, and altered shear stress (due to abnormal narrowing of blood vessels) [54]. A thorough understanding of factors affecting local distribution can be inspiring for the formulators to design responsive systems (e.g., pH-sensitive formulations, temperature-responsive platforms, and shear stress-induced nanoparticles) [54, 103].

1.3.3 What to Deliver?

The field of novel drug delivery systems aims to develop technologies for the delivery of therapeutics (including small molecules and biopharmaceuticals) to the targeted site with the ultimate goal of maximizing therapeutic effects, minimizing adverse effects, and enhancing patient compliance [104].

1.3.3.1 Small Molecules

Small molecule drugs/prodrugs are compounds with a low molecular weight that can regulate the biochemical processes of the host body to diagnose, treat, or prevent diseases. Because of their more predictable pharmacokinetics and pharmacodynamics (which frequently cause easier dosing protocols), simpler manufacturing and characterizing processes, and high stability and oral bioavailability (which usually enhances patient compliance), small molecule drugs are attractive to healthcare providers and patients [105]. In comparison to large-molecular biologics, these 'tiny creatures' can reach target sites easily, penetrate the cell membranes efficiently, and work on target tissues/cells with relatively well-known mechanisms. The main challenge of the majority of small molecules is their poor solubility and/or weak ability to permeate across the biological barriers, which collectively have been addressed in BCS (See Chapter 4). Since new drugs have recently shown a trend toward the BCS classes II and IV with low aqueous solubility and/or membrane permeability, a remarkable research and development effort has been devoted to overcoming those obstacles. Obviously, the direction of the small-molecule drugs to the intended target sites is still a very attractive goal attempted by the researchers along with those two aforementioned objectives.

1.3.3.2 Biopharmaceuticals

Biopharmaceuticals, including peptides, recombinant proteins, vaccines, hormones, enzymes, genes, monoclonal antibodies, and antibody-drug conjugates, have become more attractive therapeutic molecules than small molecules thanks to their high selectivity, specificity, efficacy, and potency. These characteristics rise from their macromolecular nature, which provides the structural complexity that is needed for specificity. Due to this structural complexity, formulation and delivery of biopharmaceuticals are challenging issues. Also, the high molecular mass of these agents causes a reduction in permeability across biological barriers such as skin and gastrointestinal tract epithelia. Because of poor stability and limited permeability, biopharmaceuticals are commonly administered by the parenteral route. To overcome these challenges, novel parenteral formulations and oral delivery strategies (such as enteric coating, enzyme inhibitors, permeation enhancers, nanoparticles, and intestinal microdevices) have been developed [106–108].

To sum up, considering how NDDSs positively affect pharmacokinetic aspects, overcome bioavailability barriers, and deliver a wide variety of cargos to their destination, while improving the stability, specificity, efficacy, and safety of drug molecules, they are considered a comprehensive and promising approach for the future of disease treatment.

References

1. Adepu, S. and Ramakrishna, S., Controlled drug delivery systems: current status and future directions. *Molecules*, 26, 19, 5905, 2021.
2. Abu-Thabit, N.Y. and Makhlouf, A.S.H., Historical development of drug delivery systems: From conventional macroscale to controlled, targeted, and

responsive nanoscale systems, in: *Stimuli Responsive Polymeric Nanocarriers for Drug Delivery Applications*, vol. 1, pp. 3–41, Elsevier, Duxford, United Kingdom, 2018.
3. Folkman, J. and Long, D.M., The use of silicone rubber as a carrier for prolonged drug therapy. *J. Surg. Res.*, 4, 3, 139–42, 1964.
4. Park, K., Controlled drug delivery systems: past forward and future back. *J. Control. Release*, 190, 3–8, 2014.
5. Hoffman, A.S., The origins and evolution of "controlled" drug delivery systems. *J. Control. Release*, 132, 3, 153–63, 2008.
6. Felice, B., Prabhakaran, M.P., Rodriguez, A.P., Ramakrishna, S., Drug delivery vehicles on a nano-engineering perspective. *Mater. Sci. Eng.: C*, 41, 178–95, 2014.
7. Tewabe, A., Abate, A., Tamrie, M., Seyfu, A., Siraj, E.A., Targeted drug delivery—from magic bullet to nanomedicine: principles, challenges, and future perspectives. *J. Multidiscip. Healthc.*, 14, 1711, 2021.
8. Bruschi, M.L. (Ed.), 2 - Modification of drug release, in: *Strategies to Modify the Drug Release from Pharmaceutical Systems*, pp. 15–28, Woodhead Publishing, Cambridge, United Kingdom, 2015, Available from: https://www.sciencedirect.com/science/article/pii/B9780081000922000023.
9. Sun, W., Hu, Q., Ji, W., Wright, G., Gu, Z., Leveraging physiology for precision drug delivery. *Physiol. Rev.*, 97, 1, 189–225, 2017.
10. Simonazzi, A., Cid, A.G., Villegas, M., Romero, A.I., Palma, S.D., Bermúdez, J.M., Chapter 3 - Nanotechnology applications in drug controlled release, in: *Drug Targeting and Stimuli Sensitive Drug Delivery Systems*, A.M. Grumezescu (Ed.), pp. 81–116, William Andrew Publishing, Oxford, United Kingdom, 2018, Available from: https://www.sciencedirect.com/science/article/pii/B9780128136898000033.
11. Vecsernyés, M., New Insights into Advanced Drug Delivery and Absorption Systems. *Sci. Pharm.*, 88, 2, 20, 2020.
12. Ritschel, W.A. and Kearns, G.L., The LADMER system: liberation, absorption, distribution, metabolism, elimination and response, in: *Handbook of Basic Pharmacokinetics Including Clinical Applications*, pp. 19–24, 1992.
13. Shargel, L., Wu-Pong, S., Yu, A.B.C., Chapter 17. Modified-Release Drug Products, in: *Applied Biopharmaceutics & Pharmacokinetics*, 6e, The McGraw-Hill Companies, New York, NY, 2012, Available from: accesspharmacy.mhmedical.com/content.aspx?aid=56605978.
14. Cid, A.G., Simonazzi, A., Palma, S.D., Bermúdez, J.M., Solid dispersion technology as a strategy to improve the bioavailability of poorly soluble drugs. *Ther. Deliv.*, 10, 6, 363–82, 2019.
15. Junyaprasert, V.B. and Morakul, B., Nanocrystals for enhancement of oral bioavailability of poorly water-soluble drugs. *Asian J. Pharm. Sci.*, 10, 1, 13–23, 2015.
16. Pillai, O., Dhanikula, A.B., Panchagnula, R., Drug delivery: an odyssey of 100 years. *Curr. Opin. Chem. Biol.*, 5, 4, 439–46, 2001.

17. Fontana, F., Figueiredo, P., Zhang, P., Hirvonen, J.T., Liu, D., Santos, H.A., Production of pure drug nanocrystals and nano co-crystals by confinement methods. *Adv. Drug Deliv. Rev.*, 131, 3–21, 2018.
18. Luo, Z., Paunović, N., Leroux, J.C., Physical methods for enhancing drug absorption from the gastrointestinal tract. *Adv. Drug Deliv. Rev.*, 175, 113814, 2021, Available from: https://www.sciencedirect.com/science/article/pii/S0169409X21001885.
19. Haque, T. and Talukder, M.M.U., Chemical enhancer: a simplistic way to modulate barrier function of the stratum corneum. *Adv. Pharm. Bull.*, 8, 2, 169, 2018.
20. Boddupalli, B.M., Mohammed, Z.N.K., Nath, R.A., Banji, D., Mucoadhesive drug delivery system: An overview. *J. Adv. Pharm. Technol. Res.*, 1, 4, 381–7, 2010, Available from: https://www.ncbi.nlm.nih.gov/pmc/articles/PMC3255397/.
21. Mueller, C.E., Prodrug approaches for enhancing the bioavailability of drugs with low solubility. *Chem. Biodivers.*, 6, 11, 2071–83, 2009.
22. Liu, M., Sun, Y., Zhao, S., Li, Y., Piao, R., Yang, Y. *et al.*, A novel prodrug strategy to improve the oral absorption of O-desmethylvenlafaxine. *Exp. Ther. Med.*, 12, 3, 1611–7, 2016.
23. Patel, M., Sheng, Y., Mandava, N.K., Pal, D., Mitra, A.K., Dipeptide Prodrug Approach to Evade Efflux Pumps and CYP3A4 Metabolism of Lopinavir. *Int. J. Pharm.*, 476, 99–107, 2014 Dec 10, Available from: https://www.ncbi.nlm.nih.gov/pmc/articles/PMC4344907/.
24. Agrawal, M., Dethe, M.R., Ahmed, H., Yadav, A., Gupta, U., Alexander, A., Nose-to-brain drug delivery for the treatment of Alzheimer's Disease: Current advancements and challenges. *Expert Opin. Drug Deliv.*, 19, 1, 87–102, 2022.
25. Alexander, A., Agrawal, M., Bhupal Chougule, M., Saraf, S., Saraf, S., Chapter 9 - Nose-to-brain drug delivery: an alternative approach for effective brain drug targeting, in: *Nanopharmaceuticals*, R. Shegokar (Ed.), pp. 175–200, Elsevier, Amsterdam, Netherlands 2020, Available from: https://www.sciencedirect.com/science/article/pii/B9780128177785000099.
26. Zhang, X.Y. and Lu, W.Y., Recent advances in lymphatic targeted drug delivery system for tumor metastasis. *Cancer Biol. Med.*, 11, 4, 247–54, 2014, Available from: https://www.ncbi.nlm.nih.gov/pmc/articles/PMC4296090/.
27. McGinnity, D.F. and Grime, K., 4.02 - ADME Optimization in Drug Discovery, in: *Comprehensive Medicinal Chemistry III*, S. Chackalamannil, D. Rotella, S.E. Ward (Eds.), pp. 34–44, Elsevier, Oxford, 2017, Available from: https://www.sciencedirect.com/science/article/pii/B9780124095472123650.
28. Alexis, F., Pridgen, E., Molnar, L.K., Farokhzad, O.C., Factors Affecting the Clearance and Biodistribution of Polymeric Nanoparticles. *Mol. Pharm.*, 5, 4, 505–15, 2008, Available from: https://doi.org/10.1021/mp800051m.

29. O'Donnell, J.P., Pharmacokinetic and Biopharmaceutical Considerations in Prodrug Discovery and Development, in: *Prodrugs and Targeted Delivery: Towards Better ADME Properties*, vol. 47, pp. 415–52, 2010.
30. Attia, M.F., Anton, N., Wallyn, J., Omran, Z., Vandamme, T.F., An overview of active and passive targeting strategies to improve the nanocarriers efficiency to tumour sites. *J. Pharm. Pharmacol.*, 71, 8, 1185–98, 2019, Available from: https://onlinelibrary.wiley.com/doi/abs/10.1111/jphp.13098.
31. Maheshwari, N., Kumar Atneriya, U., Tekade, M., Sharma, M.C., Elhissi, A., Tekade, R.K., Chapter 3 - Guiding Factors and Surface Modification Strategies for Biomaterials in Pharmaceutical Product Development, in: *Biomaterials and Bionanotechnology*, Advances in Pharmaceutical Product Development and Research, R.K. Tekade (Ed.), pp. 57–87, Academic Press, London, United Kingdom, 2019, Available from: https://www.sciencedirect.com/science/article/pii/B9780128144275000032.
32. Mahato, R., Tai, W., Cheng, K., Prodrugs for improving tumor targetability and efficiency. *Adv. Drug Deliv. Rev.*, 63, 8, 659–70, 2011, Available from: https://www.sciencedirect.com/science/article/pii/S0169409X11000184.
33. Di, L. and Kerns, E.H., Chapter 11 - Metabolic Stability, in: *Drug-Like Properties*, Second Edition, L. Di and E.H. Kerns (Eds.), pp. 161–94, Academic Press, Boston, 2016, Available from: https://www.sciencedirect.com/science/article/pii/B9780128010761000113.
34. Tillement, J.P. and Tremblay, D., 5.02 - Clinical Pharmacokinetic Criteria for Drug Research, in: *Comprehensive Medicinal Chemistry II*, J.B. Taylor and D.J. Triggle (Eds.), pp. 11–30, Elsevier, Oxford, 2007, Available from: https://www.sciencedirect.com/science/article/pii/B008045044X001176.
35. Nishimuta, H., Houston, J.B., Galetin, A., Hepatic, intestinal, renal, and plasma hydrolysis of prodrugs in human, cynomolgus monkey, dog, and rat: implications for *in vitro–in vivo* extrapolation of clearance of prodrugs. *Drug Metab. Dispos.*, 42, 9, 1522–31, 2014.
36. Blanco, E., Shen, H., Ferrari, M., Principles of nanoparticle design for overcoming biological barriers to drug delivery. *Nat. Biotechnol.*, 33, 9, 941–51, 2015, Available from: https://www.ncbi.nlm.nih.gov/pmc/articles/PMC4978509/.
37. Lima, T., Bernfur, K., Vilanova, M., Cedervall, T., Understanding the Lipid and Protein Corona Formation on Different Sized Polymeric Nanoparticles. *Sci. Rep.*, 10, 1, 1129, 2020, Available from: https://www.nature.com/articles/s41598-020-57943-6.
38. Ernsting, M.J., Murakami, M., Roy, A., Li, S.D., Factors controlling the pharmacokinetics, biodistribution and intratumoral penetration of nanoparticles. *J. Control. Release*, 172, 3, 782–94, 2013, Available from: https://www.sciencedirect.com/science/article/pii/S0168365913008018.
39. Bajaj, G. and Yeo, Y., Tumor targeted nanoparticles: state of the art and remaining challenges, in: *Nanoparticulate Drug Delivery Systems Strategies, Technologies and Applications*, Yeo, Y. (Ed.), pp. 12–21, 2013.

40. Haroon, H.B., Hunter, A.C., Farhangrazi, Z.S., Moghimi, S.M., A brief history of long circulating nanoparticles. *Adv. Drug Deliv. Rev.*, 188, 114396, 2022, Available from: https://www.sciencedirect.com/science/article/pii/S0169409X22002861.
41. Li, R., He, Y., Zhang, S., Qin, J., Wang, J., Cell membrane-based nanoparticles: a new biomimetic platform for tumor diagnosis and treatment. *Acta Pharm. Sin. B*, 8, 1, 14–22, 2018, Available from: https://www.sciencedirect.com/science/article/pii/S2211383517303283.
42. Brenner, J.S., Mitragotri, S., Muzykantov, V.R., Red blood cell hitchhiking: A novel approach for vascular delivery of Nanocarriers. *Annu. Rev. Biomed. Eng.*, 23, 225–48, 2021.
43. Brenner, J.S., Pan, D.C., Myerson, J.W., Marcos-Contreras, O.A., Villa, C.H., Patel, P. *et al.*, Red blood cell-hitchhiking boosts delivery of nanocarriers to chosen organs by orders of magnitude. *Nat. Commun.*, 9, 1, 2684, 2018, Available from: https://www.nature.com/articles/s41467-018-05079-7.
44. Zheng, J., Lu, C., Ding, Y., Zhang, J., Tan, F., Liu, J. *et al.*, Red blood cell-hitchhiking mediated pulmonary delivery of ivermectin: Effects of nanoparticle properties. *Int. J. Pharm.*, 619, 121719, 2022, Available from: https://www.sciencedirect.com/science/article/pii/S0378517322002745.
45. Zhao, D., Zhang, H., Tao, W., Wei, W., Sun, J., He, Z., A rapid albumin-binding 5-fluorouracil prodrug with a prolonged circulation time and enhanced antitumor activity. *Biomater. Sci.*, 5, 3, 502–10, 2017.
46. Huang, Y., Wang, L., Cheng, Z., Yang, B., Yu, J., Chen, Y. *et al.*, SN38-based albumin-binding prodrug for efficient targeted cancer chemotherapy. *J. Controlled Release*, 339, 297–306, 2021, Available from: https://www.sciencedirect.com/science/article/pii/S0168365921005277.
47. Yang, J., Lv, Q., Wei, W., Yang, Z., Dong, J., Zhang, R. *et al.*, Bioresponsive albumin-conjugated paclitaxel prodrugs for cancer therapy. *Drug Deliv.*, 25, 1, 807–14, 2018, Available from: https://doi.org/10.1080/10717544.2018.1451935.
48. Yoo, J.W., Chambers, E., Mitragotri, S., Factors that control the circulation time of nanoparticles in blood: challenges, solutions and future prospects. *Curr. Pharm. Des.*, 16, 21, 2298–307, 2010.
49. Barreto, E.F., Larson, T.R., Koubek, E.J., Drug Excretion, in: *Reference Module in Biomedical Sciences*, Elsevier, 2021, Available from: https://www.sciencedirect.com/science/article/pii/B9780128204726999997.
50. Skotland, T., Iversen, T.G., Llorente, A., Sandvig, K., Biodistribution, pharmacokinetics and excretion studies of intravenously injected nanoparticles and extracellular vesicles: Possibilities and challenges. *Adv. Drug Deliv. Rev.*, 186, 114326, 2022, Available from: https://www.sciencedirect.com/science/article/pii/S0169409X22002162.
51. Bose, T., Latawiec, D., Mondal, P.P., Mandal, S., Overview of nano-drugs characteristics for clinical application: the journey from the entry to the exit point. *J. Nanopart. Res.*, 16, 8, 1–25, 2014.

52. Poon, W., Zhang, Y.N., Ouyang, B., Kingston, B.R., Wu, J.L., Wilhelm, S. *et al.*, Elimination pathways of nanoparticles. *ACS Nano*, 13, 5, 5785–98, 2019.
53. Cho, S. and Yoon, Y.R., Understanding the pharmacokinetics of prodrug and metabolite. *Transl. Clin. Pharmacol.*, 26, 1, 1–5, 2018.
54. Mitchell, M.J., Billingsley, M.M., Haley, R.M., Wechsler, M.E., Peppas, N.A., Langer, R., Engineering precision nanoparticles for drug delivery. *Nat. Rev. Drug Discov.*, 20, 2, 101–24, 2021.
55. Brunner, J., Ragupathy, S., Borchard, G., Target specific tight junction modulators. *Adv. Drug Deliv. Rev.*, 171, 266–88, 2021, Available from: https://www.sciencedirect.com/science/article/pii/S0169409X21000429.
56. Lemmer, H.J. and Hamman, J.H., Paracellular drug absorption enhancement through tight junction modulation. *Expert Opin. Drug Deliv.*, 10, 1, 103–14, 2013, Available from: https://doi.org/10.1517/17425247.2013.745509.
57. Lu, R., Zhou, Y., Ma, J., Wang, Y., Miao, X., Strategies and Mechanism in Reversing Intestinal Drug Efflux in Oral Drug Delivery. *Pharmaceutics*, 14, 6, 1131, 2022.
58. Pham, Q.D., Björklund, S., Engblom, J., Topgaard, D., Sparr, E., Chemical penetration enhancers in stratum corneum — Relation between molecular effects and barrier function. *J. Control. Release*, 232, 175–87, 2016, Available from: https://www.sciencedirect.com/science/article/pii/S0168365916302449.
59. Cross, S.E. and Roberts, M.S., Physical enhancement of transdermal drug application: is delivery technology keeping up with pharmaceutical development? *Curr. Drug Deliv.*, 1, 1, 81–92, 2004.
60. Trenfield, S.J. and Basit, A.W., Chapter 6 - Modified drug release: Current strategies and novel technologies for oral drug delivery, in: *Nanotechnology for Oral Drug Delivery*, J.P. Martins and H.A. Santos (Eds.), pp. 177–97, Academic Press, London, United Kingdom, 2020, Available from: https://www.sciencedirect.com/science/article/pii/B9780128180389000065.
61. Makvandi, P., Maleki, A., Shabani, M., Hutton, A.R., Kirkby, M., Jamaledin, R. *et al.*, Bioinspired microneedle patches: Biomimetic designs, fabrication, and biomedical applications. *Matter*, 5, 2, 390–429, 2022.
62. Maheshwari, R., Kuche, K., Mane, A., Chourasiya, Y., Tekade, M., Tekade, R.K., Chapter 20 - Manipulation of Physiological Processes for Pharmaceutical Product Development, in: *Dosage Form Design Considerations*, Advances in Pharmaceutical Product Development and Research, R.K. Tekade (Ed.), pp. 701–29, Academic Press, London, United Kingdom, 2018, Available from: https://www.sciencedirect.com/science/article/pii/B9780128144237000204.
63. Rajput, A., Pingale, P., Telange, D., Chalikwar, S., Borse, V., Lymphatic transport system to circumvent hepatic metabolism for oral delivery of lipid-based nanocarriers. *J. Drug Deliv. Sci. Technol.*, 66, 102934, 2021, Available from: https://www.sciencedirect.com/science/article/pii/S1773224721006146.

64. Mittal, D., Ali, A., Md, S., Baboota, S., Sahni, J.K., Ali, J., Insights into direct nose to brain delivery: current status and future perspective. *Drug Deliv.*, 21, 2, 75–86, 2014, Available from: https://doi.org/10.3109/10717544.2013.838713.
65. Yamazoe, E., Fang, J.Y., Tahara, K., Oral mucus-penetrating PEGylated liposomes to improve drug absorption: Differences in the interaction mechanisms of a mucoadhesive liposome. *Int. J. Pharm.*, 593, 120148, 2021, Available from: https://www.sciencedirect.com/science/article/pii/S0378517320311339.
66. Ensign, L.M., Tang, B.C., Wang, Y.Y., Tse, T.A., Hoen, T., Cone, R. *et al.*, Mucus-Penetrating Nanoparticles for Vaginal Drug Delivery Protect Against Herpes Simplex Virus. *Sci. Transl. Med.*, 4, 138, 138ra79–138ra79, 2012, Available from: https://www.science.org/doi/abs/10.1126/scitranslmed.3003453.
67. Popov, A., Mucus-Penetrating Particles and the Role of Ocular Mucus as a Barrier to Micro- and Nanosuspensions. *J. Ocul. Pharmacol. Ther.*, 36, 6, 366–75, 2020.
68. Kwon, Y.B., Kang, J.H., Kim, Y.J., Kim, D.W., Lee, S.H., Park, C.W., Preparation and Evaluation of Mucus-Penetrating Inhalable Microparticles of Tiotropium Bromide Containing Sodium Glycocholate. *Pharmaceutics*, 14, 7, 1409, 2022, Available from: https://www.mdpi.com/1999-4923/14/7/1409.
69. Xia, X., Zhou, Y., Gao, H., Prodrug strategy for enhanced therapy of central nervous system disease. *Chem. Commun.*, 57, 71, 8842–55, 2021, Available from: https://pubs.rsc.org/en/content/articlelanding/2021/cc/d1cc02940a.
70. Gote, V., Ansong, M., Pal, D., Prodrugs and nanomicelles to overcome ocular barriers for drug penetration. *Expert Opin. Drug Metab. Toxicol.*, 16, 10, 885–906, 2020.
71. Mansuri, S., Kesharwani, P., Jain, K., Tekade, R.K., Jain, N.K., Mucoadhesion: A promising approach in drug delivery system. *React. Funct. Polym.*, 100, 151–72, 2016, Available from: https://www.sciencedirect.com/science/article/pii/S1381514816300104.
72. Ahadian, S., Finbloom, J.A., Mofidfar, M., Diltemiz, S.E., Nasrollahi, F., Davoodi, E. *et al.*, Micro and nanoscale technologies in oral drug delivery. *Adv. Drug Deliv. Rev.*, 157, 37–62, 2020.
73. Homayun, B., Lin, X., Choi, H.J., Challenges and recent progress in oral drug delivery systems for biopharmaceuticals. *Pharmaceutics*, 11, 3, 129, 2019.
74. Rahim, M.A., Jan, N., Khan, S., Shah, H., Madni, A., Khan, A. *et al.*, Recent Advancements in Stimuli Responsive Drug Delivery Platforms for Active and Passive Cancer Targeting. *Cancers (Basel)*, 13, 4, 670, 2021, Available from: https://www.ncbi.nlm.nih.gov/pmc/articles/PMC7914759/.
75. Peng, Y., Yu, W., Li, E., Kang, J., Wang, Y., Yang, Q. *et al.*, Discovery of an Orally Active and Liver-Targeted Prodrug of 5-Fluoro-2′-Deoxyuridine for the Treatment of Hepatocellular Carcinoma. *J. Med. Chem.*, 59, 8, 3661–70, 2016, Available from: https://doi.org/10.1021/acs.jmedchem.5b01807.

76. Hunt, N.J., Lockwood, G.P., Le Couteur, F.H., McCourt, P.A.G., Singla, N., Kang, S.W.S. *et al.*, Rapid Intestinal Uptake and Targeted Delivery to the Liver Endothelium Using Orally Administered Silver Sulfide Quantum Dots. *ACS Nano*, 14, 2, 1492–507, 2020, Available from: https://doi.org/10.1021/acsnano.9b06071.
77. Indurkhya, A., Patel, M., Sharma, P., Abed, S.N., Shnoudeh, A., Maheshwari, R. *et al.*, Influence of drug properties and routes of drug administration on the design of controlled release system, in: *Dosage Form Design Considerations*, pp. 179–223, Elsevier, London, United Kingdom, 2018.
78. Yanni, S. and Thakker, D.R., Prodrugs: Absorption, Distribution, Metabolism, Excretion (ADME) Issues, in: *Prodrugs: Challenges and Rewards Part 1*, Biotechnology: Pharmaceutical Aspects, V.J. Stella, R.T. Borchardt, M.J. Hageman, R. Oliyai, H. Maag, J.W. Tilley (Eds.), pp. 1043–81, Springer, New York, NY, 2007, Available from: https://doi.org/10.1007/978-0-387-49785-3_29.
79. Dahan, A., Zimmermann, E.M., Ben-Shabat, S., Modern Prodrug Design for Targeted Oral Drug Delivery. *Molecules*, 19, 10, 16489–505, 2014, Available from: https://www.ncbi.nlm.nih.gov/pmc/articles/PMC6271014/.
80. Rumondor, A.C., Dhareshwar, S.S., Kesisoglou, F., Amorphous solid dispersions or prodrugs: complementary strategies to increase drug absorption. *J. Pharm. Sci.*, 105, 9, 2498–508, 2016.
81. Murakami, T., A mini review: usefulness of transporter-targeted prodrugs in enhancing membrane permeability. *J. Pharm. Sci.*, 105, 9, 2515–26, 2016.
82. Aungst, B.J. and Matz, N., Prodrugs to Reduce Presystemic Metabolism, in: *Prodrugs*, pp. 339–55, Springer, 2007.
83. Zhao, Y., Sultan, D., Liu, Y., Biodistribution, excretion, and toxicity of nanoparticles, in: *Theranostic Bionanomaterials*, pp. 27–53, Elsevier, Amsterdam, Netherlands, 2019.
84. Wei, Y., Quan, L., Zhou, C., Zhan, Q., Factors relating to the biodistribution & clearance of nanoparticles & their effects on *in vivo* application. *Nanomedicine*, 13, 12, 1495–512, 2018.
85. Choi, H.S., Liu, W., Misra, P., Tanaka, E., Zimmer, J.P., Ipe, B.I. *et al.*, Renal clearance of quantum dots. *Nat. Biotechnol.*, 25, 10, 1165–70, 2007.
86. Wang, B., He, X., Zhang, Z., Zhao, Y., Feng, W., Metabolism of Nanomaterials *in vivo*: Blood Circulation and Organ Clearance. *Acc. Chem. Res.*, 46, 3, 761–9, 2013, Available from: https://doi.org/10.1021/ar2003336.
87. Tsoi, K.M., MacParland, S.A., Ma, X.Z., Spetzler, V.N., Echeverri, J., Ouyang, B. *et al.*, Mechanism of hard nanomaterial clearance by the liver. *Nat. Mater.*, 15, 11, 1212–21, 2016, Available from: https://www.ncbi.nlm.nih.gov/pmc/articles/PMC5132626/.
88. Hua, S., Advances in Oral Drug Delivery for Regional Targeting in the Gastrointestinal Tract-Influence of Physiological, Pathophysiological and Pharmaceutical Factors. *Front. Pharmacol.*, 11, 524, 2020.

89. Shukla, T., Upmanyu, N., Pandey, S.P., Sudheesh, M.S., Chapter 14 - Site-specific drug delivery, targeting, and gene therapy, in: *Nanoarchitectonics in Biomedicine*, A.M. Grumezescu (Ed.), pp. 473–505, William Andrew Publishing, Amsterdam, Netherlands, 2019, Available from: https://www.sciencedirect.com/science/article/pii/B978012816200200013X.
90. Bhawani, S.A., Ahmad, F.B., Ibrahim, M.N.M., Yakout, A.A., Chapter 3 - Polymeric micelles in biomedical science, in: *Nanoparticles in Pharmacotherapy*, A.M. Grumezescu (Ed.), pp. 45–71, William Andrew Publishing, Amsterdam, Netherlands, 2019, Available from: https://www.sciencedirect.com/science/article/pii/B9780128165041000053.
91. Griffin, B.T., Guo, J., Presas, E., Donovan, M.D., Alonso, M.J., O'Driscoll, C.M., Pharmacokinetic, pharmacodynamic and biodistribution following oral administration of nanocarriers containing peptide and protein drugs. *Adv. Drug Deliv. Rev.*, 106, 367–80, 2016.
92. Li, D., Zhuang, J., He, H., Jiang, S., Banerjee, A., Lu, Y. *et al.*, Influence of Particle Geometry on Gastrointestinal Transit and Absorption following Oral Administration. *ACS Appl. Mater. Interfaces*, 9, 49, 42492–502, 2017, Available from: https://doi.org/10.1021/acsami.7b11821.
93. Li, W., Zhang, X., Hao, X., Jie, J., Tian, B., Zhang, X., Shape design of high drug payload nanoparticles for more effective cancer therapy. *Chem. Commun.*, 49, 93, 10989–91, 2013.
94. Wang, Y., Pi, C., Feng, X., Hou, Y., Zhao, L., Wei, Y., The Influence of Nanoparticle Properties on Oral Bioavailability of Drugs. *Int. J. Nanomed.*, 15, 6295–310, 2020, Available from: https://www.ncbi.nlm.nih.gov/pmc/articles/PMC7455773/.
95. Yu, M., Xu, L., Tian, F., Su, Q., Zheng, N., Yang, Y. *et al.*, Rapid transport of deformation-tuned nanoparticles across biological hydrogels and cellular barriers. *Nat. Commun.*, 9, 1, 2607, 2018, Available from: https://www.nature.com/articles/s41467-018-05061-3.
96. Hui, Y., Yi, X., Wibowo, D., Yang, G., Middelberg, A.P.J., Gao, H. *et al.*, Nanoparticle elasticity regulates phagocytosis and cancer cell uptake. *Sci. Adv.*, 6, 16, eaaz4316, 2020.
97. Merkel, T.J., Jones, S.W., Herlihy, K.P., Kersey, F.R., Shields, A.R., Napier, M. *et al.*, Using mechanobiological mimicry of red blood cells to extend circulation times of hydrogel microparticles. *Proc. Natl. Acad. Sci.*, 108, 2, 586–91, 2011.
98. Yan, G., Dalhaimer, P., Cai, S., Tsai, R., Tewari, M., Minko, T. *et al.*, Shape effects of filaments versus spherical particles in flow and drug delivery. *Nat. Nanotechnol.*, 2, 4, 249–55, 2007.
99. Rewatkar, P., Kumeria, T., Popat, A., Size, shape and surface charge considerations of orally delivered nanomedicines, in: *Nanotechnology for Oral Drug Delivery*, pp. 143–76, Elsevier, London, United Kingdom, 2020.
100. Zein, R., Sharrouf, W., Selting, K., Physical Properties of Nanoparticles That Result in Improved Cancer Targeting. *J. Oncol.*, 2020, e5194780, 2020, Available from: https://www.hindawi.com/journals/jo/2020/5194780/.

101. Zhang, X., Chen, G., Zhang, T., Ma, Z., Wu, B., Effects of PEGylated lipid nanoparticles on the oral absorption of one BCS II drug: a mechanistic investigation. *Int. J. Nanomed.*, 9, 5503, 2014.
102. Reineke, J., Cho, D.Y., Dingle, Y.L., Cheifetz, P., Laulicht, B., Lavin, D. *et al.*, Can bioadhesive nanoparticles allow for more effective particle uptake from the small intestine? *J. Control. Release*, 170, 3, 477–84, 2013.
103. Allan-Rahill, N., Iruvanti, S., Miller, N., Sarubbe, J., Shear Stess-Induced Nanoparticle Drug Delivery in the Right Coronary Artery, 2018.
104. Mitragotri, S. and Lahann, J., Materials for drug delivery: innovative solutions to address complex biological hurdles, *Adv. Mater.*, 24, 3717–372, 2012.
105. Ngo, H.X. and Garneau-Tsodikova, S., What are the drugs of the future? *MedChemComm*, 9, 5, 757–8, 2018.
106. Mitragotri, S., Burke, P.A., Langer, R., Overcoming the challenges in administering biopharmaceuticals: formulation and delivery strategies. *Nat. Rev. Drug Discov.*, 13, 9, 655–72, 2014.
107. Zhu, Q., Chen, Z., Paul, P.K., Lu, Y., Wu, W., Qi, J., Oral delivery of proteins and peptides: Challenges, status quo and future perspectives. *Acta Pharm. Sin. B*, 11, 8, 2416–48, 2021.
108. Patel, A., Patel, M., Yang, X., K Mitra, A., Recent advances in protein and peptide drug delivery: a special emphasis on polymeric nanoparticles. *Protein Pept. Lett.*, 21, 11, 1102–20, 2014.

2

Novel Drug Delivery Systems: Approach to the Market

Maryam Motamedi* **and Mahsa Sayed Tabatabaei**

HIDA Pharmaceutical Technologies Development Center, HIDA Pharmaceuticals, Tehran, Iran

Abstract

Novel drug delivery systems (NDDSs) have created a big growing global market. Identifying the dimensions, components, and trends of this vast market will help pharmaceutical researchers, scientists, and industry owners to find a clearer horizon in front, to prepare themselves to keep pace with future changes. Along with market pull, the underlying factors that push knowledge and technology to advancements are also of particular importance. Also, finding the areas of interest for the development of NDDSs can pave the way for technological advancements and economic benefits. This chapter first provides a general analysis of the NDDS global market, suggesting comprehensive insights into essential elements and trends leading to future advancements. Several studies and projects arguing the growth rate and the specifications of this market are included with qualitative and quantitative analysis to define the future of the industry with a thorough understanding of the NDDS market and its crucial dynamics. Subsequently, the underlying factors of the emerging pharmaceutical technologies including demographic, social, economic, environmental, political, and technological factors are discussed. Finally, some of the most important development areas shaping the future of pharmaceutical technologies will be briefly introduced.

Keywords: Novel drug delivery systems, global market, pharma trends, value-added pharmaceuticals, emerging technologies

**Corresponding author*: motamedi@hidapharma.com

Mehrdad Hamidi and Mahsa Sayed Tabatabaei (eds.) Novel Drug Delivery Systems: Fundamentals and Applications, (35–66) © 2025 Scrivener Publishing LLC

2.1 Introduction

The Novel Drug Delivery System (NDDS) is a typical example of an emerging market. The main characteristics of this kind of market are, firstly showing the highly changing indices which, sometimes, makes it very difficult to analyze and recognize, secondly being affected by several internal as well as external determinants with complex interrelationships, and, thirdly being very difficult to identify and detect the real trends and signals, sometimes weak but important ones, in order to have a realistic foresight study to have a more futuristic sense even for a near future. Therefore, to do any kind of market study and analysis needed for the commercial scale decision making, both for government-based policy making and the business development plans for private enterprises, the aforementioned three faces of this sector of the pharmaceutical market should be considered very carefully to avoid any misunderstandings, misinterpretations, and inappropriate decisions. In other words, instead of relying only on figures and tables of the numbers, values, and solid indices such as compound annual growth rate (generally referred to as CAGR), some deeper analysis should be done on more 'qualitative' data related to the diseases trends, technology trends, socio/cultural trends, economical trends, political issues, and even the global education trends, e.g., the evolution of third- and fourth-generation universities. This chapter addresses three major areas of the current global market of NDDSs, the underlying factors that necessitate the development of such systems, and the new drug development areas that the world is moving toward. Obviously, a real discussion of all the factors and trends and also different scenarios and considerations is far beyond the scope and capacity of a chapter. Therefore, we have just focused on some 'more practical' issues to provide an initial space to think about this highly changing and emerging market.

2.2 Novel Drug Delivery Systems Global Market

To have a more realistic sense of the current global market of the NDDSs, some qualitative as well as quantitative indices of this market are reviewed in this section. Like any other market, with some critical data available such as market size, growth rate, and segmentation, an expert can do a series of analyses to gain an insight into the present status and future of this market.

2.2.1 Macroeconomic Indicators Affecting the Market Growth

Based the statistics issued by the Statista website, global per capita expenditure reached 1,129 U.S.D (1,482 PPP $) in 2019 and is projected to reach 1,515 U.S.D (2.050 PPP $) by 2050 (Figure 2.1) [1]. With a healthcare expenditure of USD 10, 623.85 per capita, the United States of America has been the leading country in 2018. Moreover, the estimated average annual percent change related to National Health Expenditures (NHE) in the U.S. was 5.2% in 2020, as compared to 2019 (4.5%) which is also projected to reach USD 6,192.5 Billion in 2028, a very important indicator expected to create profitable business opportunities in the coming years [2].

The international healthcare expenditure trend shows a significant change from 2020 to 2022 due to the COVID-19 pandemic but is expected to resume the same outlook as pre-COVID, afterward. The only exception is for COVID-19 vaccines (Figure 2.2). The phased integration of COVID-19 vaccine shots in the market caused a 157 Billion USD rise in the COVID vaccine incremental expenditure which resulted in a 2% rise in cumulative global spending equal to $88 billion despite the -$ 68.2 billion disruption impact during 2022 to 2025 (Figure 2.3) [3].

Both developed and pharmerging[1] countries show considerable pharmaceutical consumption variations (Figure 2.4). They also show some correlations with national income per capita. In developed countries, the privilege of single-payer healthcare programs has resulted in a weak relation between the two variables: income and pharmaceutical consumption.

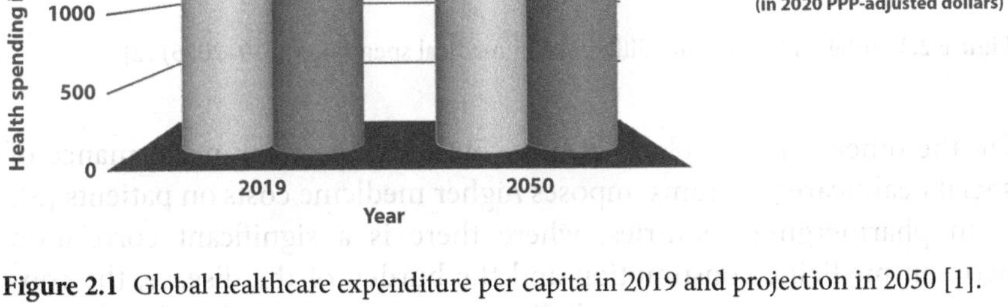

Figure 2.1 Global healthcare expenditure per capita in 2019 and projection in 2050 [1].

[1]Pharmerging market refers to countries with a GDP per capita threshold of $25,000 and more than $1 billion in spending growth.

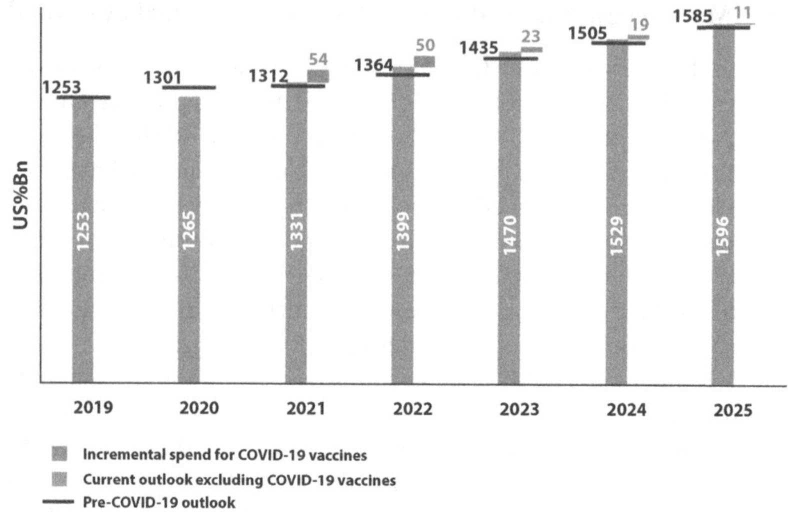

Figure 2.2 Global medical spending (2019-2025), current and pre-COVID-19 outlook [2].

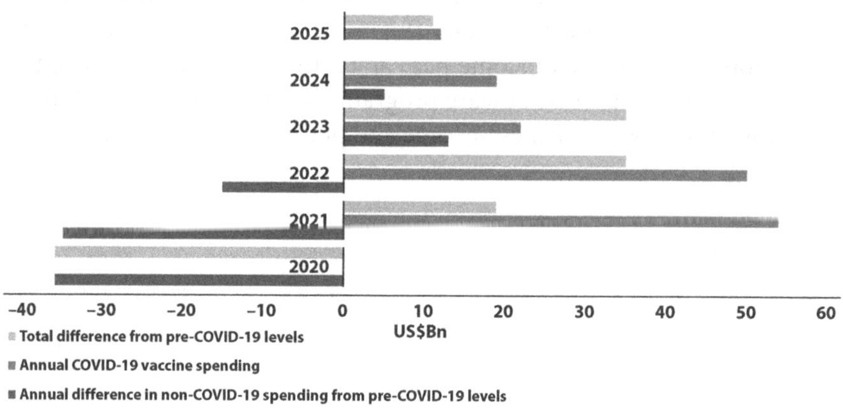

Figure 2.3 Total and segmental difference in medical spending (2019–2025) [2].

On the other hand, in pharmerging countries, the weak performance of social healthcare programs imposes higher medicine costs on patients [3].

In pharmerging countries, where there is a significant correlation between medicine consumption and the burden of the disease, the consumption rate has increased in all disease areas during the last decade, 30% of which contributed to oncology. In developed countries, chronic diseases

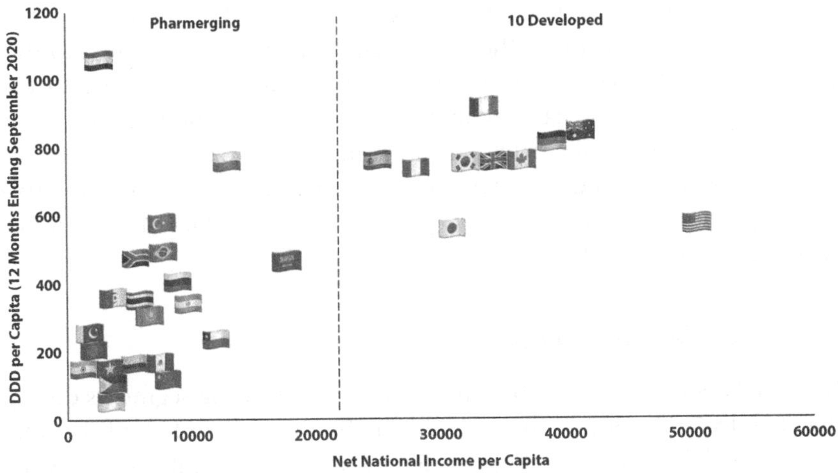

Figure 2.4 Per capita use of medicines varies by national income. The drug use in developed countries is typically higher than in pharmerging countries, data obtain from Ref. [2] (DDD: Defined Daily Dose).

contribute to the higher use of medicines, with the oncology area as the leading one, which has grown at a 14% CAGR[2] [3].

2.2.2 Novel Drug Delivery Systems

Novel drug delivery systems (NDDSs) are developed to enhance the safety/efficacy profile of drugs. The global market of NDDSs has grown rapidly over the recent years due to several factors. The growing need for a controlled-released version of the existing drugs is set to be primarily driven by the prohibiting cost of developing new drug entities, international drug patent expirations, and discovery of new materials capable of regulating the drug release thereby leading to a better drug clinical profile. New advancements in drug formulations and the technological development of NDDSs are also some of the key factors that are likely to boost market growth in the near future [4].

The global market for NDDSs was estimated at US$9 Billion in 2020, amid the COVID-19 crisis, and is predicted to reach US$28.1 Billion by 2026, with a CAGR of 20.8% over 2021 to 2026. Nanoparticles as one of the most promising segments in this market, is expected to have a 20.3% CAGR to reach US$27.2 Billion by 2026. Besides, the Embolization

[2] Compound annual growth rate

Particles segment is predicted to have a 22.9% CAGR for the 7 years. Currently, a 17.3% share of the global NDDS market is dedicated to this segment [5]. The nanoparticles segment is mainly growing due to the increasing demand for nanotechnology concepts including nanocarriers. Benefits offered by nanoparticles in 'targeting' drugs to cancer cells, lowering the side effects, and enhancing the therapeutic efficacy open up profitable opportunities to adopt these platforms for anticancer formulations currently underdevelopment.

2.2.3 Key Market Trends

Oral NDDSS is expected to be one of the fastest-growing segments during the 2022 to 2027 period. They are already building up to half of the total value-added drugs market (Figure 2.5). The market is mainly driven by the growing incidence of chronic diseases, patient compliance improvement, the potential of reformulating the existing drugs provided by advanced oral drug delivery systems, and the vast investment of major pharmaceutical companies in related technologies [4]. This market is estimated to grow with a 7.6% rate over the 2022 to 2029 period.

Injectables lead the fastest-growing market segment in value-added medicines (Figure 2.6), with a 5-year CAGR of 14%. The market size of this segment is nearly two times the segment size back in 2014 and has reached $11bn over the period. Within the segment, pre-filled syringes have recorded an outstanding growth of 24% 5y CAGR, vials with 11% 5y CAGR and infusions with 9% 5y CAGR come next. The United States accounted for more than 80% of the global value growth of injectables,

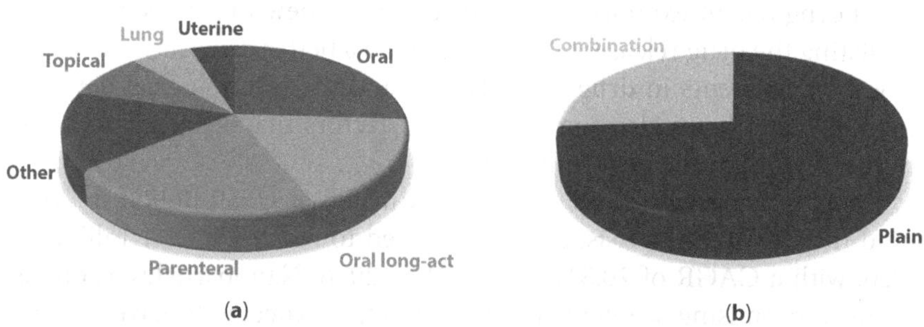

Figure 2.5 Value added medicines global market 2019, by route of administration and combination vs. plain drugs [5].

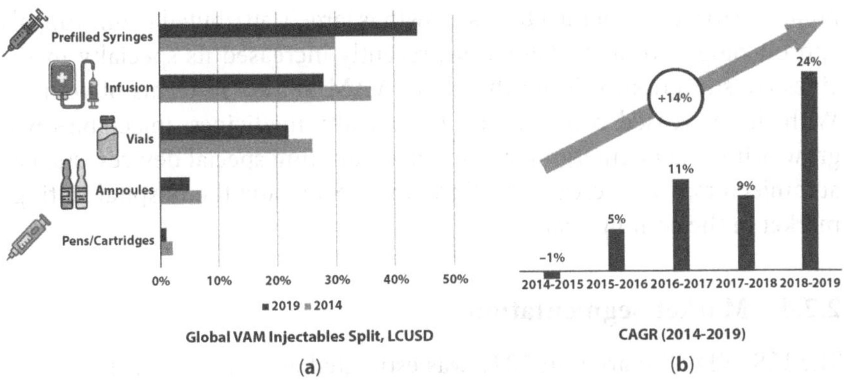

Figure 2.6 Value-added injectables global market trend (2014-2019) [5].

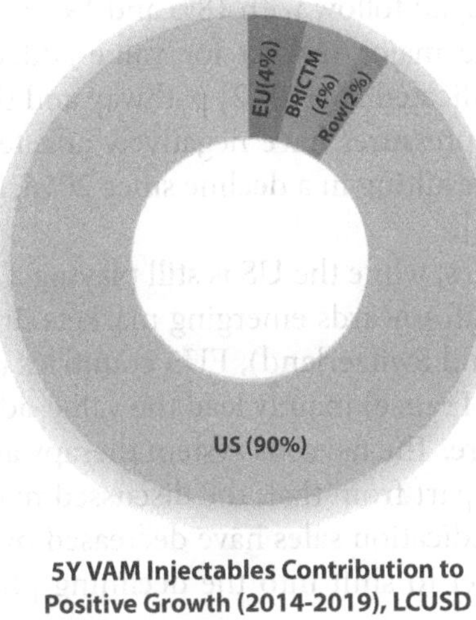

Figure 2.7 5Y VAM injectables contribution to positive growth (2014-2019), LCUSD [5].

followed by BRICTM[3] at 10% and the EU at 4% (Figure 2.7). Products, such as Alkermes' Vivitrol (naltrexone, long-acting) and Pacira's Exparel (bupivacaine, formulation) which decrease opioid usage have mainly driven the substantial growth in injectables over this period in the US [6]. Value-added medicines (VAM) provide a significant opportunity for healthcare systems to deal with substance abuse (e.g., the current opioid epidemic). In emerging economies, the VAM injectables market which

[3]Brazil, Russia, India, China, Turkey, Mexico

builds up to a tenth of injectables growth is largely attributed to antifungals and oncology products. China has recently increased its specialty medicines consumption, shifting the global VAM market in terms of volume. With the increased consumption of specialty medicines, injectables will grow primarily in emerging economies. Adopting special devices, such as autoinjectors, will increase the Euro and US outside the hospital settings market in the coming years [6].

2.2.4 Market Segmentation

The U.S. NDDS market in 2021 was estimated at $4 Billion with a 37.8% share in the international market. The second place belongs to China with a forecasted market share of $2.9 Billion by 2026 at a CAGR of 25.7%. Within Europe, Germany is the fastest-growing market at a CAGR of 19.5%. Canada and Japan follow with 18% and 19.1%, respectively [7].

The US is still the major market for value-added medicines; this is mainly due to the dedicated 505(b) [2] pathway and the flexibility of pricing policies. Pricing pressures have negatively affected the top brands in the recent few years resulting in a decline since 2016, but the US still leads global sales at 70% [6].

Regarding dynamics, while the US is still playing a key role, the growth axis will gradually shift towards emerging markets. In Europe (EU member states, Norway, and Switzerland), EU5 countries (including UK, Italy, Germany, Spain, and France) mainly lead the value-added medicines market, taking a 70% share. The nervous system therapy area has secured Italy a slight growth, but apart from that, the discussed market has declined in the last year. Pain medication sales have decreased over the last few years causing Japan's market to shift into the declining phase, already smaller than Europe [6].

More extensive coverage of US brands by the implementation of enhancements in methodology gains US $44bn or 70% of the total value of the global prescription at list price as of MAT[4] Q1 2019 (Figure 2.8). BRICTM recorded a high growth rate from 2014 and reached 15% in 2019. China and Brazil are driving the growth in the region, in the areas of Oncology and Antifungals, respectively. Value-added medicines have a small market size in emerging economies but the outlook for the future is hopeful since these markets are less payer-driven and more dependent on the patient's decision [6].

---

[4]Moving annual total

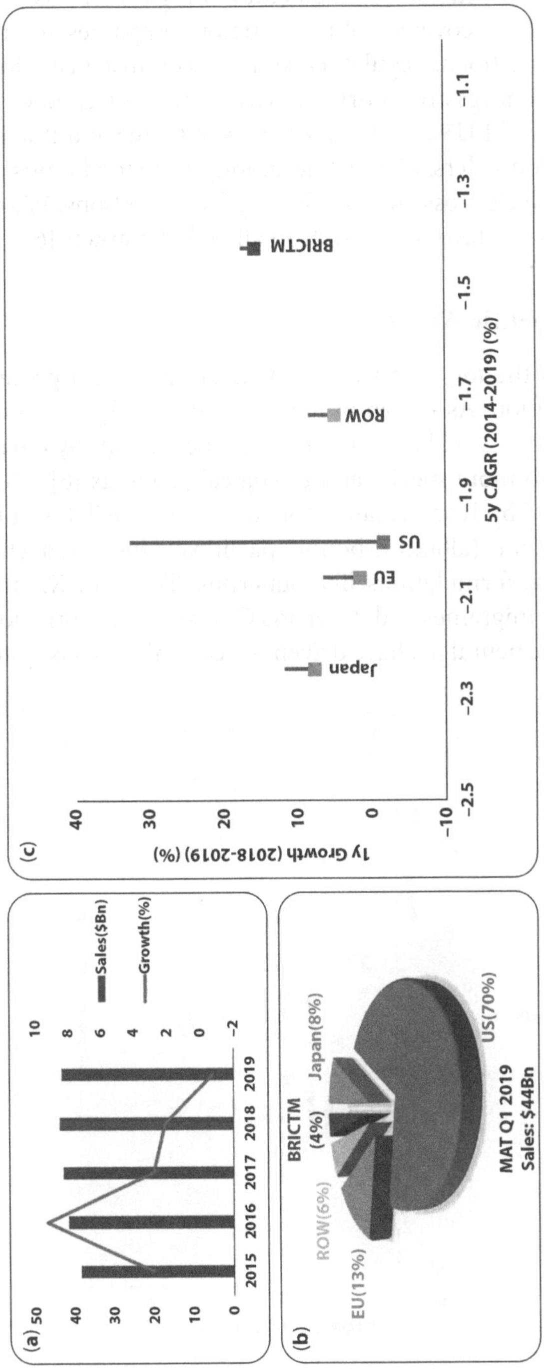

Figure 2.8 Value-added medicines global market size, by geographical segmentation and CAGR [5].

The market for value-added medicines is regionally based due to the limited geographical coverage of the initiator companies and the absence of a single international regulatory system. For instance, global brands like Mirena (levonorgestrel: a drug–device combination) have the highest sales in the US and EU5 countries, whereas in China and Brazil, domestic brands are the top sellers. Meanwhile, amongst the top brands across continents, there is little cross-over. Excluding few exceptions, VAMs sales did not record the same favorable results in all global markets [6].

2.2.5 Therapeutic Areas

Primary care is the top area where VAMs are playing a precious role in constant innovation. As shown in Figure 2.9 on the dynamics of the therapeutic domains, VAMs have caused a significant shift by developing the generic pipe into more special and biological products [6]. The oncology area is mainly led by Teva's Treanda (bendamustine HCl; formulation) and Celgene's Abraxane (albumin-bound paclitaxel; increased efficacy) [6]. Extended-release formulations like Supernus' Trokendi XR (topiramate; long-acting) for migraines and Assertio's Gralise (gabapentin; long-acting) for post-herpetic neuralgia, have driven the central nervous system market

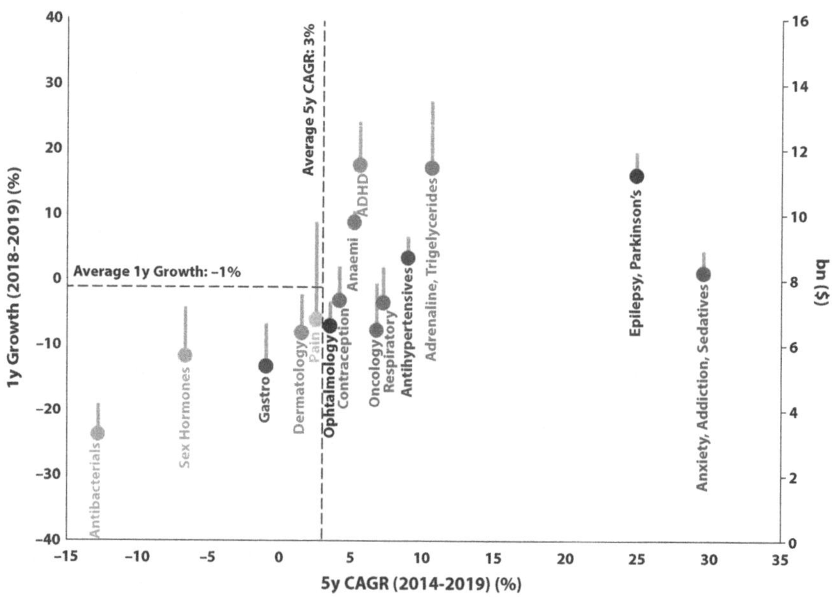

Figure 2.9 Therapeutic areas dynamics [5].

to the top in terms of growth over the studied period (MAT Q1 2014-19). The lack of innovations in the current standard pain pharmacotherapeutic area and frequent dose intervals of such regimens contribute to market differentiation through continuous innovation. Several administration modalities in the pain area such as injectables and topical formulations have resulted in the highest total relative sales. Meanwhile, there is pressure from the governments to control access to opioids. This is a great benefit for healthcare systems and a major tool to combat drug abuse [6].

Abraxane's albumin-bound formulation of paclitaxel that avoids solvents, and aerosols like Salamol (salbutamol; drug device) and Beclazone (beclometasone; drug device) have resulted in a double-digit market growth rate in Oncology and respiratory areas, respectively. Auvi-Q (adrenaline; drug device) has shown the top growth rate during the last five years. Salamol Easi-Breathe (salbutamol; drug device combination) follows next with a mild growth rate over the five years. In the analgesic area, Grünenthal's Versatis (lidocaine; drug device) patches and Mundipharma's Targin (oxycodone/naloxone; drug–drug combination) tablets led in EU5 last year while in Japan, Sector (ketoprofen; formulation, and device) and Artz (sodium hyaluronate; formulation), topped the VAMs market [2].

2.2.6 Nanotechnology-Based Formulations

Pharmaceutical nanotechnology-based products are predicted to play a major role in the international pharma market and the whole healthcare system. The US FDA (Food and Drug Administration) and EMA (European Medicines Agency) have approved around 70 nanopharmaceuticals since 1995 [8–14] (Table 2.1) and there are even more products in the pipeline yet to complete clinical trials. There are also many reports showing an annual increase in nanotechnology-based medications. Every year, to evaluate their safety/efficacy improvement, new nano-based formulations of the existing drugs undergo clinical trials [14]. Rapid growth in the R&D centers as well as the growing market demand are two major incentives. Nanoparticle-based drug products are being progressively used by patients worldwide (Table 2.1). These products, which are manufactured by different international pharmaceutical companies show the success of nanotechnology-based medications as therapeutic agents. There has been a great increase in the number of products available on the market and the investment in the development of nano-based medicines since 2010, given the healthcare system benefits provided by these formulations. 78 nanomedicines have entered the global market since 1989: 66 FDA-approved, 31 EMA-approved, and 20 nanomedicines hold both FDA and EMA

Table 2.1 List of internationally marketed nanomedicines approved by the FDA and the EMA [15].

Type	Trade name	Company	Approval	Active ingredients	Indication
Nanocrystals	Emend®	Merck & Co. Inc.	FDA (2003)	Aprepitant	Anti-emetic drug
	Ivemend®	Merck & Co. Inc.	FDA & EMA (2008)	Fosaprepitant dimeglumine (prodrug of aprepitant)	Anti-emetic drug
	Ostim®	OSARTIS GmbH & Co.	FDA (2004)	Calcium Hydroxyapatite	Bone grafting material
	Rapamune®	Wyeth Pharmaceuticals Inc. (subsidiary of Pfizer Inc.)	EMA (2001), FDA (2010)	Sirolimus (Rapamycin)	Preventing kidney transplant rejection (Immunosuppressant)
	Vitoss®	Orthovia Inc.	FDA (2003)	β-tricalcium phosphate	Bone grafting material
	Ritalin® LX	Novartis	FDA (2002)	Methylphenidate	Attention deficit hyperactivity disorder (ADHA) in children
	Avinza®	Pfizer Pharmaceuticals	FDA (2002)	Morphine sulfate	Psychostimulant
	Focalin® XR	Novartis	FDA (2008)	Dexamethylphenidate HCL	ADHD in children

(*Continued*)

Table 2.1 List of internationally marketed nanomedicines approved by the FDA and the EMA [15]. (*Continued*)

Type	Trade name	Company	Approval	Active ingredients	Indication
	Invega®	Janssen Pharmaceuticals	FDA (2009)	Paliperidone	Schizophrenia
	Invega Sustenna®	Janssen Pharmaceuticals	FDA (2009)	Paliperidone Palmitate	Schizophrenia
	Megace® ES	Par Pharmaceuticals	FDA (2005)	Megestrol acetate	Antianorexic
	NanOss®	RTI Surgical	FDA (2005)	Hydroxyapatite	Bone substitute
	Equivabone®	Zimmer Biomet	FDA (2009)	Hydroxyapatite	Bone substitute
	OsSatura®	IsoTis OrthoBiologics Inc.	FDA (2003)	Hydroxyapatite	Bone substitute
	Epaxal®	Crucell Berna Biotech	EMA (1993)	Inactivated hepatitis A virus vaccine	Preventing hepatitis A infection
	Zanaflex®	Acorda	FDA (2002)	Tizanidine HCL	Muscle relaxant
	Ryanodex®	Eagle Pharm	FDA (2014)	Dantrolene sodium	Malignant hyperthermia
	TriCor®	Abbott Laboratories	FDA (2004)	Fenofibrate	Anti-hyperlipidemia

(*Continued*)

Table 2.1 List of internationally marketed nanomedicines approved by the FDA and the EMA [15]. (*Continued*)

Type	Trade name	Company	Approval	Active ingredients	Indication
Lipid-Based nanoparticles	Doxil®	Johnson & Johnson	FDA (1995), EMA (1996)	Doxorubicin (Adriamycin)	Metastatic ovarian cancer, HIV-associated Kaposi's sarcoma
	Lipodox®	Sun Pharma Global FZE	FDA (2013)	Doxorubicin Hydrochloride	Metastatic ovarian cancer, HIV-associated Kaposi's sarcoma
	DaunoXome®	Galen ltd.	FDA, EMA (1996)	Daunorubicin	Cancers & HIV-associated Kaposi's sarcoma
	Onyvide®	Merrimack Pharmaceuticals	FDA (2015)	Irinotecan	Metastatic pancreatic cancer
	DepoCyt®	Pacira Pharmaceuticals	EMA (2002), FDA (2007)	Cytarabine	Lymphomatous meningitis
	Myocet®	Teva Pharmaceutical Industries Ltd.	EMA (2000)	Doxorubicin hydrochloride	Breast Cancer
	Caelyx®	Janssen Pharmaceuticals	EMA(1996)	Doxorubicin	Breast Cancer, Ovarian Cancer, HIV-related Kaposi's sarcoma
	Mepact®	Takeda France SAS	EMA (2009)	Mifamurtide	Osteogenic sarcoma

(*Continued*)

Table 2.1 List of internationally marketed nanomedicines approved by the FDA and the EMA [15]. *(Continued)*

Type	Trade name	Company	Approval	Active ingredients	Indication
	Marquibo®	Talon Therapeutics	FDA (2012)	Vincristine	Philadelphia chromosome-negative chronic myelogenous leukemia in adult patients
	Onpattro®	Alnylam	FDA, EMA (2018)	Patisiran	Hereditary transthyretin (TTR) mediated amyloidosis
	Lipusu®		FDA (2016)	Paclitaxel	Breast Cancer, NSCLC
	AmBisome®	NeXtar Pharmaceuticals	EMA(1990), FDA(1997)	Amphotericin B	Antifungal drug
	Vyxeos®	Jazz Pharmaceuticals	FDA (2017), EMA(2018)	Daunorubicin and Cytarabine	Acute myeloid leukemia
	Abelcet®	Defiante Farmaceutica	FDA (1995)	Amphotericin B	Antifungal drug
	DepoDur®	Skyepharma	FDA (2004), EMA (2006)	Liposomal morphine sulfate	Postoperative analgesia
	Curosurf®	Chiesi	FDA(1999)	Poractant alfa	Respiratory distress syndrome (RDS)

(Continued)

Table 2.1 List of internationally marketed nanomedicines approved by the FDA and the EMA [15]. *(Continued)*

Type	Trade name	Company	Approval	Active ingredients	Indication
	Zevalin®	Bayer Pharma	FDA (2002), EMA(2004)	90 Y – ibritumomab tiuxetan	Lymphoma
	Inflexal®	Crucell Berna Biotech	EMA (1997)	Inactivated Influenza Vaccine	Prevents Influenza Infection
	Pfizer-BioNTech Vaccine	Pfizer Pharmaceuticals	FDA(2020)	mRNA Vaccine	Prevents COVID-19 Infection
	Moderna COVID-19 Vaccine	Moderna TX Inc.	FDA (2020)	mRNA Vaccine	Prevents COVID-19 Infection
	Vysudine®	QLT Phototherapeutics	FDA & EMA (2000)	Photosensitizer (PS) Benzoporphyrine	Choroidal neovascularization caused by age-related macular degeneration

approvals. These products are categorized into nanocrystals, lipid-based nanocarriers, polymer-based nanocarriers, dendrimers, protein-based nanocarriers, and inorganic nanocarriers (Table 2.1) [15].

According to a factsheet released by World Health Organization (WHO) in 2015, Cancer is one of the major reasons of mortality worldwide, caused 14 million new cases in 2012 and 8.2 million deaths [5]. Hence, effective and modern anti-cancer drugs are urgently needed for controlling the high death rate. During the past few decades, anti-cancer nanomedicines have made significant progress with the swift advancements in pharmaceutical technologies (Table 2.2).

Table 2.2 Nano-formulated cancer therapeutics currently in the market [16].

Product name	Composition	Indications	First approval
Doxil/Caelyx	PEGylated liposomal doxorubicin	Myeloma, KS, breast and ovarian cancers	US(1995)
DaunoXome	Liposomal doxorubicin	KS	US(1996)
Myocet	Liposomal doxorubicin	Breast Cancer	EU/Canada (2000)
Abraxane	Albumin-bound paclitaxel	Breast, NSCLC, pancreatic cancers	US(2005)
Lipusu	Liposomal paclitaxel	Breast, NSCLC,	China(2006)
Nanoxel	Paclitaxel micellar	Solid Tumors	India(2006)
Oncaspar	L-Asparginase conjugate	Acute lymphoblastic leukemia	US(2006)
Depocyt	Liposomal cytarabine	Lymphoma, leukemia	US(1999)
Genexol-PM	Paclitaxel micellar	Breast Cancer, NSCLC, Ovarian and Gastric Cancers	South Korea(2007)
Mepact	Liposomal mifamurtide	Osteogenic sarcoma	EU(2009)

(*Continued*)

Table 2.2 Nano-formulated cancer therapeutics currently in the market [16]. (*Continued*)

Product name	Composition	Indications	First approval
NanoTherm	Iron Oxide nanoparticles	Brain tumors	PEU(2011)
Marquibo	Liposomal vincristine sulfate	Acute lymphoblastic leukemia	US(2012)
ONIVIDE	Liposomal irinotecan	Advanced pancreatic cancer	US(215)
DHP107	Paclitaxel lipid nanoparticles (Oral Administration)	Gastric cancer	South Korea(2016)
Vyxeos	Liposomal daunorubicin and cytarabine	High-risk acute myeloid leukemia	US(2017)
Apealea	Paclitaxel micellar	Ovarian, peritoneal, and fallopian tube cancer	EU(2018)
Hensify	Hafnium oxide nanoparticles	Locally-advanced soft tissue sarcoma	EU(2019)

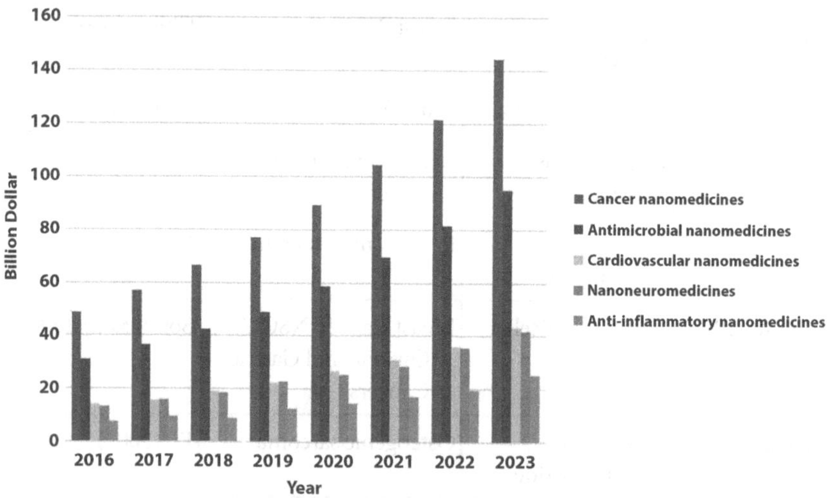

Figure 2.10 Globally marketed nanomedicines by therapeutic area [17].

To target cancerous tumors, different approaches could be adopted, including passive targeting through EPR (enhanced permeation and retention) effect, active targeting based on specific ligands, and stimuli-responsive targeting. Although today more anticancer formulations are present on the market than ever, other indications are also considered for underdevelopment formulations such as autoimmune, metabolic, inflammatory, neurological, hematological and ophthalmic diseases, among others (Figure 2.10) [17].

2.2.7 The Importance of Communicating Value

Commercialize of VAMs have historically experienced certain types of challenges from payers including the lack of recognition of the extra advantages that these formulations can provide such as the difference between the profit level (low price and/or market access) and the amount of evidence required (according to current regulations in EU: Phase III trials and results, postmarketing studies and full submission dossiers). The effective dissemination of the advantages of ongoing innovations in drug formulations to patients and regulators has yet to convince stakeholders to the adoption of value-added medicines. Payers and regulators are mainly concerned about the outcomes, whereas value-added medicines supply a variety of precious advantages to them and the whole healthcare system (Figure 2.11) [6].

Demonstration of Value-added medicines' predominant benefit in the real world as compared to the existing generics shapes what lies ahead. In the long term, the pattern will emerge using data from outside the clinical trial settings [6].

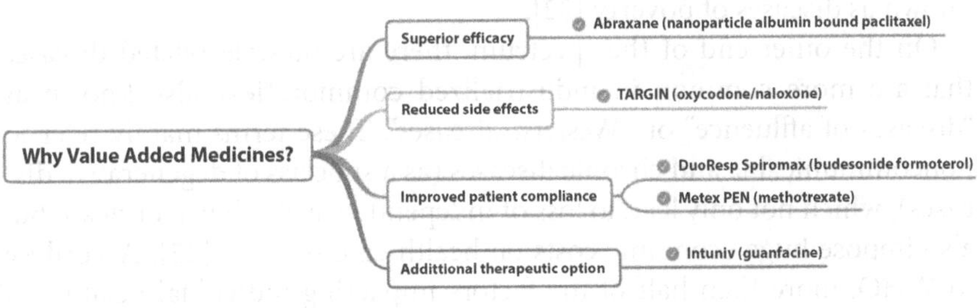

Figure 2.11 Selected benefits demonstrated by value added medicines [5].

2.3 Emerging Diseases, Emerging Technologies

New technologies are becoming more and more widespread, and, in the field of health, numerous factors necessitate such needs. Here, an attempt has been made to briefly present the most prominent aspects, and how they have led to the human need to improve diagnostic and therapeutic systems.

2.3.1 Demographic Factors

Many health-related problems remain challenging even in the 21st century. It is witnessed by the number of human beings who succumb to diseases all over the world. According to the Center for Disease Control and Prevention (CDC), chronic diseases (e.g., heart diseases, cancer, and diabetes) and infections (e.g., COVID-19, influenza, and pneumonia) are among the leading causes of death in the U.S. [18]. Given their fatality and prevalence, chronic and infectious diseases are considered major global health concerns [19].

On the other hand, life expectancy is showing an increasing trend in all countries of the world from an average of 29 in 1770 to 73 years in 2019 [20]. While satisfactory, such an increase has led to an expansion in the aging population and age-related disorders like central nervous system (CNS) diseases [21]. It seems that health-related problems are not getting smaller, but their pattern is changing [22].

2.3.2 Socioeconomic Grounds

Income and welfare state is one of the most important social determinants of health. Some infectious diseases (like malaria, tuberculosis, and HIV/AIDS) are a global dilemma, primarily in underprivileged communities where millions of people suffer from them [19]. In general, infectious diseases along with those related to malnutrition and poor health behavior are known as diseases of poverty [22].

On the other end of the spectrum, there are lifestyle-related diseases that are more common in industrialized communities—also known as "diseases of affluence" or "Western diseases". These terms mainly refer to noncommunicable and chronic diseases (as a subclass of degenerative diseases), which not only lead to loss of independence, disability, or death but also impose huge economic costs on health care services [23]. According to WHO, more than half of the factors impacting individual health and quality of life are related to lifestyle [24]. Lifestyle encompasses the daily functions and behaviors of individuals that significantly affect the physical and mental health of the human being. Today, unhealthy lifestyles such as insufficient physical activity, smoking, alcohol consumption, unhealthy diet, self-medications, and more recently, overuse and misuse of technology have become increasingly popular. This eventually leads to serious chronic problems like cardiovascular diseases (with annual deaths of > 17 million), cancer (with > 8 million deaths each year), metabolic diseases (about 1.6 million people due to diabetes), overweight, stress, etc. [25, 26].

2.3.3 Environmental Reasons

Noncommunicable diseases, one of the most important emerging health problems, are also related to environmental factors. Air pollution, for instance, which is rapidly rising in the developing world is the second leading cause of death among noncommunicable diseases [27].

2.3.4 Political Considerations

Political reasons include a wide range of policies and factors that can affect the health of communities and inequalities in health [28]. A multidimensional conceptual framework has been proposed by Navarro *et al.* [29] to schematically illustrate how politics affects health outcomes. Accordingly, power resources (politics) impact the labor market and welfare state. These factors may directly or indirectly (through impacting the socioeconomic and income inequalities) influence the health of populations.

In addition, the imposition of economic sanctions will pose serious challenges to the health and welfare of citizens, especially vulnerable groups. Economic sanctions restrict access to essential medical items, increase the cost of treatment, and challenge governments to ensure the right to health and welfare of the citizens [30]. On the other hand, sanctions lead to the growth of the black market for medicines. Supply of poor quality or outdated drugs, may, in turn, cause irreparable damage to public health [31].

2.3.5 Technological Aspects

One of the major challenges in the healthcare field is accurate and early diagnosis of diseases. In addition to health problems, late diagnosis incurs exorbitant treatment costs for end-stage diseases [19]. This is of special importance in cases where there is a poor prognosis (e.g., CNS diseases) [21]. Diagnosis often requires extensive screening strategies, wherein novel microtechnologies and nanotechnologies can worthy help by providing more functional, more accurate, and faster diagnosis equipment (e.g., lab-on-a-chip technology) and health monitoring systems (such as implantable sensors) [19, 32]. Besides, nanotechnology-enabled theranostic approaches can combine the benefits of cost-effective and safe drug delivery with high-quality data collection [33].

Another area in which the use of novel technologies is rapidly advancing is drug delivery. Novel drug delivery systems combine art and science to achieve therapeutic systems with higher efficacy and lower adverse effects. Using these technologies, it is possible to modify the physicochemical properties, pharmacokinetics, and biodistribution of drugs, increasing the efficiency of treatment by targeting, designing stimuli-responsive systems, preventing tissue damage by controlling the drug release rate and reducing the side effects of drugs [34].

2.4 New Development Areas

Today, in line with advances in medicine, biology, chemistry, and engineering, drug delivery is also evolving in many ways. Figure 2.12 illustrates new areas of progress from a macro perspective. These highlighted aspects are briefly introduced in the following.

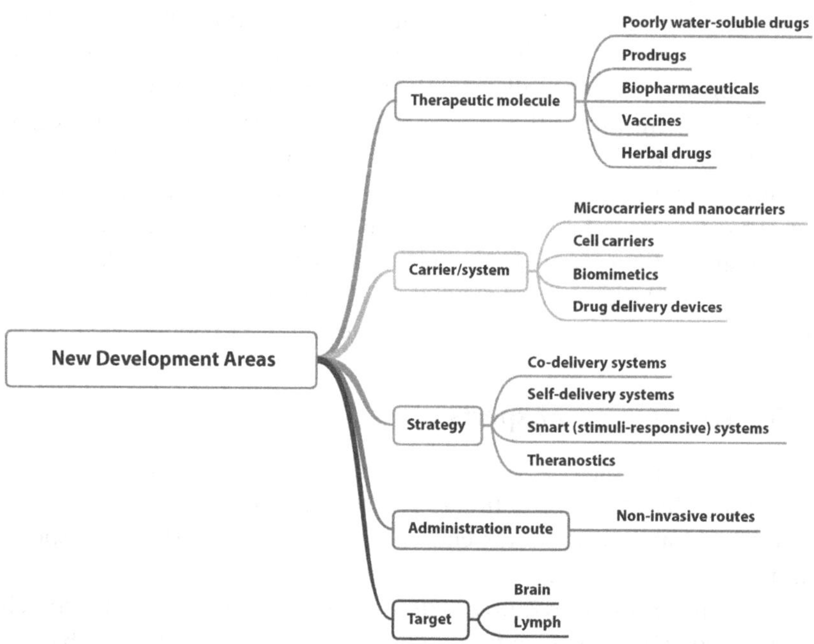

Figure 2.12 New development areas in drug delivery systems.

2.4.1 Poorly Water-Soluble Drugs

Poorly water-soluble drugs suffer from several biopharmaceutical drawbacks including insufficient and variable bioavailability, delayed onset of action, high interpersonal variability, and so on. Nevertheless, such compounds account for a large share of the market with an ever-growing trend [35]. It is reported that by 2016, about 40% of the marketed drugs and 80% of drug molecules under development were low water-soluble entities [36]. Today, many efforts are being made to convert poorly water-soluble drugs into biopharmaceutically acceptable formulations.

2.4.2 Prodrugs

Prodrugs are pharmacologically inactive entities composed of a parent active drug and a pro-moiety connected through a labile chemical bond. Although being a long-lasting approach to mask problematic functional groups of drug molecules, the prodrug strategy has recently been considered to achieve drug molecules with optimized pharmacokinetic and pharmacodynamic features. Today, about 20% of all small molecule drug approvals belong to prodrugs [37].

2.4.3 Biopharmaceuticals

Biopharmaceutical drugs refer to a wide range of drugs (e.g., peptides, proteins, enzymes, and monoclonal antibodies) that have revolutionized the pharma industry with today accounting for about 20% of all drugs in the pipeline. Due to their large size and susceptibility to degradation (i.e., permeability and stability aspects), formulation and delivery of biopharmaceuticals are challenging [38]. In parallel with recent advances in their formulation (e.g., protein modification, microsphere-based methods, etc.), efforts are being made to develop alternative noninvasive routes of administration (for example, oral, transdermal, buccal, nasal, and so on rather than injection) [38, 39].

2.4.4 Vaccines

Vaccination is the best prophylactic tool against diseases which has so far succeeded in eradicating several life-threatening diseases. In recent years, vaccination trends have shifted from the use of attenuated/inactivated pathogens to proteins and peptide antigens with higher safety and lower immunogenicity. In the development of vaccines, antigen delivery carriers (e.g., lipid-based and polymer-based systems) undertake a major role in protecting the antigens, releasing them in a controlled manner, and offering intrinsic adjuvant potential [40].

2.4.5 Herbal Drugs

Several herbal compounds are effective adjunctive therapies for combating different diseases, especially cancer. Herbal combination therapy can improve the effectiveness of commonly used chemotherapeutics through synergistic or additive effects. Today, co-encapsulation of herbal drugs with conventional anticancer drugs inside different nanocarriers to improve efficacy and safety has received growing attention [41].

2.4.6 Microcarriers and Nanocarriers

There is a world-scale attempt to develop and optimize carrier-assisted systems to deliver a wide range of drug substances in a controlled manner and ensure the desired therapeutic effect. In recent decades, a variety of microcarriers and nanocarriers consisting of different materials (e.g., lipids, polymers, proteins, and inorganic materials) and various routes of administration have been studied [42]. Today, different products of microparticles and nanoparticles have entered the market and the process of their development and commercialization is rapidly ongoing [43].

2.4.7 Cell Carriers

Cell-based drug delivery platforms are promising biological carriers, which have been considered in recent decades as a potential alternative to conventional carriers [44]. These endogenous carriers offer several advantages over synthetic microparticles and nanoparticles including high biocompatibility, nonimmunogenicity, long circulation time, and natural targeting potential (site-specific drug delivery). Besides, the existence of a variety of cell types and the diversity of cell surface chemical composition have expanded the field of exploration and development of cell carriers. Carrier erythrocytes, leucocytes, platelets, adipocytes, stem cells, and bacterial ghosts, are among the well-studied cell carriers [45]. They are prone to deliver their cargo (e.g., drugs, enzymes, and genetic materials) into specific tissues or organs in a sustained-release manner. Therefore, they are potentially applicable in the treatment of cancers, cardiovascular diseases, intracellular infections, vaccine delivery, gene therapy, and so on [44].

2.4.8 Biomimetics

Biomimetics refers to drug delivery systems that mimic the unique features of biological systems. For this reason, these systems are highly compatible and low immunogenic. Besides, they show a long circulation in the systemic blood and high targeting ability. Today, studies on the development of biomimetics revolve predominantly around cell membrane-camouflaged nanoparticles, virus-like nanoparticles, lipoprotein-coated nanoparticles, and extracellular vesicles (e.g., exosomes) [46].

2.4.9 Drug Delivery Devices

Drug delivery devices cover a wide range of topics including microfluidics (e.g., micropumps), implanted systems (e.g., drug-eluting stents), wearable systems (e.g., automated insulin delivery systems), microneedles, biodegradable systems, and so on [47]. This area is highly regarded for developing platforms for effective drug delivery. Currently, 3D printing is one of the advanced techniques that is increasingly used in the manufacturing of pharmaceutical dosage forms and biomedical devices [48].

2.4.10 Codelivery Systems

It is a usual clinical practice to treat diseases with various drugs—what is called combination therapy. However, to achieve the best synergistic effect, the pharmacokinetics of the combined drugs should be optimized. Today, various nanoparticles with well-known pharmacokinetic features are explored for the co-delivery of drugs [46].

2.4.11 Self-Delivery Systems

Despite several advantages of carrier-assisted drug delivery systems, they may face several challenging issues like low drug loading capacity, carrier-induced toxicity and/or immunogenicity, and complicated scale-up. In parallel with extensive efforts to develop modified nanocarriers, an innovative solution has been raised called "carrier-free" or "drug self-delivery" systems. In this case, the drug molecule simultaneously undertakes the functional and structural roles. So, carrier-related drawbacks are circumvented and the therapeutic effect could be achieved at a substantially lower dose. Nanocrystals and self-assembled prodrugs are well-known examples of this category [49, 50].

2.4.12 Smart (Stimuli-Responsive) Systems

Smart of stimuli-responsive systems are one of the most enchanting strategies to achieve on-demand drug release [46]. These systems are prone to lower the dosing frequency, reduce drug toxicities, improve the pharmacokinetics and therapeutic efficacy, and provide site specificity for a long period [51]. Currently, a variety of stimuli-responsive systems are expanded based on different endogenous (intrinsic) or exogenous (extrinsic) stimuli. Among the endogenous triggers, one can mention pH, ROS (reactive oxygen species), enzymes, temperature, and so on. Exogenous triggers mainly include light, temperature, magnetic field, X-ray, and ultrasound [46].

2.4.13 Theranostics

Theranostics, by the simplest definition, are the combination of diagnosis and therapy. These systems are of particular importance in the treatment of cancer. The foundation of theranostics is the concept of personalized medicine, which relies on the collection of specific health-related data of individuals (rather than population). To date, several multifunctional systems/devices have been developed to improve the quality of treatment. One of the most advanced arenas of nanomedicines is nanothernostics, where multifunctional nanoarchitectures are employed to provide the concurrent diagnosis, therapy, and monitoring of the real-time response of the drug inside the body [52]. Today, theranostics based on antibodies (e.g., nanobodies) and antibody-related therapeutics (e.g., antibody–drug conjugates, ADCs) have garnered increasing attention [53, 54].

2.4.14 Noninvasive Drug Administration Routes

Noninvasive drug delivery encompasses major routes of drug administration (like buccal, oral, inhalation, intranasal, and transdermal) excluding the parenteral route (e.g., injection through intravenous, intramuscular, subcutaneous, and intrathecal routes). Although the parenteral route has several indispensable advantages including fast systemic distribution, reproducibility, and bioavailability of 100%, the aspects of patient compliance have led to seeking alternative routes, especially for chronic diseases. Over the last several decades, the noninvasive routes have evolved in such a way that, in many cases, they can be considered the preferred option. To date, a significant number of nonparenteral products for the delivery of biologics have been approved or under clinical trials. It is an important step forward, but there is still a lot of room for improvement [55].

2.4.15 Brain Drug Delivery

The brain is a delicate vital organ that is strongly protected against detrimental external stimuli by several supportive structures including the skull, meninges, cerebrospinal fluid, and the blood–brain barrier (BBB). Such a protected structure also precludes most drugs from entering the brain [56]. Various approaches (like nanoparticulate systems, nose-to-brain delivery, etc.) have emerged recently to improve the efficiency of drug delivery to the brain [56, 57].

2.4.16 Lymphatic Drug Delivery

The lymphatic system plays a special defensive role in the body. Since most tumors metastasize through the surrounding lymphatic tissues and may eventually lead to lymphatic tumors, treatment of lymphatic metastatic tumors is of crucial importance. Common therapies (i.e., surgery, radiotherapy, and chemotherapy) have some limitations like insufficient effectiveness and low patient compliance. Under such circumstances, lymphatic targeted drug delivery systems are widely explored as a potential alternative approach to specifically deliver the drugs into the action sites [58].

References

1. Statista, Health spending per capita globally 2019 and 2050. Available from: https://www.statista.com/statistics/856380/per-person-health-spending-globally/.
2. Drug Delivery Systems Market Size, Growth, Global Report, 2029, Available from: https://www.fortunebusinessinsights.com/drug-delivery-systems-market-103070.
3. Aitken, M., Klienrock, M., Munoz, E., Global Medicine Spending and Usage Trends outlook to 2025, IQVIA, 2021, Available from: www.iqvia.com.
4. Novel Drug Delivery Systems (NDDS) Market, Industry Share, Size, Growth - Mordor Intelligence, 2022 - 27, Available from: https://www.mordorintelligence.com/industry-reports/new-drug-delivery-systems-market.
5. Inc. GIA, New Analysis from Global Industry Analysts Reveals Steady Growth for Novel Drug Delivery Systems (NDDS), with the Market to Reach $28.1 Billion Worldwide by 2026, Available from: https://www.prnewswire.com/news-releases/new-analysis-from-global-industry-analysts-reveals-steady-growth-for-novel-drug-delivery-systems-ndds-with-the-market-to-reach-28-1-billion-worldwide-by-2026–301393882.html.

6. Auras, A., A Digital Future for Value Added Medicines: Supporting Positive Patient Behaviour, 2019, Available from: www.iqvia.com.
7. Ltd R and M, Novel Drug Delivery Systems (NDDS) - Global Market Trajectory & Analytics, Available from: https://www.researchandmarkets.com/reports/5140238/novel-drug-delivery-systems-ndds-global.
8. Choi, Y.H. and Han, H.K., Nanomedicines: current status and future perspectives in aspect of drug delivery and pharmacokinetics. *J. Pharm. Invest.*, 48, 1, 43–60, 2018.
9. Ventola, C.L., Progress in nanomedicine: approved and investigational nanodrugs. *Pharm. Ther.*, 42, 12, 742, 2017.
10. Bulbake, U., Doppalapudi, S., Kommineni, N., Khan, W., Liposomal formulations in clinical use: an updated review. *Pharmaceutics*, 9, 2, 12, 2017.
11. Prasad, M., Lambe, U.P., Brar, B., Shah, I., Manimegalai, J., Ranjan, K. *et al.*, Nanotherapeutics: An insight into healthcare and multi-dimensional applications in medical sector of the modern world. *Biomed. Pharmacother.*, 97, 1521–37, 2018.
12. Anselmo, A.C. and Mitragotri, S., Nanoparticles in the clinic: An update. *Bioeng. Transl. Med.*, 4, 3, e10143, 2019, Available from: https://www.ncbi.nlm.nih.gov/pmc/articles/PMC6764803/.
13. Klein, K., Stolk, P., De Bruin, M.L., Leufkens, H.G.M., Crommelin, D.J.A., De Vlieger, J.S.B., The EU regulatory landscape of non-biological complex drugs (NBCDs) follow-on products: observations and recommendations. *Eur. J. Pharm. Sci.*, 133, 228–35, 2019.
14. Caster, J.M., Patel, A.N., Zhang, T., Wang, A., Investigational nanomedicines in 2016: a review of nanotherapeutics currently undergoing clinical trials. *Wiley Interdiscip. Rev.: Nanomed. Nanobiotechnol.*, 9, 1, e1416, 2017.
15. Halwani, A.A., Development of Pharmaceutical Nanomedicines: From the Bench to the Market. *Pharmaceutics*, 14, 1, 106, 2022.
16. Salvioni, L., Rizzuto, M.A., Bertolini, J.A., Pandolfi, L., Colombo, M., Prosperi, D., Thirty years of cancer nanomedicine: Success, frustration, and hope. *Cancers*, 11, 12, 1855, 2019.
17. Manisekaran, R., Introduction to Nanomedicine and Cancer Therapy, in: *Design and Evaluation of Plasmonic/Magnetic Au-MFe2O4 (M-Fe/Co/Mn) Core-Shell Nanoparticles Functionalized with Doxorubicin for Cancer Therapeutics*, pp. 1–36, Springer, Mexico City, Mexico, 2018.
18. Centers for Disease Control and Prevention's (CDC), National Center for Health Statistics (NCHS), FastStats, 2022, Available from: https://www.cdc.gov/nchs/fastats/leading-causes-of-death.htm.
19. Alvarez, M.M., Aizenberg, J., Analoui, M., Andrews, A.M., Bisker, G., Boyden, E.S. *et al.*, Emerging trends in micro-and nanoscale technologies in medicine: From basic discoveries to translation. *ACS Nano*, 11, 6, 5195–214, 2017.
20. Roser, M., Ortiz-Ospina, E., Ritchie, H., Life Expectancy. Our World in Data, 2019, Available from: https://ourworldindata.org/life-expectancy.

21. Cardoso, A.M., Guedes, J.R., Cardoso, A.L., Morais, C., Cunha, P., Viegas, A.T. *et al.*, Recent trends in nanotechnology toward CNS diseases: lipid-based nanoparticles and exosomes for targeted therapeutic delivery. *Int. Rev. Neurobiol.*, 130, 1–40, 2016.
22. Singh, A.R. and Singh, S.A., Diseases of Poverty and Lifestyle, Well-Being and Human Development. *Mens Sana Monogr.*, 6, 1, 187–225, 2008, Available from: https://www.ncbi.nlm.nih.gov/pmc/articles/PMC3190550/.
23. al-Maskari F. Lifestyle diseases: An economic burden on the health services. UN Chronicle. The Magazine of the United Nations. 2010. http://unchronicle.un.org/article/lifestyle-diseases-economic-burden-health-services/.
24. The WHO cross-national study of health behavior in school-aged children from 35 countries: findings from 2001-2002. *J. Sch. Health*, 74, 6, 204–6, 2004.
25. Farhud, D.D., Impact of Lifestyle on Health. *Iran J. Public Health*, 44, 11, 1442–4, 2015, Available from: https://www.ncbi.nlm.nih.gov/pmc/articles/PMC4703222/.
26. Tabish, S.A., Lifestyle diseases: consequences, characteristics, causes and control. *J. Cardiol. Curr. Res.*, 9, 3, 00326, 2017.
27. World Health Organization, Noncommunicable diseases and air pollution, 2019, Available from: https://www.euro.who.int/en/health-topics/environment-and-health/air-quality/publications/2019/noncommunicable-diseases-and-air-pollution-2019.
28. Borrell, C., Espelt, A., Rodríguez-Sanz, M., Politics and health. *J. Epidemiol. Community Health*, 61, 8, 658–9, 2007, Available from: https://www.ncbi.nlm.nih.gov/pmc/articles/PMC2652986/.
29. Navarro, V., Muntaner, C., Borrell, C., Benach, J., Quiroga, Á, Rodríguez-Sanz, M. *et al.*, Politics and health outcomes. *Lancet*, 368, 9540, 1033–7, 2006, Available from: https://www.sciencedirect.com/science/article/pii/S0140673606693410.
30. Abdollahi, M.R., The direct impact of sanctions on health, physical and mental health. *Razi J. Med. Sci.*, 28, 3, 229–45, 2021.
31. Mohammadi, D., US-led economic sanctions strangle Iran's drug supply. *Lancet*, 381, 9863, 279, 2013.
32. Kumar, N. and Kumar, R., Chapter 1 - Introduction, in: *Nanotechnology and Nanomaterials in the Treatment of Life-threatening Diseases*, N. Kumar and R. Kumar (Eds.), pp. 1–51, William Andrew Publishing, Oxford, 2014, Available from: https://www.sciencedirect.com/science/article/pii/B9780323264334000014.
33. Herrmann, I.K. and Rösslein, M., Personalized medicine: the enabling role of nanotechnology. *Nanomed. Future Med.*, 11, 1–3, 2016.
34. Nikalje, A.P., Nanotechnology and its applications in medicine. *Med. Chem.*, 5, 2, 081–9, 2015.
35. Gao, L., Liu, G., Ma, J., Wang, X., Zhou, L., Li, X. *et al.*, Application of Drug Nanocrystal Technologies on Oral Drug Delivery of Poorly Soluble Drugs.

Pharm. Res., 30, 2, 307–24, 2013, Available from: https://doi.org/10.1007/s11095-012-0889-z.
36. Nikolakakis, I. and Partheniadis, I., Self-Emulsifying Granules and Pellets: Composition and Formation Mechanisms for Instant or Controlled Release. *Pharmaceutics*, 9, 4, 50, 2017, Available from: https://www.ncbi.nlm.nih.gov/pmc/articles/PMC5750656/.
37. Chen, K.J., Plaunt, A.J., Leifer, F.G., Kang, J.Y., Cipolla, D., Recent advances in prodrug-based nanoparticle therapeutics. *Eur. J. Pharm. Biopharm.*, 165, 219–43, 2021, Available from: https://www.sciencedirect.com/science/article/pii/S0939641121001235.
38. Mitragotri, S., Burke, P.A., Langer, R., Overcoming the challenges in administering biopharmaceuticals: formulation and delivery strategies. *Nat. Rev. Drug Discov.*, 13, 9, 655–72, 2014, Available from: https://www.ncbi.nlm.nih.gov/pmc/articles/PMC4455970/.
39. Zhu, Q., Chen, Z., Paul, P.K., Lu, Y., Wu, W., Qi, J., Oral delivery of proteins and peptides: Challenges, status quo and future perspectives. *Acta Pharm. Sin. B*, 11, 8, 2416–48, 2021, Available from: https://www.sciencedirect.com/science/article/pii/S2211383521001167.
40. Cordeiro, A.S. and Alonso, M.J., Recent advances in vaccine delivery. *Pharm. Pat. Anal.*, 5, 1, 49–73, 2016.
41. Anwar, D.M., El-Sayed, M., Reda, A., Fang, J.Y., Khattab, S.N., Elzoghby, A.O., Recent advances in herbal combination nanomedicine for cancer: delivery technology and therapeutic outcomes. *Expert Opin. Drug Deliv.*, 18, 11, 1609–25, 2021.
42. Mandracchia D. and Tripodo G., In: *Silk-based Drug Delivery Systems*, ed. E. Bari, S. Perteghella, and M. L. Torre, The Royal Society of Chemistry, pp. 1–24, 2020.
43. Zhong, H., Chan, G., Hu, Y., Hu, H., Ouyang, D., A comprehensive map of FDA-approved pharmaceutical products. *Pharmaceutics*, 10, 4, 263, 2018.
44. Gutiérrez Millán, C., Colino Gandarillas, C.I., Sayalero Marinero, M.L., Lanao, J.M., Cell-based drug-delivery platforms. *Ther. Deliv.*, 3, 1, 25–41, 2012.
45. Tang, L., He, S., Yin, Y., Liu, H., Hu, J., Cheng, J. *et al.*, Combination of Nanomaterials in Cell-Based Drug Delivery Systems for Cancer Treatment. *Pharmaceutics*, 13, 11, 1888, 2021, Available from: https://www.mdpi.com/1999-4923/13/11/1888.
46. Li, C., Wang, J., Wang, Y., Gao, H., Wei, G., Huang, Y. *et al.*, Recent progress in drug delivery. *Acta Pharm. Sin. B*, 9, 6, 1145–62, 2019, Available from: https://www.sciencedirect.com/science/article/pii/S2211383519306008.
47. Chappel, E., *Drug Delivery Devices and Therapeutic Systems*, Academic Press, UK, 2020.

48. Beg, S., Almalki, W.H., Malik, A., Farhan, M., Aatif, M., Rahman, Z. *et al.*, 3D printing for drug delivery and biomedical applications. *Drug Discov. Today*, 25, 9, 1668–81, 2020, Available from: https://www.sciencedirect.com/science/article/pii/S1359644620302841.
49. Qin, S.Y., Zhang, A.Q., Cheng, S.X., Rong, L., Zhang, X.Z., Drug self-delivery systems for cancer therapy. *Biomaterials*, 112, 234–47, 2017.
50. Sayed Tabatabaei, M., Sayed Tabatabaei, F.A., Moghimi, H.R., Drug Self-Delivery Systems: A Comprehensive Review on Small Molecule Nanodrugs. *Bioimpacts*, 2024.
51. Liu, D., Yang, F., Xiong, F., Gu, N., The Smart Drug Delivery System and Its Clinical Potential. *Theranostics*, 6, 9, 1306–23, 2016, Available from: https://www.ncbi.nlm.nih.gov/pmc/articles/PMC4924501/.
52. Bai, R.G., Muthoosamy, K., Manickam, S., Chapter 12 - Nanomedicine in Theranostics, in: *Nanotechnology Applications for Tissue Engineering*, S. Thomas, Y. Grohens, N. Ninan (Eds.), pp. 195–213, William Andrew Publishing, Oxford, 2015, Available from: https://www.sciencedirect.com/science/article/pii/B9780323328890000121.
53. Moek, K.L., Giesen, D., Kok, I.C., de Groot, D.J.A., Jalving, M., Fehrmann, R.S. *et al.*, Theranostics using antibodies and antibody-related therapeutics. *J. Nucl. Med.*, 58, Supplement 2, 83S–90S, 2017.
54. Lecocq, Q., De Vlaeminck, Y., Hanssens, H., D'Huyvetter, M., Raes, G., Goyvaerts, C. *et al.*, Theranostics in immuno-oncology using nanobody derivatives. *Theranostics*, 9, 25, 7772–91, 2019, Available from: https://www.ncbi.nlm.nih.gov/pmc/articles/PMC6831473/.
55. Anselmo, A.C., Gokarn, Y., Mitragotri, S., Non-invasive delivery strategies for biologics. *Nat. Rev. Drug Discov.*, 18, 1, 19–40, 2019, Available from: https://www.nature.com/articles/nrd.2018.183.
56. Dong, X., Current Strategies for Brain Drug Delivery. *Theranostics*, 8, 6, 1481–93, 2018, Available from: https://www.ncbi.nlm.nih.gov/pmc/articles/PMC5858162/.
57. Alexander, A., Agrawal, M., Bhupal Chougule, M., Saraf, S., Saraf, S., Chapter 9 - Nose-to-brain drug delivery: an alternative approach for effective brain drug targeting, in: *Nanopharmaceuticals*, R. Shegokar (Ed.), pp. 175–200, Elsevier, 2020, Available from: https://www.sciencedirect.com/science/article/pii/B9780128177785000099.
58. Zhang, X.Y. and Lu, W.Y., Recent advances in lymphatic targeted drug delivery system for tumor metastasis. *Cancer Biol. Med.*, 11, 4, 247–54, 2014, Available from: https://www.ncbi.nlm.nih.gov/pmc/articles/PMC4296090/.

3

Intraoral Novel Drug Delivery Systems

Seyedeh Maryam Mortazavi[1]*, Leila Sayed-Tabatabaei[2], Mahsa Sayed Tabatabaei[3] and Azadeh Vaezi Moghaddam[1]

[1]*Department of Pharmaceutics and Pharmaceutical Nanotechnology, School of Pharmacy, Shahid Beheshti University of Medical Sciences, Tehran, Iran*
[2]*Department of Pharmacy, University of Copenhagen, Denmark*
[3]*HIDA Pharmaceutical Technologies Development Center, HIDA Pharmaceuticals, Tehran, Iran*

Abstract
Intraoral drug delivery systems are a promising approach to overcoming various obstacles associated with oral, or so-called per-oral, delivery of drugs. Such delivery systems provide opportunities for systemic and local treatment of diseases related to or unrelated to the oral cavity. The oral cavity is a readily accessible place for drug administration, possesses very permeable mucosal tissues, and provides precise localization. Besides, due to the presence of different anatomical sites in the oral cavity, the onset of drug action can be controlled not only by formulation modification but also by changing the administration site. However, drug delivery through the oral mucosal route suffers from certain drawbacks such as small surface area, saliva flow, jaw movement, and so forth. A variety of dosage forms have been designed for intraoral drug delivery, from orodispersible dosage forms that rapidly disintegrate on the tongue to specialized gingival and periodontal formulations targeting oral health conditions. In this chapter, we delve into the intraoral drug delivery systems, exploring the various dosage forms designed for precise drug administration within the oral cavity. Besides, the anatomy and physiology of the oral cavity, challenges and opportunities of intraoral delivery, pharmacokinetics, and commercialized products are also addressed.

Keywords: Intraoral delivery system, buccal mucosa, orodispersible, sublingual, gingival, translabial, periodontal

**Corresponding author*: mmortazavi@sbmu.ac.ir

Mehrdad Hamidi and Mahsa Sayed Tabatabaei (eds.) Novel Drug Delivery Systems: Fundamentals and Applications, (67–102) © 2025 Scrivener Publishing LLC

3.1 Introduction

The systemic delivery of some molecules through oral administration is difficult due to the possibility of being degraded and/or eliminated by hepatic first-pass metabolism, acidic condition of the stomach, gastric and pancreatic enzymes, gut-wall metabolism, and so forth. Besides, since the absorption of drugs after oral administration is not fast enough, this route of administration is not suitable for emergency cases [1]. To overcome such issues, intraoral administration is a suitable alternative route. Intraoral delivery means using the oral cavity to deliver drug molecules. Such a delivery route is not only appropriate for systemic delivery but also can be used to achieve a local effect of drug molecules for the treatment of mucosal diseases [2]. Intraoral delivery systems hold promise in enhancing patient compliance, reducing systemic exposure, and improving therapeutic outcomes. The diverse range of dosage forms and their applications underscores the importance of personalized and patient-centric approaches to healthcare. Besides, the presence of different anatomical sites including the tongue (top and bottom), soft and hard palate, gums, and cheek mucosa with different properties allows the formulator to design various delivery systems depending on the purpose of drug delivery [3]. These innovative systems have expanded the horizons of drug delivery, offering numerous advantages and solutions for a wide range of therapeutic needs. However, the small area of the oral cavity, saliva flow, loss of applied dose because of swallowing, and low permeability of some drugs through oral mucosa are important challenges of this route of administration that should be overcome [4, 5].

In this chapter, the anatomy and physiology of the oral cavity as well as the challenges and opportunities of intraoral delivery are described. A comprehensive classification of intraoral delivery systems designed for precise drug administration within the oral cavity is then presented. Additionally, the fate of the drug inside the body upon the administration of the intraoral drug delivery systems is discussed. Finally, those products that have reached the market following a long journey from laboratories to the clinic will be introduced.

3.2 Anatomy and Physiology

From an anatomical point of view, lips from the anterior part, oropharynx from the posterior, hard and soft palates from the above, some portions of tongue as well as the floor of the mouth from the inferior and cheeks from the around delimit the oral cavity [6] (Figure 3.1a). Four anatomically distinct regions with unique types of tissue are addressed for intraoral drug delivery including buccal, palatal, gingival and sublingual mucosae (Figure 3.1b). The ventral surface of the tongue as well as the floor of the mouth, lined by sublingual tissue, are the sites used for systemic drug delivery. In the case of buccal drug delivery, the buccal mucosa which lines the cheek and inside of the upper and lower lips are the sites of systemic/local delivery. The soft tissue of the upper and lower jaws, the gingival tissue that surrounds and holds the teeth is generally addressed for periodontal and gingival drug delivery. The roof of the oral cavity, which is lined by palatal tissue, is also a site for systemic/local drug delivery [3, 7].

In addition to the tissue type, the classification can be considered based on the keratinization. Keratinized and nonkeratinized epithelia have different barrier properties depending on the type and molecular weight of keratin proteins. The epithelia of the sublingual, buccal, and soft palate are not keratinized while the epithelia of the gingival and hard palate are. The presence of high molecular weight keratins makes the keratinized epithelium less flexible and similar to the upper epidermis in the skin. Due to the presence of mostly low molecular weight keratins in cells of nonkeratinized mucosa, the barrier properties of this type of mucosae are different from keratinized tissue. Another difference affected drug delivery is the various epithelial thicknesses of these different regions [8–10]. Therefore,

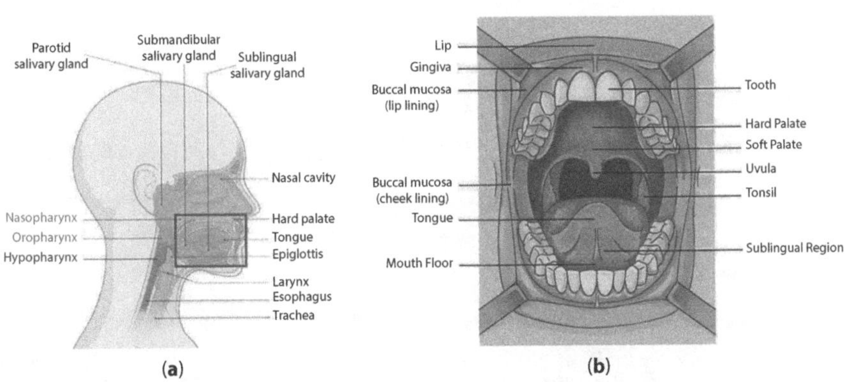

Figure 3.1 Intraoral cavity; (a) boundaries of the oral cavity (lips from the anterior part, oropharynx from the posterior, palate from the above, some portions of tongue as well as the floor of the mouth from the inferior and cheeks from the around) (created with BioRender.com), (b) different regions of the oral cavity.

given the different penetration characteristics of the oral mucosa, it is necessary to design different drug delivery systems for each area.

Due to the importance of the mucosal tissue in the barrier function of the oral cavity, it deserves to be discussed in more detail. As Figure 3.2 illustrates, oral mucosa consists of two main layers, namely the superficial stratified squamous epithelium and the deep *lamina propria*. The basement membrane is located in between. The epithelium of keratinized oral mucosa, itself, consists of four layers that from most superficial to the deepest include (i) *stratum corneum* (keratinized layer), (ii) *stratum granulosum*, (iii) *stratum spinosum*, and (iv) *stratum basale*. The *stratum corneum* consists of nonvital cells or squames. The nonkeratinized epithelium is similar to the keratinized one in the case of the deep layer but different in terms of outer layers. The intermediate and superficial layers are the outer layers of nonkeratinized epithelium. Contrary to keratinized epithelium, the superficial cells are vital and keep their nuclei and some cytoplasmic functions [11–13].

Since the nonkeratinized epithelium is more used for drug delivery and most commercialized intraoral drug delivery systems are designed for this route, the drug transport pathways from this type of epithelia, with specific attention to the buccal mucosa, are discussed here. The sublingual mucosa is considerably more permeable compared to the buccal area and can provide more rapid absorption.

Two main pathways for systemic drug delivery through oral mucosae are transcellular and paracellular. The space between epithelial cells is occupied by lipid bilayers [10]. Because of the lipophilic nature of cell membranes, the transcellular pathway will be preferred by hydrophobic

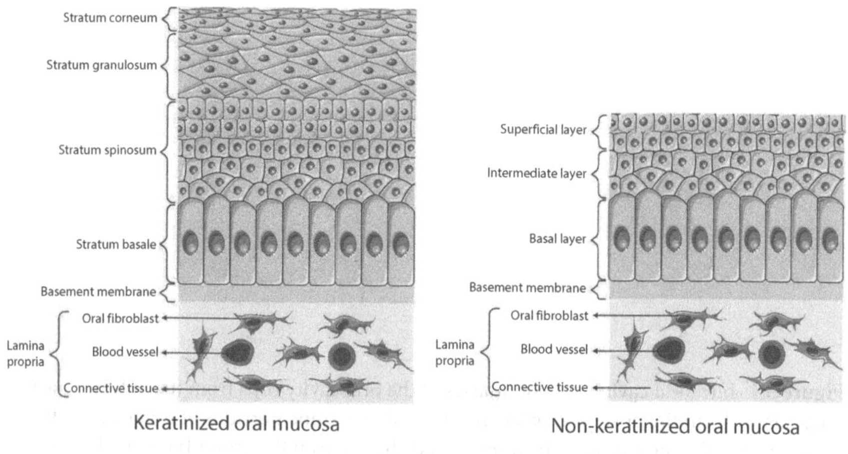

Figure 3.2 Different types of oral mucosa (created with BioRender.com).

compounds. However, hydrophilic molecules can transport across aqueous pores in the membrane. In general, the drug molecules should pass the apical membrane, then the internal space of the cell, and, subsequently, the basolateral membrane [14]. Therefore, in the case of passive transport through the epithelial cells, the balance between hydrophilicity and lipophilicity is more pronounced compared to the paracellular route.

The active transport through buccal mucosa has been reported for small molecules such as electrolytes [15], monosaccharides [16], and amino acids [17]. However, this type of transport needs to be more investigated in terms of transporter type, capacity, specificity, and so on.

3.3 Factors Affecting Intraoral Drug Delivery

Serval factors, including physiological (e.g., flow and pH of saliva), pathological (medical conditions and drug use), lifestyle-related (e.g., smoking and alcohol consumption), and pharmaceutical parameters can potentially affect intraoral drug delivery. Here, an attempt has been made to briefly describe the effect of each factor.

3.3.1 Salivary Flow

The flow of saliva can directly impact the dissolution and disintegration of formulation, and thereby the drug bioavailability. There is an optimal range of saliva, lower than which the drug absorption is low due to dryness of mouth and higher than which, there is the possibility of swallowing before absorption [18].

3.3.2 pH of Saliva

The pH of saliva could affect the drug ionization state. In the case of the transcellular pathway, drug molecules with high pKa are favored due to being relatively neutral (i.e., more lipophilic) in the saliva. Ionized (more hydrophilic) molecules, conversely, are preferred in paracellular transport [18].

3.3.3 Salivary Secretory Disorders, Medications, and Therapeutic Interventions

Some medications and medical conditions can indirectly affect intraoral drug delivery by affecting saliva secretion. For instance, dry mouth syndrome (also known as xerostomia) may be caused by taking some medications (e.g., some anticholinergics, diuretics, antidepressants, opioids analgesics, and NSAIDs) as well as some systemic disorders (e.g., Sjögren's syndrome, granulomatous diseases, and malnutrition) and therapeutic interventions (e.g., external beam radiation and internal radionuclide therapy). On the other hand, some factors may induce hypersecretion of saliva (drooling or sialorrhea). Among these, one can mention some drugs (e.g., direct cholinergic agonists, antipsychotics, and adrenergic antagonists), neurological diseases (e.g., Myasthenia gravis, seizures, and Psychosis), and systemic disorders (e.g., nasal obstruction, pancreatitis, and intestinal infections) [19].

3.3.4 Smoking

Smoking leads to vasoconstriction which, in turn, may negatively affect intraoral drug absorption [18].

3.3.5 Alcohol Consumption

Substantial acute/chronic alcohol consumption leads to thirst and xerostomia [20].

3.3.6 Residence Time of Formulation

For formulations that need considerable time for disintegration and drug release and those designed to adhere to the oral mucosa, any factor that affects the residence time is of particular importance. Therefore, it is essential to avoid food and drink intake, swallowing, and chewing the dosage form until being absorbed [18].

3.4 Challenges and Opportunities

Each administration route provides both opportunities and challenges for formulator scientists and the intraoral route is not an exception. Figure 3.3 illustrates the main challenges and opportunities associated with intraoral drug delivery. In the case of systemic delivery and compared to the oral route, intraoral delivery leads to the avoidance of hepatic first pass as well as gut-wall metabolism, which is of great importance for drugs that are substrates of enzymes. Contrary to the oral route, it is possible to remove the pharmaceutical dosage form if necessary. The formulator scientists

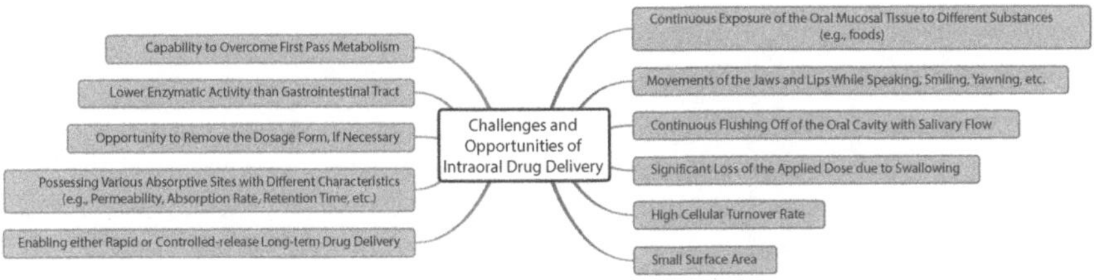

Figure 3.3 Challenges and opportunities of intraoral drug delivery.

can benefit from many capabilities of the oral cavity to design drug delivery systems for the intraoral route [2, 21]. For instance, in addition to the manipulation of dosage forms, due to the presence of different absorptive sites in the oral cavity, it is possible to adjust the retention time, absorption rate, onset of action, and permeability according to the target site where the dosage form is to be used. For instance, to achieve a faster onset, the sublingual route is preferred over the placement of dosage form on the tongue or to achieve an extended drug delivery, the buccal area is preferred over the tongue [7].

Despite all the advantages, the formulators encounter serious challenges in designing intraoral delivery systems. The movements of the lower jaw, tongue, and lips during eating, drinking, speaking, smiling, and yawning as well as the continuous secretion of saliva can be considered as the biggest challenges to be overcome. Besides, the small available surface area and the high rate of cellular turnover are other obstacles against intraoral delivery. Formulators should also consider the possibility of swallowing the dosage forms while designing the intraoral delivery systems [2, 7]. Despite all the challenges against intraoral delivery, as will be explained further, the existence of intraoral delivery systems in the pharmaceutical market indicates that the scientists have, to some extent, achieved the solutions required to conquer the mentioned challenges. However, the commercially available dosage forms are not without flaws, thus, continuous research is required.

3.5 Classification and Formulation Considerations

In this chapter, we explore the realm of intraoral drug formulations and provide a comprehensive overview of the diverse groups within this category. Intraoral drug delivery systems can be classified based on various defining features, as illustrated in Figure 3.4. However, the primary

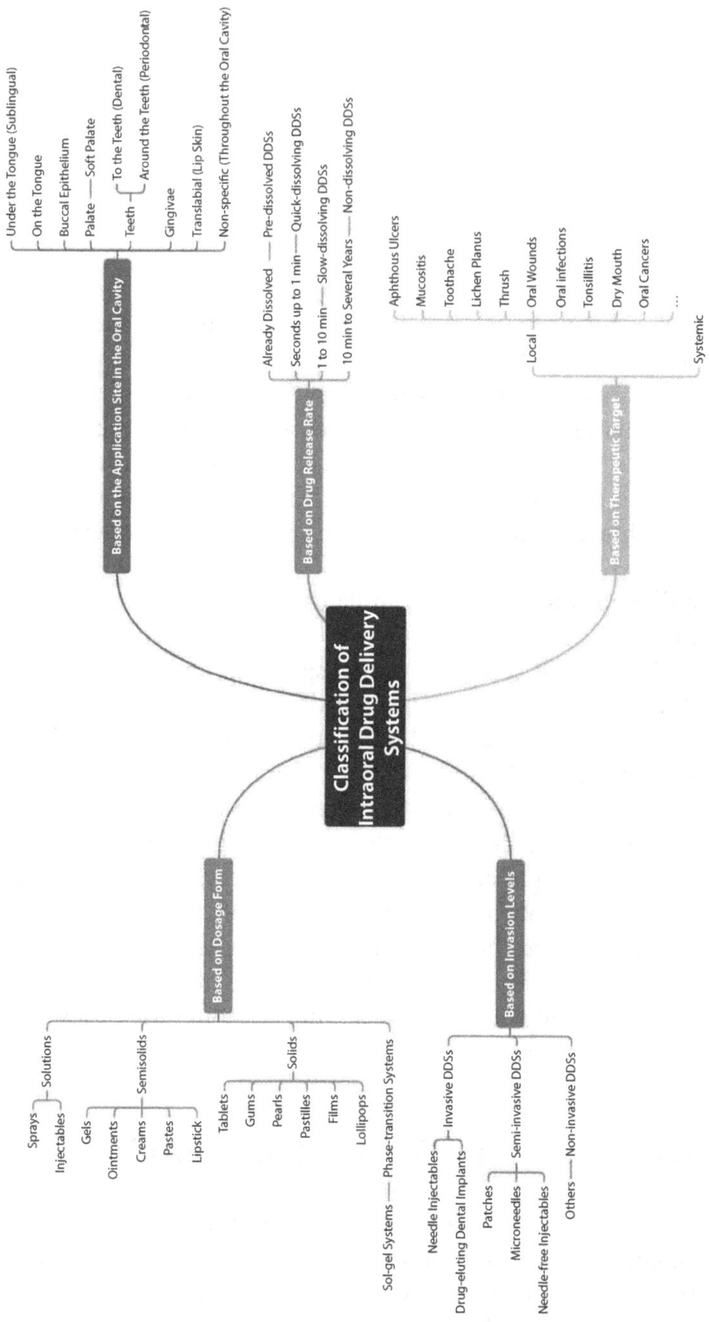

Figure 3.4 Classification of intraoral drug delivery systems.

classification within this chapter is based on the specific anatomical application sites. Consequently, intraoral drug delivery systems are categorized into the following groups:

- Orodispersible dosage forms (on the tongue)
- Sublingual dosage forms (under the tongue)
- Buccal dosage forms (buccal mucosa)
- Gingival/periodontal dosage forms
- Dental drug delivery systems
- Soft palatal platforms
- Translabial drug delivery systems
- Intraoral dosage forms with nonspecific site

Throughout each section of this chapter, we will delve into the distinctive features of each subcategory and review potential subgroups that may therein exist. By the end of this chapter, you will have gained a comprehensive understanding of the intraoral drug delivery systems and their applications.

3.5.1 Orodispersible Dosage Forms (On the Tongue)

Orodispersible dosage forms represent an innovative and patient-friendly approach to drug delivery. These solid formulations are designed to rapidly disperse in the oral cavity, typically within one minute. They offer several distinct advantages, such as high patient acceptance, ease of administration, and, in some cases, the potential to enhance drug bioavailability. These dosage forms are particularly valuable for a diverse range of patients, including children, the elderly, and individuals with swallowing difficulties. Orodispersible formulations include tablets (ODTs), films (ODFs), granules (ODGs), and oral lyophilizates (OLs) [22].

3.5.1.1 Orodispersible Tablets

Orodispersible tablets (ODTs) are one of the most recognized and widely used orodispersible dosage forms. These tablets are designed to disintegrate quickly in the presence of saliva, making them easy to swallow. The rapid disintegration is achieved through the incorporation of superdisintegrants—special ingredients that facilitate the breakdown of the tablet into fine particles. Manufacturing technologies for ODTs have evolved significantly to ensure optimal performance. Various techniques are employed, including direct compression, freeze-drying, sublimation, and 3D printing. Each method offers distinct advantages and challenges. For instance, direct compression is a widely utilized and cost-effective approach. Freeze-drying, on the other hand, is suitable for more sensitive active pharmaceutical ingredients (APIs) but requires specialized equipment and processes [23].

In cases where ODTs are intended for pediatric use, the selection of excipients is of paramount importance. It is essential to choose excipients that are safe and suitable for children. Additionally, a subcategory of ODTs, known as minitablets (ODMTs), has come to attention. These are smaller tablets with diameters under 5mm, making them particularly suitable for a wide range of patients, including neonates. The most common method for manufacturing ODMTs is direct compression [24].

3.5.1.2 Orodispersible Films

Orodispersible films (ODFs) represent an innovative approach to drug delivery. These thin and flexible sheets are designed to disperse and release the active substance rapidly when placed in the oral cavity. One notable feature of ODFs is their dose flexibility. They can be easily cut to adjust the dosage, making them suitable for a wide range of patients. The composition of ODFs primarily includes polymers, with cellulose-based polymers often forming the structural base. Plasticizers are commonly incorporated to enhance flexibility. The manufacturing of ODFs utilizes various techniques, including the solvent casting method, electrospinning, hot melt extrusion, and inkjet printing. The solvent casting method, while traditional, remains a widely used and effective approach for ODF preparation. Electrospinning is a cutting-edge technique that employs high voltage to create nanofibers from a mixture of polymer and drug. Hot-melt extrusion is particularly useful for APIs that are not stable in liquid solvents; however, it involves high temperatures, which cellulose derivatives or many drugs may not tolerate. Consequently, it is crucial to select structural polymers that are compatible with the specific conditions of the chosen method. Inkjet printing is a more recent and innovative method for producing ODFs. It involves printing an API solution or suspension onto a suitable substrate, enabling precise dosage control and customization. The versatility of inkjet printing makes it an attractive choice for ODF production [25].

3.5.1.3 Orodispersible Granules

Orodispersible granules (ODGs) are multiparticulate dosage forms usually prepared upstream for tablets and capsules, but they can also be used as they are. Granules' small size is an advantage and offers a low risk of swallowing. ODGs offer flexibility in dose, allowing dose adjustment as needed. Since ODGs are solid dosage forms, they are more stable in comparison with liquid formulations; but atmosphere humidity might affect their stability. To prevent disintegration before use, ODGs are often packaged in single-dose sachets or similar packaging, ensuring that the granules remain intact and dry until they are opened to be consumed. Any common granulation method can be used to prepare ODGs. It is worth mentioning that ODGs are not specifically categorized as individual dosage forms in the European Pharmacopoeia. Consequently, the quality control process for these formulations can be challenging, as they do not conform to the standards applied to other common formulations [26].

3.5.1.4 *Oral Lyophilizates*

Oral lyophilizates (OLs) are a unique type of orodispersible dosage form that offers rapid disintegration in the oral cavity and immediate conversion into a solution or suspension. What sets OLs apart is their manufacturing process, which involves freeze-drying a liquid or semisolid substance to create the final product. Freeze-drying, also called lyophilization, provides a porous structure with very short disintegration time, and long shelf life. Although OLs may appear similar to ODTs in their shape, packaging, and mode of administration, their production method is distinct. The main disadvantage of OLs is low mechanical strength. Also, due to the manufacturing method, the final product is highly hygroscopic. Packaging and method of preparation apply higher expenses to OLs in comparison with compressed ODTs [24, 26].

3.5.1.5 *Nanoparticles in Orodispersible Formulations*

In orodispersible formulations, taste plays an important role since the drug is exposed to the oral cavity. There are also challenges in formulating poorly soluble medications in orodispersible formulations. Nanoparticle-based orodispersible drug delivery platforms are being developed to improve the oral bioavailability and taste of poorly soluble and unpalatable medications [27].

In conclusion, orodispersible dosage forms have gained widespread recognition and application due to their remarkable advantages, including ease of administration and enhanced patient compliance. However, these formulations also present unique challenges that demand specific attention and ongoing research. As already mentioned, one of the most critical factors in the development of orodispersible formulations is taste, as they disintegrate directly in the mouth. Various approaches are employed to improve the taste, such as the use of sweeteners, flavoring agents, microencapsulation, nanocarriers, or complexation techniques. These strategies aim to ensure that patients find these dosage forms not only effective but also palatable, further enhancing their acceptance and usability in a wide range of clinical settings [22, 27, 28].

3.5.2 Sublingual Dosage Forms (Under the Tongue)

Systemic drug delivery through the sublingual route has garnered special attention for its ability to provide rapid pharmacological effects. Sublingual formulations are placed under the tongue, allowing it to be directly absorbed and enter the bloodstream through the underside of the tongue and the floor of the mouth. This method of drug absorption is notably more efficient when compared to other administration routes. Sublingual drug absorption is from 3 to 10 times higher than oral administration and is rivaled only by hypodermic injection. As a consequence, it leads to a quick onset of action but a relatively short duration. Several factors influence drug absorption through the sublingual route, including the lipid solubility and pKa of the drug, saliva pH, oral epithelium thickness, and the partition coefficient of the drug molecule, which are, collectively, pivotal in choosing the appropriate formulation [29, 30].

In this section, we will explore sublingual drug delivery through both solid and liquid dosage forms, which encompass a range of options for patients. Sublingual solid dosage forms include tablets, films, and pearls while sublingual liquid dosage forms include sprays and drops.

3.5.2.1 Sublingual Tablets

Sublingual tablets (SLTs) are solid dosage forms specifically prepared to dissolve or disintegrate rapidly when placed under the tongue. These tablets are typically small and flat, making them more convenient for patients to use. The quick disintegration of SLTs allows for the rapid absorption of the drug through the highly vascularized sublingual mucosa. Various methods are employed in the preparation of SLTs, including direct compression, compression molding, freeze-drying, and hot melt extrusion. Commercially available SLTs have gained popularity for their role in providing fast-acting medications which, eventually, contributes to the improvement of patients' quality of life [29].

3.5.2.2 *Sublingual Films or Strips*

Sublingual films or strips belong to the category of orodispersible films. These films share similarities with sublingual tablets in their ability to dissolve easily in the mouth and disintegrate rapidly. The preparation methods for sublingual films are similar to those used for orodispersible tablets and films in general [31].

3.5.2.3 *Sublingual Solutions*

Sublingual sprays and drops are concentrated drug-containing solutions designed to be administered under the tongue. These formulations offer a notable advantage in that the drug is predissolved and, when administered as a solution, can provide rapid absorption. The ease of use and rapid onset of action make sublingual sprays and drops a good choice for patients seeking quick drug effects [32, 33].

3.5.2.4 *Sublingual Soft Gelatine*

Sublingual soft gelatine formulations, also known as pearls, are formulations containing drugs in liquid or semisolid phase. The soft gel capsule format has a shell that encapsulates drug-loaded formulation, which provides easy administration and accuracy in dose adjustment. Compared to sublingual tablets, sublingual pearls offer faster absorption and onset of action [34].

3.5.2.5 *Nanoparticles in Sublingual Formulations*

Nanoparticles are employed in sublingual formulations to enhance bioavailability. These nanoparticle-based sublingual formulations have demonstrated improvements in drug permeability across the epithelium, modification of drug release kinetics, solubilization, and protection of compounds sensitive to degradation [18].

Overall, sublingual dosage forms, either solid or liquid, represent a valuable option for patients who require fast-acting medication. The choice between sublingual tablets, sprays, drops, pearls, or films often depends on patient preferences, the specific drug being administered, and the particular clinical requirements for quick onset of action. The efficient and rapid absorption characteristics of the sublingual route make it an attractive option for drugs that necessitate a quick response, such as medications for acute pain relief or emergencies. However, it is essential to carefully select the appropriate dosage form and formulation to ensure optimum safety and efficacy.

3.5.3 Buccal Dosage Forms

The buccal area, which is covered by a mucosal layer, serves as a valuable drug delivery site, for an array of dosage forms. These formulations, available in solid, semi-solid, and liquid forms, present a multitude of advantages, such as convenient application, high patient compliance, direct absorption into the bloodstream, and protection of drugs from gastrointestinal degradation. Despite these benefits, buccal drug administration faces some challenges, including limited surface area for application and absorption, a short duration of medication effectiveness due to continuous saliva flow, and the risk of drug destabilization attributed to buccal pH variations. Additionally, achieving high therapeutic dosages with buccal formulations can be challenging due to their limited size [35, 36]. In this section, we will address buccal formulations, including tablets, patches/films, microneedle platforms, needle-free injectors, and buccal solutions.

3.5.3.1 Buccal Tablets

Buccal tablets have emerged as a prominent choice among buccal dosage forms. These solid formulations are designed to provide drug release, either unidirectional, (i.e., in direct contact with the buccal mucosa) or multidirectional (i.e., in saliva). To achieve unidirectional release, buccal tablets often feature a backing layer that inhibits drug release from the back side. The incorporation of mucoadhesive polymers into the tablet structure ensures that these tablets remain in the buccal cavity for an extended duration. The primary method for manufacturing buccal tablets is direct compression. It is worth mentioning that buccal tablets may not be suitable for certain patient populations, particularly children and the elderly. Discomfort during application, the potential for mucosal irritation, the risk of tablet detachment from the buccal mucosa, subsequent swallowing, and adhesion to the esophagus, are factors that merit consideration [37, 38].

3.5.3.2 Buccal Films/Patches

Buccal films and patches represent another set of buccal dosage forms, characterized by their thin, flat, and flexible structures. These formulations are designed to provide unidirectional or multidirectional drug release within the oral cavity. Buccal patches are typically multilayered; though simpler unilayer designs are also employed. Mucoadhesion is a prevalent feature of these formulations, achieved by incorporating mucoadhesive polymers into the primary structure or as a separate layer covering the drug-loaded section. In unidirectional patches, a hydrophobic backing layer, covering the drug-loaded section, inhibits release from one side, adds protection, and facilitates application. Various methods are utilized in the preparation of buccal films and patches, including solvent casting, hot melt extrusion, freeze drying, and electrospinning. Patients generally find these formulations highly acceptable due to their flexibility and structure. They are particularly well suited for small children and the elderly. Nonetheless, a significant challenge associated with the use of mucoadhesive films/patches is achieving high therapeutic dosages of drugs, which is influenced by formulation structure and size [39, 40].

3.5.3.3 *Microneedle Platforms*

Microneedle devices are designed for active transmembrane drug delivery, capable of penetrating both skin and mucosal layers, including the buccal region. These microneedles, typically ranging in height from 25 to 2000 microns, are considered one of the future platforms for transmucosal drug delivery. Microneedles offer a host of advantages that position them as strong candidates for drug and vaccine delivery. To mention some, bypassing the liver and gastrointestinal system, reducing pain when compared to traditional hypodermic needles, simple application, and overcoming the barrier posed by upper dermal layers that can hinder drug absorption. Microneedles can be prepared from silicon, metal, or polymers, and their manufacturing processes include 3D printing, injection molding, wet chemical or reactive ion etching, laser drilling, hot embossing, drawing lithography, or lithography plus electroforming, and two-photon polymerization. Different microneedle designs have been developed, including solid, hollow, coated, and dissolvable microneedles. While microneedle platforms offer numerous advantages, they are not without limitations and challenges. Some of these include patient acceptance, manufacturing complexities, and specific considerations related to drug delivery [41–43].

3.5.3.4 *Needle-Free Injectors*

Needle-free injectors, also known as "jet-injectors," provide an alternative to traditional needles. These devices employ pressure to propel liquid formulations, including vaccines and drugs, through a small orifice into the skin or mucosal area. This needle-free approach reduces discomfort and has gained popularity due to its convenience and ease of use [44].

3.5.3.5 *Buccal Solutions*

Buccal solutions are liquid dosage forms designed for buccal administration. These solutions are typically concentrated and presolubilized, allowing for rapid absorption and drug delivery through the buccal mucosa. Buccal solutions are being used for many purposes, in patients of a wide range of ages, even while unconscious. These formulations are easy to use and able to contain high dosages of drugs; but considering the liquid base, they cannot remain for a long time in a buccal area, as they are washed out or swallowed [38, 45].

3.5.3.6 *Nanoparticles in Buccal Formulations*

Similar to the application of nanoparticles in sublingual formulations, they are also utilized in buccal formulations to improve bioavailability. Nanoparticle-containing buccal formulations have shown enhancement in drug permeability, drug release modification, solubilization, and protection of sensitive compounds [18].

Buccal drug delivery continues to evolve, offering various formulations to cater to the unique needs of patients and healthcare providers. As understanding and advancements in buccal drug delivery continue to develop, the potential for enhancing drug administration and patient care remains promising.

3.5.4 Gingival/Periodontal Dosage Forms

Gingival and periodontal drug delivery is a specialized and targeted approach in the field of intraoral drug administration. These formulations are tailored for oral health conditions, particularly those related to the gingival and periodontal tissues. Gingiva, the soft tissue surrounding the teeth, and periodontal structures, including the gums and supporting bone, are prone to various oral diseases, infections, and inflammations. Localized drug delivery to these areas offers a means to manage and treat such conditions effectively. Here, we will address gingival/periodontal formulations, including gels, chips, rinse solutions, microspheres, and nanoparticles.

3.5.4.1 *Gingival/Periodontal Gels*

Gingival/periodontal gels are topical dosage forms designed to be applied directly to the affected areas within the oral cavity. These gels are composed of a semi-solid or viscous base that provides contact with the gingival and periodontal tissues thereby enhancing drug absorption. They are particularly useful for localized treatment of periodontal diseases, including gingivitis and periodontitis. The active ingredients in gingival/periodontal gels can vary, including antibiotics and anti-inflammatory agents. The use of gels allows for precise and targeted drug delivery, thus reducing systemic exposure and potential side effects. *In situ* gel-forming systems and mucoadhesive formulations demonstrate promising potential in localized drug delivery for gingival and periodontal applications [46, 47].

3.5.4.2 Gingival/Periodontal Chips

Gingival/periodontal chips are small, solid, and often biodegradable units that are inserted directly into periodontal pockets or gingival crevices. These chips serve as reservoirs for drug release, providing prolonged therapeutic effects at the local level. The biodegradable nature of these chips is advantageous, as they gradually dissolve or disintegrate, eliminating the need for removal and minimizing patient discomfort. Gingival/periodontal chips are designed to deliver antibiotics, antiseptics, and other medications locally [48].

3.5.4.3 Gingival/Periodontal Rinse Solutions

Gingival/periodontal rinse solutions are designed for topical application and provide a convenient means of delivering medication to the gingival and periodontal tissues. These solutions can contain antimicrobial agents, antiseptics, or other active ingredients that help maintain oral health and manage oral conditions such as plaque and gingivitis. Patients can easily incorporate gingival/periodontal rinse solutions into their daily oral hygiene routines. These solutions can be an integral part of managing gum diseases or as a preventive measure for individuals at risk of periodontal issues [49].

3.5.4.4 Biodegradable Microspheres and Nanoparticles

Innovations in drug delivery have brought about the development of biodegradable microspheres and nanoparticles for gingival and periodontal applications. These drug carriers can be loaded with various active ingredients and precisely targeted to the affected gingival or periodontal sites. They provide controlled drug release over an extended period, ensuring consistent therapeutic concentrations at the treatment site. Biodegradable microspheres and nanoparticles offer advantages such as sustained drug delivery, reduced systemic exposure, and improved patient compliance. They are particularly useful in the management of chronic periodontal conditions and for localized therapy in periodontal pockets [50].

Gingival and periodontal dosage forms play an important role in managing a range of oral health issues. The localized nature of drug delivery in these formulations allows for effective treatment of gingival and periodontal diseases while minimizing the risk of systemic side effects. Additionally, the ease of administration and patient compliance make these formulations valuable tools in the field of oral healthcare. In the realm of intraoral drug delivery, gingival and periodontal dosage forms play a pivotal role in improving oral health, managing periodontal diseases, and enhancing patient outcomes. Advances in the development of these formulations continue to expand the possibilities for more effective and patient-friendly treatments.

3.5.5 Dental Drug Delivery Systems

Dental drug delivery systems represent a specialized and evolving branch of intraoral drug administration that focuses on oral health management, localized treatment of oral diseases, and targeted delivery of therapeutic agents within the oral cavity. These systems include a range of innovative approaches designed to improve the effectiveness of oral healthcare and therapeutic outcomes.

3.5.5.1 Drug-Eluting Implants and Screws

Dental drug-eluting implants and screws are innovative devices designed to provide localized drug delivery within the dental area. These implants or screws are typically composed of biocompatible materials and are loaded with therapeutic agents. Upon placement, they release the drug gradually, ensuring continuous exposure to the treatment site. One of the primary applications of drug-eluting implants in dentistry is for the management of periodontal diseases. These implants can be placed in periodontal pockets or other sites to deliver antimicrobial agents, anti-inflammatories, or other medications directly to the affected area. This targeted approach minimizes systemic exposure while maximizing the efficacy of treatment [51].

3.5.5.2 Smart Dental Implants

Smart dental implants are an important advancement in dental drug delivery. These implants are equipped with sensors or responsive elements that can detect changes in the oral environment, such as pH levels, bacterial activity, or the presence of specific biomarkers. In response to these cues, smart dental implants can release drugs or therapeutic agents as needed, providing personalized and on-demand treatment. For instance, smart dental implants can be designed to release antimicrobial agents when bacterial overgrowth is detected, thereby preventing infection or complications after oral surgery. These implants can also assist in the management of conditions like peri-implantitis by releasing anti-inflammatory medications when signs of inflammation are detected [52].

3.5.5.3 Dental Patches

Dental patches are adhesive and flexible systems that adhere to the oral mucosa, offering localized drug delivery to the gums, cheeks, or other areas within the oral cavity. These patches can be loaded with therapeutic agents, designed for specific dental conditions or provide localized anesthesia. One common application of dental patches is for the treatment of oral aphthous ulcers, also known as canker sores. These patches are designed to adhere to the affected area and release therapeutic agents, such as pain relievers or anti-inflammatory drugs, directly to the ulcer site, providing relief and expediting healing [53].

3.5.5.4 Nanoparticles for Dental Drug Delivery

Nanoparticles have made significant inroads in the field of dental drug delivery. These submicron-sized carriers, often made of biocompatible materials, can be loaded with therapeutic agents and utilized to manage various dental conditions. The versatility of nanoparticles allows for the time-controlled release of drugs, ensuring that therapeutic concentrations are maintained in the oral cavity. One of the key applications of nanoparticles in dental drug delivery is the management of dental caries, by providing sustained release of therapeutic agents. Additionally, nanoparticles can be used in the treatment of oral infections, tooth remineralization, oral biofilm and periodontal disease management, root canal disinfection, and other dental health issues [50].

In the evolving landscape of dental drug delivery systems, these innovations are transforming the way oral health conditions are managed. By offering targeted and localized drug delivery, reducing systemic exposure, and providing on-demand treatment, these systems hold great promise for improving patient care and overall oral health outcomes. The continued development of dental drug delivery technologies ensures that patients receive effective and personalized treatment while minimizing side effects and discomfort associated with traditional approaches.

3.5.6 Soft-Palatal Platforms

Soft-palatal platforms represent a specialized area in intraoral drug delivery, focusing on the soft and delicate tissue of the palate. The palate, located at the roof of the mouth, plays an important role in speech, swallowing, and sensory functions. The soft-palatal area, specifically, is an intricate region with unique drug delivery challenges and opportunities.

3.5.6.1 Soft-Palatal Solutions

Soft-palatal liquid dosage forms are a group of intraoral drug delivery systems designed for delicate and sensitive tissues of the soft palate. These liquid formulations are manufactured to be applied directly to the soft-palatal mucosa, offering advantages in both systemic and local drug administration. Soft-palatal sprays are a group of formulations, designed to be used on palatal areas, with applications in a range of oral health issues. These formulations are particularly valuable in cases of palatal inflammation, ulceration, or discomfort, offering rapid relief and promoting the healing process. Furthermore, soft-palatal liquid dosage forms can be used as carriers for antimicrobial agents, providing effective solutions for infections in this specific area. They also can be used in allergies to reduce symptoms. Their ease of application, quick onset of action, and localized drug delivery make soft-palatal liquid dosage forms good choices in maintaining oral health and managing conditions that affect the soft palate [54].

3.5.6.2 Soft-Palatal Patches

Soft-palatal patches are usually adhesive and flexible systems that attach to the soft-palatal tissue, providing localized drug delivery to this specific area. These patches are designed to address conditions that affect the soft palate, such as palatal cysts or postoperative management after palatal surgeries. One notable application of soft-palatal patches is in the management of obstructive sleep apnea. These patches can be loaded with medications or therapeutic agents aimed at reducing palatal tissue vibrations and airway obstructions, offering potential relief for individuals with this condition [55, 56].

Innovations in soft-palatal platforms are ongoing, and they continue to play a valuable role in specific conditions and concerns related to this unique region within the oral cavity. As the field of intraoral drug delivery advances, these specialized platforms contribute to improving patient care and the management of soft-palatal issues.

3.5.7 Translabial Drug Delivery Systems

Translabial drug delivery systems are a novel and specialized approach for drug administration that specifically targets the labial, or lip, region. Although less common than other intraoral delivery methods, translabial drug delivery systems have unique applications and advantages in specific healthcare conditions.

One of the most accessible and widely recognized forms of translabial drug delivery is through the use of medicated lip balms. Lip balms, a common cosmetic and therapeutic product, are frequently used to deliver therapeutic agents to the lips. These lip balms can contain a variety of ingredients, such as sunscreens, moisturizers, or medications for conditions like cold sores or lip chapping. The topical application of lip balm provides a straightforward way of delivering medications to the labial tissues [57, 58].

Transdermal patches designed for labial drug delivery represent another innovative option in this field. These adhesive patches are specifically designed to adhere to the lips and gradually release medications over time. This approach ensures a continuous and controlled drug delivery to the labial tissues. Labial transdermal patches have found applications in pain management, particularly for oral lesions or cold sores. These patches can contain local anesthetics or antiviral medications, offering relief from pain and expediting the healing process. They are especially useful when systemic drug exposure is not required or when other oral administration routes may not be practical [57, 59].

Using liposomal formulations has also garnered attention not only in general drug delivery but also in cosmetics, including translabial formulations. Liposomes are small, spherical vesicles composed of amphiphilic lipids that can encapsulate drugs or other therapeutic agents. This encapsulation enhances the delivery of these agents to the tissues. Liposomal formulations are particularly beneficial for conditions that affect the lips, such as *herpes labialis* (cold sores) or angular cheilitis, as they can provide controlled and localized drug release and reduce the risk of systemic exposure [60, 61].

Challenges are an integral part of drug delivery systems, including translabial formulations. The labial tissues are highly sensitive, and any formulations or medications applied to this area should be well tolerated and free from irritants. Ensuring that the medications or therapeutic agents do not cause discomfort or allergic reactions is essential to the success of translabial drug delivery. Moreover, given the proximity of the labial tissues to the oral cavity, there is a potential for oral ingestion of the delivered drug, which should be considered and managed, to avoid unintended systemic exposure.

In general, translabial drug delivery systems offer a unique approach for specific conditions or issues affecting the labial region. Their localized and controlled delivery mechanisms make them a valuable option. As advancements in the formulation and drug delivery technologies continue, translabial drug delivery systems may find expanded applications, providing solutions to a range of lip-related health concerns.

3.5.8 Intraoral Dosage Forms with Nonspecific Site

Intraoral drug delivery systems encompass a diverse range of dosage forms tailored for various anatomical sites within the oral cavity. Among these, there are dosage forms that do not target specific anatomical sites but instead provide versatile and convenient administration options. Chewable dosage forms (such as pastilles and tablets), lollipops, and lozenges, fall into this category. These formulations offer a unique set of advantages and applications in the field of intraoral drug delivery.

3.5.8.1 Chewable Dosage Forms: Pastilles and Tablets

Chewable dosage forms have gained popularity for their ease of administration and pleasant taste, making them suitable for patients of all ages. Pastilles and tablets are common representatives of this category and are designed to be disintegrated and dissolved by chewing within the oral cavity.

Pastilles are soft, typically gelatine-based dosage forms that can be easily broken down by chewing in the mouth. These formulations are favored for their ability to mask the taste of unpalatable medications, making them a suitable choice for pediatric and geriatric patients who may have difficulty swallowing. Pastilles are being used for both systemic and localized drug administration [62, 63].

Chewable tablets are solid dosage forms, similar to conventional tablets, but formulated to disintegrate rapidly when chewed. They can be used for a wide range of therapeutic purposes. These tablets are popular for delivering vitamins, minerals, analgesics, and antacids. Their chewable nature enhances patient compliance, as they are easier to ingest.

3.5.8.2 Lollipops

Lollipops have found a unique role in the field of drug delivery. These confectionery-like dosage forms are designed to be slowly dissolved in the mouth, allowing for the controlled release of medication. Lollipops offer several advantages, including ease of administration, and a pleasant taste, making them particularly well suited for pediatric patients or individuals who have difficulty in swallowing traditional medications. Lollipops can be used for various therapeutic purposes. They are frequently used to deliver medications for pain management, such as analgesics or local anesthetics. Additionally, lollipops can serve as an effective vehicle for antitussive medications to alleviate coughing or sore throat. These formulations are also being used for systemic drug delivery through absorption through the oral mucosa. The unique aspect of lollipops in drug delivery is their ability to provide a consistent and controlled release of medication as the candy slowly dissolves. This controlled release can be advantageous for maintaining therapeutic drug levels over an extended period [64].

3.5.8.3 Lozenges

Lozenges are solid dosage forms designed to be held in the mouth to allow them to be dissolved slowly. They can also be used for various therapeutic applications, including relieving sore throats, alleviating coughs, and administering antimicrobial agents for local oral infections. The slow dissolution of lozenges allows prolonged drug exposure to the oral mucosa and provides a soothing effect in cases of irritation or inflammation. They can also include local anesthetics or analgesics. Lozenges are particularly beneficial for patients who prefer not to swallow medications or those with localized oral conditions [65].

In conclusion, dosage forms with nonspecific site targeting, offer versatile and patient-friendly options for intraoral drug delivery. Their ease of administration, pleasant taste, and controlled drug release characteristics make them valuable choices in various healthcare settings, particularly for pediatric and geriatric patients and those with difficulty swallowing conventional medications. These formulations showcase the adaptability and innovation in the field of intraoral drug delivery, focusing on the diverse needs and preferences of patients of all ages.

3.6 Pharmacokinetics

The pharmacokinetics of intraorally administered drugs is influenced by (i) the physiological aspects of the application site, (ii) the physicochemical properties of the drug molecule, and (iii) characteristics of the formulation and dosage form. Here, an attempt has been made to briefly discuss the role of each factor.

The permeability of nonkeratinized oral mucosa (i.e., the floor of the mouth, sublingual, and buccal regions) is higher than the keratinized parts (i.e., hard palate and gingiva). Therefore, nonkeratinized oral regions are of great interest for both systemic and topical drug delivery [66], while keratinized mucosa is generally considered a topical route [67, 68].

Buccal mucosa undertakes a barrier function against xenobiotics. So, as a foreign substance, the drug molecules should overcome several obstacles. The first barrier is the epithelial layer of the buccal region, particularly its upper third part, which resists drug permeation. Then, the drug encounters an enzymatic barrier. Fortunately, however, this does not create a critical challenge for drug delivery for three reasons; (i) drug degradation by salivary enzymes is negligible, (ii) compared to all mucosal enzymes, the activity of buccal enzymes is the lowest; and (iii) the activity of the buccal enzymes, if needed, could be inhibited using different permeation enhancers. Another barrier that is often overlooked in studies is the lamina propria, which can provide a hydrophilicity-related barrier against drug absorption. Generally, permeation of highly lipophilic drugs through the lamina propria is challenging [69].

The sublingual region is more permeable than buccal, due to its less thickness and more blood supply. So, this route is preferred when a rapid onset of drug action is required. However, the residence time of the drug in this region is restricted due to salivation and tongue movement [67]. The buccal mucosa is smoother, more immobilized, and resistant to irritation, damage, and repeated food exposure, compared to the sublingual region [67, 70]. Here again, the salivary flow is the main challenge to be overcome [67]. Additionally, due to the higher thickness of the buccal compared to the sublingual mucosa, the absorption of some drugs may be too slow [68].

In general, the permeation molecular weight cut off in the buccal mucosa is considerably higher than the skin. So, buccal mucosa is an interesting region for the delivery of proteins, peptides, and genes. Additionally, a large number of chemical penetration enhancers that are generally used to improve skin permeability (e.g., surfactants, fatty acids, terpenes, etc.) are also prone to improve buccal permeation [66]. Also, it is possible to further improve the buccal permeability by optimizing the physicochemical properties of drug molecules (e.g., molecular weight, lipophilicity, and ionization) [67]. Generally, drug molecules with aqueous solubility of > 1mg/ml, molecular weight of less than 500 Da, and 1 < logP < 3 are well absorbed from the oral mucosa [68].

Nanoparticles are of great interest for sublingual and buccal drug delivery owing to several capabilities: (i) modifying the kinetic of drug release, (ii) enhancing the permeability of drugs, (iii) increasing the solubility of low water-soluble drugs, and (iv) improving the stability of susceptible drugs to degradation (e.g., proteins and peptides). For effective sublingual or buccal delivery, there are two main issues to be considered. Firstly, the physicochemical properties of nanoparticles should be optimized. Unfortunately, there is a lack of comprehensive studies about the optimal size and charge of nanoparticles. However, most experiments in this field have studied the size range of 100 to 300 nm. It has been shown that the surface charge of nanoparticles affects their interaction with the oral mucosa as well as the depth of penetration. Nevertheless, there are contradictory results about the optimum charge of nanoparticles. The second matter to be considered is the interaction of nanoparticles with the pharmaceutical base that incorporates such. Nanoparticles should be stable within the formulation base, and the base should ensure the sufficient residence time of embedded nanoparticles in the intended region and provide an appropriate concentration gradient for a long term [18].

When a formulation is placed in the intraoral cavity, nanoparticles may have different fates (Figure 3.5). The first fate is the washing-out

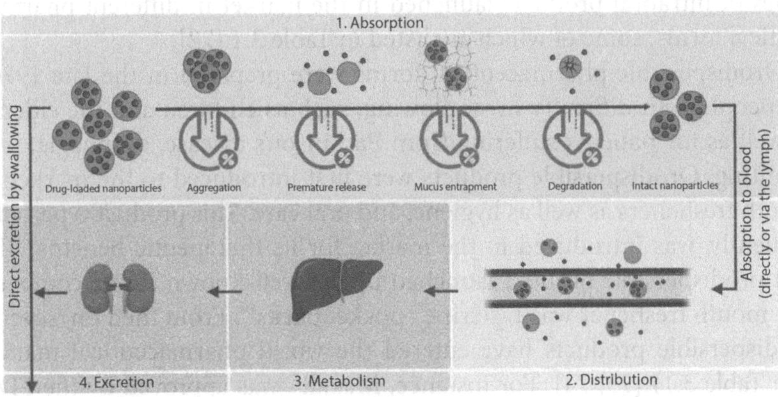

Figure 3.5 Different fates of intraoral nanoparticles (pink circles) containing drug molecules (cyan circles) according to the pharmacokinetic stage (Absorption, Distribution, Metabolism, and excretion) (created with BioRender.com).

elimination; nanoparticles may swallowed and enter the gastrointestinal tract. In this case, nanoparticles may be excreted or degraded, and drugs released in the lumen. The second scenario is the failure of intraoral drug delivery due to premature release of drugs from nanoparticles in the oral cavity, early pH- or enzymatic-induced degradation, burst release of drugs, and aggregation of nanoparticles. If nanoparticles pass these challenges safely, they reach the oral mucosa. Mucosal surfaces may prevent the permeation of nanoparticles. In this case, however, the drug absorption may be improved due to the enhancement of the residence time of the nanoparticles. Another possibility is the absorption of "intact" nanoparticles into the blood circulation and thereby, enhancement of the plasma half-life of drugs. Then, absorbed nanoparticles may be degraded in the liver or undergo the lymphatic pathway. Based on the physicochemical properties (e.g., size, surface charge, composition, morphology, and surface functionalization), the absorbed nanoparticles may perform passive or active targeting [71].

3.7 Products in the Market

In general, there is a good acceptance of intraoral drug delivery by patients. As mentioned earlier, many studies have been conducted on intraoral drug delivery with different pharmaceutical forms, some of which have entered the pharmaceutical market. The first product of intraoral drug delivery that entered the market was nitroglycerin sublingual tablets, which were introduced in the early 1960s for treating angina pectoris. Since then, a series of intraoral products launched in the market in different pharmaceutical forms, some of which are listed in Table 3.1 [72].

Orodispersible pharmaceutical forms were prepared in the late 1970s for people with difficulty in swallowing, such as children and the elderly, as well as for patients suffering from Parkinson's disease, dysphagia, and vomiting. Orodispersible products were first introduced to the market as mouth fresheners as well as hygiene, and oral care. This product type, subsequently, was introduced in the market for its therapeutic benefits. The first orodispersible product launched by the well-known Pfizer company as a mouth freshener was Listerine® pocket packs™. From then on, several orodispersible products have entered the world pharmaceutical market (see Table 3.1) [73, 74]. For instance, Maxalt® was approved by the FDA as an orodispersible tablet in 1998. At present, this product has been discontinued and replaced by FDA-approved rizatriptan orodispersible film under the brand name Rizafilm® [75].

Table 3.1 Some commercial intraoral products.

Anatomical region	Dosage form	Brand name	Active ingredient	Manufacturer
On the tongue	Orodispersible Tablets	Alavert®	loratadine	Wyeth Consumer Healthcare
		Remeron SolTab®	Mirtazapine	Organon Inc
		Fazaclo®	Clozapine	Alamo Pharmaceuticals
		NuLev®	Hyoscyamine	Alaven Pharmaceutical LLC
		Zyprexa Zydis®	Olanzapine	Eli Lilly Nederland B.V
		Staxyn®	Vardenafil	GlaxoSmithKline
		Zelapar®	Selegiline hydrochloride	Valeant Pharmaceuticals International
		Orapred ODT®	Prednisolone sodium phosphate	BioMarin Pharmaceutical Inc./Alliant Pharmaceuticals, Inc
	Orodispersible Films	Zuplenz®	Ondansetron	Aquestive Therapeutics
		RizaFilm®	Rizatriptan	IntelGenx Corp
Sublingual	Tablet	Abstral®	Fentanyl Citrate	Orexo AB
		Subutex®	Buprenorphine	Reckitt Benckiser
		Edluar®	Zolpidem tartrate	Orexo and Meda AB
		Isordil®	Isosorbide dinitrate	BIOVAIL
		Nitrostat®	Nitroglycerin	Parke-Davis Pharmaceutical Research
		Zubsolv®	Buprenorphine and naloxone	Orexo AB
		Intermezzo®	Zolpidem tartrate	Purdue Pharma LP
		Saphris®	Asenapine	Allergan plc
		Ismigen®	Polyvalent mechanical bacterial lysate (biological)	Bruschettini S.R.L.

(*Continued*)

Table 3.1 Some commercial intraoral products. (*Continued*)

Anatomical region	Dosage form	Brand name	Active ingredient	Manufacturer
		Dsuvia®	sufentanil	AcelRx Pharmaceuticals,Inc
		Nocdurna®	Desmopressin acetate	Ferring Pharmaceuticals Inc
		DesmoMelt®	Desmopressin acetate	Ferring Pharmaceuticals Ltd
	Film	Kynmobi®	Apomorphine hydrochloride	Sunovion Pharmaceuticals Inc.
		Suboxone®	buprenorphine and naloxone	Reckitt Benckiser Pharmaceuticals Inc.
	Spray	Subsys®	Fentanyl	INSYS Therapeutics, Inc.
		Glytrin®	Glyceryl trinitrate	Multiple international companies e.g. Sanofi-aventis, Surry, Ayrton Saunders, Wirral, AFT Pharmaceuticals, Aukland
		NitroMist®	Nitroglycerin	NovaDel Pharma
	Powder	GoNitro®	Nitroglycerin	Espero Biopharma
Buccal	Film	Bunavail®	Buprenorphine and naloxone	BioDelivery Sciences International, Inc.
		Belbuca®	buprenorphine	Endo Pharmaceuticals Inc.
	Tablets	Fentora®/ Effentora®	fentanyl	Cephalon, Teva Pharma B.V
		Oravig®	Miconazole Lauriad	BioAlliance Pharma SA
		Loramyc®	Miconazole Lauriad	Strativa Pharmaceuticals Inc, Woodcliff Lake, Therabel Lucien Pharma Laboratories, Levallois-Perre
		Buccastem M®	Prochlorperazine maleate	Alliance Pharmaceuticals
		Sitavig®	Acyclovir	BioAlliance Pharma SA

(*Continued*)

Table 3.1 Some commercial intraoral products. (*Continued*)

Anatomical region	Dosage form	Brand name	Active ingredient	Manufacturer
		Striant®	Testosterone	Columbia Laboratories
	Paste	Aphthasol®	Amlexanox	Discus Dental, Inc
	Patch	OraDisc A®	Amlexanox	Access Pharmaceuticals, Inc
Gingival/ periodontal	Gel	Atridox®	Doxycycline hyclate	Tolmar Inc. Fort Collins
		Biotene OralBalance®	enzymes (lactoperoxidase, lactoferrin, lysozyme)	EKR Therapeutics, Inc., Bedminster, N
		Chlo-Site®	chlorhexidine gluconate	Ghimas
		Elyzol®	Metronidazole benzoate	Dumex, Copenhagen
		Emdogain®	Hydrophobic enamel	Institute Straumann AG, Basel
		Dentomcin®	Minocycline hydrochloride	Henry Schein UK Holdings Ltd.
		Aftamed®	Hyaluronic acid	BioPlax Limited
		Aminogam®	Hyaluronic acid	Errekappa Euroterapici
	Microspheres	Arestin®	Minocycline	OraPharma, Inc., Warminster
	Fiber	Actisite®	Tetracycline	Alza Corporation, Palo Alto, CA
	Ointment	Periocline®	Minocycline	Sunstar Corp
	Mouthwash	Periogard®/ Peridex®	Chlorhexidine gluconate mouth rinse	Procter and Gamble, Cincinatti, and Colgate-Palmolive Company
		Listerine®	Antiseptic essential oil	McNeill-PPC, Inc. Skillman
		Amicar®	Aminocaproic acid syrup	VersaPharm Inc., Marietta, GA
	Insert, or Chip	PerioChip®/ PerioCol-CG®	chlorhexidine gluconate	DEXCEL PHARMA/ Eucare Pharmaceuticals Ltd

(*Continued*)

Table 3.1 Some commercial intraoral products. (*Continued*)

Anatomical region	Dosage form	Brand name	Active ingredient	Manufacturer
	Dye Solution	HELBO®	Photosensitiser	Bredent Medical
		Fotosan®	Photosensitiser	CMS Dental, Copenhagen
Dental	Paste	Kenalog in Orabase®	Triamcinalone acetonide	Bristol-Myers Squibb Co., Princeton, Taro Pharmaceuticals, Hawthorne
		Oralone®	Triamcinalone acetonide	Bristol-Myers Squibb Co, Princeton,Taro Pharmaceuticals, Hawthorne
	Gel	Corsodyl®	Chlorhexidine Digluconate	GlaxoSmithKline Consumer Healthcare Trading Limited, Brentford, TW8 9GS, and GlaxoSmithKline Consumer Healthcare Limited
Non-specific site	Chewable gum/ lozenge	Nicorette®	Tobacco cessation	GlaxoSmithKlein, Bentford, Middlesex
	Lozenge/ lollipops	Actiq®	Fentanyl citrate	Anesta Corporation
	Throat spray and lozenge	Benactiv®	Flurbiprofen	Reckitt Benckiser H.C.
	Chewable tablet	Singulair®	Montelukast Sodium	Organon
		Tegretol®	Carbamazepine	Novartis
		Zyrtec®	Cetirizine	Pfizer Inc
	Oral Spray	Zolpimist®	Zolpidem tartrate	NovaDel Pharma, Inc.
	Oromucosal solution	Buccolam®	Midazolam	ViroPharma
	Oromucosal Spray	Sativex®	delta-9-tetrahydrocannabinol, cannabidiol	GW Pharma
	Gel	Orabase-B Gel®	Benzocaine	Rochem International Inc. LGM Pharma

The first commercialized buccal formulation was fentanyl Oralet®, which was approved by the FDA in 1996. Effentora®/Fentora®, Breakyl®/Onsolis®, and Abstral® are other commercialized fentanyl products that are used buccally and/or sublingually and have been approved in the US and/or Europe [75].

Reports show that the risk of periodontitis and thereby, the global market for periodontal drug therapy has an ever-growing trend due to an increase in the population of smokers and people with diabetes [76]. Atridox® is a couple-syringe of controlled-release product, which is applied to gingival infections. Syringe A contains Atrigel® (a bioabsorbable liquid system) and Syringe B contains doxycycline hyclate [77, 78].

The journey of innovation in intraoral drug delivery systems seems far from over, offering continuous opportunities for the development of various intraoral products to introduce to the market.

References

1. Homayun, B., Lin, X., Choi, H.J., Challenges and Recent Progress in Oral Drug Delivery Systems for Biopharmaceuticals. *Pharmaceutics*, 11, 3, 129, 2019.
2. Liu, Y., Guo, H., Xu, R., Chen, W., Smart intraoral systems for advanced drug delivery. *MedComm – Biomater. Appl.*, 1, e19, 2022.
3. Ghosh, T.K. and Pfister, W.R., Intraoral delivery systems: An overview, current status, and future trends, in: *Drug Delivery to the Oral Cavity, Molecules to Market*, T.K. Ghosh and W.R. Pfister (Eds.), pp. 1–40, CRC Press, Boca Raton, 2005.
4. Abdul Rasool, B.K. and Shahiwala, A., Buccal and Intraoral Drug Delivery: Potential Alternative to Conventional Therapy, in: *Novel Drug Delivery Technologies: Innovative Strategies for Drug Re-positioning*, A. Misra and A. Shahiwala (Eds.), pp. 29–71, Springer, Singapore, 2019.
5. Hearnden, V., Sankar, V., Hull, K., Juras, D.V., Greenberg, M., Kerr, A.R. *et al.*, New developments and opportunities in oral mucosal drug delivery for local and systemic disease. *Adv. Drug Deliv. Rev.*, 64, 1, 16–28, 2012.
6. Kamrani, P. and Sadiq, N.M., Anatomy, Head and Neck, Oral Cavity (Mouth), in: *StatPearls*, StatPearls Publishing, Treasure Island (FL) ineligible companies, 2023.
7. Madhav, N.V., Shakya, A.K., Shakya, P., Singh, K., Orotransmucosal drug delivery systems: A review. *J. Control. Release*, 140, 1, 2–11, 2009.
8. Bing, L. and JR, R., Preclinical Assessment of Oral Mucosal Drug Delivery Systems, in: *Drug Delivery To The Oral Cavity, Molecules To Market*, G. TK and P. WR (Eds.), CRC Press, Boca Raton, 2005.

9. Shetty, S., Keratinization and its disorders. *Oman Med. J.*, 27, 5, 348–57, 2012.
10. Squier, C.A., The permeability of oral mucosa. *Crit. Rev. Oral. Biol. Med.*, 2, 1, 13–32, 1991.
11. Brizuela, M. and Winters, R., Histology, Oral Mucosa, in: *StatPearls*, StatPearls Publishing, Treasure Island (FL) ineligible companies, 2023, Disclosure: Ryan Winters declares no relevant financial relationships with ineligible companies.
12. Groeger, S. and Meyle, J., Oral Mucosal Epithelial Cells. *Front. Immunol.*, 10, 2019.
13. Presland, R. and Jurevic, R., Making Sense of the Epithelial Barrier: What Molecular Biology and Genetics Tell Us About the Functions of Oral Mucosal and Epidermal Tissues. *J. Dent. Educ.*, 66, 564–74, 2002.
14. Rathbone, M.J. and Tucker, I.G., Mechanisms, barriers and pathways of oral mucosal drug permeation. *Adv. Drug Deliv. Rev.*, 12, 1, 41–60, 1993.
15. Orlando, R.C., Tobey, N.A., Schreiner, V.J., Readling, R.D., Active electrolyte transport in mammalian buccal mucosa. *Am. J. Physiol.*, 255, 3 Pt 1, G286–91, 1988.
16. Oyama, Y., Yamano, H., Ohkuma, A., Ogawara, K., Higaki, K., Kimura, T., Carrier-mediated transport systems for glucose in mucosal cells of the human oral cavity. *J. Pharm. Sci.*, 88, 8, 830–4, 1999.
17. Vadgama, J.V. and Evered, D.F., Absorption of amino acids from the human mouth. *Amino Acids*, 3, 3, 271–86, 1992.
18. Hua, S., Advances in Nanoparticulate Drug Delivery Approaches for Sublingual and Buccal Administration. *Front. Pharmacol.*, 10, 1328, 2019.
19. Miranda-Rius, J., Brunet-Llobet, L., Lahor-Soler, E., Farré, M., Salivary Secretory Disorders, Inducing Drugs, and Clinical Management. *Int. J. Med. Sci.*, 12, 10, 811–24, 2015.
20. Inenaga, K., Ono, K., Hitomi, S., Kuroki, A., Ujihara, I., Thirst sensation and oral dryness following alcohol intake. *Jpn. Dent. Sci. Rev.*, 53, 3, 78–85, 2017.
21. Kaneko, K., Osman, N., Carini, V., Scagnetti, G., Saleem, I., Overview of the Advantages and Disadvantages of Different Mucosal Sites for the Delivery of Nanoparticles, in: *Mucosal Delivery of Drugs and Biologics in Nanoparticles*, P. Muttil and N.K. Kunda (Eds.), pp. 61–82, Springer International Publishing, Cham, 2020.
22. Cilurzo, F., Musazzi, U.M., Franzé, S., Selmin, F., Minghetti, P., Orodispersible dosage forms: biopharmaceutical improvements and regulatory requirements. *Drug Discov. Today*, 23, 2, 251–9, 2018.
23. Dey, P. and Maiti, S., Orodispersible tablets: A new trend in drug delivery. *J. Nat. Sci. Biol. Med.*, 1, 1, 2–5, 2010.
24. Cornilă, A., Iurian, S., Tomuţă, I., Porfire, A., Orally Dispersible Dosage Forms for Paediatric Use: Current Knowledge and Development of Nanostructure-Based Formulations. *Pharmaceutics*, 14, 8, 1621, 2022.

25. Turković, E., Vasiljević, I., Drašković, M., Parojčić, J., Orodispersible films — Pharmaceutical development for improved performance: A review. *J. Drug Deliv. Sci. Technol.*, 75, 103708, 2022.
26. Slavkova, M. and Breitkreutz, J., Orodispersible drug formulations for children and elderly. *Eur. J. Pharm. Sci.*, 75, 2–9, 2015.
27. Deng, Y., Shen, L., Yang, Y., Shen, J., Development of nanoparticle-based orodispersible palatable pediatric formulations. *Int. J. Pharm.*, 596, 120206, 2021.
28. Dziemidowicz, K., Lopez, F.L., Bowles, B.J., Edwards, A.J., Ernest, T.B., Orlu, M. *et al.*, Co-Processed Excipients for Dispersible Tablets-Part 2: Patient Acceptability. *AAPS PharmSciTech*, 19, 6, 2646–57, 2018.
29. Prathusha, P., Kamarapu, P., A review on sublingual tablets. *J. Formul. Sci. Bioavailab.*, 1, 103, 2019.
30. Narang, N., and Sharma, J., Sublingual mucosa as a route for systemic drug delivery. *Int. J. Pharm. Pharm. Sci.*, 3, 18–22, 2011.
31. Prabhu, S.C., Parsekar, S.D., Shetty, A., Monteiro, S.S., Azharuddin, M., Shabaraya, A.R., A review on fast dissolving sublingual films for systemic drug delivery. *Int. J. Pharm. Chem. Sci.*, 3, 2, 501–11, 2014.
32. Baltzley, S., Malkawi, A.A., Alsmadi, M., Al-Ghananeem, A.M., Sublingual spray drug delivery of ketorolac-loaded chitosan nanoparticles. *Drug Dev. Ind. Pharm.*, 44, 9, 1467–72, 2018.
33. Rawat, I., Kandwal, M., Patil, S., A comprehensive review on sublingual drug delivery system. *Eur. Chem. Bull.*, 12, s3, 10, 2023.
34. Naharros-Molinero, A., Caballo-González, M.Á., de la Mata, F.J., García-Gallego, S., Shell Formulation in Soft Gelatin Capsules: Design and Characterization. *Adv. Healthcare Mater.*, 13, 1, 2302250, 2024.
35. Barua, S., Kim, H., Jo, K., Seo, C.W., Park, T.J., Lee, K.B. *et al.*, Drug delivery techniques for buccal route: formulation strategies and recent advances in dosage form design. *J. Pharm. Invest.*, 46, 7, 593–613, 2016.
36. Mohit, and Hussain, M.S., A brief review on buccal drug delivery system: advantages, limitations, and impact on healthcare system. *World J. Pharm. Res.*, 10, 558–76, 2021.
37. Montero-Padilla, S., Velaga, S., Morales, J.O., Buccal Dosage Forms: General Considerations for Pediatric Patients. *AAPS PharmSciTech*, 18, 2, 273–82, 2017.
38. Chinna Reddy, P., Chaitanya, K.S., Madhusudan Rao, Y., A review on bioadhesive buccal drug delivery systems: current status of formulation and evaluation methods. *Daru*, 19, 6, 385–403, 2011.
39. Bruschi, M.L. and de Freitas, O., Oral Bioadhesive Drug Delivery Systems. *Drug Dev. Ind. Pharm.*, 31, 3, 293–310, 2005.
40. Stie, M.B., Öblom, H., Hansen, A.C.N., Jacobsen, J., Chronakis, I.S., Rantanen, J. *et al.*, Mucoadhesive chitosan- and cellulose derivative-based nanofiber-on-foam-on-film system for non-invasive peptide delivery. *Carbohydr. Polym.*, 303, 120429, 2023.

41. Faraji Rad, Z., Prewett, P.D., Davies, G.J., An overview of microneedle applications, materials, and fabrication methods. *Beilstein J. Nanotechnol.*, 12, 1034–46, 2021.
42. Creighton, R.L. and Woodrow, K.A., Microneedle-Mediated Vaccine Delivery to the Oral Mucosa. *Adv. Healthcare Mater.*, 8, 4, e1801180, 2019.
43. Aldawood, F.K., Andar, A., Desai, S., A Comprehensive Review of Microneedles: Types, Materials, Processes, Characterizations and Applications. *Polym. (Basel)*, 13, 16, 2815, 2021.
44. Ravi, A.D., Sadhna, D., Nagpaal, D., Chawla, L., Needle free injection technology: A complete insight. *Int. J. Pharm. Investig.*, 5, 4, 192–9, 2015.
45. Ludwig, C. and Fisher, L., Buccal Midazolam Solution for the Management of Prolonged Acute Convulsive Seizures: A Cost Analysis. *Pharmacoeconomics*, 4, 1, 171–9, 2020.
46. Yadav, R., Kanwar, I.L., Haider, T., Pandey, V., Gour, V., Soni, V., *In situ* gel drug delivery system for periodontitis: an insight review. *Future J. Pharm. Sci.*, 6, 1, 33, 2020.
47. Aslani, A. and Malekpour, N., Design, formulation, and physicochemical evaluation of periodontal propolis mucoadhesive gel. *Dent. Res. J. (Isfahan)*, 13, 6, 484–93, 2016.
48. Al-Bayaty, F.H., Kamaruddin, A.A., Ismail, M.A., Abdulla, M.A., Formulation and Evaluation of a New Biodegradable Periodontal Chip Containing Thymoquinone in a Chitosan Base for the Management of Chronic Periodontitis. *J. Nanomater.*, 2013, 397308, 2013.
49. Davies, R.M., Rinses to control plaque and gingivitis. *Int. Dent. J.*, 42, 4 Suppl 1, 276–80, 1992.
50. Basudan, A.M., Nanoparticle based periodontal drug delivery - A review on current trends and future perspectives. *Saudi Dent. J.*, 34, 8, 669–80, 2022.
51. Johnson, A.R., Forster, S.P., White, D., Terife, G., Lowinger, M., Teller, R.S. *et al.*, Drug eluting implants in pharmaceutical development and clinical practice. *Expert Opin. Drug Deliv.*, 18, 5, 577–93, 2021.
52. Gaobotse, G., Mbunge, E., Batani, J., Muchemwa, B., Non-invasive smart implants in healthcare: Redefining healthcare services delivery through sensors and emerging digital health technologies. *Sensors Int.*, 3, 100156, 2022.
53. Ou, Y.H., Ou, Y.H., Gu, J., Kang, L., Personalized anesthetic patches for dental applications. *Int. J. Bioprinting*, 5, 2.1, 203, 2019.
54. Thosar, M.M., Intra oral sprays-an overview. *Int. J. Pharm. Life Sci.*, 11, 1235–1246, 2011.
55. Shakya, P., Madhav, N.V.S., Shakya, A.K., Singh, K., Palatal mucosa as a route for systemic drug delivery: A review. *J. Control. Release*, 151, 1, 2–9, 2011.
56. Huang, T.W. and Young, T.H., Novel porous oral patches for patients with mild obstructive sleep apnea and mouth breathing: a pilot study. *Otolaryngol. - Head Neck Surg.*, 152, 2, 369–73, 2015.
57. Mehrotra, S. and Pathak, K., Translabial drug delivery: potential and possibilities. *Ther. Deliv.*, 11, 11, 673–6, 2020.

58. Kokil, S., Kadu, M., Vishwasrao, D., Singh, D., Review on Natural Lip Balm. *Int. J. Res. Cosmet. Sci.*, 5, 1, 2014.
59. Kang, S., Jang, E.J., Jo, H.M., Kang, S.S., Lee, M.S., Yun, S.Y. *et al.*, Effects of a Topically Applied Oral Wound Dressing Film on Intra-oral Wound Healing in Rabbits. *In Vivo*, 36, 4, 1745–52, 2022.
60. Ashtiani, H., Bishe, P., Lashgari, N.-A., Nilforoushzadeh, M., Zare, S., Liposomes in Cosmetics. *J. Skin Stem Cell*, e65815, 2016.
61. Seth, A.K., Misra, A., Umrigar, D., Topical liposomal gel of idoxuridine for the treatment of herpes simplex: pharmaceutical and clinical implications. *Pharm. Dev. Technol.*, 9, 3, 277–89, 2004.
62. Silva, F.C., Marto, J.M., Salgado, A., Machado, P., Silva, A.N., Almeida, A.J., Nystatin and lidocaine pastilles for the local treatment of oral mucositis. *Pharm. Dev. Technol.*, 22, 2, 266–74, 2017.
63. Lankalapalli, S. and Tenneti, V.S., Formulation and Evaluation of Rifampicin Liposomes for Buccal Drug Delivery. *Curr. Drug Deliv.*, 13, 7, 1084–99, 2016.
64. Sangle, V. and Tathe, P., A review on medicated lollipop: A novel dosage form. *Int. J. Sci. Res. Arch.*, 8, 565–70, 2023.
65. Rathod, M., Poharkar, S., Pandhre, Y., Muneshwar, M., Sul, S., Medicated lozenges as an easy to use dosage form. *World J. Pharm. Res.*, 7, 16, 305–22, 2018.
66. Wertz, P.W., Roles of Lipids in the Permeability Barriers of Skin and Oral Mucosa. *Int. J. Mol. Sci.*, 22, 10, 5229, 2021.
67. Mazzinelli, E., Favuzzi, I., Arcovito, A., Castagnola, R., Fratocchi, G., Mordente, A. *et al.*, Oral Mucosa Models to Evaluate Drug Permeability. *Pharmaceutics*, 15, 5, 1559, 2023.
68. Rathbone, M.J., Pather, I., S.Ş., Overview of Oral Mucosal Delivery, in: *Oral Mucosal Drug Delivery and Therapy*, M.J. Rathbone, S. Senel, I. Pather (Eds.), pp. 17–29, Springer US, Boston, MA, 2015.
69. Guo, Y.-G. and Singh, A.P., Emerging strategies for enhancing buccal and sublingual administration of nutraceuticals and pharamaceuticals. *J. Drug Deliv. Sci. Technol.*, 52, 440–51, 2019.
70. Colombo, P., Cagnani, S., Buttini, F., Santi, P., Sonvico, F., Russo, P. *et al.*, Biological *In Vitro* Models for Absorption by Non-Oral Routes, in: *Comprehensive Medicinal Chemistry II*, pp. 279–99, 2006.
71. Feitosa, R.C., Geraldes, D.C., Beraldo-de-Araújo, V.L., Costa, J.S.R., Oliveira-Nascimento, L., Pharmacokinetic Aspects of Nanoparticle-in-Matrix Drug Delivery Systems for Oral/Buccal Delivery. *Front. Pharmacol.*, 10, 1057, 2019.
72. Ghosh, T.K. and Pfister, W.R., *Drug delivery to the oral cavity: molecules to market*, CRC Press, Boca Raton, 2005.
73. Bala, R., Pawar, P., Khanna, S., Arora, S., Orally dissolving strips: A new approach to oral drug delivery system. *Int. J. Pharm. Investig.*, 3, 2, 67, 2013.

74. Heer, D., Aggarwal, G., Kumar, S.H., Recent trends of fast dissolving drug delivery system–an overview of formulation technology. *Pharmacophore*, 4, 1, 1–9, 2013.
75. Şenel, S., Rathbone, M.J., Cansız, M., Pather, I., Recent developments in buccal and sublingual delivery systems. *Expert Opin. Drug Deliv.*, 9, 6, 615–28, 2012.
76. Tan, O.L., Safii, S.H., Razali, M., Commercial local pharmacotherapeutics and adjunctive agents for nonsurgical treatment of periodontitis: A contemporary review of clinical efficacies and challenges. *Antibiotics*, 9, 1, 11, 2019.
77. Joshi, D., Garg, T., Goyal, A.K., Rath, G., Advanced drug delivery approaches against periodontitis. *Drug Deliv.*, 23, 2, 363–77, 2016.
78. Steinberg, D. and Friedman, M., Sustained-release delivery of antimicrobial drugs for the treatment of periodontal diseases: Fantasy or already reality? *Periodontol. 2000*, 84, 1, 176–87, 2020.

4

Oral Novel Drug Delivery Systems

Mahsa Sayed Tabatabaei

HIDA Pharmaceutical Technologies Development Center, HIDA Pharmaceuticals, Tehran, Iran

Abstract
Oral (or more accurately, peroral) route of drug administration is a well established, popular, and oftentimes, the first choice for drug delivery. This is due to the outstanding advantages of oral administration such as high patient compliance, ease of administration, feasibility for different dosage forms, high dosing flexibility, and ease of scale-up. Moreover, the gastrointestinal (GI) tract is, physiologically, a specialized absorption site. Despite these advantages, the high absorption of drugs is hampered by different physical, chemical, and metabolic barriers of the GI. In addition, a large number of orally administered drug molecules suffer from inherent difficulties (such as inadequate solubility and permeability), which ultimately leads to poor bioavailability. To conquer the aforementioned challenges, and take the opportunities that GI provides for absorption, oral novel drug delivery systems have emerged. This chapter first overviews the principles of oral drug delivery (including anatomy and physiology of the GI tract, bioavailability barriers, and factors affecting oral drug delivery). Subsequently, we will have a brief overview of the opportunities that the GI tract offers, and the challenges that must be overcome. In the following, oral novel drug delivery systems based on the anatomical site (including regional drug delivery to the esophagus, stomach, small intestine, and colon) are presented. Then, the principles of oral pharmacokinetics are briefly discussed. And finally, some of the novel oral products that have entered the market are introduced.

Keywords: Oral drug delivery, novel drug delivery systems, oral drug delivery barriers, nanocarriers, gastroretentive systems, colon drug delivery, pharmacokinetics

Email: tabatabaei@hidapharma.com

Mehrdad Hamidi and Mahsa Sayed Tabatabaei (eds.) Novel Drug Delivery Systems: Fundamentals and Applications, (103–166) © 2025 Scrivener Publishing LLC

4.1 Introduction

Among the various routes of drug administration, the oral (or more accurately, peroral) route is the most common and oftentimes the first choice [1]. The reason for this is to be found in the salient advantages of oral drug delivery, including the convenience of administration, high patient compliance, feasibility for a variety of dosage forms, high degree of dosing flexibility, the capability of uniform delivery, ease of transport, storage, production and scale-up, and economic feasibility [2–5].

Besides, the GI tract is physiologically specialized for absorption; hence, the drug can also be well absorbed by taking advantage of the available absorptive facilities (such as the very large surface area provided by the abundance of enterocytes throughout the GI tract) [2].

However, oral delivery is not that easy either, since the GI tract creates several obstacles to the drug. Anatomical heterogeneity and diversity of variables throughout the GI tract (e.g., pH, enzymatic activity, and transit time are the most important physiological barriers challenging oral drug delivery [4].

On the other hand, it has been reported that about 50% of orally administered drugs suffer from poor bioavailability, arising from low solubility, inadequate mucosal and/or cellular permeability, limited physical/chemical stability, short residence time, or narrow absorption window [6]. Moreover, the proportionality of the number of drug molecules in development belonging to different BCS classes is changing remarkably in such a way that the share of BCS class I drugs (with ideal physicochemical properties of high solubility and permeability) is changing dramatically in favor of class II and IV which are poorly water-soluble and/or low permeable candidates (Figure 4.1). Also, many biopharmaceuticals have a large molecular weight and restricted lipophilicity (class III); again leading to poor bioavailability [7]. Consequently, adopting new approaches to the successful delivery of drug candidates with undesirable physicochemical feature(s) is of crucial importance [8].

Conventional drug delivery systems are not fortunate enough to conquer the obstacles listed. Therefore, in recent years, an upsurge of research has been carried out by researchers from different fields to develop advanced systems to conquer the aforementioned challenges. This goal could be achieved by designing novel drug entities (e.g., prodrugs), formulations (e.g., nanoparticles), and devices (e.g., intestinal patches) [2, 4].

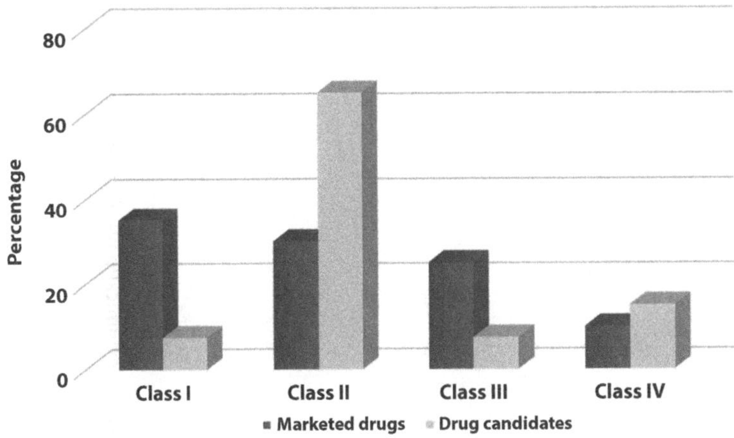

Figure 4.1 How the share of drugs belonging to each class of BCS is changing.

The design and development of oral novel delivery systems require a full understanding of the structure and function of the GI tract as well as all the formulation-related factors involved in drug release and absorption in the GI tract. In this context, a realistic consideration of the pharmacokinetic profile of the drug is of key importance. This chapter, accordingly, aims to encompass all of those prerequisites in brief.

4.2 Anatomy and Physiology of Gastrointestinal Tract

The majority of the GI tract segments are of importance from the drug delivery perspective. In this chapter, we will focus on the esophagus, stomach, small intestine, and colon as organs involved in oral drug delivery. Other parts of the GI tract, including the oral cavity and rectum, have been discussed in detail as separate chapters in this book (chapters 3 and 5, respectively).

The GI tract—also called the alimentary canal—is a hollow muscular tube starting from the mouth and ending at the anal opening, through which foods and drinks are imported, mechanically and chemically processed, extracted, and finally disposed of [9, 10]. It consists of the mouth, esophagus, stomach, small intestine, and finally the large intestine. Understanding the heterogeneous characteristics of these segments is a prerequisite for efficient drug delivery to each region. Different segments of the GI tract along with their anatomical and physiological features are presented in Figure 4.2.

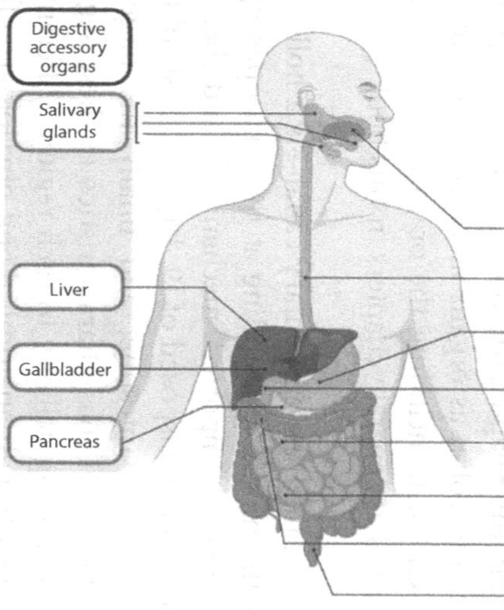

Organ	Length (m)	Surface area (m^2)	pH	Transit time	Epithelial type	Major enzymatic activity
Oral cavity	-	0.01	6.5	-	stratified squamous	polysaccharidase
Esophagus	0.20-0.25	0.02	-	4-8 s	stratified squamous	-
Stomach	0.25	3.5	1.0-3.0	0-2 h (fast) 2-6 h (fed)	secretory columnar	proteases, lipases
Duodenum	0.35	1.9	4.0-5.5	30-40 min	simple columnar	poly- and oligosaccharidase, proteases, peptidases, lipases
Jejunum	2.80	184	5.5-7.0	1.5-2 h	simple columnar	oligosaccharidase, peptidases, lipases
Ileum	4.20	276	7.0-7.5	5-7	simple columnar	oligosaccharidases and peptidases (non-secreted)
Colon	1.50	1.3	7.5-8.0	16-35 h	simple columnar	broad spectrum of bacterial enzymes
Rectum	0.12	-	7.0	-	simple columnar	-

Figure 4.2 A scheme of the main and accessory organs of the human digestive system along with the major anatomical/physiological features of the gastrointestinal tract (created with BioRender.com).

The oral cavity is the first host for foods and drinks, wherein food is retained for a short time, partly digested mechanically (chewed), mixed with salivary secretions, and pushed toward the pharynx by the tongue [9]. Once food (or pharmaceutical) is swallowed, peristaltic contractions of the esophagus direct it to the stomach to undergo enzymatic and chemical processing [11]. Short-term residence time, despite the proper accessibility, has excluded the esophagus from the spotlight of drug delivery for years [12]. Nevertheless, some efforts have been made recently towards targeted, long-lasting, and local treatment of esophagus disorders [13]. The stomach is not used to be conventionally considered as the final destination for most drugs (due to its limited surface area as well as the limited/variable residence time) but a temporary reservoir that enters its contents into the initial part of the small intestine at a certain rate [11]. Today, however, the stomach has attracted much more attention for drug delivery [12].

The small intestine or small bowel anatomically consists of the duodenum, jejunum, and ileum. It is the lengthiest organ in the GI tract with a proximity of 6 m in length, whose responsibility is absorption, digestion, and propulsion of the ingested mass. Owing to its high length as well as enormously increased surface area, the small intestine is considered the specialized absorption site not only for food and nutrients but also for many drugs [9, 11, 14]. This key feature relies on the presence of the extensive wavy mucosal folds (known as circular folds or Kerckring valves), protruding from which are densely-packed finger-like structures, called villi (singular villus) and microvilli. Such structural adaptions collectively expand the effective absorption area about 600 to 1000 times, which means a total surface area of 250 to 400 m^2. Besides, these adaptions facilitate absorption owing to being well-supplied by blood vessels and capillaries [11, 15]. The main structural features of the small intestine are shown in Figure 4.3.

The large intestine successively comprises the caecum and appendix, colon (which is further subdivided into ascending, transverse, descending, and sigmoid segments), rectum, and anal canal [16]. As its functional role, the large intestine absorbs the remained nutrients and the water to ensure homeostasis. From a pharmaceutical point of view, the colon plays a crucial role, both in systemic and local drug delivery. The absence of villi in the smooth mucosa of the colon has drastically reduced its surface area compared with the small intestine. Nevertheless, this shortcoming has been somewhat compensated by tube-like crypt structures. Throughout the colon, there is a wide range of bacteria, collectively called "microbiota" (with a much larger population compared to the human body cells and weighing more than 1 kilogram), whose presence is essential for normal bowel function [9, 11].

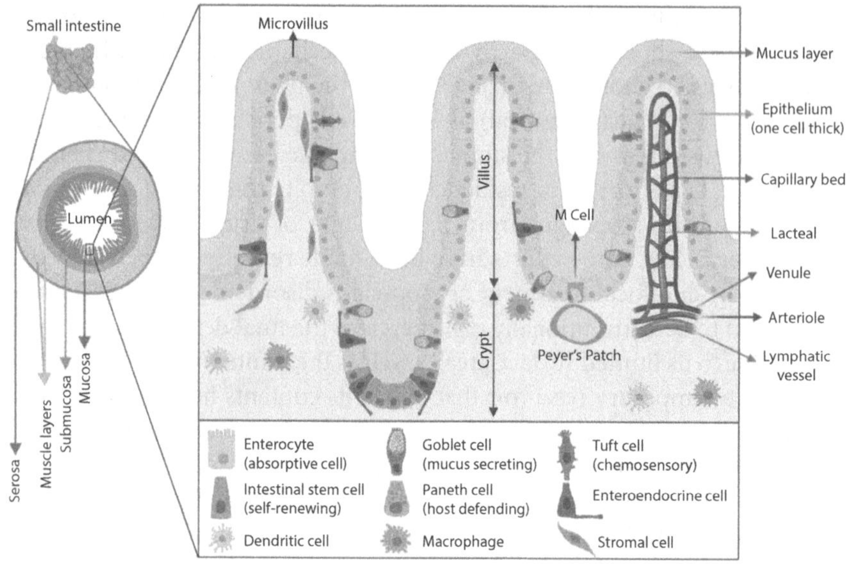

Figure 4.3 Key structural features of the small intestine (created with BioRender.com).

4.3 Barriers to Oral Drug Delivery

As already discussed, the GI tract is specialized for absorption. However, being located in a strategic position with constant contact with the external environment, the GI lumen has been equipped with highly efficient barriers against xenobiotics, thereby serving as the first protecting mechanism against the oral entry of harmful substances [9, 17]. As an external substance, the drug is no exception to being exposed to these obstacles. The overall efficiency of these barriers is such that minimizes the bioavailability of a wide range of drugs. This value is reported to be about 1% (or even less) for intact peptides and proteins. In a general classification, the GI barriers against xenobiotics may be divided into two classes: (i) bioavailability barriers (including biochemical, physical, and cellular barriers) and (ii) immunological/anti-microbial defense barriers [18]. From the drug delivery point of view, the former is of crucial importance and will be discussed below (see Figure 4.4).

4.3.1 Biochemical (Stability) Barriers

Enzymatic degradation and pH are the main biochemical barriers hindering the bioavailability of orally administered drugs. The harsh acidic condition inside the stomach created by hydrochloric acid is the first barrier that decreases the efficiency of many drug molecules (particularly

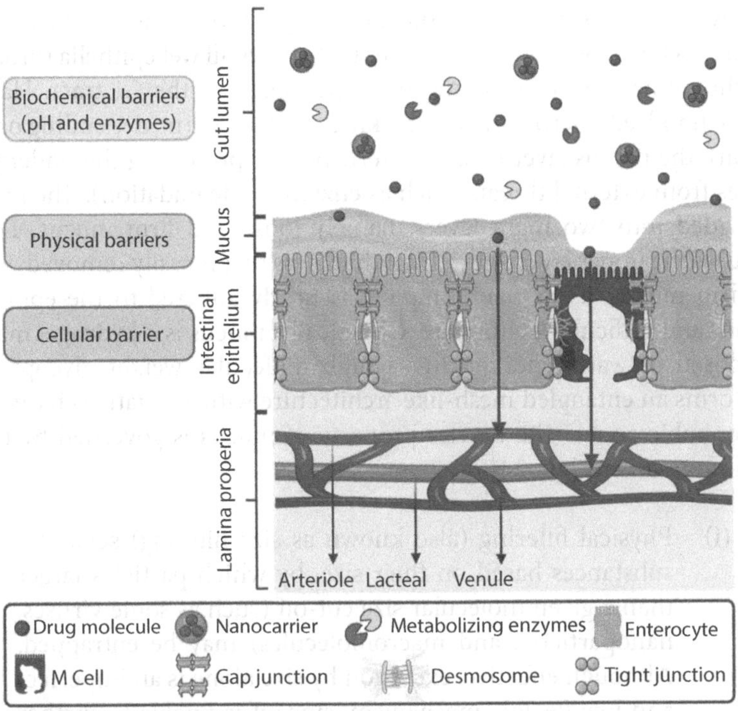

Figure 4.4 Main bioavailability barriers to oral drug delivery (created with BioRender.com).

biopharmaceuticals) by degradation/denaturation [2]. Additionally, pH variability throughout the GI may induce precipitation of drugs with pH-dependent solubility [19]. Besides, metabolizing enzymes within the stomach (e.g., proteolytic pepsins), as well as the duodenum lumen (e.g., trypsin, chymotrypsin, carboxypeptidase, and elastase), undertake a major role in drug degradation. In the colon, however, the enzymatic activity is minimized, providing the desired condition to avoid metabolic degradation [2]. Also, the surfactant-like bile salts may negatively affect the structural integrity of drugs. In addition to all these, even if the obstacles listed and those that will follow are overcome safely, the drug is further threatened by the microsomal enzymes located inside the cytoplasm of the liver cells (in particular, the cytochrome P450 family) [20].

4.3.2 Physical (Permeability) Barriers

Once a drug overcomes the lumen challenges, it faces several physical barriers before reaching the enterocytes (e.g., unstirred water layer, mucus, and glycocalyx). Among these, the mucus layer is the major challenge. The mucus is a tenacious viscoelastic layer that covers all wet epithelia throughout the human body, including the inner space of the GI tract. Having several hundred micrometers in thickness (100–800 μm, depending on the region), the mucus layer creates a steric barrier protecting the underlying tissues from external threats (such as enzymatic degradation). The mucus is divided into two main layers; namely loose and firm mucus. Loose mucus is the upper layer with a high thickness that is easily removed; while the firm mucus, as its name implies, is firmly adhered to the epithelial surface and difficult to eliminate. Chemically, mucus is a hydrogel mainly composed of water and mucin—a high molecular weight glycoprotein that forms an entangled mesh-like architecture with a negative charge and hydrophobic zones. The barrier property of mucus is governed by three main mechanisms [20–24]:

(i) Physical filtering (also known as size filtering) separates substances based on their size, by which particles larger than a given molecular size cut-off (such as some viruses, nanoparticles, and macromolecules) may be entrapped. Although effective, exclusion by size alone is an imperfect and low flexible mechanism, since it is unable to further differentiate between the below-cut-off-sized particles. This drawback has been compensated by the second filtering mechanism, as follows.

(ii) Chemical filtering (or interaction filtering), is based on the physicochemical interactions of particles (smaller than mesh pore) with mucus slowing down the diffusion of particles. This pathway affects molecules regardless of their size. Drugs or ions with low molecular weight and positive charge are electrostatically attached to negatively charged components in the mucosa.

(iii) Dynamically rebuilt barrier; because mucus is a dynamic layer and constantly renewed, it creates a time limit for the entry of drugs.

4.3.3 Cellular (Epithelial) Barrier

Intestinal epithelium comprises a single layer of columnar epithelial cells (mainly including absorptive cells or enterocytes). On the apical side of enterocytes, there is a highly dense and ordered brush border as a fundamental barrier [20, 25]. In general, there are two main pathways for a drug molecule to cross the intestinal epithelial barrier and reach blood circulation including passive and active transportation [26].

The term "passive" means that there is no energy expenditure by the system during the permeation process. Rather, the concentration gradient is the driving force for the moving down of drugs [27]. Diffusion refers to the random motion of solute molecules leading to transportation from one point to another. Paracellular and transcellular diffusion are two main mechanisms associated with passive diffusion. In the paracellular route, the drug bypasses the cellular barrier through the water-filled pores of tight junctions located between the enterocytes through passive diffusion; while in the transcellular pathway, the drug navigates across the cells through passive diffusion or carrier-mediated transportation [28].

Carrier-mediated active transportation is a type of facilitated diffusion, in which the movement requires energy and may be carried out bidirectional (i.e., the drug is also capable of transport against an opposite concentration gradient). The involved transporters are usually called active- or concentrating transporters [27]. The gut lumen active transporters undertake different roles based on their location and function [28]. Depending on the direction of movement, i.e. into or out of the cell, they are categorized into the influx (uptake) and efflux transporters, respectively [27]. Uptake transporters have an absorptive role and commonly direct the drug into the cell according to the concentration gradient. Conversely, during efflux, the drug molecule is thrown out of the cell. This process is usually against the gradient of concentration and is known as secretion [28]. Figure 4.5 schematically illustrates the intestinal drug absorption pathways.

Transporters are asymmetrically expressed on the apical and basolateral sides of enterocytes. They may be selective for only one substrate or recognize many different compounds [28]. The main barrier to the transcellular

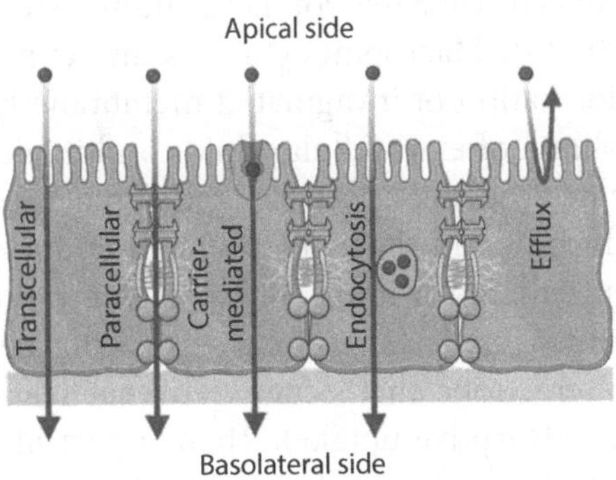

Figure 4.5 Intestinal drug absorption pathways (created with BioRender.com).

pathway is efflux pumps located in the apical membrane of the intestine. Among the most important highly expressed intestinal transporters are ATP-binding cassette transporters (known as ABCs), including P-gp, BCRP, and MRP2. From the proximal to the distal part of the small intestine, the abundance of P-gp increases, whilst the population of CYP3A4 enzymes plummets. Given the synergistic overlapping of their substrate spectrum in some cases, the drugs are prone to extensive pre-systemic intestinal metabolism [29]. For instance, through a concerted performance, the apical efflux transporters and intra-enterocytic CYP450s may reduce the systemic absorption of some orally administered drugs [27].

In general, it is challenging to overcome the epithelial barrier, for low-soluble and low-permeable drugs (i.e., BCS Class IV) and those with high molecular weights (e.g., proteins) [26, 30]. The absorption of a drug is a consequence of its ability to penetrate the mucosal layer, cross intercellular tight junctions, and also the ability to conquer epithelial transporters and enzymes [29].

What has been discussed so far is the transfer and absorption of drugs in the molecular form. Due to the fundamental differences between the molecules and nanocarriers, mainly the size, these drug delivery vehicles follow other mechanisms to enter the cells, generally known as bulk transport or endocytosis where the drug-loaded carrier is "uptaken" by the epithelial cells *via* transformation in cell membrane surface ended by budding a membrane-surrounded vesicle containing the whole body of the carrier [31, 32].

Endocytosis is the major pathway for nanocarriers entrance, which is divided into four main mechanisms including (i) phagocytosis, (ii) macropinocytosis, (iii) clathrin-dependent (receptor-mediated) endocytosis, and (iv) caveolae-dependent endocytosis [33]. Phagocytosis is the preferential route for particles larger than 500 nm, which is merely performed by a specialized type of cells (like macrophages, monocytes, neutrophils, and dendritic cells) [20, 33]. Macropinocytosis is an actin-regulated process that involves the formation of invaginated membrane ruffles, the engulfment of a large volume of extracellular fluids and its debris, and the closure to form large heterogeneous vesicles called macropinosomes [20, 33, 34]. Clathrin-mediated endocytosis is the main endocytic pathway for a cell to take its requirements. It is also of particular interest to viruses and nanoparticles. Clathrin-mediated endocytosis is commenced with uptake (either through interactions that occur between ligands and receptors or *via* non-specific adsorptive uptake). Then, a part of the guest mass is recycled and the remainder is transferred into lysosomes. This process is followed by the scission of the budding and degradation of sequestered cargo by lysosomal enzymes [20, 33]. Caveolin-mediated endocytosis is a clathrin-independent procedure, wherein plasma membrane invaginates to form flask-shaped caveolae (literally means small caves) with an average diameter of 60 to 80 nm [35]. This endocytic route provides an opportunity for internalized mass to escape from lysosomal degradation; therefore, caveolin-mediated endocytosis is advantageous for the delivery of proteins and genes [20].

4.3.4 Hepatic First-Pass Metabolism (Metabolic Barrier)

Once a drug molecule or a nanocarrier successfully conquers the barriers listed, it has two absorptive pathways including systemic and lymphatic circulations. Most orally administered drug molecules enter the former pathway through direct permeation from the enterocytes into the mesenteric blood capillaries, followed by movement towards the liver and being exposed to hepatic metabolism, the so-called first-pass effect [30, 36]. During the first-pass (presystemic) metabolism, a considerable extent of the drug may undergo biotransformation and get out of reach. Therefore, bioavailability may be significantly reduced [37].

The lymphatic circulation is an additional absorption pathway that could be adopted by certain drugs [36]. Being more permeable than blood vessels, lymphatic vessels are the preferred pathway for chylomicrons with a diameter of 100 to 1000 nm. Besides, the entry of chylomicrons is aided by specialized openings in the lacteals (blind lymphatic capillaries projecting into the villus). Following the distribution of chylomicrons in the mesenteric lymph vessels, the lymph enters the thoracic duct and, then, joins the general circulation through the subcla*via*n vein. Today, the lymphatic pathway is of special importance due to the development of new drug molecules with high efficiency but low aqueous solubility, as well as prodrug- and lipid-based nanoparticles. Advantageously, upon entering the lymph vessels, the portal pathway is virtually bypassed. So, there will be a high drug concentration in lymphatic ducts and nodes, which could be a potential site for drug actions. Any factor influencing the production and secretion of chylomicrons can also impact the lymphatic uptake of the drug and its bioavailability [30, 38].

4.4 Factors Affecting Oral Drug Delivery

From a macro perspective, factors affecting oral drug delivery can be divided into four categories: (i) physiological factors, (ii) pathological conditions, (iii) demographic characteristics, and (iv) physicochemical/pharmaceutical parameters. Here, the major factors are described.

4.4.1 Physiological Factors

Physiological factors varying across the GI tract can significantly affect the bioavailability of the drug by changing its solubility, stability, retention time at the absorption site, and so on [9].

Gastric emptying rate

Gastric emptying rate is one of the most important variables affecting the bioavailability of drugs, which, in turn, is influenced by various other factors (such as fed- or fasted-state, calorie intake, volume, osmolality, and viscosity of the gastric fluid, gender, and stress). In addition, the emptying rate could be accelerated using some drugs (e.g., erythromycin). Through an individual or collective effect, these factors markedly impact the gastric residence time, and consequently the bioavailability of a vast variety of drugs [26, 30].

Such time limitation added by the small absorption area of the stomach makes gastric absorption practically insignificant compared to intestinal absorption. Most importantly, the rate at which particles leave the stomach depends on their size. For instance, in the fasted-state, suspensions are expelled at the same time as the water, while tablets, granules, and non-disintegrating dosage forms are emptied more slowly [30]. Solid particles smaller than 2 mm, similar to liquids, are capable of passing through the pyloric sphincter even in the fed-state [39].

Gastric pH

The gastric pH is highly dependent on food intake. In the fasted-state, the median pH is reported to be around the baseline level of < 2. Meanwhile, this value increases significantly in the fed-state. Such variability is especially important for the pharmacokinetics of weakly basic drugs, whose solubility is the rate-limiting step of absorption [30]. Besides, a majority of proteins and peptides are, at least partially, hydrolyzed during passing through such a harsh condition, and their biological activity is lost [40].

Enzymatic degradation in the stomach

Although the enzymatic diversity of the stomach is much lower than that of the small intestine, it contains a very important protein-specific enzyme. Pepsin, as the major enzyme found in the stomach, breaks down the proteins and peptides into smaller absorbable parts (i.e., peptide fragments and amino acids). This enzymatic activity is highly dependent on the acidic conditions of the stomach (pH < 5, created by hydrochloric acid); by which the inactive form of pepsin, pepsinogen, is activated [41]. Unexpectedly, pepsin is also capable of digesting nucleic acids [42]. Therefore, the stomach is a major barrier to the oral administration of peptides, proteins, and genes.

Small intestinal transit time

The small intestinal transit time is roughly estimated to be 3-4 hours; but highly depends on various factors such as intestinal motility, individual dietary habits, expression of efflux and uptake transporters and metabolizing enzymes throughout the GI tract, and drug/dosage form characteristics. In this context, some drugs can affect GI motility and thereby, change the small intestinal transit time of themselves or any other concurrently used medication. In general, intestinal transit time is extremely important for three classes of drugs: (i) low-permeable ones (BCS classes III and IV), (ii) those with limited absorption in the colon, (iii) those with preferential absorption zone (absorption window) in the small intestine. The major part of the drug absorption occurs in the upper small intestine at a golden opportunity when the drug in the gastric chime is still supersaturated and has not yet had enough time to nucleate or precipitate [30].

Intestinal pH

During the movement of partially digested food through the intestine, the pH of the stomach contents starts to gradually increase from about 2 to reach 4–5.5 in the duodenum, 5.5–7.0 in the jejunum, and 7.0–7.5 in the ileum before delivery to the colon. This pH gradient, along with other factors, strongly affects the pharmacokinetics of oral drugs. Nevertheless, such pH diversity can be considered an opportunity for controlled-release drug delivery [17].

Enzymatic degradation in the small intestine

The small intestine is home to a wide range of enzymes including amylases, lipases, proteases, and so on. They can alter the bioavailability of drugs through destabilization, degradation, and decreasing the net absorption [26].

Colon transit time

The colon transit time depends on several factors including formulation-related parameters (e.g., size, shape, and surface properties), demographic characteristics (e.g., age and gender), individual physiological conditions (e.g., circadian rhythm, body position, and sleeping), food intake, dosing conditions, properties of luminal contents (e.g., viscosity and osmolality), and pathological conditions. For instance, it has been shown that the transfer time of smaller dosage forms (e.g., dispersed particles) is slower than that of larger ones such as capsules [30, 43]. Also, patients with inflammatory bowel disease (IBD) usually experience a delayed transit time than healthy subjects. Conversely, this time is reduced when there is a dysbiotic condition (i.e., bacterial overgrowth) [11].

Colon transit time is especially important for colon-targeted and extended-release formulations that require a sufficiently long time to be dissolved and absorbed. In addition to the early mass movement in the colon leading to a short-term drug residence, this causes the premature entry of contents from the ileum into the cecum and thus an incomplete absorption in the small intestine and limited bioavailability [30].

Colon fluid volume

The water-absorption capacity of the colon is very high, as it resorbs about 90% of the water in the contents and leaves an average of only about 13 ml. Therefore, no wonder why the dissolution of drugs in this area is so challenging. The physiologically optimal region of the colon from a drug delivery perspective is the ascending colon –the only part where there is a sufficient volume of free fluid [43].

Colonic pH and enzymatic degradation in the colon

The colonic pH first decreases relative to the distal part of the small intestine and then gradually increases. The effect of pH on pharmacokinetics is manifested through its impact on the solubility of drugs or other ingredients of the formulation (such as coatings). One of the most influential factors on colonic pH is the sugar-rich diet, which is fermented by colon bacteria to form short fatty acid chains. Accordingly, formulations containing polysaccharides can have a similar effect. Some disease states, such as ulcerative colitis, also change the pH of the colon [43].

The microbiota

The colon, as mentioned earlier, is home to several aerobic and anaerobic bacteria that produce a wide range of metabolizing enzymes. Therefore, the colon is where a part of the biotransformation process of xenobiotics (e.g., drugs) and some internal substances (like bile salts) as well as the inactivation of harmful substances takes place [43]. The composition of the microbiome is unique to individuals leading to diversity in drug metabolism. It could be affected by age, lifestyle, and medications [30]. A drug in the colon can become activated, deactivated, and even toxic [30, 43]. However, the effect of microbiota is not limited to drugs that reach directly to the colon. Rather, it has been shown that microbiota can indirectly impact the pharmacokinetics of drugs administered from other routes (even injected) by altering the hepatic metabolism of drugs, affecting the pH, concentration, and composition of bile acids, and changing the expression of drug transporters in the epithelium [30, 44]. One of the designing approaches of controlled-release and colon-targeted systems is to use polysaccharides resisting the GI enzymes but vulnerable to anaerobic microflora of the colon [43]. Additionally, it is possible to design prodrugs metabolized by the gut microbiota [44].

4.4.2 Pathological Conditions

Depending on the duration and severity of the disease, the physiological condition of some or all parts of the GI tract may change. Acute diseases usually cause temporary GI dysfunction. For instance, increased motility in diarrhea reduces the residence time of the dosage form and the chance for complete drug release and absorption; acute GI infections disrupt the microbiome; toxins secreted by GI pathogens cause intestinal inflammation leading to enhanced permeability. Chronic diseases, on the other hand, usually have significant long-term effects. For example, in IBD, depending on the sub-type of the disease, inflammation develops in some areas of the GI mucosa. This can lead to increased intestinal epithelial permeability as well as changes in the motility, pH, volume of GI contents, microbiota, and enzymatic activities. Nevertheless, disease conditions affecting oral drug delivery are not limited to GI disorders, but a range of non-digestive diseases (such as cystic fibrosis, diabetes, Parkinson's disease, HIV, and COVID-19) may anatomically/functionally change the GI tract leading to pharmacokinetic alteration of drugs [11, 45]. It has even been shown that pain can also impact motility, secretion, epithelial permeability, mucosal perfusion, and the microbiome [11].

In addition to all these, a problem like esophageal dysphagia resulting from benign or malignant causes makes it difficult to swallow the medicine and fails drug administration from the beginning [46]. Besides, any drug that stimulates GI motility, induces diarrhea/constipation, alters pH, or causes dysbiosis (imbalance of microflora) is potentially prone to affect the fate of the orally co-administered drug [11].

4.4.3 Demographic Characteristics

By classifying based on certain demographic characteristics, one can find specific populations where physiological conditions are naturally different from healthy adults; the most prominent of these factors are age [47], gender, and ethnicity [30]. Besides, there are other potential factors prone to affect the bioavailability of drugs (e.g., diet, obesity, and pregnancy).

4.4.4 Physicochemical/Pharmaceutical Parameters

In addition to the physiological and anatomical aspects of the barriers, the physicochemical and pharmaceutical characteristics of the drugs are also of crucial importance. To be orally bioavailable, a drug molecule must first be molecularly dissolved in the GI fluid and then cross the intestinal membrane. Inherently, many drugs lack such competence (i.e., concurrent high solubility and permeability); hence, they are challenged in at least one of these steps [6].

The Biopharmaceutics Classification System (BCS) is a guide developed by Amidon *et al.* [48] and then promulgated by the US FDA to predict the oral absorption of drugs. This system classifies drug molecules based on indicators of solubility, dissolution rate, and intestinal permeability [49]. In the BCS, solubility boundaries are determined based on the FDA solubility which is the highest-dose strength of an immediate-release product in an aqueous media with a maximum volume of 250 ml over a pH range of 1 to 7.5 at 37°C [50]. To more accurately simulate the GI conditions, the pH range is recently adjusted to be 1 to 6.8 [51]. It is worth noting that FDA solubility is not an API intrinsic property, but a formulation-related factor [49]. An immediate-release product refers to a dosage form with rapid dissolution, in which at least 85% of the labeled amount of the drug content is dissolved within 30 minutes using USP Apparatus I at 100 rpm or USP Apparatus II at 50 rpm in one of the reference environments [52]. Immediate-release is important since a high dissolution rate will guarantee that, in the case of highly soluble and permeable drugs, the gastric emptying rate will not be the rate-limiting step of absorption [50]. The classification of permeability is either directly based on drug absorption from the human intestine, indirectly by measurement of mass transfer rate through the human intestinal membrane, or by absorption prediction using animal/membrane models. A drug would be highly permeable if at least 90% of its oral dose, compared with the intravenous dose, is absorbed [52]. When the permeability is high, the short intestinal transit time does not prevent maximum absorption [49].

More recently, a revised version of BCS has been developed by Butler and Dressman [53]; the so-called Developability Classification System (DCS), in which biopharmaceutics concepts have been included to improve oral product development. This system was, then, modified as the Refined Developability Classification System [54, 55]. Although the framework remained the same for the BCS, significant differences were made in the following three cases [53, 54]:

(i) Solubility of drugs was applied in the fasted state simulated intestinal fluid (FaSSIF) rather than simple buffers of BCS
(ii) Permeability was determined as human jejunal permeability instead of the absorbed dose fraction in BCS
(iii) High permeable-low soluble drugs (BCS class II) were classified into two main subcategories: dissolution rate-limited drugs (DCS class IIa) and solubility-limited ones (DCS class IIb).

How drugs are classified in BCS and DCS is shown in Figure 4.6. In addition to providing a useful guide to determine appropriate drug candidates for oral delivery, such classification is also beneficial to selecting the optimum formulation based on the drug characteristics [6]. The most suitable approaches for improving the oral absorption of different BCS classes are illustrated in Figure 4.7.

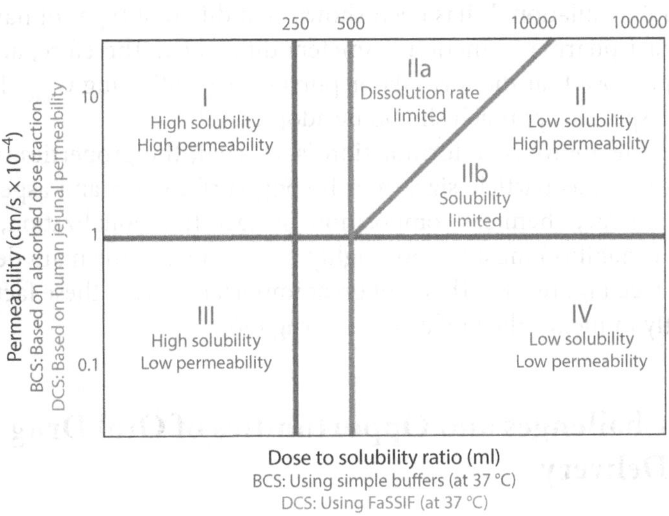

Figure 4.6 Classification of drugs based on biopharmaceutics classification system (BCS) and developability classification system (DCS).

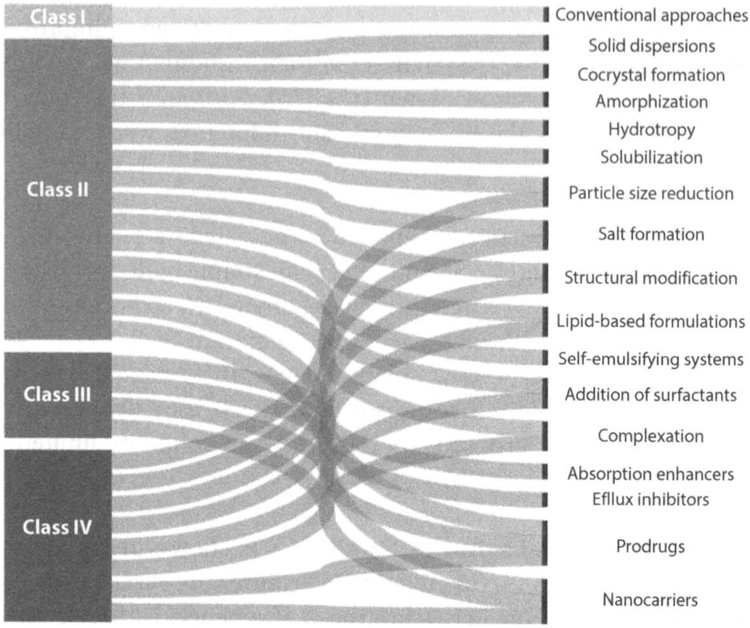

Figure 4.7 Common approaches to improve oral absorption of drugs belonging to different classes of biopharmaceutics classification system (BCS).

In addition to the properties of the drug, the selection of the appropriate carrier is also determined by the expected *in vivo* performance of the nanoformulation. It has been shown that different types of nanocarriers impact pharmacokinetic parameters differently. Therefore, according to the purpose that the formulator pursues (e.g., affecting C_{max}, T_{max}, and AUC), a specific approach should be adopted [6].

Once the appropriate formulation is selected, the properties of nanocarriers (such as particle size, morphology, surface area and aspect ratio, surface charge, chemical composition, surface functionalization, roughness, and stability) must be thoroughly investigated as the main designing aspects (see Figure 4.8). This is of great importance since these factors significantly influence the biofate of the drug [56].

4.5 Challenges and Opportunities of Oral Drug Delivery

The specific anatomical and physiological characteristics of the GI tract create numerous barriers to drug delivery, while also providing opportunities.

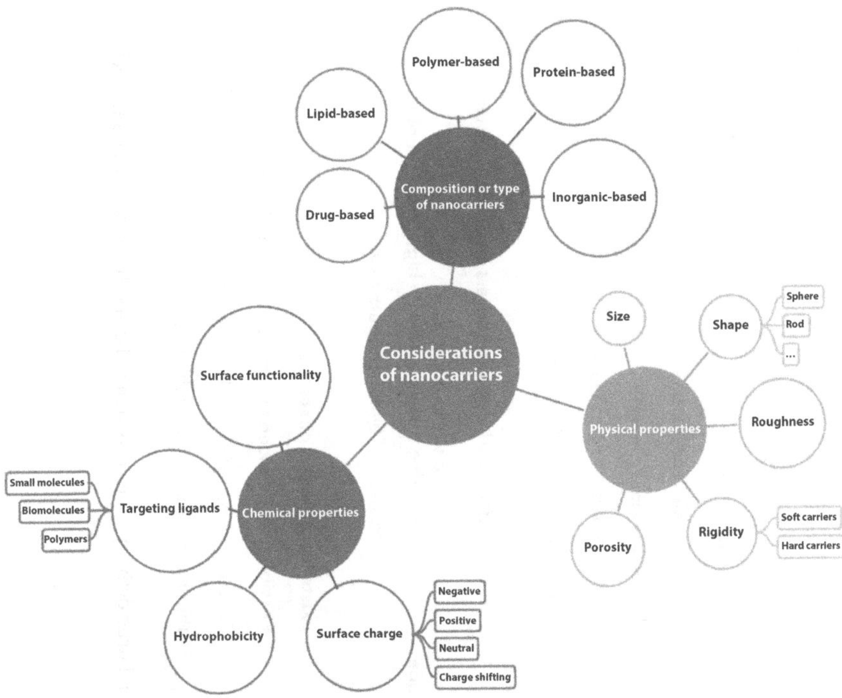

Figure 4.8 Considerations to be observed in designing of nanocarriers.

Figure 4.9 shows an organ-based classification of the main challenges and opportunities of the oral NDDSs.

4.6 Classification of Oral Novel Drug Delivery Systems

4.6.1 Regional Drug Delivery to the Esophagus

As mentioned at the beginning of the chapter, the esophagus has been long overlooked as a site of drug delivery compared to other parts of the GI tract. The key restriction is the transient nature of the esophagus which leads to a very short residence time of dosage forms. However, the growing number of patients with esophageal disorders (e.g., cancer, fungal infections, gastric reflux, and motility dysfunctions) has increased the need for local and long-lasting esophageal delivery systems [57]. A large number of oral cavity targeting strategies (e.g., gels, films, lozenges, and chewing gums) may be translated to the esophagus targeting; so that, the drug released

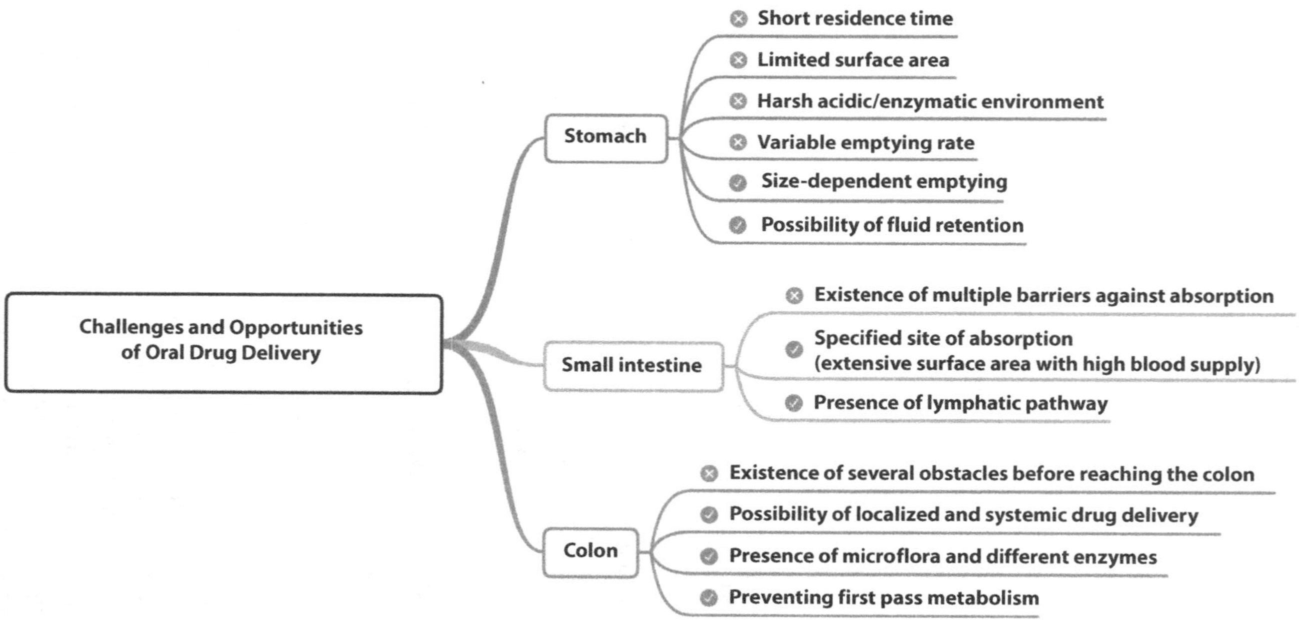

Figure 4.9 Corresponding challenges and opportunities of each organ for oral drug delivery (The cross sign indicates the challenge and the tick mark indicates the opportunity).

into saliva is constantly swallowed over a long period [58, 59]. If the viscosity-enhancing polymers are also included (e.g., in the erodible lozenges), the residence time of the drug will be even longer [58]. Bioadhesive dosage forms (e.g., bioadhesive viscous liquids, bioadhesive films, etc.) are among the most promising systems in this regard, improving the esophageal mucosa adherence and consequently, prolonging the residence time of dosage form [57, 58].

4.6.1.1 *EsoCap Technology*

EsoCap drug delivery system is an innovative platform for the treatment of esophageal diseases that enables an effective local, targeted, long-lasting, and painless drug administration to the esophageal mucosa [60]. Its design principle is based on a thin hydrophilic polymer film that absorbs water and becomes sticky upon contacting the mucous membranes. Figure 4.10 schematically illustrates the design of EsoCap and what happens during its administration. Structurally, the EsoCap system is composed of a hard capsule (inside which there is a rolled-up drug-loaded mucoadhesive film and a sinker to enhance the density) and a soluble thin thread –the so-called retainer. To be administrated, the whole system is housed inside

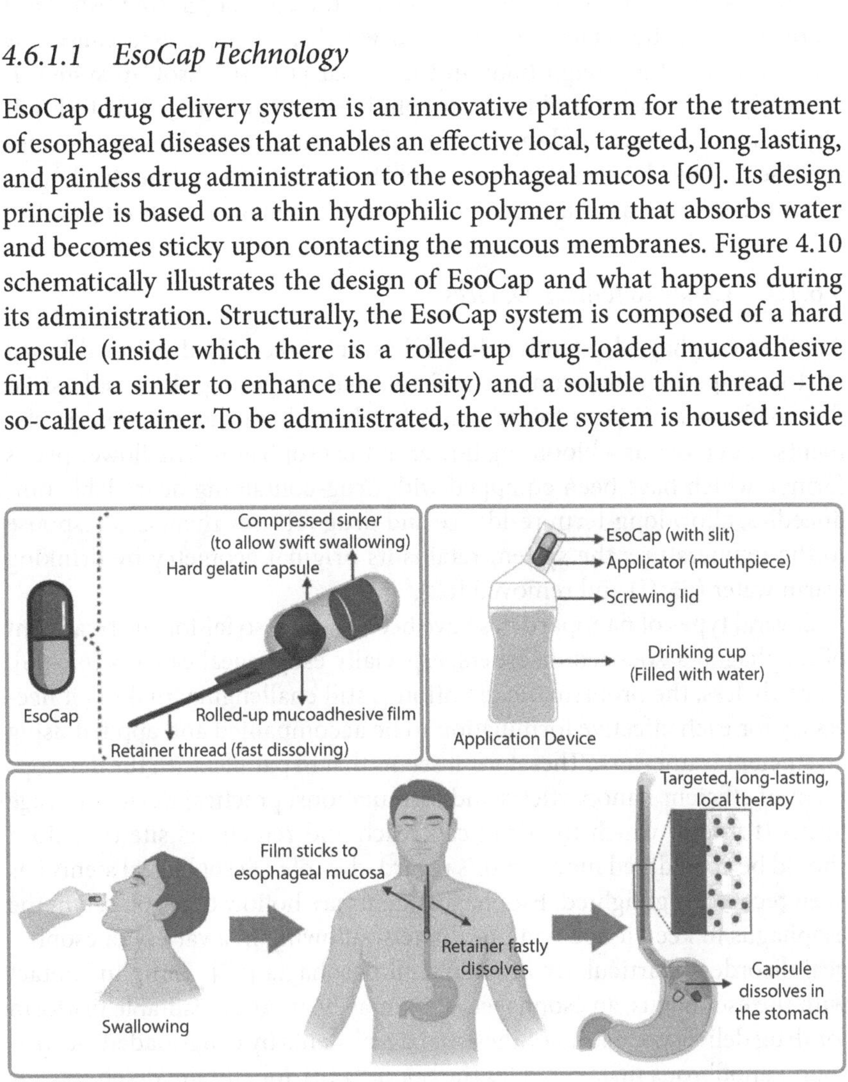

Figure 4.10 Schematic illustration of EsoCap system; structural design, how to use, and mechanism of action (according to Ref. [13, 59, 60]) (created with BioRender.com).

a mouthpiece applicator that is screwed onto a water-filled drinking cup. The two ends of the retainer are connected to the polymeric film and the applicator [13, 61].

The patient inserts the mouthpiece of the device and drinks the water. Immediately after the capsule is swallowed (endowed by the sinker enabling swift swallowing and comfortable drinking), the film is dragged out from the capsule by the retainer. As the capsule passes through the esophagus, the mucoadhesive film unrolls and sticks to the esophageal wall. After just mere seconds, the retainer is dissolved which reduces nausea caused by the sensation of a foreign body in the throat [13, 61]. EsoCap system is currently in a phase 2 trial for eosinophilic esophagitis [61]. This technology seems a forward-looking platform for a variety of GI disorders since provides a high local concentration of drugs for a long time, leading to enhanced efficacy, as well as decreased dose frequency and side effects [13].

4.6.1.2 Other Retentive Devices

Another esophageal retentive device has been recently developed based on the shape-memory concept and thermal triggering. A capsule-sized flower-like system (in a folded state) is taken; thanks to elastomeric elements, it deploys as a blooming flower in the esophagus. The flower petals (arms), which have been equipped with drug-containing degradable millineedles, allow long-term residence and drug release. Then, as a response to the temperature, the system retains its original geometry by drinking warm water (55°C) and removed [62].

Several types of nanoparticles have been studied so far for the treatment of esophagus-associated disorders, especially esophageal cancers [63–66]. Nevertheless, the problematic retention is still challenging; making it necessary for each effective formulation to be accompanied and applied using an efficient framework. Therefore, it seems that in parallel with the development of efficient nanoparticles and formulations, practical devices/dosage forms (through which the drug can reach and remain its site of action) should be considered more. In this regard, the role of esophageal stents has been recently highlighted. Esophageal stents are hollow tubes placed in the esophagus to keep it open and facilitate swallowing in a variety of esophageal disorders, particularly in malignant dysphagia [67]. Being in contact with the esophagus, an esophageal stent may be used as a suitable platform for drug delivery. Coating of the esophageal stents by drug-loaded electrospun nanofibrous mats provides the opportunity for sustained delivery for several days [68].

4.6.2 Regional Drug Delivery to the Stomach

As already discussed, one of the most challenging physiological barriers ahead of oral drug delivery is the unpredictable and high emptying rate of the stomach, leading to insufficient gastric residence time (normally 2-3 h). This restricts and fluctuates the bioavailability of drugs administered as conventional delivery systems. Therefore, in recent years, numerous studies have focused on the systems remaining for a desired period of time in the stomach (entitled gastroretentive systems) to achieve a prolonged/controlled release profile. When the residence time is long enough, there will be also an opportunity to achieve sufficient drug concentration to overcome local gastric disorders [4].

The gastroretention approach is especially helpful for drugs whose efficiency is hindered by dissolution, absorption, or stability challenges [69–71]. However, it is not the option of choice for drugs with limited solubility or stability in harsh gastric conditions (e.g., peptides and proteins), those causing gastric lesions upon slow liberation (e.g., NSAIDs), drugs extensively affected by the first-pass metabolism, and those with specific absorption in the colon [69, 72]. Gastroretentive systems are designed in two main platforms of single- and multiple units. Since the single-unit systems suffer from the all-or-nothing risk during the stomach emptying, multiple-unit systems have been proposed to, in addition to compensating for this defect, provide a uniform distribution in the stomach, gradual emptying, (probable) longer residence time, and lower inter-subject variations [73]. In addition to the unit-based classification, gastroretentive systems can be categorized based on the strategy used to increase their retention time; as follows:

4.6.2.1 Floating (Low-Density) Gastroretentive Systems

The concept of floating for pharmaceutical dosage forms was first introduced in 1968 by Davis [74]. According to Archimedes' principle, for an object to stay afloat, the buoyancy force exerted by the fluid vertically must be greater than the oppositely exerted gravity. This can be achieved by adjusting the density of the object relative to the fluid on which it is floating [75]. Therefore, floating gastroretentive systems are dosage forms with a lower bulk density than that of gastric fluid (reported as 1.004-1.01 g/cm^3) that keep them buoyant in the stomach [4]. Such a strategy improves the gastric residence time of the drug, independently of the emptying rate; hence, the fluctuation of bioavailability significantly decreases. Apart from the mentioned advantages, there are also limitations. For instance, an acceptable level of fluid is required to float; likewise, the floating systems are highly affected by the change of body position in the fed state [72].

Raft-forming systems are a subclass of floating systems which is considered a revolution in controlled release oral delivery systems. They are effervescent-containing liquid formulations with *in-situ* gelling capability, providing sustained drug delivery for more than 48 hours. The mechanism of gelation may be physical (swelling or diffusion), chemical (ion-induced cross-linking), or physiological stimuli (pH or temperature). Swelling of each portion of the liquid upon contacting with the gastric fluid leads to the formation of a continuous cohesive gel layer with high viscosity but low density (created by the formation of CO_2) (see Figure 4.11). Without being affected by the emptying rate of the stomach, these layers remain buoyant on the gastric fluid for a long time and release the drug slowly; then, the remained mass is emptied. In addition to being afloat on the gastric content, the gel with an adhesive nature may adhere to the gastric mucosa to create a physical barrier between the esophagus and stomach and hence, prevent acid reflux. The main components of a raft system are a gel-forming material (e.g., polymers) and carbonate/bicarbonate to provide the low density required to float. Appropriate polymers should possess biocompatibility and pseudo-plastic (shear thinning) behavior. Various natural and synthetic polymers such as alginic acid, chitosan, pectin, xyloglucan, gellan gum, and carbopol could be employed for the preparation of raft-forming systems [76]. Ease of manufacturing and administration, highly extended

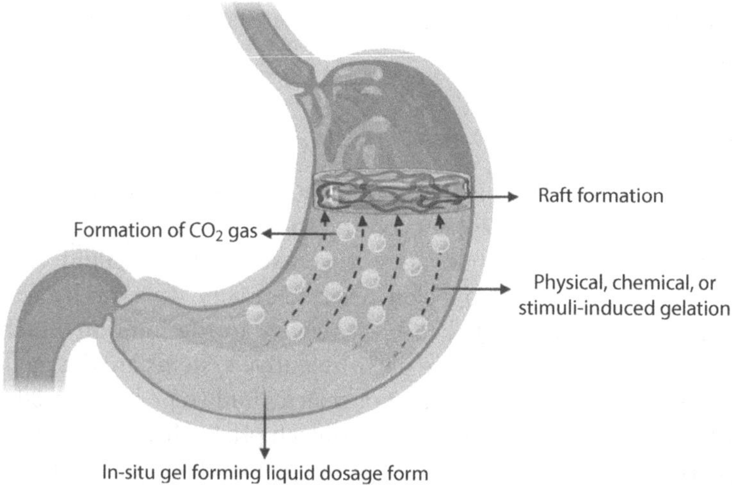

Figure 4.11 Schematic illustration of the raft (floating *in-situ* gel) forming system (created with BioRender.com).

residence time, and fast performance are the most important advantages of raft-forming systems [77].

4.6.2.2 *Sinking (High-Density) Gastroretentive Systems*

Sinking systems refer to dosage forms with a higher density than that of normal gastric contents (around 1.004 g/cm^3); hence, they are capable of withstanding peristalsis and settling down in the stomach [69]. The first-ever study on these systems dates back to 1993 when Clarke *et al.* [78] investigated the effect of material density on gastric residence time. In a subsequent study, a critical range of 2.4 to 2.8 g/cm^3 was determined to improve the residence time [79]. The major drawback of sinking systems is their problematic formulation; because it is difficult to achieve such a density while containing a high amount of drug (> 50% by weight). Besides, there is a lack of animal and clinical studies on these systems [4]. Usually, barium sulfate, iron powder, zinc oxide, or titanium oxide are accompanied by polymers as density-increasing agents [69]. High-density gastric-resident approach has been so far employed to produce a wide range of dosage forms such as matrix tablets [80], multi-particulate pulsatile tablets [81], and microspheres [82].

4.6.2.3 *Expandable (Size-Increasing) Gastroretentive Systems*

As its name implies, expandable systems achieve prolonged residence time in the stomach through an increase in their volume and/or change in shape; which could be based on swelling or unfolding. Regardless of the mechanism involved in expansion, it must be assured that any change is in line with the requirements of each of the following three steps: (i) the system size is small enough to be swallowed easily, (ii) once inserted into the stomach and opened, it gains a larger size than the pyloric sphincter (with an approximate diameter of 11-13 mm in the wide-open position) to be able to withstand stomach waves to remove the contents, and (iii) after the programmed drug release, the size of the remaining structure is small enough to be evacuated [74]. The elimination phase could be aided either by product breaking due to its reduced resistance against gastric contractions or being exposed to orally administered specific substances [83]. Since the expandable systems are prone to block the pyloric sphincter in their expanded state, they are also called plug-type systems [70].

In the swelling systems, size-increasing occurs through hydration in contact with gastric fluid. Here, hydrophilic polymers have a major role (e.g., polyacrylate polymers, polycarbonates, polyethylene oxides, polyvinyl acetate, hydroxypropyl methylcellulose (HPMC), carbopol, sodium alginate, and agar), since they undergo several modifications upon water absorption [5]. Swelling, plasticization (reduction of glass transition temperature), chain disentanglement (which in turn leads to enhanced diffusion coefficient), and erosion are among the effects of water absorption on polymers. Nowadays, intelligent polymers in which the swelling behavior and drug release characteristics are under the influence of special stimuli (such as pH, temperature, and solvent) have received much attention to design novel swelling gastroretentive systems [4]. Besides, the role of superporous hydrogels with great capability in water absorption and swelling has been recently highlighted.

In the case of unfolding systems, the product (composed of biodegradable polymers, inside which the drug is dispersed) is collapsed by force (e.g., through folding) and placed in a dosage form for ease of swallowing. Then, in the stomach, the carrier is degraded and the product deploys and regains its initial bulky configuration [4, 83]. The variety of polymers with different properties (e.g., erodible, non-erodible, *in-situ* forming gel polymers, etc.) allows for achieving different drug release manners (such as biphasic and sustained release) [4, 5]. Several geometries have been reported hitherto for unfolding expandable systems including ring-shaped, tetrahedron, disc, lobed, planar, etc. [5]. Some of these geometries are represented in Figure 4.12.

4.6.2.4 *Bio/Mucoadhesive Systems*

Bioadhesive systems, as an efficient platform for oral-controlled drug delivery, were first introduced in 1984 by Park and Robinson [84]. From

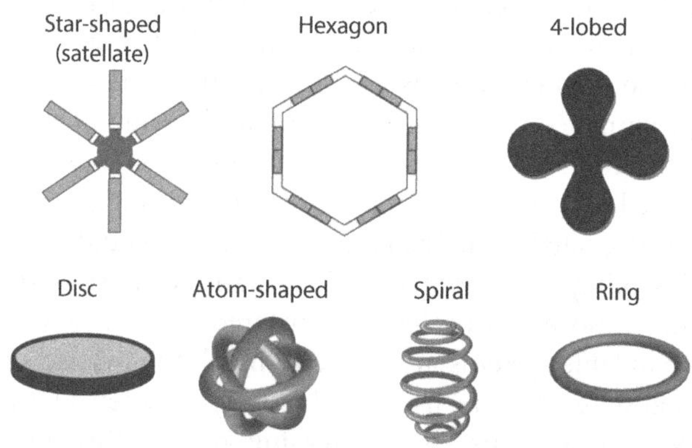

Figure 4.12 Examples of expandable systems' geometries (based on [83, 84]) (created with BioRender.com).

then on, bioadhesive polymers were considered the point of convergence to create more efficient controlled systems in several routes of administration including oral, ocular, nasal, and vaginal [4]. Bioadhesive and mucoadhesive oral dosage forms are designed to adhere inside the gastric lumen and protect the drug from removal by peristaltic flow for a longer time. Also, these site-specific systems are beneficial to local absorption of drugs; for instance, where there is an infected region in the stomach [85].

Ideally, a mucoadhesive polymer should be inert, non-toxic, and non-irritating with site-specificity and adherence to the mucosal surface. How strength a polymer interacts with mucin depends on its chemical structure, molecular weight, charge, chain flexibility, hydrogen-bonding capacity, cross-linking density, and degree of hydration [70]. Examples of commonly used bioadhesives are chitosan, sodium alginate, tragacanth, cholestyramine, dextran, PEG, HPMC, poly(acrylic acid), poly(alkyl cyanoacrylate), polylactic acid, Sephadex, and sucralfate [4, 70]. Mucoadhesive systems may be formulated in different forms such as microspheres, beads, films, capsules, and tablets. Among the main challenges with bioadhesive systems are the continuous turnover of mucus resulting in dosage form removal, a high degree of hydration in the stomach leading to a reduction in bioadhesion, and a high risk of dosage from adhesion to the esophagus which may cause collateral lesions [70].

4.6.2.5 Magnetic Field-Assisted Systems

The basis of magnetic field-assisted systems is the attraction of two magnets: (i) the internal magnet in a small amount which is accompanied by active ingredients and excipients to form a dosage form, and (ii) an external (extracorporeal) magnet located on the abdomen over the stomach (Figure 4.13) [4]. The magnetic field intensity and position of the external magnet impact the gastroretention. This strategy can improve the bioavailability of drugs by positively affecting the residence time of dosage forms. However, since it is difficult to precisely locate the external magnet at the specific position in the case of clinical research, the patients may feel discomfort, and thereby, the compliance decreases [4, 70].

4.6.2.6 Nanocarriers

Since small particles (i.e., below 7 mm in diameter) are quickly evacuated from the stomach, the use of nanoparticles as gastroretentive systems is challenging. However, their advantages (e.g., optimal distribution throughout the stomach, enhanced mucosal interaction, and high cellular uptake)

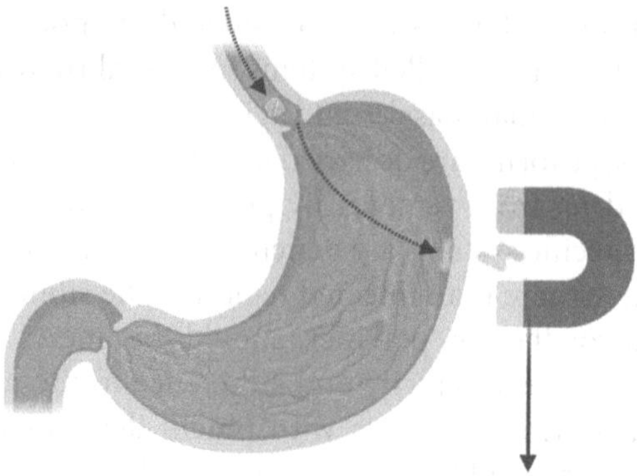

Figure 4.13 A schematic illustration of a magnetic field-assisted system (created with BioRender.com).

are highly attractive. Particularly, for the treatment of local gastric diseases and drugs with preferential absorption in the stomach, nanoparticles cannot be ignored or excluded [11]. So far, some of the previously mentioned approaches such as mucoadhesion have been incorporated to provide gastroretentive nanoparticulate systems like polymeric nanoparticles [86–88], liposomes [89], and porous silica nanoparticles [90]. Although promising data have been obtained from both *in vitro* and *ex vivo* experiments of gastroretentive nanoparticles, their extrapolation is challenging concerning a large number of gastric variables. Also, in the case of mucoadhesive nanosystems, the problems with the mucus (e.g., fast turnover and being hydrated) remain. To sum up, more *in vivo* investigations are needed to determine how translatable and reproducible gastroretentive nanoparticles are [11].

A potential alternative for gastroretentive nanoparticles is nanofibers. Nanofibers are defined as structures with (i) cylindrical shape, (ii) outer diameter of less than 1000 nm, and (iii) an aspect ratio (i.e., length-to-width ratio) of larger than 50 [91]. Nanofibrous mats are an emerging class of gastroretentive systems [92–95]. Very low density along with high surface hydrophilicity, adsorption properties, significant porosity, and a great surface-to-volume ratio have made nanofibrous systems a superior option as floating dosage forms [92].

4.6.3 Regional Drug Delivery to the Small Intestine

The main goal of novel intestinal drug delivery systems is to achieve optimal bioavailability. This goal is usually accomplished by structural modification of drugs (i.e., prodrugs), or using controlled-release systems, mucoadhesive dosage forms, and nanotechnology-based carriers.

4.6.3.1 Prodrugs

A prodrug is an inert, non-toxic, and pharmacologically inactive derivative of the main drug, which converts *in vivo* to the active parent drug through enzymatic/chemical transformation [96]. Prodrug formation is commonly used to resolve one or more challenging properties of the drug (e.g., low aqueous solubility, high lipophilicity, low permeability, insufficient systemic/local absorption, chemical instability, low half-life, and toxicity) [97]. Also, lipophilic prodrugs may enhance lymphatic absorption. In particular, polar/charged drugs with low intestinal absorption will benefit the best from prodrug formation [98].

Prodrugs are divided into different categories depending on the type of chemical bond used; the most common of which are ester, amide, carbamate, and azo bonds. In addition, polar moieties (e.g., PEG) may be included to achieve prodrugs. In this case, not only will the benefits of prodrugs be achieved but also the absorption will be further enhanced *via* the bypassing efflux pumps (mainly P-gp) and CYPs metabolism. Worth considering, prodrugs may also provide site-specific drug delivery by targeting special biotransformation mechanisms [98].

4.6.3.2 Hydrogels

Hydrogels are three-dimensional structures of physically/chemically bonded polymeric networks imbibing a large amount of water and entrapping it within the intermolecular spaces while remaining undissolved [99, 100]. High affinity to water absorption, soft consistency, high biocompatibility, tunable properties, and the ability of site-specific and controlled release of both small molecules and biotherapeutics, have granted ideal features to hydrogels as drug delivery systems [100, 101]. In particular, "superporous" hydrogels with high mechanical strength and swelling rate are excellent candidates for oral drug delivery [101]. Although hydrogels are typically employed for the delivery of hydrophilic compounds and particularly hydrophilic macromolecules, many efforts (e.g., incorporation of hydrophobic domains within the matrix network, and adding cyclodextrins to the hydrogel structure) are being made today to effectively load lipophilic drugs as well [102]. Nowadays, several types of hydrogels are being developed as oral formulations, the most common of which are pH-responsive and enzymatically sensitive systems [100].

By adjusting the particle size of the hydrogels, targeted drug delivery to different sites of the GI tract could be achieved (e.g., 1–1000 μm for the small intestine, 50–200 μm for intracellular vesicles, and < 30 μm for lymphatic vessels). For intestinal drug delivery, anionic hydrogels are generally used since, in the pH values higher than the polymer chain pKa, they are ionized/swollen, thus providing the drug release. For the delivery of drugs in the stomach and intracellular environments, cationic hydrogels are more suitable because they are ionized at the lower pH ranges [101].

4.6.3.3 Osmotic Pumps

Osmotic pumps are promising drug delivery systems providing controlled release of drugs (e.g., prolonged, pulsatile, etc.) with uniform blood concentrations. The key advantages of osmotic pumps are their independence from the pH and hydrodynamics of the surrounding environment, thereby providing a robust system with exceptional *in vitro/in vivo* correlation, as well as being applicable for a variety of drugs with wide ranges of water solubility. Furthermore, osmotic pumps are easy to formulate, scale up, and operate [103]. Structurally, osmotic pumps are similar to reservoir devices except that the driving force of release is osmosis [104]. Osmosis refers to a process through which the solvent molecules pass through a semipermeable membrane from a less concentrated site to a more concentrated one. Osmotic pumps can be administered orally or implanted [103].

For oral purposes, an osmotic pump generally consists of a compressed osmotic core (containing a drug, an osmotic agent (so-called osmogen), and a water-swellable polymer) which is caged by a semipermeable membrane coating. As the core imbibes the water, the volume is expanded and pushes the drug out through the delivery port(s). The rate of drug release depends on the permeability of the coating as well as the osmotic pressure produced by the core [104]. There are several types of oral osmotic pump systems. They may be formulated as single-layer tablets, bilayer tablets, sandwiched systems, or soft gel capsules; the drug content may be as solution or suspension; and the drug may be expelled through orifice(s), channels, or *in situ* formed cracks [103, 104]. ALZA Corporation (Mountain View, CA, USA) is the pioneer of the use of osmotic pumps orally under the famous trademark of OROS® (osmotic controlled release oral delivery system).

4.6.3.4 Mucoadhesive Systems

By prolonging the intestinal residence time, mucoadhesive dosage forms not only facilitate targeting and localization of the drug at the proper site but also increase the chance of intimate contact with the mucosa which, in turn, leads to a higher flux of the drug through the intended GI segment [105]. GI mucoadhesive patches are unique dosage forms composed of several layers that prolong drug release (including both small and large therapeutics) by adhering to the mucosa [106]. They could reduce luminal loss and/or degradation of the drug which makes them a potential alternative for painful injections. This delivery system can potentially improve drug bioavailability by allowing unidirectional diffusion toward the intestinal tissue [2, 107]. GI patches may widely differ in diameter (micrometers to millimeters) and surface area. Also, they may possess one or more laminated layer(s) [106, 108]. Usually, each mucoadhesive patch consists of at least two layers including a backing layer and a drug-laden mucoadhesive layer. The backing layer is composed of water-impermeable polymers covering all systems but one side. Such a design not only leads to the proximal and unidirectional release of the drug into the mucosal surface but also precludes enzymatic degradation of the drug. This is especially important for drugs susceptible to proteolytic enzymes such as peptides and proteins. The mucoadhesive layer provides a drug depot and also assists the drug transport. Upon contact with the intestinal fluids, the patch adheres to the mucosal layer. Charge-based interactions (between the cationic polymers of the formulation and negative groups of the mucin), hydrogen bonding, and polymer chain entanglement, collectively, ensure the effective adhesion of the patch to the intestinal mucosa [106].

To achieve optimal performance of patches, the inclusion of permeation enhancers is recently highly regarded [109]. The main challenge of GI patches is the threat of being detached from the lumen wall and washed away by gastric boluses. So, a strong attachment to the mucus is of crucial importance. Even so, the high turnover rate of the intestinal mucus layer impedes the real application of intestinal patches [2]. Therefore, unlike transdermal patches, which are designed for multi-day release, intestinal patches aim to release drugs over several hours [110].

4.6.3.5 Geometrically Modified Dosage Forms

The geometry of a dosage form strongly affects the drug release profile. Therefore, the geometric principles have been widely employed by researchers and companies to achieve devices in various forms (e.g., spherical, cylindrical, biconvex, etc.) with modified-release behaviors –from zero-order release kinetics to bimodal release model and non-zero-order scheduled chronotherapy. To realize zero-order kinetics as an optimal goal for many therapies, the geometry is commonly changed to provide a constant surface area for drug release. Multilayered tablets and donut-shaped tablets are examples designed for this purpose. Geometrically modified dosage forms have evolved thanks to the presence of various polymers with different degrees of swellability, porosity, and erosion. In general, swelling, dissolution, and erosion are the main mechanisms by which different polymers may play their roles in this context [111].

4.6.3.6 Nanocarriers

To date, several types of nanostructures have been widely studied as oral delivery systems with the eventual purpose of improving bioavailability and therapeutic outcomes of already marketed drugs and providing an efficient platform for the delivery of recently developed compounds or those administrated in other routes. Possessing several advantages (such as modifiable size, geometry, composition, coating, and surface characteristics like charge, coating, and targetability), nanocarriers can undertake different roles in improving bioavailability [112]. For instance, nanocarriers can preserve their susceptible cargo (particularly, biomacromolecules) against destabilizing factors in the GI tract. Also, nanocarriers can enable the delivery of lipophilic molecules by increasing their apparent aqueous solubility, improving the permeability of drugs *via* several mechanisms (e.g., improving lymphatic transport and reducing efflux), and reducing the systemic side effects [26, 113]. Meanwhile, thanks to their targeting capability, nanocarriers can allow drug delivery to particular regions of the GI tract (e.g., enterocytes and M cells). Therefore, nanocarriers may be used for oral immunization as well [114]. Furthermore, endogenous triggers of the GI tract (such as pH) provide opportunities for the development of stimuli-responsive nanocarriers [112].

According to the ever-growing trend of published research on oral nanoformulations from 2000-2018, the largest share is respectively allocated to lipid-based nanocarriers (>50%), polymer-based nanosystems (>25%), nanocrystals, and inorganic nanoparticles [112]. Following on, we merely introduce some types of nanocarriers used hitherto as oral delivery systems and the readers are referred to the original literature for more details.

Liposomes

Liposomes are renowned lipid-based bilayer nanostructures mainly composed of phospholipids and cholesterol. Owing to mimicking the cell membranes, they are basically considered biocompatible and safe. Moreover, liposomes have a large capacity for the entrapment of lipophilic, hydrophilic, and also amphiphilic drugs and can, then, gradually release them [6]. In particular, liposomes have appropriate potential for oral administration; for instance, they have a suitable adherence to the biological membranes. Also, they are prone to form mixed systems with the bile salts which improves the solubility of lipophilic drugs. Besides, liposomes are proper candidates to be uptaken into the lymphatic circulation [115]. Despite their success in the parenteral route, the use of liposomes *via* the oral route is hampered by several GI barriers. The most challenging obstacle against oral delivery of liposomes is their instability in the GI tract (due to aggregation, fusion, or hydrolysis of phospholipids) [116, 117]. Today, by modulation of the composition of the lipid bilayer and surface modifications by polymeric coatings, surface charging, and ligand attachment, this challenge has been considerably conquered [117]. To date, no liposomes have been marketed as oral pharmaceutical products [117]. Nevertheless, liposomes are widely used to carry supplements and nutraceuticals (e.g., Sideral® and Sucrosomial® iron) as oral products [118].

Niosomes

Niosomes, having a bilayer vesicular self-assembled structure are considered as the structural analogs of liposomes. The difference is that niosomes are mainly composed of non-ionic single-chain surfactants, whereas liposomes are formed from neutral/charged double-chain phospholipids. Similar to liposomes, niosomes can improve the oral bioavailability of drugs by extending the residence time and improving drug stability. Also, they are suitable carriers for the targeted delivery of small molecules [119]. The main superiority of niosomes over liposomes is the higher stability leading to a longer shelf-life and enhanced stability in the GI tract. Also, their scale-up is cost-effective with no need to provide special storage conditions. Besides, the encapsulation efficiency in the niosomes is generally higher, by the presence of less cholesterol in their structure. By overcoming several drawbacks of liposomes, niosomes follow a growing trend as a potential alternative delivery system [119, 120].

Solid lipid nanoparticles (SLNs) and nanostructured lipid carriers (NLCs)

SLNs are promising lipid-based nanostructures that have garnered more attention in recent years for the delivery of lipophilic cargos. They possess several advantages like biocompatibility, enhanced lymphatic transport, and, thereby, reduced first-pass metabolism [6]. In addition, they have higher stability than liposomes (due to their rigid lipid matrix) and more feasible mass production [121]. However, SLNs suffer from limited loading capacity and potential leakage of the drug. Owing to the special structure of NLCs, containing a mixture of solid and liquid lipids, these shortcomings have been compensated. NLCs have a greater payload (due to the less-ordered structure of the lipid matrix) and a much-lowered cargo expulsion during storage [6].

Nanoemulsions and self-nanoemulsifying drug delivery systems (SNEDDSs)

Nanoemulsions are thermodynamically, but not kinetically, unstable, transparent, or translucent nano-sized dispersions of oil droplets in the aqueous media stabilized by an interfacial layer of surfactants and co-surfactants [122]. SNEDDSs are anhydrous homogenous systems consisting of oils, surfactants, and co-surfactants, which spontaneously form oil-in-water nanoemulsions upon dilution with water under gentle agitation [123, 124]. Both nanoemulsions and SNEDDSs have the benefits of lipid-based systems. Additionally, they are easy to prepare and feasible for mass production. Compared with the ready-to-use nanoemulsions, SNEDDSs offer additional benefits like higher storage stability, improved palatability, and the possibility to be carried with gelatin capsules [6, 123]. The aforementioned lipid-based nanocarriers used for oral drug delivery are graphically illustrated in Figure 4.14.

Cyclodextrin-based nanostructures

An inclusion complex refers to a complex in which a chemical compound (the guest) occupies the cavity of another (the host). The most famous

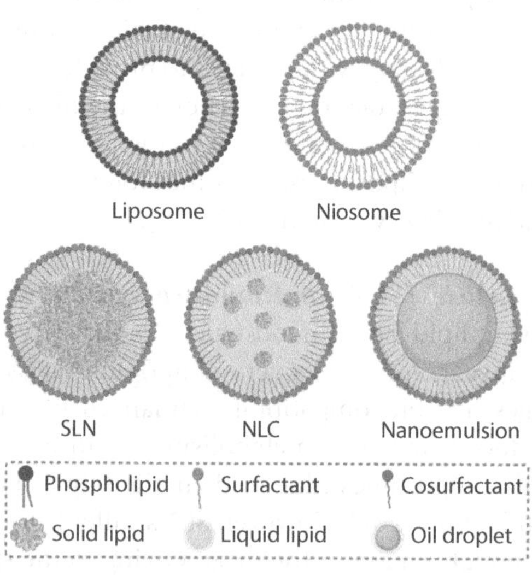

Figure 4.14 Some lipid/surfactant-based nanocarriers used in oral drug delivery (created with BioRender.com).

example of host-guest inclusion complexes in the pharmaceutical field is cyclodextrins. Cyclodextrins are truncated cones generally made of 6, 7, or 8 glucose units connected as closed-rings *via* the glycosidic bonds. The outside of the cyclodextrins is hydrophilic due to the presence of free hydroxyls; while several oxygen atoms in the glycosidic bonds and the hydrogen atoms make the cavity hydrophobic. Concerning this structure, cyclodextrins are widely used to solubilize drugs with insufficient aqueous solubility and improve their bioavailability. Due to their bulky and hydrophilic nature, cyclodextrins are negligibly absorbed orally and generally considered safe [125].

Polymeric Micelles

Micelles are dynamic aggregates of amphiphilic molecules that form above a certain concentration called CMC (critical micelle concentration). In the aqueous media, amphiphilic molecules are constantly disintegrating and self-assembling to form aggregates with the headgroups outwards and central tails (a hydrophobic core). Owing to such structure, they can solubilize poorly water-soluble cargos [126]. CMC is an indicator of micellization ability as well as micellar stability. The less CMC, the micelles are more stable and easier to form [127]. Micelles made of nonionic surfactants are extensively used as an ingredient in various drug delivery systems. However, using them as independent carriers is challenging. This is because surfactant-based micelles generally have a high CMC; therefore, they are very susceptible to dilution which may lead to their breaking apart and premature drug release [128].

Polymeric micelles, made of amphiphilic block copolymers, having about a thousand times lower CMC than surfactant micelles, have remarkably higher thermodynamic stability [127]. In addition, the preparation condition of polymeric micelles is generally mild, which is especially advantageous for labile drugs [6]. Upon oral administration, polymeric micelles withstand dissociation and drug leakage. Additionally, certain polymers can improve oral bioavailability, either through enhancing the permeability of the membrane to drug and/or carrier, and inhibition of efflux transporters or first-pass effect. Besides, using bioadhesive polymers in the micellar structure can extend the residence time of the system in the GI tract. Also, pH-sensitive micelles can improve site-specific absorption [129]. Today, polymeric micelles are gaining much attention in the delivery of macromolecules [130]. If more than one type of polymer is involved in the formation of micelles, then the "mixed micelles" are achieved which offer additional advantages like improved kinetic and thermodynamic stability (lower CMC), easier modification, increased loading capacity, and more precise size control [131].

Polymeric nanoparticles

Polymeric nanoparticles are promising nanocarriers for controlled and targeted drug delivery. They have been widely evaluated for improving the pharmacokinetic characteristics of different drug molecules (e.g., anti-cancers, siRNA, and insulin). High stability, protecting cargos against pH and enzymatic degradation in the GI tract, controlled drug release, and capability of encapsulating small molecules and biopharmaceuticals are among the crucial advantages that make them appropriate to be administered orally [132]. In addition, they are generally made of biocompatible and biodegradable polymers that are "generally regarded as safe (referred to as GRAS)" by the US FDA [133]. Polymeric nanoparticles encompass both nanocapsules and nanospheres, with different morphologies. Nanocapsules are reservoir systems composed of a drug-containing oily core surrounded by a polymeric shell controlling the drug release. Nanospheres are made of a continuous polymeric network to form a matrix system; drug molecules may be adsorbed on the shell or entrapped inside [132]. Compared with micelles, polymeric nanoparticles are anticipated to be superior concerning their higher loading efficiency and stability [134]. Surface modification (by PEG, for instance) facilitates the penetration of nanoparticles through the mucus barrier. Besides, conjugation of the nanoparticles by targeting ligands can improve drug absorption through receptor-mediated transcytosis. Also, mucoadhesive polymers can increase the drug concentration at the absorption site by extending the residence time and increasing the contact of nanoparticles with the epithelium. In addition, M cell transport of polymeric nanoparticles could be enhanced by adjusting the particle size and hydrophobicity. Although this pathway is closely associated with the immune system and, therefore, not of choice for long-term oral delivery, it can be counted on for oral mucosal vaccination [132].

Mucoadhesive, muco-penetrating, and muco-inert nanoparticles

Similar to the other mucoadhesive systems, mucoadhesive nanoparticles show a prolonged residence time at the site of absorption. If cell-penetrating peptides are combined, the permeability of the cargo through both the intestinal mucus layer and epithelium would be significantly improved [135]. Among the most studied mucoadhesive materials are some kinds of polysaccharides (e.g., carboxymethyl cellulose, chitosan, alginate, and pectin). They are of interest not only due to their mucoadhesion but also to forming hydrogels. Another approach to facilitate mucoadhesion is the modification of polymers (e.g., thiol-functionalization) to be able to be covalently bound with mucin [136].

As a main concern, mucoadhesive nanoparticles may be immobilized within the mucus layer and subsequently cleared away. So, another approach (the so-called mucopenetration) has been developed to make particles slip through the mucus. As their name implies, mucopenetrating particles can penetrate the mucus layer, reach the proximity of the cells, and release the drug in a targeted/sustained manner [137, 138]. It should be noted, however, that the superiority of mucopenetrating particles compared with mucoadhesive ones has not yet been proven empirically [137, 139].

Coating particles with several charged groups to obtain high charge density but overall neutral surfaces is one of the most regarded approaches [136]. The development of these particles is inspired by nature. Some viruses with a sufficiently small size and a hydrophilic net-neutral surface (densely coated with both positively and negatively charged groups) can freely spread through mucus as water [138]. Therefore, they are called virus-like or virus-mimicking particles.

Besides, charge-shifting particles could address the challenge of entrapment in the mucus. The net charge of the mucus is negative due to the presence of several sialic and sulfonic acid residues. So, positively charged particles electrostatically interact and get stuck. On the other hand, in the vicinity of enterocytes, positively charged particles show a higher cellular uptake than that of their negative counterparts [136]. Therefore, the concept of zeta-potential shifting delivery has been proposed by the Bernkop-Schnürch group [140]. Accordingly, phosphate-functionalized nanoparticles with a negative charge penetrate the mucus. Then, the alkaline phosphatase adjacent to epithelial cells cleaves these groups leading to a surface charge shift to neutral or positive. The negative charge of epithelium attracts and thereby, facilitates the absorption of positively-charged particles.

Another strategy is coating particles with a non-interacting polymer to achieve muco-inert formulations. PEG is an optimal coat in this context providing an electrically neutral, hydrophilic surface with minimized electrostatic and hydrophobic interactions with the mucus [136]. The polymer-based systems discussed in this section are illustrated in Figure 4.15.

Nanocrystals (nanosuspensions)

The solubility of a drug is intrinsically affected by its particle size. As a general rule, the smaller the particle, the higher the surface-to-volume ratio, and consequently the greater interaction with solvent, and eventually

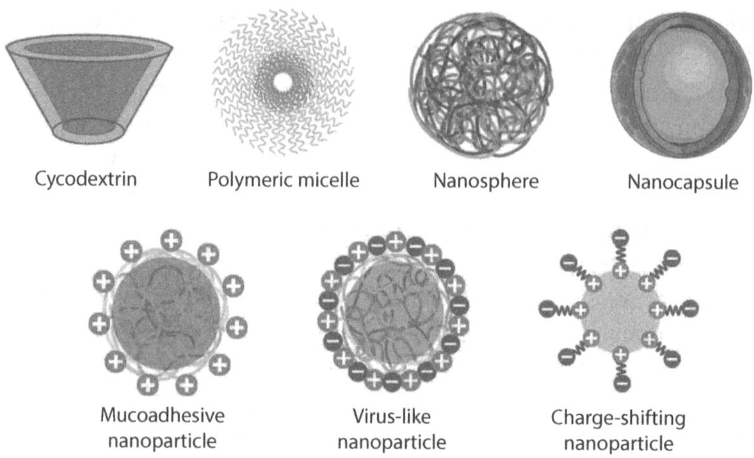

Figure 4.15 Some polymer-based nanocarriers used in oral drug delivery (created with BioRender.com).

the higher solubility [141]. Micronization is a size-reduction method that is capable of enhancing the dissolution- and diffusion rate by improving the surface area. However, it cannot increase the equilibrium solubility. Therefore, micronization is not the option of choice for drugs with very low solubility and/or high doses [141]. Further diminution in the particle size (i.e., nanocrystals) is an influential, simple, and applicable strategy for enhancing both dissolution rate and saturation solubility, and thus, the oral bioavailability of a wide range of drugs [142].

Nanocrystals are self-delivery (carrier-free) systems made of pure active pharmaceutical ingredient(s) with no or minimal non-therapeutic agents that provide a unique approach for the delivery of poorly water-soluble drugs. Being composed of up to 100% drug, nanocrystals allow excellent drug loading capacity; so, the therapeutic effect could be achieved in a much lower dose. Besides, the tedious carrier preparation processes are avoided. In addition, the total cost of the formulation will be significantly reduced due to the exclusion of excipients. Moreover, the carrier-related issues (e.g., toxicity and immunogenicity) will no longer be a concern [143]. Nanocrystals acquire several advantages through downsizing [144], which is schematically illustrated in Figure 4.16:

(i) Increased saturated solubility owing to the remarkable enhancement of particle surface area along with significant changes in the basic particles (e.g., molecules) arrangement within the solid bulk;

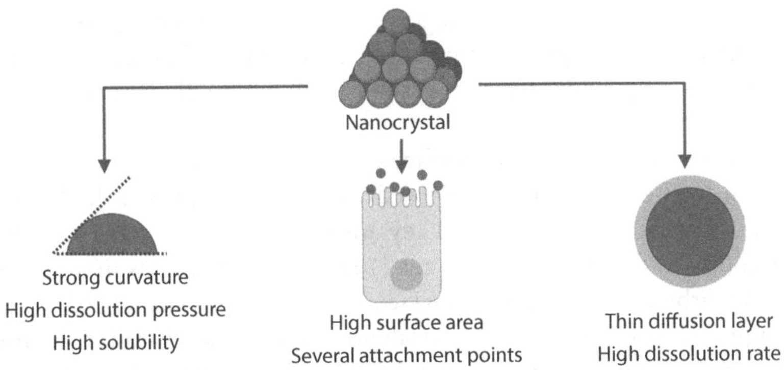

Figure 4.16 How do nanocrystals improve drug absorption? (created with BioRender.com).

(ii) Enhanced dissolution rate due to the expansion of surface area and a thinned diffusion layer surrounding each nanoparticle;

(iii) Improved membrane adhesion as a result of increased surface area and the attachment points number; so, the retention time of nanoparticle is extended, which in turn, causes improved oral bioavailability of the drug.

From a pharmacokinetic point of view, nanocrystals not only improve the bioavailability of drugs *via* acquired features but also reduce intra- and inter-subject pharmacokinetic variations (e.g., fasted/fed states). This is because the absorption of nanocrystals is permeation-limited, while the absorption of larger particles (i.e., microcrystals) is a dissolution-limited process [145].

Silica-based nanoparticles

Silicon dioxide (SiO_2), also known as silica, is a widely abundant element in nature and diverse living organisms. In recent years, silica-based nanomaterials have been in the spotlight as drug delivery systems more than ever. Possessing excellent biocompatibility and biodegradability, silica has been recognized as safe by the US FDA [146]. Besides, silica nanoparticles have high stability, rigid skeleton, large surface area for adsorption of drug molecules, high porosity, and controllable morphology. Also, they could be easily functionalized owing to the presence of several easily reachable silanol groups (Si-OH) on their inner and outer surfaces [147, 148]. Despite the safety of nonporous silica as oral delivery systems, most studies on silica-based nanocarriers have been performed on the mesoporous variant which confers additional advantages such as highly ordered pore channels with tunable size range resulting in an excellent loading capacity [148]. The term "mesoporous" refers to systems with a pore diameter of 2-50 nm [146].

The size, morphology, and surface characteristics of the pores could be tuned to load either hydrophobic or hydrophilic drugs with high encapsulation efficiency. So, one of the key advantages of silica nanoparticles is their ability to deliver drug cocktails [149]. Drug molecules may be physically adsorbed on the internal surface of mesoporous silica (and released through diffusion) or attached covalently (and released under the influence of internal or external triggers) [147]. Mesoporous silica nanoparticles also provide an excellent platform for the delivery of macromolecules with different size ranges. For instance, when the surface is appropriately functionalized, proteins strongly interact with the pore channels through electrostatic force. So, an increased protein amount is loaded while released slowly [148]. Oral delivery of silica nanoparticles not only is capable of managing the poor solubility challenges of BCS II and IV drugs but also improves intestinal permeability *via* enhancing cellular uptake and decreasing efflux [147]. Despite the advantages mentioned, given the novelty of these systems, their biofate and pharmacokinetic aspects are needed to be comprehensively examined [149].

4.6.4 Regional Drug Delivery to the Colon

Colon-targeted drug delivery is an attractive research field to treat local diseases (e.g., colorectal cancer, ulcerative colitis, Crohn's disease, and IBD), as well as systemic delivery of biopharmaceuticals such as proteins, peptides, and vaccines [150]. The main strategies for the delivery of drugs to the colon include pH-responsive systems, delayed-release dosage forms, osmotically-controlled systems, and biodegradable formulations. Also, nanocarriers are promising in colonic drug delivery due to their enhanced accumulation and uptake in the affected tissues. Due to the high inter- and intra-subject variations throughout the GI tract, single-unit dosage forms oftentimes lead to an unpredictable drug release. Therefore, multi-unit systems (including granules, pellets, beads, tablets-in-capsule, minitablets, multiparticulate tablets, etc.) are usually preferred [151].

4.6.4.1 pH-Responsive Systems

Since the pH of the colon is higher than the upper GI, pH responsiveness has been widely studied as a targeting strategy. Accordingly, pH-dependent polymers such as acrylate resins, e.g., polymethacrylate-based copolymers (Eudragit®), cellulose acetate phthalates (CAPs), and hydroxypropyl methyl-cellulose phthalate (HPMCP) derivatives are commonly employed, whether alone or in combination, to form colon-targeted systems. Ideally, the polymer used should have selective solubility, such that withstands the acidic pH of the stomach and the proximal part of the small intestine while dissolving in the distal part of the small intestine and the colon. The pH range of 6.0-7.0 is usually considered optimal for a pH-targeted colonic system to hinder premature release. However, the disintegration of dosage form and drug release may vary due to the presence of several inter/intra-subject GI variables like pH, transit time, fluid volume, and motility. Also, it may be affected by diet, disease conditions, and co-administered drugs. These may lead to the lack of site-selective drug release from the dosage form [11, 152]. As an alternative approach, it has been suggested that pH-dependent systems be combined with other strategies (e.g., time-dependent or enzyme-triggered systems) to ensure consistent drug release [152]. In this regard, multi-unit formulations in which different mechanism-based systems are integrated with pH-dependent coating are being developed to improve the targeting efficiency [153].

4.6.4.2 *Time-Dependent (Delayed-Release) Dosage Forms*

Time-dependent systems rely on the rather constant transit time before reaching the colon (around 6 h in the upper GI tract in the fed state) [154]. Usually, hydrophilic polymers (such as ethylcellulose and HPMC) are employed as a matrix or coating to provide a delayed phase through gradual swelling before drug release. Upon contact with aqueous fluids, a gel layer forms on the polymer surface which controls the rate of drug release [11]. It should be noted, however, that although the transit is relatively consistent in the small intestine, the variable gastric emptying rate may lead to drug release in an inaccurate location. Moreover, the transit time is simply affected by diseases like IBD. Therefore, a time-dependent strategy is not reliable enough if used alone [154].

4.6.4.3 *Osmotically Controlled Systems*

Osmotically controlled-release systems, generally known as osmotic pumps, benefit from the osmotic pressure gradient to release the drug. An enteric-coated polymer protects the drug-containing core over the harsh gastric environment. Once it is degraded in the colon, water diffuses through the semi-permeable membrane of the core and generates hydrostatic pressure, which in turn, pushes the drug out in a controlled (nearly zero-order) manner. This strategy is independent of pH and time. However, like the previous two systems, the therapeutic efficiency of osmotic pumps can also be circumscribed by the physio/pathological variables [155].

4.6.4.4 Biodegradable Formulations

Biodegradable formulations, also known as enzyme-sensitive or microbiota-controlled systems, are one of the most promising approaches for colon-targeted drug delivery. This is because they are independent of the dynamic pH and transit time of the GI tract [156]. Microbiota-activated systems (e.g., polysaccharide coatings) have been developed based on the fact that a series of polysaccharides are degraded enzymatically and metabolized in the lower GI tract. This category encompasses polysaccharide-based conjugates, drug-polymer matrices, and prodrugs. The major enzymes secreted by the colonic anaerobic microbiota include azoreductases, nitroreductases, glucuronidases, glycosidases, esterases, and amidases [156].

4.6.4.5 Nanocarriers

Upon accumulation in the diseased regions, colon-targeted nanosystems not only enable localized drug delivery and enhance the efficacy of therapeutics but also minimize the systemic side effects. However, considering the anatomical location and physiological condition of the colon, a nanocarrier has to take a demanding path. Before reaching the colon, it may be endangered by the burst release, acidic/enzymatic degradation in the stomach, systemic absorption in the small intestine, mucus entrapment, and so on. Therefore, advanced nanosystems are being developed that can address the aforementioned challenges. Ideally, a nanocarrier should efficiently overcome the biological barriers, selectively target the colon and diseased tissue, and release a certain dose of drug on demand [150].

Dual-sensitive nanocarriers

The combination of colon-targeted strategies with nanotechnology can lead to the creation of promising drug delivery systems. Acid-sensitive polymers may be integrated with time-sensitive [157] or enzyme-sensitive [158] components to form dual-responsive nanocarriers. Compared to single-responsive systems, dual-responsive nanocarriers have succeeded in delivering the drug to inflamed tissue in sufficient quantities and with selective accumulation. In addition, the side effects and clinical complications could also be reduced.

Redox-sensitive nanocarriers

In active inflammation, reactive oxygen species (ROS) and reactive nitrogen species (RNS) are increasingly produced by inflammatory cells, which adversely affect the surrounding tissues. In this case, redox-sensitive nanoparticles can effectively scavenge ROS/RNS and thereby help to treat inflammation. To date, several types of redox-active nanoparticles including inorganic (e.g., metals and metal oxides) and organic systems (like bilirubin-based and polyphenols-based nanoparticles) have been studied to treat IBD [159].

Mucoadhesive nanoparticles

It has been shown that cationic nanoparticles can improve mucoadhesion to the inflamed tissues of the colon. However, there is a risk of their non-specific interaction with the mucus (through electrostatic binding of cationic particles to anionic mucus surface) in the upper GI. To improve site-selective delivery, charge-reversal nanoparticles have been proposed, whose cationic surface is protected before reaching the colon, followed by un-shielding in the colon [160].

Ligand-mediated nanocarriers

To improve the site-specificity of nanoparticles, they can be decorated with different ligands. These ligands are selected based on the expression profile of certain receptors or proteins in the colon and mainly include antibodies, peptides, folic acid, and hyaluronic acid. To maximize their efficiency, ligand-decorated nanoparticles could be combined with pH-responsive systems [152].

4.7 Pharmacokinetics

As an extravascular route, oral administration delays and also decreases the climax of the plasma concentration-time curve than that of intravascular (Figure 4.17). During the administration of a single oral dose, there is initially an ascending phase corresponding to the drug absorption. At this stage, there is the maximum absorption rate, while the elimination rate is minimal. As drug absorption continues, the absorption rate –which is a function of the amount of remaining drug– plummets, while the elimination rate begins to increase. This process continues until the two rates

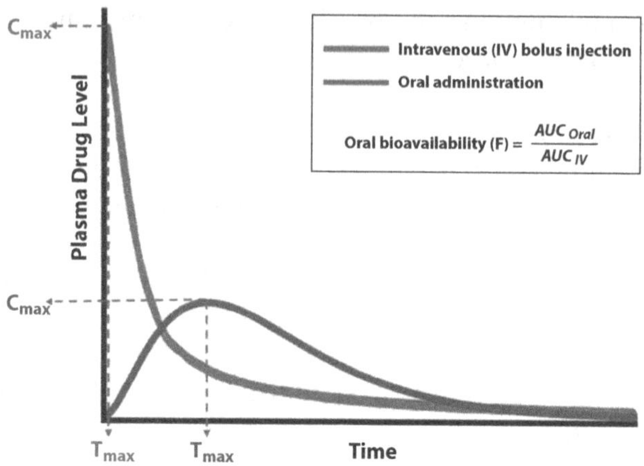

Figure 4.17 Comparative plasma concentration-time curves of a model drug administered intravenously and orally (the bioavailability formula applies to the same dose).

become equal to each other (at the time and concentration called T_{max} and C_{max}, respectively). From then on, a decreasing phase (i.e., elimination phase) begins in which the elimination rate prevails over the absorption rate [161].

Even when the drug is administered in solution, oral bioavailability is typically less than 1 or 100%. The reason for this is the anatomical path the drug takes before entering the systemic circulation (Figure 4.18). In general, an unabsorbed fraction of the drug remains in the GI lumen and is then excreted through the feces (F_F); a fraction is degraded within the enterocytes or returned to the lumen by efflux pumps (F_G); and the remainder will undergo hepatic first-pass loss (F_H). Accordingly, the overall oral bioavailability (F) is equal to $F_F \times F_G \times F_H$ [162].

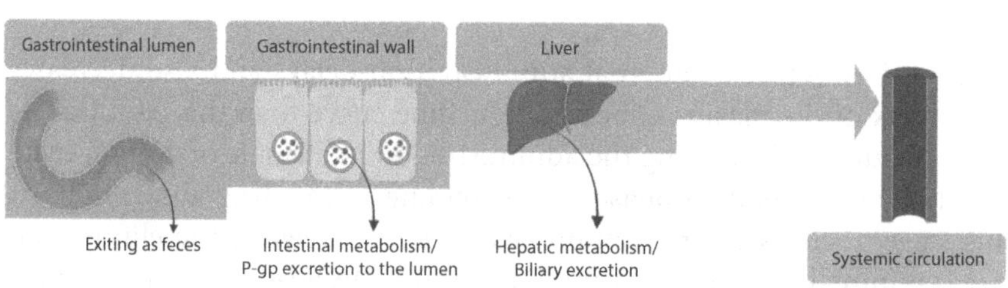

Figure 4.18 Why oral bioavailability is generally lower than 1? (created with BioRender.com).

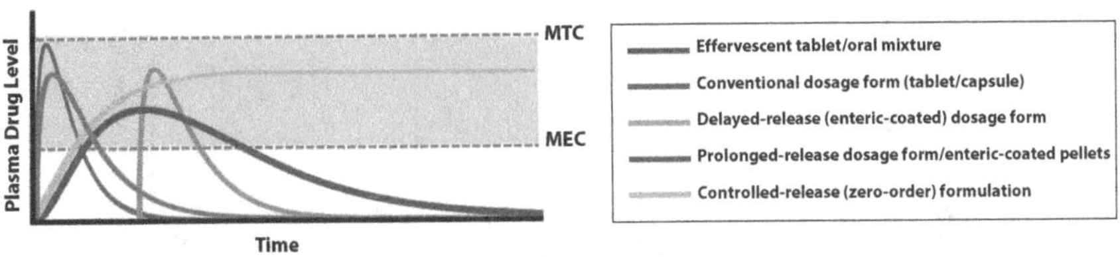

Figure 4.19 Schematic plasma concentration-time curves of different single-dose oral formulations.

Sudden ascent of the plasma concentration-time curve following oral administration (due to rapid drug release) and a consequent abrupt decline, are pharmacokinetic characteristics of the immediate-release formulations. In an attempt to achieve more constant plasma concentrations over time, avoid excessive concentration fluctuations, reduce dose frequency, and eventually increase patient compliance and drug effectiveness when multiple dosing is required, a variety of modified-release systems have been developed. In these systems, the input rate is adjusted so that the drug release rate is the limiting step of absorption. Accordingly, it is possible to achieve zero-order absorption [163]. Figure 4.19 comparatively illustrates plasma concentration-time curves of different single-dose oral formulations.

The five-letter acronym LADME (Liberation, Absorption, Distribution, Metabolism, and Excretion) describes the biofate of orally administrated drugs [164]. The pharmacokinetics of oral drugs can be studied in separate areas based on whether the drug is in the molecular form or particulate. In the case of free drugs, their biofate generally depends on their physicochemical characteristics, mainly including molecular weight, diffusivity, solubility, lipophilicity, and stability. Noteworthy, this is especially true for passively diffused drugs, and those with carrier-mediated absorption are generally considered poor candidates for controlled release systems [165].

For nanoparticles, there are two scenarios: (i) the drug is released from the nanoparticle in the GI tract (whose fate is similar to what happens for a free drug), and (ii) the nanoparticle enters intact into the systemic circulation. In the latter case, a similar path is taken with nanoparticles that are administered intravascularly [166].

4.8 Products in the Market

The number of oral novel products that have entered the market is very large. Table 4.1 represents just a few of them as examples, based on the strategy used in their manufacturing.

Table 4.1 Some commercialized oral novel drug delivery systems.

Strategy	Brand name	Active pharmaceutical ingredient(s)	Manufacturer
Bilayer floating capsule	Cytotec®	Misoprostol	Pharmacia/Pfizer Inc., USA
HBS floating capsule-controlled release	Madopar HBS®	Levodopa and benserzide	Roche, UK
	Prolopa HBS®	Levodopa and benserzide hydrochloride	Roche, UK
	Valrelease®	Diazepam	Roche, UK
Floating capsule	Madopar®	Levodopa and benserazide	Roche, UK
Foam-based floating tablet	Inon Ace Tablets®	Simethicone, aluminum-magnesium salts	Sato Pharma, Japan
Floating and swelling system	Minextab Floating®-Metformin HCL	Metformin hydrochloride	Galanix, France
	Minextab Floating®-Cefaclor LP	Cefaclor	Galanix, France
	Minextab Floating®-Tramadol LP	Tramadol	Galanix, France
Swelling tablet	Cetapin® XR	Metformin hydrochloride	Sanofi, India

(Continued)

Table 4.1 Some commercialized oral novel drug delivery systems. (*Continued*)

Strategy	**Brand name**	**Active pharmaceutical ingredient(s)**	**Manufacturer**
Polymer-based swelling tablet (AcuForm ™ technology)	Metformin GR™	Metformin HCl	Depomed, USA
	Proquin® XR	Ciprofloxacin	Depomed, USA
	Gralise®	Gabapentin	Depomed, USA
	Nucynta® ER	Tapentadol	Depomed, USA
	Glumetza®	Metformin HCl	Salix Pharmaceuticals, USA
	Janumet® XR	Sitagliptin and Metformin HCl	Merck, USA
Multiple-unit floating and swelling system	Baclofen GRS®	Baclofen	Sun Pharma, India
Effervescent floating tablet	Prazopress® XL	Prazosin hydrochloride	Sun Pharma, India
Gas-generating and swelling capsule	Liofen® XL	Baclofen	Sun Pharma, India
Effervescent floating system (Film-coated tablet)	Zanocin OD®	Ofloxacin	Ranbaxy, India
	Riomet OD®	Metformin hydrochloride	Ranbaxy, India
	Cifran OD®	Ciprofloxacin	Ranbaxy, India
Colloidal gel-forming floating system	Conviron®	Ferrous sulfate	Ranbaxy, India

(*Continued*)

Table 4.1 Some commercialized oral novel drug delivery systems. (*Continued*)

Strategy	**Brand name**	**Active pharmaceutical ingredient(s)**	**Manufacturer**
Raft forming system	Topalkan®	Aluminum-magnesium salts	Pierre Fabre Medicament, France
	Almagate Flatcoat®	Aluminum-magnesium salts	Pierre Fabre Medicament, France
	Liquid Gaviscon®	Calcium bicarbonate, sodium bicarbonate	Reckitt Benckiser Healthcare, UK
	Algicon®	Potassium bicarbonate	Rhone-Poulenc Rorer, UK
	Gastrocote®	Magnesium Trisilicate, Sodium Bicarbonate	Boehringer Mannheim, UK
	Peptac Liquid®	Calcium carbonate, sodium bicarbonate	Baker Norton, UK
	Mylanta® Heartburn Relief	Aluminum hydroxide, calcium carbonate, magnesium hydroxide, sodium bicarbonate	Warner-Lambert, Australia
	Bisodol®	Calcium carbonate, sodium bicarbonate, magnesium carbonate	Forest Laboratories, UK
	Rennie® Duo	Calcium carbonate, magnesium carbonate	Roche, UK
Erodible matrix system	Cipro® XR	Ciprofloxacin hydrochloride and betaine	Bayer, USA

(*Continued*)

Table 4.1 Some commercialized oral novel drug delivery systems. (*Continued*)

Strategy		Brand name	Active pharmaceutical ingredient(s)	Manufacturer
Bioadhesive system		Xifaxan®	Rifaximin	Lupin, India
Nanocrystal	Tablet	Rapamune®	Sirolimus	Wyeth/Pfizer Inc.
		Tricor®	Fenofibrate	Abbott
		Triglide™	Fenofibrate	SkyePharma AG
		Gris-PEG®	Griseofulvin	Novartis
	Capsule	Ritalin® LA	Methyl Phenidate HCl	Novartis
		Focalin®XR	Dexmethylphenidate hydrochloride	Novartis
		Emend®	Aprepitant	Merck
		Cesamet®	Nabilone	Eli Lilly
		Avinza®	Morphine sulfate	Pfizer Inc.
		Zanaflex™	Tizanidine HCl	Acorda
	Oral suspension	Megace® ES	Megestrol acetate	Par Pharmaceutical

(*Continued*)

Table 4.1 Some commercialized oral novel drug delivery systems. (*Continued*)

Strategy		Brand name	Active pharmaceutical ingredient(s)	Manufacturer
SEDDS	Soft gelatin capsule	Sandimmune®	Cyclosporin A	Novartis
		Neoral®	Cyclosporin A	Novartis
		Norvir®	Ritonavir	AbbVie
		Depakene®	Valproic acid	AbbVie
		Fortovase®	Saquinavir	Roche
		Vesanoid®	Tretinoin	Roche
		Accutane®	Isotretinoin	Roche
		Rocaltrol®	Calcitriol	Roche
		Agenerase®	Amprenavir	GlaxoSmithKline
		Targretin®	Bexarotene	Ligand
		Aptivus®	Tipranavir	Boehringer Ingelheim
		Panimun Bioral®	Cyclosporin A	Panacea Biotec
	Hard gelatin capsule	Gengraf®	Cyclosporin A	AbbVie
		Solufen®	Ibuprofen	Sanofi-Aventis
		Lipirex®	Fenofibrate	Sanofi-Aventis

(*Continued*)

Table 4.1 Some commercialized oral novel drug delivery systems. *(Continued)*

Strategy		Brand name	Active pharmaceutical ingredient(s)	Manufacturer
Prodrug	Soft capsule	Vepesid®	Etoposide phosphate	Xediton
Osmotic Pump	Capsule	Coreg® CR	Carvedilol phosphate	GlaxoSmithKline, UK
	Tablet	Ditropan XL®	Oxybutynin chloride	Janssen, USA
		Altoprev®	Lovastatin	Covis Pharma
Solid dispersion	Capsule	Cesamet®	Nabilone	Eli Lilly
	Tablet	Grispeg®	Griseofulvin	Pendinal pharm inc
		Afeditab®	Nifedipine	Elan corp, Ireland
		Rezulin®	Troglitazone	Pfizer
		Nimotop®	Nimodipine	Bayer
pH-responsive system	Tablet	Mesasal®	Mesalazine	Sanofi

References

1. Pereira de Sousa, I. and Bernkop-Schnürch, A., Pre-systemic metabolism of orally administered drugs and strategies to overcome it. *J. Control. Release*, 192, 301–9, 2014.
2. Homayun, B., Lin, X., Choi, H.J., Challenges and recent progress in oral drug delivery systems for biopharmaceuticals. *Pharmaceutics*, 11, 3, 129, 2019.
3. Pinto, J.F., Site-specific drug delivery systems within the gastro-intestinal tract: From the mouth to the colon. *Int. J. Pharm.*, 395, 1, 44–52, 2010, Available from: https://www.sciencedirect.com/science/article/pii/S0378517310003467.
4. Lopes, C.M., Bettencourt, C., Rossi, A., Buttini, F., Barata, P., Overview on gastroretentive drug delivery systems for improving drug bioavailability. *Int. J. Pharm.*, 510, 1, 144–58, 2016.
5. Bhatt, P., Patel, D., Patel, A., Patel, A., Nagarsheth, A., Oral Controlled Release Systems: Current Strategies and Challenges, in: *Novel Drug Delivery Technologies: Innovative Strategies for Drug Re-positioning*, A. Misra and A. Shahiwala (Eds.), pp. 73–120, Springer, Singapore, 2019, Available from: https://doi.org/10.1007/978-981-13-3642-3_4.
6. Babadi, D., Dadashzadeh, S., Osouli, M., Abbasian, Z., Daryabari, M.S., Sadrai, S. *et al.*, Biopharmaceutical and pharmacokinetic aspects of nanocarrier-mediated oral delivery of poorly soluble drugs. *J. Drug Deliv. Sci. Technol.*, 102324, 102324, 2021.
7. Benet, L.Z., Hosey, C.M., Ursu, O., Oprea, T.I., BDDCS, the Rule of 5 and Drugability. *Adv. Drug Deliv. Rev.*, 101, 89–98, 2016, Available from: https://www.ncbi.nlm.nih.gov/pmc/articles/PMC4910824/.
8. Nikolakakis, I. and Partheniadis, I., Self-Emulsifying Granules and Pellets: Composition and Formation Mechanisms for Instant or Controlled Release. *Pharmaceutics*, 9, 4, 50, 2017, Available from: https://www.ncbi.nlm.nih.gov/pmc/articles/PMC5750656/.
9. Carlson, B.M., Chapter 12 - The Digestive System, in: *The Human Body: Linking Structure and Function*, B.M. Carlson (Ed.), pp. 321–55, Academic Press, London, UK, 2019, Available from: https://www.sciencedirect.com/science/article/pii/B9780128042540000120.
10. Reinus, J. and Simon, D., *Gastrointestinal Anatomy and Physiology*, Wiley Online Library, UK, 2014.
11. Hua, S., Advances in Oral Drug Delivery for Regional Targeting in the Gastrointestinal Tract-Influence of Physiological, Pathophysiological and Pharmaceutical Factors. *Front. Pharmacol.*, 11, 524, 2020.
12. Durán-Lobato, M., Niu, Z., Alonso, M.J., Oral delivery of biologics for precision medicine. *Adv. Mater.*, 32, 13, 1901935, 2020.
13. Rosenbaum, C., Grimm, M., Krause, J., Rump, A., Kessler, R., Hosten, N. *et al.*, Functionality and Acceptance of the EsoCap System—A Novel Film-Based

Drug Delivery Technology: Results of an *In Vivo* Study. *Pharmaceutics*, 13, 6, 828, 2021.
14. Mahadevan, V., Anatomy of the small intestine. *Surg. (Oxford)*, 35, 8, 407–12, 2017, Available from: https://www.sciencedirect.com/science/article/pii/S0263931917301187.
15. Campbell, J., Berry, J., Liang, Y., Chapter 71 - Anatomy and Physiology of the Small Intestine, in: *Shackelford's Surgery of the Alimentary Tract, 2 Volume Set*, Eighth Edition, C.J. Yeo (Ed.), pp. 817–41, Elsevier, Philadelphia, 2019, Available from: https://www.sciencedirect.com/science/article/pii/B9780323402323000716.
16. Mahadevan, V., Anatomy of the caecum, appendix and colon. *Surg. (Oxford)*, 35, 3, 115–20, 2017, Available from: https://www.sciencedirect.com/science/article/pii/S0263931917300145.
17. Brown, T.D., Whitehead, K.A., Mitragotri, S., Materials for oral delivery of proteins and peptides. *Nat. Rev. Mater.*, 5, 2, 127–48, 2020.
18. Oh, A. and Lee, C.H., Physiological Barriers in Advanced Drug Delivery: Gastrointestinal Barrier, in: *Advanced Drug Delivery*, A.K. Mitra, C.H. Lee, K. Cheng (Eds.), Wiley, New Jersey, USA, 2013.
19. Carlert, S., Investigation and prediction of small intestinal precipitation of poorly soluble drugs: a study involving in silico, *in vitro* and *in vivo* assessment [PhD Thesis], Acta Universitatis Upsaliensis, 2012.
20. Wang, Y., Pi, C., Feng, X., Hou, Y., Zhao, L., Wei, Y., The Influence of Nanoparticle Properties on Oral Bioavailability of Drugs. *Int. J. Nanomed.*, 15, 6295–310, 2020, Available from: https://www.ncbi.nlm.nih.gov/pmc/articles/PMC7455773/.
21. Boegh, M. and Nielsen, H.M., Mucus as a barrier to drug delivery–understanding and mimicking the barrier properties. *Basic Clin. Physiol. Pharmacol.*, 116, 3, 179–86, 2015.
22. Murgia, X., Loretz, B., Hartwig, O., Hittinger, M., Lehr, C.M., The role of mucus on drug transport and its potential to affect therapeutic outcomes. *Adv. Drug Deliv. Rev.*, 124, 82–97, 2018, Available from: https://www.sciencedirect.com/science/article/pii/S0169409X17302338.
23. Sigurdsson, H.H., Kirch, J., Lehr, C.M., Mucus as a barrier to lipophilic drugs. *Int. J. Pharm.*, 453, 1, 56–64, 2013, Available from: https://www.sciencedirect.com/science/article/pii/S0378517313004572.
24. Lieleg, O. and Ribbeck, K., Biological hydrogels as selective diffusion barriers. *Trends Cell Biol.*, 21, 9, 543–51, 2011.
25. Vancamelbeke, M. and Vermeire, S., The intestinal barrier: a fundamental role in health and disease. *Expert Rev. Gastroenterol. Hepatol.*, 11, 9, 821–34, 2017, Available from: https://www.ncbi.nlm.nih.gov/pmc/articles/PMC6104804/.
26. Pathak, K. and Raghuvanshi, S., Oral bioavailability: issues and solutions *via* nanoformulations. *Clin. Pharmacokinet.*, 54, 4, 325–57, 2015.

27. Tozer, T.N. and Rowland, M., Chapter 4- Membranes: A Determining Factor, in: *Essentials of pharmacokinetics and pharmacodynamics*, 2016.
28. Park, M.S. and Chang, J.H., Absorption of Drugs *via* Passive Diffusion and Carrier-Mediated Pathways, in: *Oral Bioavailability: Basic Principles, Advanced Concepts, and Applications*, M. Hu, and X. Li. (Eds.), pp. 63–75, Wiley Online Library, New Jersey, USA, 2011.
29. Vertzoni, M., Augustijns, P., Grimm, M., Koziolek, M., Lemmens, G., Parrott, N. *et al.*, Impact of regional differences along the gastrointestinal tract of healthy adults on oral drug absorption: An UNGAP review. *Eur. J. Pharm. Sci.*, 134, 153–75, 2019.
30. Vinarov, Z., Abdallah, M., Agundez, J., Allegaert, K., Basit, A.W., Braeckmans, M. *et al.*, Impact of gastrointestinal tract variability on oral drug absorption and pharmacokinetics: An UNGAP review. *Eur. J. Pharm. Sci.*, 76, 105812, 2021.
31. Hayward, C.E., Jones, R.L., Sibley, C.P., 12 - Mechanisms of Transfer Across the Human Placenta, in: *Fetal and Neonatal Physiology*, Fifth Edition, R.A. Polin, S.H. Abman, D.H. Rowitch, W.E. Benitz, W.W. Fox (Eds.), pp. 121–133.e5, Elsevier, Philadelphia, 2017, Available from: https://www.sciencedirect.com/science/article/pii/B9780323352147000123.
32. Flaherty, D.K. (Ed.), Chapter 12 - Phagocytosis and Intracellular Killing, in: *Immunology for Pharmacy*, pp. 97–101, Mosby, Saint Louis, 2012, Available from: https://www.sciencedirect.com/science/article/pii/B9780323069472100124.
33. Zhao, J. and Stenzel, M.H., Entry of nanoparticles into cells: The importance of nanoparticle properties. *Polym. Chem.*, 9, 3, 259–72, 2018.
34. Su, M.J., Parayath, N.N., Amiji, M.M., 11 - Exosome-Mediated Communication in the Tumor Microenvironment, in: *Diagnostic and Therapeutic Applications of Exosomes in Cancer*, M. Amiji and R. Ramesh (Eds.), pp. 187–218, Academic Press, UK, 2018, Available from: https://www.sciencedirect.com/science/article/pii/B9780128127742000110.
35. Manzanares, D. and Ceña, V., Endocytosis: the nanoparticle and submicron nanocompounds gateway into the cell. *Pharmaceutics*, 12, 4, 371, 2020.
36. Brocks, D.R. and Davies, N.M., Lymphatic Drug Absorption *via* the Enterocytes: Pharmacokinetic Simulation, Modeling, and Considerations for Optimal Drug Development. *J. Pharm. Pharm. Sci.*, 21, 1s, 254s–70s, 2018.
37. Maheshwari, R., Kuche, K., Mane, A., Chourasiya, Y., Tekade, M., Tekade, R.K., Chapter 20 - Manipulation of Physiological Processes for Pharmaceutical Product Development, in: *Dosage Form Design Considerations*, Advances in Pharmaceutical Product Development and Research, R.K. Tekade (Ed.), pp. 701–29, Academic Press, London, UK, 2018, Available from: https://www.sciencedirect.com/science/article/pii/B9780128144237000204.
38. Yáñez, J.A., Wang, S.W., Knemeyer, I.W., Wirth, M.A., Alton, K.B., Intestinal lymphatic transport for drug delivery. *Adv. Drug Deliv. Rev.*, 63, 10–11, 923–42, 2011.

39. Newton, J.M., Gastric emptying of multi-particulate dosage forms. *Int. J. Pharm.*, 395, 1, 2–8, 2010, Available from: https://www.sciencedirect.com/science/article/pii/S0378517310003443.
40. Cao, S.J., Xu, S., Wang, H.M., Ling, Y., Dong, J., Xia, R.D. *et al.*, Nanoparticles: Oral Delivery for Protein and Peptide Drugs. *AAPS PharmSciTech*, 20, 5, 190, 2019, Available from: https://doi.org/10.1208/s12249-019-1325-z.
41. Heda, R., Toro, F., Tombazzi, C.R., Physiology, Pepsin, in: *StatPearls*, StatPearls Publishing, Treasure Island, FL, 2021, Available from: http://www.ncbi.nlm.nih.gov/books/NBK537005/.
42. Zhang, Y., Li, C., Liu, Y., Wang, X., Dong, P., Liang, X., Mechanism of extraordinary DNA digestion by pepsin. *Biochem. Biophys. Res. Commun.*, 472, 1, 101–7, 2016.
43. Amidon, S., Brown, J.E., Dave, V.S., Colon-Targeted Oral Drug Delivery Systems: Design Trends and Approaches. *AAPS PharmSciTech*, 16, 4, 731–41, 2015, Available from: https://doi.org/10.1208/s12249-015-0350-9.
44. Zhang, X., Han, Y., Huang, W., Jin, M., Gao, Z., The influence of the gut microbiota on the bioavailability of oral drugs. *Acta Pharm. Sin. B*, 11, 7, 1789–812, 2021, Available from: https://www.sciencedirect.com/science/article/pii/S2211383520307279.
45. Vinarov, Z., Abrahamsson, B., Artursson, P., Batchelor, H., Berben, P., Bernkop-Schnürch, A. *et al.*, Current challenges and future perspectives in oral absorption research: An opinion of the UNGAP network. *Adv. Drug Deliv. Rev.*, 171, 289–331, 2021, Available from: https://www.sciencedirect.com/science/article/pii/S0169409X21000351.
46. Wf, K., Pd S, A.B., Fp, V., Challenges in oral drug delivery in patients with esophageal dysphagia. *Expert Opin. Drug Deliv.*, 13, 5, 645–658, 2016, Available from: https://pubmed.ncbi.nlm.nih.gov/26781167/.
47. Stillhart, C., Vučićević, K., Augustijns, P., Basit, A.W., Batchelor, H., Flanagan, T.R. *et al.*, Impact of gastrointestinal physiology on drug absorption in special populations—An UNGAP review. *Eur. J. Pharm. Sci.*, 147, 105280, 2020.
48. Amidon, G.L., Lennernäs, H., Shah, V.P., Crison, J.R., A theoretical basis for a biopharmaceutic drug classification: the correlation of *in vitro* drug product dissolution and *in vivo* bioavailability. *Pharm. Res.*, 12, 3, 413–20, 1995.
49. Benet, L.Z., The role of BCS (biopharmaceutics classification system) and BDDCS (biopharmaceutics drug disposition classification system) in drug development. *J. Pharm. Sci.*, 102, 1, 34–42, 2013, Available from: https://onlinelibrary.wiley.com/doi/abs/10.1002/jps.23359.
50. Chen, M.L., Amidon, G.L., Benet, L.Z., Lennernas, H., Yu, L.X., The BCS, BDDCS, and Regulatory Guidances. *Pharm. Res.*, 28, 7, 1774–8, 2011, Available from: https://doi.org/10.1007/s11095-011-0438-1.
51. Hosey, C.M., Chan, R., Benet, L.Z., BDDCS Predictions, Self-Correcting Aspects of BDDCS Assignments, BDDCS Assignment Corrections, and Classification for more than 175 Additional Drugs. *AAPS J.*, 18, 1,

251–60, 2015, Available from: https://www.ncbi.nlm.nih.gov/pmc/articles/PMC4706282/.
52. Yu, L.X., Amidon, G.L., Polli, J.E., Zhao, H., Mehta, M.U., Conner, D.P. *et al.*, Biopharmaceutics classification system: the scientific basis for biowaiver extensions. *Pharm. Res.*, 19, 7, 921–5, 2002.
53. Butler, J.M. and Dressman, J.B., The developability classification system: application of biopharmaceutics concepts to formulation development. *J. Pharm. Sci.*, 99, 12, 4940–54, 2010.
54. Rosenberger, J., Butler, J., Dressman, J., A refined developability classification system. *J. Pharm. Sci.*, 107, 8, 2020–32, 2018.
55. Rosenberger, J., Butler, J., Muenster, U., Dressman, J., Application of a refined developability classification system. *J. Pharm. Sci.*, 108, 3, 1090–100, 2019.
56. Gatoo, M.A., Naseem, S., Arfat, M.Y., Mahmood Dar, A., Qasim, K., Zubair, S., Physicochemical Properties of Nanomaterials: Implication in Associated Toxic Manifestations. *Biomed. Res. Int.*, 2014, e498420, 2014, Available from: https://www.hindawi.com/journals/bmri/2014/498420/.
57. Batchelor, H., Bioadhesive dosage forms for esophageal drug delivery. *Pharm. Res.*, 22, 2, 175–81, 2005.
58. Batchelor, H. and Conway, B., Targeting Infections within the Gastrointestinal Tract, in: *Advanced Drug Formulation Design to Optimize Therapeutic Outcomes*, CRC Press, New York, USA, pp. 237–264, 2008.
59. Zhang, L., Russell, D., Conway, B.R., Batchelor, H., Strategies and therapeutic opportunities for the delivery of drugs to the esophagus. *Crit. Rev. Ther. Drug Carrier Syst.*, 25, 3, 259–304, 2008.
60. Krause, J., Rosenbaum, C., Grimm, M., Rump, A., Keßler, R., Hosten, N. *et al.*, The EsoCap-system – An innovative platform to drug targeting in the esophagus. *J. Control. Release*, 327, 1–7, 2020, Available from: https://www.sciencedirect.com/science/article/pii/S0168365920304491.
61. Building a smart drug delivery technology for the esophagus. Available from: https://www.nature.com/articles/d43747-021-00083-1.
62. Babaee, S., Pajovic, S., Kirtane, A.R., Shi, J., Caffarel-Salvador, E., Hess, K. *et al.*, Temperature-responsive biometamaterials for gastrointestinal applications. *Sci. Transl. Med.*, 11, 488, eaau8581, 2019.
63. Xu, R., Zhang, K., Liang, J., Gao, F., Li, J., Guan, F., Hyaluronic acid/polyethyleneimine nanoparticles loaded with copper ion and disulfiram for esophageal cancer. *Carbohyd. Polym.*, 261, 117846, 2021.
64. Chang, Y., Wu, F., Pandey, N.K., Chudal, L., Xing, M., Zhang, X. *et al.*, Combination of Disulfiram and Copper–Cysteamine Nanoparticles for an Enhanced Antitumor Effect on Esophageal Cancer. *ACS Appl. Bio Mater.*, 3, 10, 7147–57, 2020, Available from: https://doi.org/10.1021/acsabm.0c00949.
65. Fu, Q., Wang, J., Liu, H., Chemo-immune synergetic therapy of esophageal carcinoma: trastuzumab modified, cisplatin and fluorouracil co-delivered lipid–polymer hybrid nanoparticles. *Drug Deliv.*, 27, 1, 1535–43, 2020, Available from: https://doi.org/10.1080/10717544.2020.1837294.

66. Wang, J.Y., Song, Y.Q., Peng, J., Luo, H.L., Nanostructured Lipid Carriers Delivering Sorafenib to Enhance Immunotherapy Induced by Doxorubicin for Effective Esophagus Cancer Therapy. *ACS Omega*, 5, 36, 22840–6, 2020.
67. Adler, D., Esophageal Stents: Placement, Complications, Tips, and Tricks. *Video J. Encycl. GI Endosc.*, 1, 1, 66–8, 2013, Available from: https://www.sciencedirect.com/science/article/pii/S221209711370030X.
68. Park, C.G., Kim, M.H., Park, M., Lee, J.E., Lee, S.H., Park, J.H. *et al.*, Polymeric nanofiber coated esophageal stent for sustained delivery of an anticancer drug. *Macromol. Res.*, 19, 11, 1210–6, 2011.
69. Ramanathan, M., Manikandan, S., Subramanian, L., Solairaj, P., Gastro retentive non-floating drug delivery approach: A review. *J. Drug Deliv. Ther.*, 8, 4, 20–3, 2018.
70. Tripathi, J., Thapa, P., Maharjan, R., Jeong, S.H., Current state and future perspectives on gastroretentive drug delivery systems. *Pharmaceutics*, 11, 4, 193, 2019.
71. Patel, J.K., Dalvadi, H.P., Shah, D.P., Time and/or site specific drug delivery of floating pulsatile release delivery system. *Syst. Rev. Pharm.*, 2, 1, 59–65, 2011.
72. Javadzadeh, Y. and Hamedeyazdan, S., Novel drug delivery systems for modulation of gastrointestinal transit time, in: *Recent Advances in Novel Drug Carrier Systems*, Sezer, A.D. (Ed.), IntechOpen, Rijeka, Croatia, 2012.
73. Streubel, A., Siepmann, J., Bodmeier, R., Floating microparticles based on low density foam powder. *Int. J. Pharm.*, 241, 2, 279–92, 2002, Available from: https://www.sciencedirect.com/science/article/pii/S0378517302002417.
74. Davis, D.W., *Method of swallowing a pill*, Google Patents, 1968.
75. Reddy Dumpa, N., Bandari S, A., Repka, M., Novel gastroretentive floating pulsatile drug delivery system produced *via* hot-melt extrusion and fused deposition modeling 3D printing. *Pharmaceutics*, 12, 1, 52, 2020.
76. Prajapati, V.D., Jani, G.K., Khutliwala, T.A., Zala, B.S., Raft forming system—An upcoming approach of gastroretentive drug delivery system. *J. Control. Release*, 168, 2, 151–65, 2013, Available from: https://www.sciencedirect.com/science/article/pii/S0168365913001417.
77. Abou Youssef, N.A.H., Kassem, A.A., EL-Massik, M.A.E., Boraie, N.A., Development of gastroretentive metronidazole floating raft system for targeting Helicobacter pylori. *Int. J. Pharm.*, 486, 1, 297–305, 2015. Available from: https://www.sciencedirect.com/science/article/pii/S0378517315003117.
78. Clarke, G.M., Newton, J.M., Short, M.D., Gastrointestinal transit of pellets of differing size and density. *Int. J. Pharm.*, 100, 1–3, 81–92, 1993.
79. Clarke, G.M., Newton, J.M., Short, M.B., Comparative gastrointestinal transit of pellet systems of varying density. *Int. J. Pharm.*, 114, 1, 1–11, 1995.
80. Ashoka, C.M. and Hiraman, S.K.G.H., Development and *in Vitro* Evaluation of Gastroretentive High Density Tablet of Propafenone HCL. *Asian J. Pharm. Res. Health Care*, 5, 2, 89–99, 2013.

81. Desai, N. and Purohit, R., Development of novel high density gastroretentive multiparticulate pulsatile tablet of clopidogrel bisulfate using quality by design approach. *AAPS PharmSciTech*, 18, 8, 3208–18, 2017.
82. Ahad, H.A., Sreenivasulu, R., Mallapu, Rani, E., Reddy, B.V., Preparation and evaluation of famotidine high density gastro retentive microspheres with synthetic and natural polymers. *J. Pharm. Res.*, 2, 1, 2011.
83. Melocchi, A., Uboldi, M., Inverardi, N., Briatico-Vangosa, F., Baldi, F., Pandini, S. *et al.*, Expandable drug delivery system for gastric retention based on shape memory polymers: Development *via* 4D printing and extrusion. *Int. J. Pharm.*, 571, 118700, 2019.
84. Park, K. and Robinson, J.R., Bioadhesive polymers as platforms for oral-controlled drug delivery: method to study bioadhesion. *Int. J. Pharm.*, 19, 2, 107–27, 1984.
85. Mandal, U.K., Chatterjee, B., Senjoti, F.G., Gastro-retentive drug delivery systems and their *in vivo* success: A recent update. *Asian J. Pharm. Sci.* , 11, 5, 575–84, 2016, Available from: https://www.sciencedirect.com/science/article/pii/S1818087616300320.
86. Umamaheshwari, R.B., Ramteke, S., Jain, N.K., Anti-Helicobacter pylori effect of mucoadhesive nanoparticles bearing amoxicillin in experimental gerbils model. *AAPS PharmSciTech*, 5, 2, 60–8, 2004, Available from: https://doi.org/10.1208/pt050232.
87. Ramteke, S., Ganesh, N., Bhattacharya, S., Jain, N.K., Amoxicillin, clarithromycin, and omeprazole based targeted nanoparticles for the treatment of H. pylori. *J. Drug Targeting*, 17, 3, 225–34, 2009, Available from: https://doi.org/10.1080/10611860902718649.
88. Modi, J., Joshi, G., Sawant, K., Chitosan based mucoadhesive nanoparticles of ketoconazole for bioavailability enhancement: formulation, optimization, *in vitro* and *ex vivo* evaluation. *Drug Dev. Ind. Pharm.*, 39, 4, 540–7, 2013.
89. Jain, P., Jain, S., Prasad, K.N., Jain, S.K., Vyas, S.P., Polyelectrolyte Coated Multilayered Liposomes (Nanocapsules) for the Treatment of Helicobacter pylori Infection. *Mol. Pharm.*, 6, 2, 593–603, 2009, Available from: https://doi.org/10.1021/mp8002539.
90. Sarparanta, M.P., Bimbo, L.M., Mäkilä, E.M., Salonen, J.J., Laaksonen, P.H., Helariutta, A.M.K. *et al.*, The mucoadhesive and gastroretentive properties of hydrophobin-coated porous silicon nanoparticle oral drug delivery systems. *Biomaterials*, 33, 11, 3353–62, 2012, Available from: https://www.sciencedirect.com/science/article/pii/S0142961212000610.
91. Wilson, A., 1 - The formation of dry, wet, spunlaid and other types of nonwovens, in: *Applications of Nonwovens in Technical Textiles, Woodhead Publishing Series in Textiles*, R.A. Chapman (Ed.), pp. 3–17, Woodhead Publishing, Cambridge, UK, 2010, Available from: https://www.sciencedirect.com/science/article/pii/B978184569437150001X.
92. Darbasizadeh, B., Motasadizadeh, H., Foroughi-Nia, B., Farhadnejad, H., Tripolyphosphate-crosslinked chitosan/poly (ethylene oxide) electrospun

nanofibrous mats as a floating gastro-retentive delivery system for ranitidine hydrochloride. *J. Pharm. Biomed. Anal.*, 153, 63–75, 2018, Available from: https://www.sciencedirect.com/science/article/pii/S073170851732887X.
93. Anup, N., Chavan, T., Chavan, S., Polaka, S., Kalyane, D., Abed, S.N. *et al.*, Reinforced electrospun nanofiber composites for drug delivery applications. *J. Biomed. Mater. Res. Part A*, 109, 10, 2036–64, 2021, Available from: https://onlinelibrary.wiley.com/doi/abs/10.1002/jbm.a.37187.
94. Nangare, S., Dugam, S., Patil, P., Tade, R., Jadhav, N., Silk industry waste protein: isolation, purification and fabrication of electrospun silk protein nanofibers as a possible nanocarrier for floating drug delivery. *Nanotechnology*, 32, 3, 035101, 2020, Available from: https://doi.org/10.1088/1361-6528/abb8a9.
95. Malik, R., Garg, T., Goyal, A.K., Rath, G., Polymeric nanofibers: targeted gastro-retentive drug delivery systems. *J. Drug Targeting*, 23, 2, 109–24, 2015.
96. Cho, S. and Yoon, Y.R., Understanding the pharmacokinetics of prodrug and metabolite. *Transl. Clin. Pharm.*, 26, 1, 1–5, 2018.
97. Jornada, D.H., dos Santos Fernandes, G.F., Chiba, D.E., de Melo, T.R.F., dos Santos, J.L., Chung, M.C., The Prodrug Approach: A Successful Tool for Improving Drug Solubility. *Molecules*, 21, 1, 42, 2015, Available from: https://www.ncbi.nlm.nih.gov/pmc/articles/PMC6273601/.
98. Alqahtani, M.S., Advances in oral drug delivery. *Front. Pharmacol.*, 12, 62, 2021.
99. Batista, R.A., Otoni, C.G., Espitia, P.J.P., Chapter 3 - Fundamentals of chitosan-based hydrogels: elaboration and characterization techniques, in: *Materials for Biomedical Engineering*, A.M. Holban and A.M. Grumezescu (Eds.), pp. 61–81, Elsevier, Amsterdam, Netherlands, 2019, Available from: https://www.sciencedirect.com/science/article/pii/B9780128169018000031.
100. Sharpe, L.A., Daily, A.M., Horava, S.D., Peppas, N.A., Therapeutic applications of hydrogels in oral drug delivery. *Expert Opin. Drug Deliv.*, 11, 6, 901–15, 2014.
101. Jacob, S., Nair, A.B., Shah, J., Sreeharsha, N., Gupta, S., Shinu, P., Emerging role of hydrogels in drug delivery systems, tissue engineering and wound management. *Pharmaceutics*, 13, 3, 357, 2021.
102. Hoare, T.R. and Kohane, D.S., Hydrogels in drug delivery: Progress and challenges. *Polymer*, 49, 8, 1993–2007, 2008, Available from: https://www.sciencedirect.com/science/article/pii/S0032386108000487.
103. Keraliya, R.A., Patel, C., Patel, P., Keraliya, V., Soni, T.G., Patel, R.C. *et al.*, Osmotic Drug Delivery System as a Part of Modified Release Dosage Form. *ISRN Pharm.*, 2012, 528079, 2012, Available from: https://www.ncbi.nlm.nih.gov/pmc/articles/PMC3407637/.
104. Qiu, Y. and Lee, P.I., Chapter 19 - Rational Design of Oral Modified-Release Drug Delivery Systems, in: *Developing Solid Oral Dosage Forms*, Second Edition, Y. Qiu, Y. Chen, G.G.Z. Zhang, L. Yu, R.V. Mantri (Eds.), pp. 519–54, Academic Press, Boston, 2017, Available from: https://www.sciencedirect.com/science/article/pii/B9780128024478000194.

105. Boddupalli, B.M., Mohammed, Z.N.K., Nath, R.A., Banji, D., Mucoadhesive drug delivery system: An overview. *J. Adv. Pharm. Technol. Res.*, 1, 4, 381–7, 2010, Available from: https://www.ncbi.nlm.nih.gov/pmc/articles/PMC3255397/.
106. Banerjee, A. and Mitragotri, S., Intestinal patch systems for oral drug delivery. *Curr. Opin. Pharmacol.*, 36, 58–65, 2017.
107. Ahadian, S., Finbloom, J.A., Mofidfar, M., Diltemiz, S.E., Nasrollahi, F., Davoodi, E. *et al.*, Micro and nanoscale technologies in oral drug delivery. *Adv. Drug Deliv. Rev.*, 157, 37–62, 2020.
108. Rajput, G., Majmudar, F., Patel, J., Formulation and evaluation of mucoadhesive glipizide films. *Acta Pharm.*, 61, 2, 203–16, 2011.
109. Maher, S., Brayden, D.J., Casettari, L., Illum, L., Application of Permeation Enhancers in Oral Delivery of Macromolecules: An Update. *Pharmaceutics*, 11, 1, 41, 2019, Available from: https://www.ncbi.nlm.nih.gov/pmc/articles/PMC6359609/.
110. Banerjee, A., Lee, J., Mitragotri, S., Intestinal mucoadhesive devices for oral delivery of insulin. *Bioeng. Transl. Med.*, 1, 3, 338–46, 2016.
111. Moodley, K., Pillay, V., Choonara, Y.E., Du Toit, L.C., Ndesendo, V.M., Kumar, P. *et al.*, Oral drug delivery systems comprising altered geometric configurations for controlled drug delivery. *Int. J. Mol. Sci.*, 13, 1, 18–43, 2012.
112. Bertoni, S., Passerini, N., Albertini, B., Nanomaterials for oral drug administration, in: *Nanotechnology for Oral Drug Delivery*, Martins, J.P. and Santos, H.A. (Eds.), pp. 27–76, Elsevier, London, UK, 2020.
113. Agrawal, U., Sharma, R., Gupta, M., Vyas, S.P., Is nanotechnology a boon for oral drug delivery? *Drug Discov. Today*, 19, 10, 1530–46, 2014.
114. Jazayeri, S.D., Lim, H.X., Shameli, K., Yeap, S.K., Poh, C.L., Nano and Microparticles as Potential Oral Vaccine Carriers and Adjuvants Against Infectious Diseases. *Front. Pharmacol.*, 12, 1399, 2021, Available from: https://www.frontiersin.org/article/10.3389/fphar.2021.682286.
115. Daeihamed, M., Dadashzadeh, S., Haeri, A., Akhlaghi, M.F., Potential of Liposomes for Enhancement of Oral Drug Absorption. *Curr. Drug Deliv.*, 14, 2, 289–303, 2017.
116. Potluri, P. and Betageri, G.V., Mixed-micellar proliposomal systems for enhanced oral delivery of progesterone. *Drug Deliv.*, 13, 3, 227–32, 2006.
117. He, H., Lu, Y., Qi, J., Zhu, Q., Chen, Z., Wu, W., Adapting liposomes for oral drug delivery. *Acta Pharm. Sin. B*, 9, 1, 36–48, 2019, Available from: https://www.sciencedirect.com/science/article/pii/S2211383517307311.
118. Gómez-Ramírez, S., Brilli, E., Tarantino, G., Muñoz, M., Sucrosomial® Iron: A New Generation Iron for Improving Oral Supplementation. *Pharmaceuticals (Basel)*, 11, 4, 97, 2018, Available from: https://www.ncbi.nlm.nih.gov/pmc/articles/PMC6316120/.
119. Ibrahim Bekraki, A., Chapter 7 - Liposomes-and niosomes-based drug delivery systems for tuberculosis treatment, in: *Nanotechnology Based Approaches*

for Tuberculosis Treatment, P. Kesharwani (Ed.), pp. 107–22, Academic Press, London, UK, 2020, Available from: https://www.sciencedirect.com/science/article/pii/B9780128198117000072.
120. Marianecci, C., Di Marzio, L., Rinaldi, F., Celia, C., Paolino, D., Alhaique, F. *et al.*, Niosomes from 80s to present: The state of the art. *Adv. Colloid Interface Sci.*, 205, 187–206, 2014, Available from: https://www.sciencedirect.com/science/article/pii/S0001868613001711.
121. Duan, Y., Dhar, A., Patel, C., Khimani, M., Neogi, S., Sharma, P. *et al.*, A brief review on solid lipid nanoparticles: part and parcel of contemporary drug delivery systems. *RSC Adv.*, 10, 45, 26777–91, 2020, Available from: https://pubs.rsc.org/en/content/articlelanding/2020/ra/d0ra03491f.
122. Nagalingam, A., Chapter 15 - Drug Delivery Aspects of Herbal Medicines, in: *Japanese Kampo Medicines for the Treatment of Common Diseases: Focus on Inflammation*, S. Arumugam and K. Watanabe (Eds.), pp. 143–64, Academic Press, London, UK, 2017, Available from: https://www.sciencedirect.com/science/article/pii/B9780128093986000159.
123. Date, A.A., Desai, N., Dixit, R., Nagarsenker, M., Self-nanoemulsifying drug delivery systems: formulation insights, applications and advances. *Nanomedicine (Lond.)*, 5, 10, 1595–616, 2010.
124. Li, L., Hui Zhou, C., Ping Xu, Z., Chapter 14 - Self-Nanoemulsifying Drug-Delivery System and Solidified Self-Nanoemulsifying Drug-Delivery System, in: *Nanocarriers for Drug Delivery*, Micro and Nano Technologies, S.S. Mohapatra, S. Ranjan, N. Dasgupta, R.K. Mishra, S. Thomas (Eds.), pp. 421–49, Elsevier, Amsterdam, Netherlands, 2019, Available from: https://www.sciencedirect.com/science/article/pii/B978012814033800014X.
125. Carneiro, S.B., Costa Duarte, F.Í., Heimfarth, L., Siqueira Quintans J de, S., Quintans-Júnior, L.J., da Veiga Júnior, V.F. *et al.*, Cyclodextrin–Drug Inclusion Complexes: *In Vivo* and *In Vitro* Approaches. *Int. J. Mol. Sci.*, 20, 3, 642, 2019, Available from: https://www.ncbi.nlm.nih.gov/pmc/articles/PMC6387394/.
126. Patel, D.K., Kesharwani, R., Kumar, V., 7 - Nanoparticles: an emerging platform for medical imaging, in: *Nanoparticles in Analytical and Medical Devices*, S.C.B. Gopinath and F. Gang (Eds.), pp. 113–26, Elsevier, Amsterdam, Netherlands, 2021, Available from: https://www.sciencedirect.com/science/article/pii/B9780128211632000078.
127. Kapse, A., Anup, N., Patel, V., Saraogi, G.K., Mishra, D.K., Tekade, R.K., Chapter 6 - Polymeric micelles: a ray of hope among new drug delivery systems, in: *Drug Deliv. Syst.*, Advances in Pharmaceutical Product Development and Research, R.K. Tekade (Ed.), pp. 235–89, Academic Press, London, UK, 2020, Available from: https://www.sciencedirect.com/science/article/pii/B9780128144879000065.
128. Francis, M.F., Cristea, M., Winnik, F.M., Polymeric micelles for oral drug delivery: Why and how. *Pure Appl. Chem.*, 76, 7–8, 1321–35, 2004.

129. Gaucher, G., Satturwar, P., Jones, M.C., Furtos, A., Leroux, J.C., Polymeric micelles for oral drug delivery. *Eur. J. Pharm. Biopharm.*, 76, 2, 147–58, 2010.
130. Patel, G. and Misra, A., 10 - Oral Delivery of Proteins and Peptides: Concepts and Applications, in: *Challenges in Delivery of Therapeutic Genomics and Proteomics*, A. Misra (Ed.), pp. 481–529, Elsevier, London, 2011, Available from: https://www.sciencedirect.com/science/article/pii/B9780123849649000104.
131. Cagel, M., Tesan, F.C., Bernabeu, E., Salgueiro, M.J., Zubillaga, M.B., Moretton, M.A. *et al.*, Polymeric mixed micelles as nanomedicines: Achievements and perspectives. *Eur. J. Pharm. Biopharm.*, 113, 211–28, 2017, Available from: https://www.sciencedirect.com/science/article/pii/S0939641116306944.
132. Zielińska, A., Carreiró, F., Oliveira, A.M., Neves, A., Pires, B., Venkatesh, D.N. *et al.*, Polymeric Nanoparticles: Production, Characterization, Toxicology and Ecotoxicology. *Molecules*, 25, 16, 3731, 2020, Available from: https://www.ncbi.nlm.nih.gov/pmc/articles/PMC7464532/.
133. Pridgen, E.M., Alexis, F., Farokhzad, O.C., Polymeric Nanoparticle Technologies for Oral Drug Delivery. *Clin. Gastroenterol. Hepatol.*, 12, 10, 1605–10, 2014, Available from: https://www.ncbi.nlm.nih.gov/pmc/articles/PMC4171204/.
134. Cho, H.K., Cheong, I.W., Lee, J.M., Kim, J.H., Polymeric nanoparticles, micelles and polymersomes from amphiphilic block copolymer. *Korean J. Chem. Eng.*, 27, 3, 731–40, 2010.
135. Sheng, J., He, H., Han, L., Qin, J., Chen, S., Ru, G. *et al.*, Enhancing insulin oral absorption by using mucoadhesive nanoparticles loaded with LMWP-linked insulin conjugates. *J. Control. Release*, 233, 181–90, 2016, Available from: https://www.sciencedirect.com/science/article/pii/S0168365916302784.
136. Taipaleenmäki, E. and Städler, B., Recent advancements in using polymers for intestinal mucoadhesion and mucopenetration. *Macromol. Biosci.*, 20, 3, 1900342, 2020.
137. Cheng, H., Cui, Z., Guo, S., Zhang, X., Huo, Y., Mao, S., Mucoadhesive versus mucopenetrating nanoparticles for oral delivery of insulin. *Acta Biomater.*, 135, 506–19, 2021, Available from: https://www.sciencedirect.com/science/article/pii/S1742706121005791.
138. Lai, S.K., Wang, Y.Y., Hanes, J., Mucus-penetrating nanoparticles for drug and gene delivery to mucosal tissues. *Adv. Drug Deliv. Rev.*, 61, 2, 158–71, 2009.
139. Akiyama, Y., Nagahara, N., Nara, E., Kitano, M., Iwasa, S., Yamamoto, I. *et al.*, Evaluation of oral mucoadhesive microspheres in man on the basis of the pharmacokinetics of furosemide and riboflavin, compounds with limited gastrointestinal absorption sites. *J. Pharm. Pharmacol.*, 50, 2, 159–66, 1998.
140. Akkus, Z.B., Nazir, I., Jalil, A., Tribus, M., Bernkop-Schnürch, A., Zeta potential changing polyphosphate nanoparticles: a promising approach to overcome the mucus and epithelial barrier. *Mol. Pharmaceutics*, 16, 6, 2817–25, 2019.

141. Savjani, K.T., Gajjar, A.K., Savjani, J.K., Drug solubility: importance and enhancement techniques. *Int. Sch Res. Notices*, 2012.
142. Junyaprasert, V.B. and Morakul, B., Nanocrystals for enhancement of oral bioavailability of poorly water-soluble drugs. *Asian J. Pharm. Sci.*, 10, 1, 13–23, 2015.
143. Qin, S.Y., Zhang, A.Q., Cheng, S.X., Rong, L., Zhang, X.Z., Drug self-delivery systems for cancer therapy. *Biomaterials*, 112, 234–47, 2017.
144. Chogale, M.M., Ghodake, V.N., Patravale, V.B., Performance parameters and characterizations of nanocrystals: A brief review. *Pharmaceutics*, 8, 3, 26, 2016.
145. Gao, L., Liu, G., Ma, J., Wang, X., Zhou, L., Li, X. *et al.*, Application of Drug Nanocrystal Technologies on Oral Drug Delivery of Poorly Soluble Drugs. *Pharm. Res.*, 30, 2, 307–24, 2013, Available from: https://doi.org/10.1007/s11095-012-0889-z.
146. Steven, C.R., Busby, G.A., Mather, C., Tariq, B., Briuglia, M.L., Lamprou, D.A. *et al.*, Bioinspired silica as drug delivery systems and their biocompatibility. *J. Mater. Chem. B*, 2, 31, 5028–42, 2014, Available from: https://pubs.rsc.org/en/content/articlelanding/2014/tb/c4tb00510d.
147. Diab, R., Canilho, N., Pavel, I.A., Haffner, F.B., Girardon, M., Pasc, A., Silica-based systems for oral delivery of drugs, macromolecules and cells. *Adv. Colloid Interface Sci.*, 249, 346–62, 2017.
148. Xu, C., Lei, C., Yu, C., Mesoporous Silica Nanoparticles for Protein Protection and Delivery. *Front. Chem.*, 7, 290, 2019, Available from: https://www.frontiersin.org/article/10.3389/fchem.2019.00290.
149. Zhang, J. and Rosenholm, J.M., The *via*bility of mesoporous silica nanoparticles for drug delivery. *Ther. Deliv.*, 6, 8, 891–3, 2015.
150. Naeem, M., Awan, U.A., Subhan, F., Cao, J., Hlaing, S.P., Lee, J. *et al.*, Advances in colon-targeted nano-drug delivery systems: challenges and solutions. *Arch. Pharmacal Res.*, 43, 1, 153–69, 2020.
151. Fassihi, S.C., Talukder, R., Fassihi, R., Colon-Targeted Delivery Systems for Therapeutic Applications: Drug Release from Multiparticulate, Monolithic Matrix, and Capsule-Filled Delivery Systems, in: *Targeted Nanosystems for Therapeutic Applications: New Concepts, Dynamic Properties, Efficiency, and Toxicity*, Sakurai, K. and Ilies, M.A. (Eds), pp. 309–38, ACS Publications, USA, 2019.
152. Lee, S.H., Bajracharya, R., Min, J.Y., Han, J.W., Park, B.J., Han, H.K., Strategic approaches for colon targeted drug delivery: an overview of recent advancements. *Pharmaceutics*, 12, 1, 68, 2020.
153. Park, H.J., Jung, H.J., Ho, M.J., Lee, D.R., Cho, H.R., Choi, Y.S. *et al.*, Colon-targeted delivery of solubilized bisacodyl by doubly enteric-coated multiple-unit tablet. *Eur. J. Pharm. Sci.*, 102, 172–9, 2017.
154. Yang, L., Chu, J.S., Fix, J.A., Colon-specific drug delivery: new approaches and *in vitro*/*in vivo* evaluation. *Int. J. Pharm.*, 235, 1–2, 1–15, 2002.

155. Verma, R.K., Mishra, B., Garg, S., Osmotically controlled oral drug delivery. *Drug Dev. Ind. Pharm.*, 26, 7, 695–708, 2000.
156. Kotla, N.G., Rana, S., Sivaraman, G., Sunnapu, O., Vemula, P.K., Pandit, A. *et al.*, Bioresponsive drug delivery systems in intestinal inflammation: State-of-the-art and future perspectives. *Adv. Drug Deliv. Rev.*, 146, 248–66, 2019, Available from: https://www.sciencedirect.com/science/article/pii/S0169409X18301601.
157. Naeem, M., Bae, J., Oshi, M.A., Kim, M.S., Moon, H.R., Lee, B.L. *et al.*, Colon-targeted delivery of cyclosporine A using dual-functional Eudragit® FS30D/PLGA nanoparticles ameliorates murine experimental colitis. *Int. J. Nanomed.*, 13, 1225, 2018.
158. Naeem, M., Cao, J., Choi, M., Kim, W.S., Moon, H.R., Lee, B.L. *et al.*, Enhanced therapeutic efficacy of budesonide in experimental colitis with enzyme/pH dual-sensitive polymeric nanoparticles. *Int. J. Nanomed.*, 10, 4565, 2015.
159. Ren, Q., Sun, S., Zhang, X.D., Redox-active nanoparticles for inflammatory bowel disease. *Nano Res.*, 14, 8, 2535–57, 2021.
160. Naeem, M., Oshi, M.A., Kim, J., Lee, J., Cao, J., Nurhasni, H. *et al.*, pH-triggered surface charge-reversal nanoparticles alleviate experimental murine colitis *via* selective accumulation in inflamed colon regions. *Nanomed. Nanotechnol. Biol. Med.*, 14, 3, 823–34, 2018, Available from: https://www.sciencedirect.com/science/article/pii/S1549963418300145.
161. Tozer, T.N. and Rowland, M., Chapter 7- Quantifying Events Following an Extravascular Dose, in: *Essentials of pharmacokinetics and pharmacodynamics*, 2016.
162. Tozer, T.N. and Rowland, M., Chapter 8- Physiologic and Physicochemical Determinants of Drug Absorption, in: *Essentials of pharmacokinetics and pharmacodynamics*, 2016.
163. Yáñez, J.A., Brocks, D.R., Forrest, L.M., Davies, N.M., Pharmacokinetic behaviors of orally administered drugs, in: *Oral bioavailability: Basic principles, advanced concepts, and applications*, pp. 183–219, 2011.
164. Caldwell, J., Gardner, I., Swales, N., An introduction to drug disposition: the basic principles of absorption, distribution, metabolism, and excretion. *Toxicol. Pathol.*, 23, 2, 102–14, 1995.
165. Indurkhya, A., Patel, M., Sharma, P., Abed, S.N., Shnoudeh, A., Maheshwari, R. *et al.*, Influence of drug properties and routes of drug administration on the design of controlled release system, in: *Dosage Form Design Considerations*, R.K. Tekade (ed.), pp. 179–223, Elsevier, London, UK, 2018.
166. Tekade, R.K., Rewatkar, P., Kumeria, T., Popat, A., Size, shape and surface charge considerations of orally delivered nanomedicines, in: *Nanotechnology for Oral Drug Delivery*, J.P. Martins and H.A. Santos (Eds.), pp. 143–76, Elsevier, London, UK, 2020.

5

Rectal Novel Drug Delivery Systems

Mahsa Sayed Tabatabaei and Taraneh Gazori*

HIDA Pharmaceutical Technologies Development Center, HIDA Pharmaceuticals, Tehran, Iran

Abstract

Noticed from ancient times, drug administration through the rectum (rectal drug delivery) is still used as an effective pathway for local and systemic therapeutic purposes. The rectal route is commonly regarded as an alternative to other routes of administration (particularly the oral route). However, recent studies have revealed the potential of this route to increase bioavailability owing to the unique conditions of the rectum (such as the physiologically stable environment and minimal first-pass metabolism). Additionally, a high drug accumulation could be achieved for the management of local colorectal conditions. Despite these advantages, the wide use of rectal dosage forms is hampered by several drawbacks like poor patient compliance, low residence time, and variable bioavailability. Therefore, novel rectal drug delivery systems have been developed to take advantage of the potential benefits of this approach, while conquering the limitations. In this chapter, we will first review the general aspects of rectal drug delivery including anatomy, physiology, existing barriers, factors affecting rectal drug delivery, formulation considerations, and challenges and opportunities associated with rectal drug delivery. Then, a comprehensive classification of novel rectal formulations, including modified suppositories, advanced hydrogels, carrier-based systems, carrier-in-gel composites, and liquid crystalline gels is presented. Finally, the pharmacokinetics of rectally administered drugs and novel formulations being considered for entering the market will be briefly reviewed.

***Keywords*:** Rectal drug delivery, rectal drug administration, rectal dosage form, rectal novel drug delivery system, rectal pharmacokinetics

**Corresponding author*: gazori@hidapharma.com

Mehrdad Hamidi and Mahsa Sayed Tabatabaei (eds.) Novel Drug Delivery Systems: Fundamentals and Applications, (167–198) © 2025 Scrivener Publishing LLC

5.1 Introduction

Rectal drug delivery is generally referred to as the administration of drug molecules using a suitable dosage form through the rectal mucosa to achieve systemic or local therapeutic effects [1]. As one of the oldest routes of drug administration, the history of using the rectal systems for therapeutic purposes dates back to antiquity [2]. Having several advantages, rectal systems are still used both for local and systemic drug delivery. Typically, rectal drug delivery is considered an effective alternative for other routes of drug administration, particularly the oral route. However, several physiological opportunities in this pathway can promote its position as an effective route of administration [3].

In rectal drug delivery, the harsh environment of the stomach and extensive enzymatic degradation of the gastrointestinal (GI) tract are avoided. Additionally, the hepatic first-pass metabolism could be partially bypassed which leads to improved bioavailability. Furthermore, compared to the greatly variable situations in the upper GI, the conditions in the rectum are relatively constant. Also, rectal administration is suitable in special populations and situations such as people with difficulty in swallowing solid dosage forms (e.g., infants, children, and the elderly), nausea, vomiting, GI disorders (e.g., malabsorption syndromes), or lack of consciousness. In addition, in rectal drug delivery, the organoleptic issues (e.g., taste and color) that limit patient compliance (particularly to children) would not be a matter of concern. Finally, in the case of overdosing, the drug absorption could be discontinued [1, 2]. In particular, rectal drug administration is a suitable solution for drugs with limited solubility and permeability in the upper GI, susceptibility to the acidic pH or enzymatic degradation, extensive first-pass metabolism, irritation for the gastric mucosa, localized effect in the end parts of the GI, and those whose administration through other routes is challenging [4, 5].

However, rectal drug administration is not that simple. Rectal, compared to oral, is an unpleasant, discomfort, and less popular route; so, many patients resist rectal administration. Also, there are some biological limitations against rectal drug absorption. For instance, the small surface area and low volume of rectal fluid alongside the absence of anatomical adaption for absorption (i.e., villi and microvilli) reduce the absorption capacity of the rectum [1]. In addition, drug absorption from the rectum is irregular which may lead to variable bioavailability. There is also the possibility of poor mucoadhesion or expulsion of the dosage form before absorption of drug (e.g., following the defecation). Furthermore, there is the possibility of drug degradation and reduction of bioavailability due to the microbiota metabolism [6].

Despite these challenges, the rectal pathway is still considered an important drug delivery route, and several efforts are being made to improve rectal drug delivery and overcome its challenges using novel drug delivery systems. In this chapter, various aspects of rectal drug delivery will be reviewed, from the anatomy and physiology of the rectum and barriers against rectal drug delivery to formulation considerations, classification of novel rectal delivery systems, and pharmacokinetics of the rectally administered drugs. Finally, some novel rectal formulations with the potential to enter the market will be reviewed.

5.2 Anatomy and Physiology

The rectum, as the last absorptive section of the GI, is anatomically placed between the sigmoid colon and the anal duct with an S shape and about 15 cm in length [1]. Primarily, the rectum acts passively as a space for transporting and temporarily storing the stool and does not play a significant role in the reabsorption of water and electrolytes [4]. From an anatomical and functional point of view, the rectum is divided into the upper (pelvic) and the lower (anal) parts. The upper part is located in the pelvic cavity and undertakes the role of stool propelling while the lower part is placed in the lower wall of the pelvis. As shown in Figure 5.1, three veins drain the rectum including superior, middle, and inferior rectal (haemorrhoidal) veins.

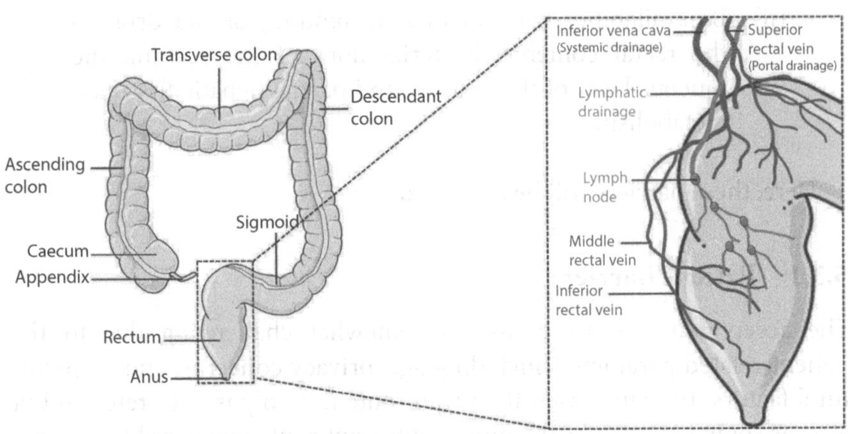

Figure 5.1 Anatomy of the rectum as well as its venous and lymphatic drainage.

The upper rectum is drained by the superior rectal vein that supplies the blood to the portal vein, while the lower rectum is drained by the middle and lower rectal veins that, at least, partially bypass the first-pass metabolism of the liver and directly transported into the systemic circulation. So, it is generally recommended that the rectal dosage forms are placed in the lower rectum to improve systemic absorption and bioavailability of drugs. Additionally, the lymphatic drainage supports the rectum which could be advantageous for absorption of highly lipophilic compounds [1, 2]. Due to the presence of both systemic and portal pathways, drug absorption from the rectum is commonly variable [2].

Histologically, the rectal epithelia are similar to the upper GI and therefore, have the capability of drug absorption. However, they also have remarkable differences such as a considerably smaller surface area (about 200–400 cm^2) and the absence of villi and microvilli. Nevertheless, it is worth noting that, the absorptive surface area of the rectum is larger than that of the buccal and nasal [2]. In contrast to the upper parts of the GI, there is a relatively stable situations in the empty rectum. The average volume of the rectal fluid is only about 1 to 3 ml with an approximately neutral pH (about 7–8) and low buffering capacity [4].

5.3 Barriers to Rectal Drug Delivery

There are two major types of barriers against rectal drug delivery including:

- (i) Patient barrier, that is the resistance of patients against rectal use;
- (ii) Bioavailability barriers include binding of the drug to the rectal contents, bacterial flora of the rectum, the mucous layer of the rectum, and partial hepatic first-pass metabolism

Here, these barriers will be discussed.

5.3.1 Patient Barrier

The acceptability of rectal use is somewhat challenging due to the patient-related parameters including age, privacy concerns, and sociocultural factors. In many cases, the rectal route not only is not preferred but also, as shown in a study, the most unpleasant route compared to the oral and injection. In addition to patients or their parents (in the case of infants and children), the attitudes of the treatment staff can also impact the rectal administration. It seems that the underlying factors for reluctance to use rectal dosage forms are more acquired than intrinsic. Therefore, increasing the knowledge of prescribers and consumers about the benefits of the rectal route can lead to improved acceptance [2, 6].

5.3.2 Rectal Contents

The presence of feces in the rectum can alter the viscosity of the rectal content, and thereby, negatively affect the bioavailability of drugs (by impacting the stability, dissolution, and contact of drugs with the rectal mucosa for absorption). Such interactions may lead to a non-specific and irregular absorption from the rectum [4].

5.3.3 Time Barrier

To be well absorbed, the drug needs sufficient residence time in the rectal space. So, in the case of early expulsion of the rectal contents (for example, following defecation), the dosage form/drug may be removed before complete absorption [4].

5.3.4 Colonic Microbiota

Apart from the host cell enzymes, various metabolic enzymes produced by the gut microbiota can be also involved in drug metabolism, and hence, impact the bioavailability of drugs [7]. Rectally administered drugs are prone to undergo microbiota-mediated metabolism due to their close contact with the colonic bacteria [8].

5.3.5 Mucus Layer

To reach the rectal epithelial cells, drug molecules have to overcome the mucous layer. As mentioned in Chapter 4 (Oral Novel Drug Delivery Systems), the mucus layer is a major permeability barrier that acts as a physical, chemical, and time filter. The mucus layer of the rectum with an approximate thickness of 150 μm resists drug penetration. Also, due to the turnover rate of the rectal mucus (about every 3–4 hours), there is a limited time for rectally administered drug molecules to be absorbed [4].

5.3.6 Partial First-Pass Effect

Avoiding extensive first-pass metabolism (in both the liver and GI) is a remarkable advantage of rectal drug delivery. However, in the case of dosage forms that reach the upper part of the rectum, drugs may partially undergo first-pass metabolism, leading to reduced bioavailability [9]. It has been reported that about 50% of the rectally administered drug can bypass the liver [6].

5.4 Factors Affecting Rectal Drug Delivery

Several factors can potentially affect drug delivery through the rectum and thereby, bioavailability, efficacy, and safety of the rectally administered drugs. These factors include physiological (e.g., volume and pH of the rectal fluid), demographic (e.g., age), pathological (e.g., diseases and taking medications), pharmaceutical (e.g., composition and type of the vehicle and presence of excipients), and physicochemical properties (e.g., solubility, lipophilicity, particle size, and ionization state), which will be reviewed in the following.

5.4.1 Physiological Factors

Different physiological factors can impact rectal drug delivery including the volume and pH of the rectum, the presence of the stool, and the rectal mucous layer. The volume of the rectal fluid (1–3 ml) is significantly lower than the upper GI, which may hamper the dissolution of the drug in the rectum. This is especially important for drugs with slight solubility. As already mentioned, the rectum has a neutral pH which is favored for drugs whose pKa is almost equal to or above the physiological pH. However, due to the low buffering capacity of the rectum, any change in the pH can alter the degree of drug ionization and hence, the absorption. In addition, the change in pH may adversely affect the mucus layer (e.g., cause irritation), which, in turn, can impact drug absorption. Another influencing factor on drug absorption is whether the rectum is empty or full. Generally, drugs will be more absorbed when the rectum is empty. When stool is present in the rectum, it can interfere with the drug in terms of dissolution, stability, and availability for contact with the mucous layer. In addition to the mentioned factors, the residence time of the drug in the intended site is of crucial importance. The presence of mucus layer can shorten the drug retention time and reduce its absorption. Also, the bowel movement can impact the retention time of the dosage form and the opportunity for the drug to be completely released and absorbed. Bowel movements should be considered when dosing the drug to the patient. Since, for instance, the dosage form may be prematurely excreted following the defecation. In addition to the intrasubject variability of the bowel movement that could make the absorption hardly predictable, the intrasubject variation should also be noticed [1, 4, 5].

5.4.2 Demographic Factors

Age is a determining factor to be considered for rectal drug delivery since it can impact some anatomical factors related to drug efficacy and safety. For instance, the size of the rectum is considerably different between children and adults. Also, in some populations, rectal administration is precluded or should be done with caution. For example, rectal administration is not common for neonates since it may lead to variable absorption. In addition, the delicate and thin rectal lining of the neonates may be damaged and subsequently infected during the rectal administration [4].

5.4.3 Pathological Factors

Pathological conditions can affect the absorption of drugs from several aspects. For instance, in situations where tissue integrity and barrier functions of the rectum mucosa are disturbed (e.g., in inflammatory bowel disease (IBD), hemorrhoids, and long-term administration of some rectal drugs), drug absorption may increase [5, 10]. On the other hand, some diseases can impact bowel motility, and through it, the transit time of the dosage form and drug absorption. For example, accelerated colonic motility (in diarrhea, for instance, caused by GI infections, fecal incontinence, or some chronic diseases like IBD) can reduce the time available for the dosage form for disintegration, dissolution, and drug absorption [5]. On the other hand, diseases that cause constipation (e.g., irritable bowel syndrome (IBS) and hypothyroidism) can prolong the retention time of the drug in the GI [4]. In addition to diseases, taking medicines can also impact the GI motility. Among the main classes of drugs causing constipation, one can mention opioids, anticholinergics, antihypertensives, antipsychotic agents, antiepileptic drugs, antidepressants, and antihistamines [11]. Conversely, some drugs can lead to diarrhea such as laxatives, cholinergics, and serotonergics [12].

5.4.4 Pharmaceutical Factors

In addition to the mentioned factors, formulation-related parameters have an undeniable effect on rectal drug delivery efficiency. Rectal dosage forms are divided into three categories based on the physical state of their base (vehicle), including solids, semi-solids, and liquids; the formulation of each has specific considerations that will be discussed in the next section. One or more pharmaceutical ingredients may be incorporated in the dosage form, as dissolved or dispersed [13]. The rate of drug release from the base may vary depending on the composition of the formulation as well as the physical state of the base. Commonly, in the liquid dosage form, there is no liberation phase which makes the drug absorption faster. In the solid dosage form, conversely, the vehicle should first undergo melting or dissolution (in the case of fatty-based and water-soluble bases, respectively). The spreadability of the base (determined by its viscosity and bioadhesion) can impact the efficacy of rectal drugs for local effects. Solid dosage forms usually remain in the rectum, while the liquids may reach the proximal parts of the colon, depending on their volume. Additionally, the administered volume affects the drug retention, so that the lower volumes have a higher retention while the higher volumes (e.g., > 80 ml) induce defecation [14]. Different formulation modifiers such as diluents, lubricants, and preservatives may be added to the formulation [13]. In addition, the incorporation of absorption enhancers (e.g., enamine derivatives and salicylates) can provide brilliant opportunities for rectal delivery. For instance, the low bioavailability of rectally administered peptides and proteins could be improved using absorption enhancers [15].

Nanotechnology has emerged as a powerful tool to aid drug delivery through a variety of administration routes, namely the rectal pathway. For

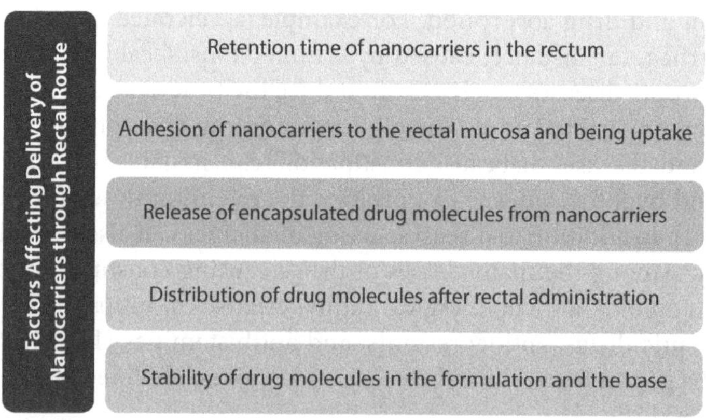

Figure 5.2 Factors affecting rectal delivery of nanocarriers.

example, considering the specific conditions of the rectum with low content of fluid for drug dissolution, using nanosystems to increase drug solubility can be helpful. Additionally, nanocarriers could be engineered to improve mucoadhesion, retention time, accumulation, and permeability of the rectally administered drugs [3]. The general factors to be considered for rectal delivery of nanocarriers are listed in Figure 5.2. In the following sections, the types of rectal nanocarriers and how they help to improve rectal drug delivery will be discussed.

5.4.5 Physicochemical Factors

Absorption of a drug is highly dependent on its physicochemical properties such as water solubility, partition coefficient, molecular weight, and ionization state. When a drug molecule is released from the vehicle, its solubility in the rectal fluid determines its concentration as the driving force for absorption. The higher the solubility, the higher the rate of dissolution and hence, the faster absorption. Optimum logP, in such a way that the drug has enough hydrophilicity to be dissolved in the rectal fluid and, at the same time, adequate lipophilicity to cross the epithelial membrane is another important factor. Since a significant proportion of drug molecules lack such ideal solubility properties, different techniques could be employed to improve the solubility of drugs (e.g., particle size reduction, salt formation, and encapsulation in the nanocarriers). Regardless of readily water-soluble drugs whose particle size does not considerably affect their absorption, as a general rule, the smaller the particle size, the higher the ratio of surface area to volume, the higher interaction with the dissolving medium, and thereby, the higher the rate of absorption. A range of 50 to 100 μm is commonly considered ideal for the particle size for rectal administration [4]. A majority of the rectally administered drugs have a molecular weight of < 500 Da and logP > 2.0 [16]. Last but not least, the ionization state of the drug in the rectal media impacts its absorption. The ionized form of drugs has a lower lipophilicity than the non-ionized form. So, the absorption of the drug is highly dependent on its pKa; if near or above the physiological pH, the drug will be in its non-ionized form and is predicted to have a better absorption [4].

5.5 Formulation Considerations

As already mentioned, based on the physical state of the base, there are three main rectal dosage forms including liquids (solutions, emulsions, and suspensions; that depending on their volume are generally called enema or microenema), semisolids (gels, creams, ointment and foams) and solids (e.g., suppositories, tablets, and capsules) [14]. Several parameters should be considered in the formulation of each type. These parameters are summarized in Table 5.1. Additionally, there are some general formulation considerations for all dosage forms. Among these, the investigation of drug-excipient and excipient-excipient interactions is of crucial importance since they can considerably impact the characteristics of the final product. It has been shown that mucoadhesion, gelling strength, thermo-gelling behavior, and melting point could be affected by formulation component interactions. For instance, carbopol and polycarbophil can lower the gelation temperature, while acetaminophen can decrease this value and lower the gel strength [16].

Table 5.1 Formulation considerations of rectal dosage forms.

Type of rectal dosage form	Formulation	Ideal drug characters	Main formulation considerations
Liquid	Solution	• High water solubility • High mucus permeability • High stability in the solution state • Compatibility with formulation ingredients	• Solubility of the drug in the solution • Stability of the drug in the solution state • Isotonicity • pH of formulation • Volume of formulation (impacting retention time)
	Suspension	• Low water solubility • High mucus permeability • High stability in aqueous media • Compatibility with formulation ingredients	• Size of the dispersed phase • Long-term formulation stability • pH of formulation • Volume of formulation

(*Continued*)

Table 5.1 Formulation considerations of rectal dosage forms. (*Continued*)

Type of rectal dosage form	Formulation	Ideal drug characters	Main formulation considerations
	Emulsion	• High lipophilicity • High mucus permeability • High stability in the emulsion matrix • Compatibility with formulation ingredients	• Size of the dispersed phase • Viscosity of formulation • Long-term formulation stability • pH of formulation • Volume of formulation
Semi-solid	Gel Cream Ointment Foam	• Local effect • High mucus permeability • High stability in the formulation matrix • Compatibility with formulation ingredients	• Solubility of the drug in the base • Disintegration/dispersion • Rheological behavior • Vehicle-to-mucus partitioning • Long-term formulation stability • pH of formulation • Volume of formulation
Solid	Suppository Tablet Capsule	• Appropriate solubility/dispersibility in the base • High mucus permeability • High stability in the formulation matrix • Compatibility with formulation ingredients	• Appropriate drug release profile • Particle size of drug • Drug/base hydrophilicity ratio (for suppositories) • Melting point of the base (for suppositories) • Softening time of the base (for suppositories) • Wettability of the base • Homogeneity of the base • Disintegration of the base • Dissolution of drug

5.6 Challenges and Opportunities of Rectal Drug Delivery

Due to its special anatomical and physiological characteristics, the rectal area provides special possibilities for drug absorption, and on the other hand, creates challenges in the way of drug delivery. These opportunities and challenges are summarized in Figure 5.3.

5.7 Classification of Rectal Novel Drug Delivery Systems

Generally, the number of studies of rectal systems is very small compared to the systems for other routes of drug delivery. This lack of interest may be probably due to the lack of patient compliance. However, in recent years, several efforts have been made to improve rectal systems. As already mentioned, conventional rectal drug delivery systems include solid, semisolid, and liquid dosage forms. Advancements in rectal formulations have been mainly focused on improving the existing dosage forms and overcoming their drawbacks. In general, novel rectal drug delivery systems could be classified into five main categories (Figure 5.4): (i) modified suppositories, (ii) hydrogel-based systems, (iii) carrier-based systems, (iv) carrier-in-gel systems, and (v) surfactant-based liquid crystalline gels. In the following, these categories and their subcategories are introduced.

5.7.1 Modified Suppositories

Suppositories, as traditional dosage forms, could be modified in different ways to be converted into novel systems. Generally, they can be (i) geometrically modified (to achieve a controlled drug release), (ii) formulated in some way to form self-emulsifying suppositories (to increase bioavailability), (iii) coated (to overcome instability issues), or (iv) formulated as non-dissolving dosage forms (to conquer the challenges of melting bases).

5.7.1.1 Geometrically Modified Suppositories

Conventional rectal suppositories are commonly referred to as homogeneous solid bodies containing one active ingredient which are disable to provide a constant or extended drug release over time [17]. In recent years, several efforts have been made for the advancement of this dosage

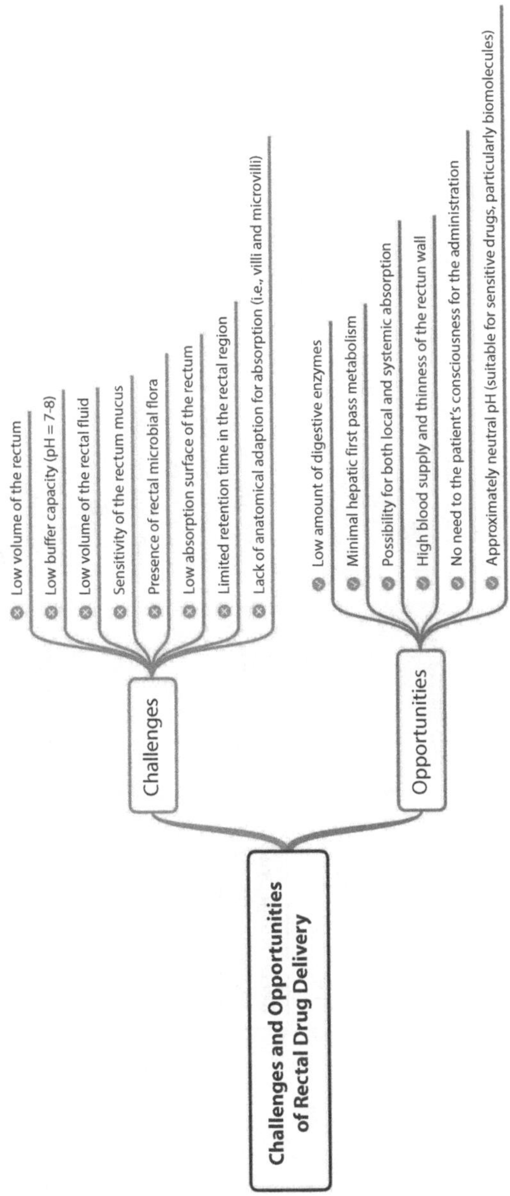

Figure 5.3 Challenges and opportunities of the rectal drug delivery.

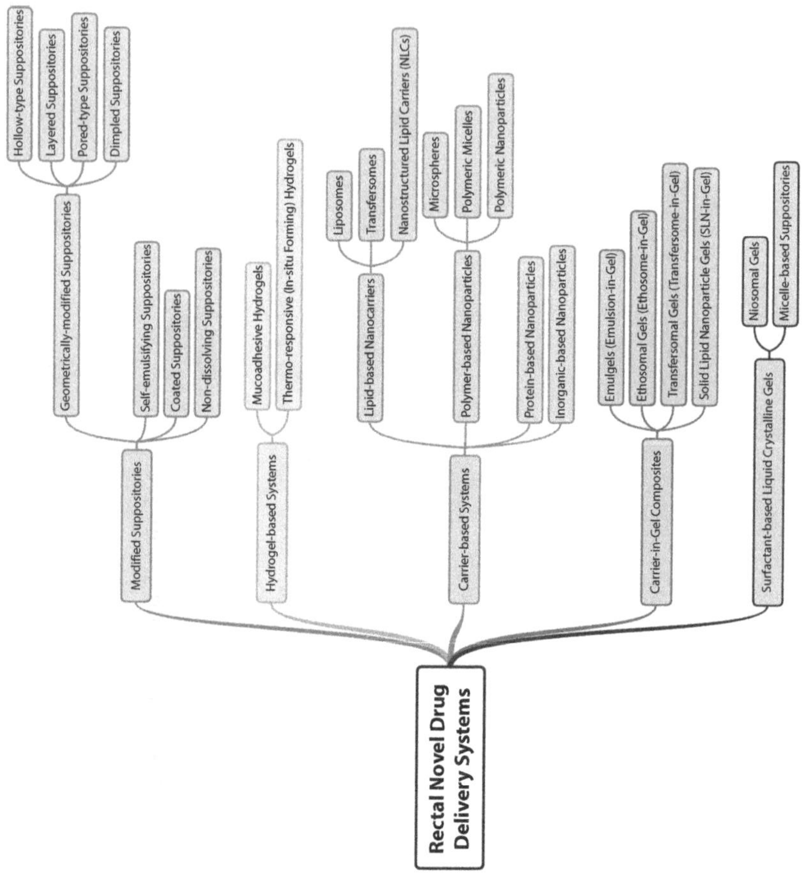

Figure 5.4 Classification of rectal novel drug delivery systems.

form. For instance, the design and geometry of the suppositories have been changed to add new capabilities to these conventional systems (e.g., controlled drug release and combinational drug delivery). In this regard, novel fabrication techniques, in particular 3D printing [18–21], have received much attention due to their ability to make different structures. Such dosage forms that are built with no need for molds are sometimes called "self-supporting" suppositories [21]. Here, we briefly overview different structurally modified suppositories.

5.7.1.1.1 Hollow-Type Suppositories

Structurally, hollow-type suppositories consist of a shell with a hollow cavity inside, which can be filled with solid (e.g., powders), semisolids (e.g., gels), and liquid (e.g., solutions) dosage forms to achieve different drug release profiles (including immediate and controlled-release). With this strategy, it is also possible to deliver high-molecular-weight drug molecules. To obtain the optimum drug delivery, different properties of the surrounding shell (e.g., thickness and hardness) as well as the inner structure characteristics could be modified [18]. Additionally, for drug release control, it is possible to add a gelling agent to the oily shell or the core part of the suppository [22]. The main advantages of hollow-type suppositories are (i) ease of drug interchangeability, (ii) controlled drug dose, and (iii) avoiding drug-base interactions [4].

5.7.1.1.2 Layered Suppositories

Bilayered suppositories, as their name implies, are composed of two different layers that can be along (i.e., as front and terminal layers) or inside each other (i.e., as inner and outer layers). Such a design provides a suitable platform for (i) fixed-dose drug combination, (ii) delivery of incompatible drug pairs (through incorporating them in the physically separated layers), and (iii) adjustable drug release patterns (e.g., combined immediate and sustained release) [23]. Recently, a multilayered suppository (called Matryoshka-type) composed of multiple-layered shells has been designed to achieve a controllable pulsatile release profile. To obtain the optimum drug release, the inner/outer space volume as well as drug concentration could be adjusted [19].

5.7.1.1.3 Pored-Type Suppositories

The idea of preparing suppositories with holes has been proposed by Tagami *et al.* [19]. They made holes with different numbers, diameters, and positions in the suppository shells to control the drug dissolution from the dosage form. They concluded that there is a good relation between the drug release profile from the suppositories and these three factors. So, the number, dimension, and location of the holes could be modified to obtain the desired release profile.

5.7.1.1.4 Dimpled Suppositories

In dimple-type suppositories, one or more dimples are created in the body, in which the drug is embedded. When rectally administered, such drug-condensed areas provide a higher drug release and absorption. This strategy could be especially helpful for drugs with poor absorption (e.g., biomolecules) [24].

5.7.1.2 Self-Microemulsifiyng Suppositories

Self-microemulsifying bases are isotropic mixtures of lipids, surfactants, and cosurfactants that spontaneously form small oil-in-water droplets upon contact with water. Drug molecules with sufficient lipophilicity can be encapsulated in this base and administered rectally. Especially, drug molecules with low bioavailability, caused by inadequate dissolution rate, will benefit from this strategy since the formed microemulsion droplets can improve the rate and extent of drug absorption [25]. The self-emulsifying systems could be administered in different forms. For instance, in a study, the indomethacin-loaded self-emulsifying system has been incorporated into the cavity of the hollow-type suppositories [26].

5.7.1.3 Coated Suppositories

In the cases where the oily base of the suppository may undergo oxidative degradation (e.g., bases with high amounts of free fatty acids), coating the surface not only postpones the instability and extends the shelf-life but also reduces the rectal irritation caused by acids. Furthermore, a non-greasy smooth surface is provided which is generally more favorable for the patients [27].

5.7.1.4 Non-Dissolving Suppositories

Conventional suppositories may face several challenges. For instance, the residue of the melted suppository may migrate upwards or leak out from the site of administration. Also, extended drug release may fail. To overcome these challenges, non-dissolving suppositories have been proposed.

Silicone elastomers, due to their biocompatibility but not biodegradability, can be a good alternative as suppository bases. In the body, non-biodegradable materials are potentially neither melted nor displaced while they can provide long-term drug delivery [28].

5.7.2 Hydrogel-Based Systems

Hydrogels are of great interest for various routes of drug delivery (including rectal), due to their brilliant advantages such as adjustable swelling, biodegradability, and high water content absorption. To date, different forms of hydrogels have been investigated for rectal purposes which will be reviewed.

5.7.2.1 Mucoadhesive Hydrogels

Mucoadhesion is a desirable feature for rectal dosage forms since it can prevent drug leakage after administration and extend the residence time of the drug in the rectum. To realize this purpose, mucoadhesive hydrogels are appropriate candidates. They can improve the efficacy of rectal drug delivery by extending the duration of action and increasing the drug bioavailability [9, 29].

5.7.2.2 Thermoresponsive (In situ Forming) Hydrogels

In situ gelling is a phase transition change of the sol form of the formulation to the gel state in the body, in response to different stimuli such as temperature and pH [30]. So, it is considered a powerful strategy to improve the residence of formulation in the intended site [31]. In thermoresponsive hydrogels, the sol-gel transition is triggered by the temperature change [32]. The main reason for the development of thermoresponsive suppositories for rectal application was the drawbacks of conventional solid suppositories namely unpleasant feeling and discomfort of the patients after use. Since thermoresponsive suppositories are liquid at room temperature, they have a simple administration and do not cause mucosal damage. Then, in the body, they convert to a mucoadhesive gel, thus preventing drug leakage while not creating the feeling of a foreign substance for the patient [33].

5.7.3 Carrier-Based Systems

There are several aspects of differentiation between conventional rectal dosage forms and nanocarrier systems including formulation-related characteristics (e.g., spreadability), residence time, as well as release/pharmacokinetic profiles. Colloidal carriers can be well dispersed on the rectal mucosa, improve the residence time of drugs, and increase their absorption. Generally, drug molecules are encapsulated within the nanocarriers and then, the nanocarriers are dispersed into the rectal dosage form [3, 5].

5.7.3.1 *Lipid-Based Nanocarriers*

Liposomes are spherical lipid bilayers with a versatile application in drug delivery. However, there are only a few studies on rectal administration of liposomes. For instance, liposomes have been studied for mucosal siRNA delivery against TNF-α for the treatment of IBD [34, 35].

Transfersomes are a unique type of liposomes with a significantly higher deformability. In addition to the phosphatidylcholine, they have edge activators in their structure, which endows them an elastic nature and facilitates their deep transportation across the biological membranes. Due to their high deformability, this platform has been widely explored for transdermal drug delivery to overcome the skin barrier. Recently, transferosomes have been employed for rectal delivery to improve the absorption and bioavailability of cannabidiol as a candidate small molecule to improve stability and permeation [36].

Solid lipid nanoparticles (SLNs) have also gained attention as rectal delivery systems, as will be discussed in the following section. SLNs are sub-micron-sized solidified lipid-based drug carriers. They are considered a powerful alternative for liposomes due to outstanding advantages such as higher drug-loading capacity and improved physical stability [37].

Nanostructured lipid carriers (NLCs) are the next generation of SLNs. They are lipid-based systems composed of a mixture of solid and liquid lipids. They can encapsulate both hydrophilic and lipophilic drugs and compared to SLNs, they have a higher loading capacity, higher water content, and minimized drug leakage [38]. In this regard, in a 2021 study, NLC-based suppositories have been studied. First, ondansetron was encapsulated in the NLCs, and then, NLCs were incorporated into a lipophilic suppository base. Compared to the plain suppository, the NLC-based suppository showed a faster onset, extended drug release, and improved bioavailability [39].

5.7.3.2 *Polymer-Based Carriers*

Polymeric mucoadhesive microspheres are good candidates for rectal drug delivery since they can be uniformly distributed in the rectum, adhere to the rectal mucosa, improve the therapeutic effect of the drug, and reduce drug fluctuation and intrasubject variability [40]. In addition to polymeric microspheres, polymeric nanoparticles have also gained attention for rectal drug delivery. So far, different biocompatible polymers (including synthetic and natural ones) have been used to prepare nanoparticles. Among the synthetic polymers, PLGA is one of the most widely used. The efficacy and selectivity of PLGA-based nanoparticles could be further improved by surface decoration for targeting (for instance, by peanut agglutinin). Chitosan and its derivatives are natural polymers that have garnered special interest as rectal nanoparticles, particularly for the delivery of biopharmaceuticals [3].

Polymeric micelles are another polymer-based platform to enhance the solubility of poorly water-soluble drugs in the physiological fluids. It has been shown that the incorporation of micelles into a lipophilic suppository base can positively affect the dissolution of the drug from the base [41]. In another study, docetaxel-loaded thermoresponsive and bioadhesive nanomicelles have been studied to improve the bioavailability of the drug and its chemotherapeutic effect. The results showed an enhanced antitumor efficiency and reduced toxicity with no mucosal irritation [42].

5.7.3.3 *Protein-Based Nanoparticles*

To date, only a few studies have been conducted on protein-based nanoparticles for rectal drug delivery [43, 44]. In this regard, silk fibroin as a natural and biocompatible protein with anti-inflammatory effects has gained attention. For instance, silk fibroin nanoparticles were proposed for the controlled release of resveratrol in animal models for IBD. The *in vivo* results demonstrated immunomodulatory and anti-inflammatory effects [43].

5.7.3.4 *Inorganic-Based Nanoparticles*

Like the previous systems, the number of studies on inorganic-based rectal drug delivery is also very small. Among inorganic nanoparticles, calcium phosphate as a naturally occurring biomineral with a high degree of biodegradability and biocompatibility has been investigated for rectal gene delivery. Nucleic acid-loaded calcium phosphate nanoparticles were prepared and dispersed in a hard fat base to form a particulate-containing suppository. At the body temperature, nanoparticles are released from the molten base and dispersed in the medium. *In vitro* cell culture results showed that this system is promising for intestinal transfection and gene silencing [45]. Other recently proposed inorganic carriers for local/systemic rectal drug delivery are drug-loaded gold nanoparticles and gold–protein nanoclusters. These nanocarriers have been dispersed in solid fat suppositories for colon targeting. The cell culture experiments demonstrated that these nanocarriers have a good uptake by the HeLa cells. This could encourage researchers to consider these systems as an option for further research on rectal delivery of various drugs, including proteins [46].

5.7.4 Carrier-in-Gel Composites

Incorporation of carriers into a hydrogel base can improve the delivery system by combining the advantages of both. For instance, drug carriers can improve the solubility/dissolution rate of drugs and hydrogel base provides mucoadhesion and long-term drug release.

5.7.4.1 Emulgel (Emulsion-in-Gel)

Emulgel is an advanced semi-solid formulation for topical drug delivery, in which the emulsion droplets are embedded in a continuous gel phase [47]. Owing to this unique structure, emulgels combine the advantages of gels and emulsions [48]. Recently, a patent has been published on the preparation of an oil-in-water emulsion in a temperature-sensitive gel to form an *in situ* gel-forming suppository to improve the drug release performance [49].

5.7.4.2 Ethosomal Gel (Ethosome-in-Gel)

Ethosomes are ethanol-rich lipid vesicles that enhance membrane fluidity and drug permeation. If ethosomes are incorporated into the hydrogels, the advantages of both (such as transmucosal permeation, bioadhesion, and stability) are combined [50].

5.7.4.3 Transfersomal Gel (Trnaferosome-in-Gel)

As previously mentioned, transferosomes have been recently used for rectal drug delivery. In one study, transferosomes were loaded in a thermosensitive gel for rectal drug delivery of tizanidine HCl to bypass the first-pass metabolism of the liver. Nanotransferosomes with a high encapsulation efficiency were first prepared followed by incorporated in a thermosensitive gel. The results showed an extended drug release and improved bioavailability compared to the oral solution [51].

5.7.4.4 *Solid Lipid Nanoparticle Gel (SLN-in-Gel)*

For rectal administration, SLNs could be embedded within a hydrogel matrix. For instance, in an innovative study, a dual thermoresponsive system of SLN-in-hydrogel was developed to enhance bioavailability and prevent the burst effect of flurbiprofen. First, the drug-loaded SLNs were prepared with a melting point of about 32 °C; which is solid at room temperature but liquid in the body. The thermosensitive SLN was then incorporated into a polymeric thermosensitive hydrogel with a gelation point of about 35 °C. So, it is liquid in the room but gelled at the physiological situation; an opposite reversible property compared to the SLNs. Such a dual-reverse thermoresponsive SLN-loaded hydrogel possesses easy rectal administration and rapid gelation in the body. This system showed a considerably increased dissolution rate, compared to the non-encapsulated drug, but retarded, compared to the plain hydrogel. Also, compared to the drug, this system led to an increase in C_{max} and AUC, but reduced initial concentration and C_{max} compared to the hydrogel. Altogether, due to no mucosal irritation, enhanced bioavailability, and lowered burst effect, this dual system seems a potential novel rectal drug delivery system [52]. The same group, more recently, proposed a similar double system for rectal administration of irinotecan [53].

5.7.5 Surfactant-Based Liquid Crystalline Gels

Liquid crystals are mesophases between crystalline solids and isotropic liquids which are broadly categorized into lyotropic and thermotropic systems. In the thermotropic liquid crystals, the phase transition is dependent on the temperature, while the formation of lyotropic systems is a function of both temperature and solvent [54]. As a subclass of lyotropic systems, surfactant-based liquid crystal gels are defined as soft materials that combine the properties of lyotropic liquid crystals with the consistency of physical gels [55, 56].

5.7.5.1 *Niosomal Gel*

Proniosomes of jelly-like liquid crystals are compact niosomes that could be easily converted into niosomal dispersion upon water addition [57]. Several drawbacks of aqueous niosomal dispersion could be overcome by proniosomes such as physical and chemical stability issues (e.g., aggregation and hydrolysis, respectively). Also, different constituents of proniosomes, such as non-ionic surfactants may act as drug permeation enhancers. In this regard, a proniosomal gel was formulated for the rectal delivery of rutin (a polyphenolic compound with an anti-inflammatory effect) for the treatment of hemorrhoids. *In vitro* results showed that this formulation can provide a high drug release as well as a high local accumulation of drugs, which could be a driving force for drug permeation [58].

5.7.5.2 Micelle-Based Suppository

Solidified reverse micellar solutions (SRMSs) are solid lipid dispersions made up of phospholipids (mainly lecithin) and triglycerides or hard fats [59]. In contact with water, this base melts and transforms into a semi-solid liquid crystal enriched with microstructures. This system has been used for the preparation of controlled-release rectal suppositories with an improved residence time in the rectum [60].

5.8 Pharmacokinetics

Liberation of a drug from its vehicle is a prerequisite for drug absorption. In the case of solid dosage forms (mainly suppositories), the base should be first molten or dissolved in the rectum. Since the rectal water content is small, it will be difficult for the dosage form to be completely dissolved in the absence of additional water. When the dissolving dosage form absorbs the low water, an unpleasant fill may occur for the patient which, in turn, may reduce compliance. Regardless of the dosage form, if the drug is molecularly dissolved in the base, it starts to diffuse from the base to the rectal mucosa. But, if the drug is suspended in the vehicle, it should first leave its base and then start dissolving [13]. The rectal absorption steps of a suspended drug within a suppository are illustrated in Figure 5.5.

After the drug release, the physicochemical properties of the drug undertake a major role in its biofate. Typically, drug molecules with low logP, high molecular weight, high charge, and capability of forming hydrogen bonds are prone to low absorption [15]. The maximum available concentration of the drug in the rectum is determined by its solubility. Generally, the higher the solubility, the higher the dissolution rate. It is worth noting that, for effective rectal delivery, an optimum hydrophilicity-lipophilicity balance (i.e., logP) is necessary. In general, the drug must be hydrophilic

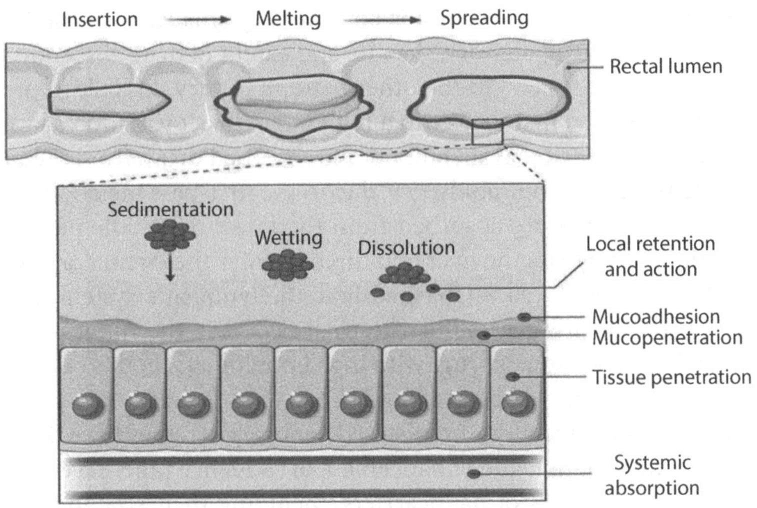

Figure 5.5 Drug absorption from a drug-suspended suppository in rectal area.

enough to be dissolved in the rectal fluid and then, be sufficiently lipophilic to cross the membrane. However, many drug molecules, particularly a major part of new chemical entities, lack this feature. In this regard, solubility improvement strategies (e.g., encapsulation) are helpful [4].

Then, to create a local or systemic effect, the mucus barrier should be overcome. Therefore, the leakage problem of traditional formulations should first be overcome (for example, using hydrogels or *in situ* gelling systems). Also, the mucus barrier should be overcome. This layer, with approximately 150 μm thickness and 3 to 4 hours turnover, is considered a serious barrier to rectal bioavailability [4]. Here, mucoadhesive systems can extend the residence time of formulation and prolong its contact with the rectal mucosa.

The next step is absorption; that is diffusion of drug molecules through the mucus layer followed by rectal epithelium penetration. In rectal absorption, the passive diffusion pathway (either paracellular or transcellular) is predominantly involved and, unlike upper GI, no evidence of the presence of active transport has been found [13]. The paracellular pathway is more favorable for hydrophilic, ionized, and high molecular weight drugs, though narrow tight junctions restrict drug transport. The transcellular pathway is considered the main transportation pathway in the rectal region, which is preferred by more lipophilic and non-ionized compounds (e.g., basic drugs with a pKa approximately equal to or higher than rectal pH) [4].

The surface area of the rectum is considerably smaller than the GI; so, compared to oral drug delivery, the rate and extent of drug absorption are lower in the rectal route. On the other hand, it is very difficult to predict rectal bioavailability, mainly due to a high intersubject variation as well as venous drainage. The bioavailability of the absorbed drugs from the upper part of the rectum is commonly low, due to the first-pass metabolism. On the contrary, if drugs are absorbed from the lower or middle part of the rectum, they can bypass the liver and directly enter the circulatory system [13]. It has not been well established how the lymphatic system impacts rectal absorption. However, similar to what occurs in the gut, it may contribute to the absorption of drugs with high lipophilicity. In this case, again, first-pass metabolism will be avoided [4].

In the case of the rectally administered nanocarriers, they have the capability of self-distribution and enhancement of residence time. Several physicochemical parameters of nanocarriers, including their size and surface properties can affect their biodistribution and biofate. Following administration, nanocarriers may be prematurely eliminated from the rectum due to rectal motility [3]. This is one of the factors that have led researchers to prepare nanocarrier-in-gel composites to improve their residence time. Moreover, nanocarriers may have retrograde transportation into the upper parts of the colon. Another pathway is to remain in the rectum to create a local effect or pass through the mucus layer to provide a systemic effect. Different pathways of rectally administered nanocarriers are illustrated in Figure 5.6.

5.9 Products in the Market

There are several novel rectal drug delivery systems in different phases of clinical trials. Rathi *et al.* [5] have comprehensively reviewed these novel formulations (including suppositories, tablets, enemas, and nanocarriers) and their clinical trial outcomes. Here, we only introduce some micro- and nano-based systems whose clinical trials have been completed (Table 5.2). The reader is referred to the cited reference for further information.

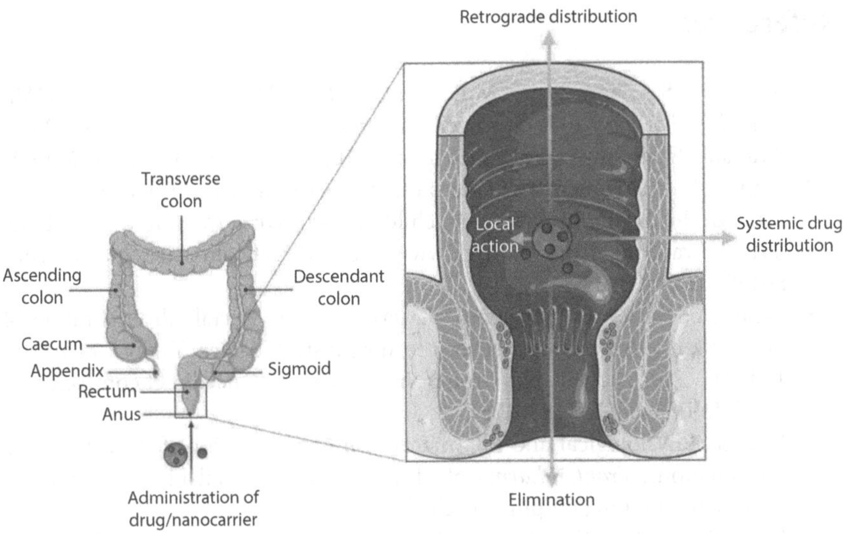

Figure 5.6 Anatomy of rectum and drug distribution in this part (created with BioRender.com).

Table 5.2 Nano-based rectal novel drug delivery systems with completed clinical trial.

Platform	Drug(s)	Indication
Microsphere	Irinotecan	Colorectal Carcinoma Neoplasm Metastasis
	FOLFOX	Colorectal Cancer Colorectal Carcinoma Liver Metastases
	Regorafenib	Colorectal Neoplasms
	FOLFOX 6m	Colorectal Cancer Metastatic
Nanoparticle	TKM-080301	Colorectal Cancer with Hepatic Metastasis
Liposome	Irinotecan HCl Floxuridine	Colorectal Neoplasms
	SN-38	Colorectal Cancer
	Promitil Capecitabine Bevacizumab	Colorectal Cancer

References

1. Montoto, S.S. and Ruiz, M.E., Rectal Drug Delivery, in: *The ADME Encyclopedia*, pp. 1–7, Springer International Publishing, Cham, 2021, Available from: https://link.springer.com/10.1007/978-3-030-51519-5_13-1.
2. Jannin, V., Lemagnen, G., Gueroult, P., Larrouture, D., Tuleu, C., Rectal route in the 21st Century to treat children. *Adv. Drug Deliv. Rev.*, 73, 34–49, 2014, Available from: https://www.sciencedirect.com/science/article/pii/S0169409X14001069.
3. Melo, M., Nunes, R., Sarmento, B., das Neves, J., Rectal administration of nanosystems: from drug delivery to diagnostics. *Mater. Today Chem.*, 10, 128–41, 2018, Available from: https://www.sciencedirect.com/science/article/pii/S2468519418301757.
4. Hua, S., Physiological and Pharmaceutical Considerations for Rectal Drug Formulations. *Front. Pharmacol.*, 10, 1196, 2019, Available from: https://www.ncbi.nlm.nih.gov/pmc/articles/PMC6805701/.
5. Rathi, R., Sanshita, Kumar, A., Vishvakarma, V., Huanbutta, K., Singh, I. *et al.*, Advancements in Rectal Drug Delivery Systems: Clinical Trials, and Patents Perspective. *Pharmaceutics*, 14, 10, 2210, 2022, Available from: https://www.mdpi.com/1999-4923/14/10/2210.
6. Lam, J.K., Cheung, C.C., Chow, M.Y., Harrop, E., Lapwood, S., Barclay, S.I. *et al.*, Transmucosal drug administration as an alternative route in palliative and end-of-life care during the COVID-19 pandemic. *Adv. Drug Deliv. Rev.*, 160, 234–43, 2020, Available from: https://www.sciencedirect.com/science/article/pii/S0169409X20301654.
7. Dhurjad, P., Dhavaliker, C., Gupta, K., Sonti, R., Exploring drug metabolism by the gut microbiota: modes of metabolism and experimental approaches. *Drug Metab. Dispos.*, 50, 3, 224–34, 2022, Available from: https://dmd.aspetjournals.org/content/50/3/224.abstract.
8. Sousa, T., Paterson, R., Moore, V., Carlsson, A., Abrahamsson, B., Basit, A.W., The gastrointestinal microbiota as a site for the biotransformation of drugs. *Int. J. Pharm.*, 363, 1–2, 1–25, 2008, Available from: https://www.sciencedirect.com/science/article/pii/S0378517308004985.
9. Ramadan, A.A., Elbakry, A.M., Esmaeil, A.H., Khaleel, S.A., Pharmaceutical and pharmacokinetic evaluation of novel rectal mucoadhesive hydrogels containing tolmetin sodium. *J. Pharm. Investig.*, 48, 6, 673–83, 2018, Available from: http://link.springer.com/10.1007/s40005-017-0365-1.
10. Malamed, S.F. (Ed.), Chapter 8 - Rectal Sedation, in: *Sedation*, Sixth Edition, pp. 120–4, Mosby, Missouri, 2018, Available from: https://www.sciencedirect.com/science/article/pii/B9780323400534000081.
11. Valladales-Restrepo, L.F., Paredes-Mendoza, M., Machado-Alba, J.E., Potentially inappropriate prescriptions for anticholinergic medications for

patients with constipation. *Dig. Dis.*, 38, 6, 500–6, 2020, Available from: https://karger.com/ddi/article-abstract/38/6/500/96264.
12. Abraham, B.P. and Sellin, J.H., 23Drug-Induced Diarrhea. *Diarrhea*, 393, 365–72, 2010, Available from: https://onlinelibrary.wiley.com/doi/10.1002/9781119211419
13. Dodou K. Chapter 44- Rectal and vaginal drug delivery. In: Aulton's pharmaceutics: the design and manufacture of medicines. 6th Edition. Elsevier; 2021. p. 735–51.
14. Nunes, R., Sarmento, B., das Neves, J., Formulation and delivery of anti-HIV rectal microbicides: advances and challenges. *J. Control. Release*, 194, 278–94, 2014, Available from: https://www.sciencedirect.com/science/article/pii/S0168365914006427.
15. Prasanna, J.L., Deepthi, B., Rao, N.R., Rectal drug delivery: A promising route for enhancing drug absorption. *Asian J. Res. Pharm. Sci.*, 2, 4, 143–9, 2012, Available from: https://www.indianjournals.com/ijor.aspx?target=ijor:ajrps&volume=2&issue=4&article=006.
16. Purohit, T.J., Hanning, S.M., Wu, Z., Advances in rectal drug delivery systems. *Pharm. Dev. Technol.*, 23, 10, 942–52, 2018, Available from: https://www.tandfonline.com/doi/full/10.1080/10837450.2018.1484766.
17. Kumar, A., Kolay, A., Havelikar, U., Modern Aspects of Suppositories: A Review. *Eur. J. Pharm. Res.*, 3, 4, 23–9, 2023, Available from: https://ej-pharma.org/index.php/pharma/article/view/68.
18. Tagami, T., Ito, E., Hayashi, N., Sakai, N., Ozeki, T., Application of 3D printing technology for generating hollow-type suppository shells. *Int. J. Pharm.*, 589, 119825, 2020, Available from: https://www.sciencedirect.com/science/article/pii/S0378517320308103.
19. Tagami, T., Hayashi, N., Sakai, N., Ozeki, T., 3D printing of unique water-soluble polymer-based suppository shell for controlled drug release. *Int. J. Pharm.*, 568, 118494, 2019, Available from: https://www.sciencedirect.com/science/article/pii/S0378517319305368.
20. Awad, A., Hollis, E., Goyanes, A., Orlu, M., Gaisford, S., Basit, A.W., 3D printed multi-drug-loaded suppositories for acute severe ulcerative colitis. *Int. J. Pharm.: X*, 5, 100165, 2023, Available from: https://www.sciencedirect.com/science/article/pii/S2590156723000099.
21. Seoane-Viaño, I., Ong, J.J., Luzardo-Álvarez, A., González-Barcia, M., Basit, A.W., Otero-Espinar, F.J. *et al.*, 3D printed tacrolimus suppositories for the treatment of ulcerative colitis. *Asian J. Pharm. Sci.*, 16, 1, 110–9, 2021, Available from: https://www.sciencedirect.com/science/article/pii/S1818087620302920.
22. Shiohira, H., Fujii, M., Koizumi, N., Kondoh, M., Watanabe, Y., Novel chronotherapeutic rectal aminophylline delivery system for therapy of asthma. *Int. J. Pharm.*, 379, 1, 119–24, 2009, Available from: https://www.sciencedirect.com/science/article/pii/S0378517309004086.

23. Ali, M., A review on bi-layered suppositories. *Eur. J. Sci. Res.*, 146, 45–54, 2017, Available from: https://www.researchgate.net/profile/Muaadh-Ali/publication/320280536_A_Review_on_Bi-layered_Suppositories/links/59db0830aca272e6096db806/A-Review-on-Bi-layered-Suppositories.pdf.
24. Matsumoto, A., Murakami, K., Watanabe, C., Murakami, M., Improved systemic delivery of insulin by condensed drug loading in a dimpled suppository. *Drug Discov. Ther.*, 11, 6, 293–9, 2017, Available from: https://www.jstage.jst.go.jp/article/ddt/11/6/11_2017.01072/_article/-char/ja/.
25. Gugulothu, D., Pathak, S., Suryavanshi, S., Sharma, S., Patravale, V., Self-Microemulsifiyng Suppository Formulation of β-Artemether. *AAPS Pharm. Sci. Tech.*, 11, 3, 1179–84, 2010, Available from: http://link.springer.com/10.1208/s12249-010-9478-9.
26. Kim, J.Y. and Ku, Y.S., Enhanced absorption of indomethacin after oral or rectal administration of a self-emulsifying system containing indomethacin to rats. *Int. J. Pharm.*, 194, 1, 81–9, 2000, Available from: https://www.sciencedirect.com/science/article/pii/S0378517399003671.
27. Loftsson, T., Coated suppositories, WO2013035113A1, 2013, Available from: https://patents.google.com/patent/WO2013035113A1/en.
28. Sun, Y., Ruan, X., Li, H., Kathuria, H., Du, G., Kang, L., Fabrication of non-dissolving analgesic suppositories using 3D printed moulds. *Int. J. Pharm.*, 513, 1, 717–24, 2016, Available from: https://www.sciencedirect.com/science/article/pii/S0378517316309188.
29. Xu, J., Tam, M., Samaei, S., Lerouge, S., Barralet, J., Stevenson, M.M. *et al.*, Mucoadhesive chitosan hydrogels as rectal drug delivery vessels to treat ulcerative colitis. *Acta. Biomater.*, 48, 247–57, 2017, Available from: https://www.sciencedirect.com/science/article/pii/S1742706116305530.
30. Kolawole, O.M. and Cook, M.T., *In situ* gelling drug delivery systems for topical drug delivery. *Eur. J. Pharm. Biopharm.*, 184, 36–49, 2023, Available from: https://www.sciencedirect.com/science/article/pii/S0939641123000073.
31. Mandal, S., Thimmasetty, M.K., Prabhushankar, G., Geetha, M., Formulation and evaluation of an *in situ* gel-forming ophthalmic formulation of moxifloxacin hydrochloride. *Int. J. Pharm. Investig.*, 2, 2, 78–82, 2012, Available from: https://www.ncbi.nlm.nih.gov/pmc/articles/PMC3482769/.
32. Chatterjee, S., Hui, P.C.L., Kan, C.W., Thermoresponsive Hydrogels and Their Biomedical Applications: Special Insight into Their Applications in Textile Based Transdermal Therapy. *Polym. (Basel)*, 10, 5, 480, 2018, Available from: https://www.ncbi.nlm.nih.gov/pmc/articles/PMC6415431/.
33. Bialik, M., Kuras, M., Sobczak, M., Oledzka, E., Achievements in Thermosensitive Gelling Systems for Rectal Administration. *Int. J. Mol. Sci.*, 22, 11, 5500, 2021, Available from: https://www.ncbi.nlm.nih.gov/pmc/articles/PMC8197127/.

34. Zhang, Y., Cristofaro, P., Silbermann, R., Pusch, O., Boden, D., Konkin, T. *et al.*, Engineering Mucosal RNA Interference *in Vivo*. *Mol. Ther.*, 14, 3, 336–42, 2006, Available from: https://www.sciencedirect.com/science/article/pii/S1525001606001389.
35. Ocampo, S.M., Romero, C., Aviñó, A., Burgueño, J., Gassull, M.A., Bermúdez, J. *et al.*, Functionally Enhanced siRNA Targeting TNFα Attenuates DSS-induced Colitis and TLR-mediated Immunostimulation in Mice. *Mol. Ther.*, 20, 2, 382–90, 2012, Available from: https://www.sciencedirect.com/science/article/pii/S1525001616304907.
36. Moqejwa, T., Marimuthu, T., Kondiah, P.P., Choonara, Y.E., Development of stable nano-sized transfersomes as a rectal colloid for enhanced delivery of cannabidiol. *Pharm.*, 14, 4, 703, 2022, Available from: https://www.mdpi.com/1999-4923/14/4/703.
37. Mukherjee, B., Chakraborty, S., Mondal, L., Satapathy, B.S., Sengupta, S., Dutta, L. *et al.*, Chapter 7 - Multifunctional drug nanocarriers facilitate more specific entry of therapeutic payload into tumors and control multiple drug resistance in cancer, in: *Nanobiomaterials in Cancer Therapy*, A.M. Grumezescu (Ed.), pp. 203–51, William Andrew Publishing, Norwich, NY, 2016, Available from: https://www.sciencedirect.com/science/article/pii/B9780323428637000074.
38. Elmowafy, M. and Al-Sanea, M.M., Nanostructured lipid carriers (NLCs) as drug delivery platform: Advances in formulation and delivery strategies. *Saudi Pharm. J.*, 29, 9, 999–1012, 2021, Available from: https://www.sciencedirect.com/science/article/pii/S1319016421001420.
39. Alotaibi, B.S., Pervaiz, F., Buabeid, M., Ashames, A., Fahelelbom, K.M., Siddique, S. *et al.*, Nanostructured lipid carriers based suppository for enhanced rectal absorption of ondansetron: *In vitro* and *in vivo* evaluations. *Arabian J. Chem.*, 14, 12, 103426, 2021, Available from: https://www.sciencedirect.com/science/article/pii/S187853522100441X.
40. Havaldar, V.D., Yadav, A.V., Dias, R.J., Mali, K.K., Kale, S.S., Pujari, P.P., Rectal Suppository of Mucoadhesive Microspheres of Alverine Citrate for İrritable Bowel Disease: *In vitro* Evaluation. *Res. J. Pharm. Technol.*, 11, 7, 3091–8, 2018, Available from: https://www.indianjournals.com/ijor.aspx?target=ijor:rjpt&volume=11&issue=7&article=067.
41. Siriratawan, W., Narapanich, K., Tungthanabordee, A., Ratanasawetasak, T., Chantasart, D., Suksiriworapong, J., Effect of polymeric micelles on the characteristics and stability of diazepam-loaded hydrophilic and lipophilic suppositories. *Mahidol. Univ. J. Pharm. Sci.*, 43, 1, 28–36, 2016
42. Seo, Y.G., Kim, D.W., Yeo, W.H., Ramasamy, T., Oh, Y.K., Park, Y.J. *et al.*, Docetaxel-Loaded Thermosensitive and Bioadhesive Nanomicelles as a Rectal Drug Delivery System for Enhanced Chemotherapeutic Effect.

Pharm. Res., 30, 7, 1860–70, 2013, Available from: https://doi.org/10.1007/s11095-013-1029-0.

43. Lozano-Pérez, A.A., Rodriguez-Nogales, A., Ortiz-Cullera, V., Algieri, F., Garrido-Mesa, J., Zorrilla, P. *et al.*, Silk fibroin nanoparticles constitute a vector for controlled release of resveratrol in an experimental model of inflammatory bowel disease in rats. *Int. J. Nanomed.*, 9, 4507–20, 2014, Available from: https://www.ncbi.nlm.nih.gov/pmc/articles/PMC4181735/.
44. Rodriguez-Nogales, A., Lozano-Pérez, A.A., Aznar-Cervantes, S.D., Algieri, F., Garrido-Mesa, J., Garrido-Mesa, N. *et al.*, Effect of aqueous and particulate silk fibroin in a rat model of experimental colitis. *Int. J. Pharm.*, 511, 1, 1–9, 2016, Available from: https://www.sciencedirect.com/science/article/pii/S037851731630610X.
45. Hosseini, S. and Epple, M., Suppositories with bioactive calcium phosphate nanoparticles for intestinal transfection and gene silencing. *Nano Select*, 2, 3, 561–72, 2021, Available from: https://onlinelibrary.wiley.com/doi/abs/10.1002/nano.202000150.
46. Hosseini, S., Wetzel, O., Kostka, K., Heggen, M., Loza, K., Epple, M., Pathways for Oral and Rectal Delivery of Gold Nanoparticles (1.7 nm) and Gold Nanoclusters into the Colon: Enteric-Coated Capsules and Suppositories. *Molecules*, 26, 16, 5069, 2021, Available from: https://www.mdpi.com/1420-3049/26/16/5069.
47. Wang, Y., Liu, Q., Yang, Y., Qiu, C., Jiao, A., Jin, Z., Fabrication of a double-network high internal phase emulsion gel stabilized by bacterial cellulose nanofibrils: Enhancement of heat stability and 3D printing. *Food Hydrocolloids*, 143, 108872, 2023, Available from: https://www.sciencedirect.com/science/article/pii/S0268005X23004186.
48. Jain, A., Kumar, P., Verma, A., Mohanta, B.C., Ashique, S., Pal, R. *et al.*, Emulgel: A Cutting Edge Approach for Topical Drug Delivery System. *Curr. Drug Res. Rev.*, 2024, Available from: https://europepmc.org/article/med/38303537.
49. 袁民政,侯海婷, Levoornidazole/ornidazole oil-in-water emulsion temperature-sensitive gel suppository and preparation method thereof. CN112089684A, 2020, Available from: https://patents.google.com/patent/CN112089684A/en.
50. Sakran, W., Abdel-Rashid, R.S., Saleh, F., Abdel-Monem, R., Ethosomal gel for rectal transmucosal delivery of domperidone: design of experiment, *in vitro*, and *in vivo* evaluation. *Drug Deliv.*, 29, 1, 1477–91, 2022, Available from: https://www.ncbi.nlm.nih.gov/pmc/articles/PMC9103489/.
51. Moawad, F.A., Ali, A.A., Salem, H.F., Nanotransfersomes-loaded thermosensitive *in situ* gel as a rectal delivery system of tizanidine HCl: preparation, *in vitro* and *in vivo* performance. *Drug Deliv.*, 24, 1, 252–60, 2017, Available from: https://www.tandfonline.com/doi/full/10.1080/10717544.2016.1245369.

52. Din, F.U., Mustapha, O., Kim, D.W., Rashid, R., Park, J.H., Choi, J.Y. *et al.*, Novel dual-reverse thermosensitive solid lipid nanoparticle-loaded hydrogel for rectal administration of flurbiprofen with improved bioavailability and reduced initial burst effect. *Eur. J. Pharm. Biopharm.*, 94, 64–72, 2015, Available from: https://www.sciencedirect.com/science/article/pii/S093964111500199X.
53. Din, F.U., Choi, J.Y., Kim, D.W., Mustapha, O., Kim, D.S., Thapa, R.K. *et al.*, Irinotecan-encapsulated double-reverse thermosensitive nanocarrier system for rectal administration. *Drug Deliv.*, 24, 1, 502–10, 2017, Available from: https://doi.org/10.1080/10717544.2016.1272651.
54. Garidel, P., Kaconis, Y., Heinbockel, L., Wulf, M., Gerber, S., Munk, A. *et al.*, Self-organisation, thermotropic and lyotropic properties of glycolipids related to their biological implications. *Open Biochem. J.*, 9, 49, 2015, Available from: https://www.ncbi.nlm.nih.gov/pmc/articles/PMC4598379/.
55. Dieterich, S., Sottmann, T., Giesselmann, F., Gelation of Lyotropic Liquid-Crystal Phases—The Interplay between Liquid Crystalline Order and Physical Gel Formation. *Langmuir*, 35, 51, 16793–802, 2019, Available from: https://pubs.acs.org/doi/10.1021/acs.langmuir.9b02621.
56. Steck, K., Dieterich, S., Stubenrauch, C., Giesselmann, F., Surfactant-based lyotropic liquid crystal gels–the interplay between anisotropic order and gel formation. *J. Mater. Chem. C*, 8, 16, 5335–48, 2020, Available from: https://pubs.rsc.org/en/content/articlehtml/2020/tc/d0tc00561d.
57. Khatoon, M., Shah, K.U., Din, F.U., Shah, S.U., Rehman, A.U., Dilawar, N. *et al.*, Proniosomes derived niosomes: recent advancements in drug delivery and targeting. *Drug Deliv.*, 24, 2, 56–69, 2017, Available from: https://www.ncbi.nlm.nih.gov/pmc/articles/PMC8812579/.
58. Kamel, R., Basha, M., Abd El-Alim, S.H., Development of a novel vesicular system using a binary mixture of sorbitan monostearate and polyethylene glycol fatty acid esters for rectal delivery of rutin. *J. Liposome Res.*, 23, 1, 28–36, 2013, Available from: http://www.tandfonline.com/doi/full/10.3109/08982104.2012.727422.
59. Uronnachi, E.M., Attama, A., Kenechukwu, F., Umeyor, C., Gugu, T., Nwakile, C. *et al.*, Solidified Reverse Micellar Solution- (SRMS-) Based Microparticles for Enhanced Oral Bioavailability and Systemic Antifungal Efficacy of Miconazole Nitrate in Immunocompromised Mice. *Biomed. Res. Int.*, 2022, 8930709, 2022, Available from: https://www.ncbi.nlm.nih.gov/pmc/articles/PMC8807030/.
60. Schneeweis, A. and Müller-Goymann, C.C., Controlled release of solid-reversed-micellar-solution (SRMS) suppositories containing metoclopramide-HCl. *Int. J. Pharm.*, 196, 2, 193–6, 2000, Available from: https://www.sciencedirect.com/science/article/pii/S0378517399004196.

6

Injectable Novel Drug Delivery Systems

Payam Khazaeli* **and Marzieh SajadiBami**

Department of Pharmaceutics, Faculty of Pharmacy, Kerman University of Medical Sciences, Kerman, Iran

Abstract

Literally, parenteral means non-oral route of administration but parenteral drug delivery is commonly referred to as drug administration by injection. In this administration route, the physiological body barriers such as the *stratum corneum* and mucous are circumvented. The three most frequently used injection methods are intravenous (IV), subcutaneous (SC), and intramuscular (IM). Parenteral drug delivery possesses major advantages such as high bioavailability and rapid onset of action. For specific therapeutic agents, such as peptides and proteins, administering through injection is frequently considered the most effective administration route. However, the possibility of infection, pain, and poor compliance of the patient make the parenteral route challenging. Today, with the advancement of technology and sciences, to increase effectiveness and reduce side effects and disadvantages, novel parenteral systems have been developed. In this chapter, after the initial introduction of injectable drug delivery systems, the anatomy and physiology of the skin have been discussed to understand its different layers as the application sites in the different injection methods. Then, the usual and special injection routes, classification of injectable formulations emphasizing novel delivery systems, formulation considerations, and parenteral pharmacokinetics have been reviewed. In the end, some examples of commercial products of novel parenteral drug delivery are provided.

Keywords: Parenteral drug delivery, injection, intravenous, bioavailability, *in situ* forming systems, infusion device

**Corresponding author*: pkhazaeli@kmu.ac.ir

Mehrdad Hamidi and Mahsa Sayed Tabatabaei (eds.) Novel Drug Delivery Systems: Fundamentals and Applications, (199–234) © 2025 Scrivener Publishing LLC

6.1 Introduction

Parenteral is a term of Greek origin consisting of the Greek words para (beside) and enteron (the intestine), and literally means *to avoid the intestine* [1]; however, this term is usually used for the injection administration routes. Since the anatomical and physiological barriers (mainly the skin, mucous layers and GI trace) are bypassed by the process of parenteral administration, there can be a significant risk to the patient. The most common and efficient route for the delivery of drugs with inadequate bioavailability and a narrow therapeutic window is parenteral administration [2–4]. In hospitals, injectables account for a considerable share of all administered drugs (about 40%) [5]. Today, the term parenteral mainly refers to dosage forms that deliver the drug into the body through injection or infusion [1, 5]. The bioavailability of drugs administered from transdermal and mucosal routes is hampered due to the presence of absorption barriers. For instance, the stratum corneum and variable situations of the gastrointestinal tract hinder the high bioavailability. The obstacles are bypassed by injection and infusion (i.e., continuous injection) [3, 6, 7]. The most common injection types are intravenous (IV), subcutaneous (SC), and intramuscular (IM). Infusions are given by intravenous route, such that the drug molecules can directly reach the target site. In fact, in the current healthcare paradigm, one of the key elements of inpatient therapy is the administration of parenteral products [8]. Also, the injection route provides excellent drug bioavailability. This route is advantageous for patients who cannot tolerate the oral administration and the rapid onset of action is required, like unconscious patients, inpatients, and bedridden patients who are completely dependent on parenteral nutrients such as water, electrolytes, and nutrients via parenteral routes [7, 8]. By injecting drugs intravenously, subcutaneously, or intramuscularly, the drug molecules reach the systemic circulation with rapid absorption or even without absorption step (in the case of IV injection). As a result of rapid absorption, the concentration of drugs in the systemic circulation may rapidly decline. Affected persons who receive daily or more than one weekly injection for years or even their entire lives are suffering terrible compliance with treatment [9]. Hence, though parenteral administration is beneficial, it suffers from several shortcomings, such as the possibility of infections, if the aseptic conditions during manufacturing are not fully observed. Additionally, if the injection devices are used inappropriately, there is a possibility of contracting life-threatening diseases (e.g., hepatitis C and AIDS). The invasive nature of injection is another drawback [10, 11]. In addition, parenteral preparations are more expensive than alternative preparations. Parenteral formulations must be prepared and administered using specialized personnel, equipment, tools, and techniques [5]. Therefore, this route of administration is saved for urgent situations. Furthermore, there are certainly many technical problems involved with even the conventional injection or infusion formulations, such as drug aqueous solubility, product stability (especially with biopharmaceuticals), and cross-compatibility between different coadministered medications. Today, novel parenteral systems such as implants, colloidal carriers such as and depot injections are evolving to play a vital role in different therapeutic protocols [12]. The possibility of reducing the dosing frequency is an opportunity for so-called long-acting products particularly for patients who use these products chronically and mainly as self-administered (such as insulin) [13].

In the context just described, there are formulations available with modified-release profiles with a variety of modalities of actions suitable for different categories of patients. These novel products may provide patients with time-controlled and/or targeted drug delivery [14]. Nowadays, some technological progress has been employed to develop sophisticated systems for special patient populations, e.g., portable injection pumps. Generally, it is desired to maintain the drug level in the therapeutic window for an extended time.

The two common ways of achieving a constant drug level in circulation are controlling the absorption rate (through modification of formulation) and controlling the elimination rate (through modification of the body situations). Obviously, it is more feasible to modify the dosage form.

Sustained drug levels within the safe therapeutic window provide the effective treatment that one needs. From another point of view, there are specific health risks associated with continued hospitalization and in-depth medical supervision, which can be minimized using the "future parenterals." Recent advances in pharmaceutical technology allow the development of a wide spectrum of specialized injectables tailor-made for a special mode of pharmacological action [15, 16]. Despite all the issues accompanied by the practical use of injections, parenteral preparations still occupy a place of choice for hospitalized cases. Since parenteral products are intended to be administered directly into the tissue or bloodstream, to protect patients from physical, chemical, and microbial contaminants, high quality/purity standards should be observed during the manufacturing of these dosage forms [17, 18].

6.2 Anatomy and Physiology

The skin is a protective barrier against external threatening factors. As shown in Figure 6.1, the skin has three main layers including epidermis, dermis, and hypodermis. The epidermis is mainly composed of keratinocytes, which produce keratin, a long, threadlike protein necessary for the skin's function. The dermis, the middle layer, is composed mainly of collagen, a fibrillary structural protein. Subcutaneous fat cells called lipocytes are small lobes found in the dermis and atop the panniculus, or

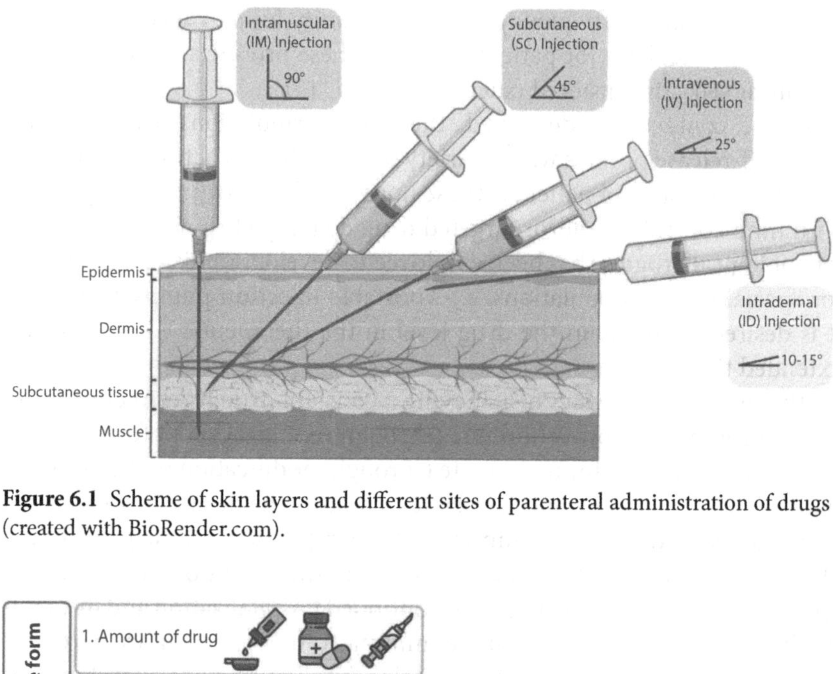

Figure 6.1 Scheme of skin layers and different sites of parenteral administration of drugs (created with BioRender.com).

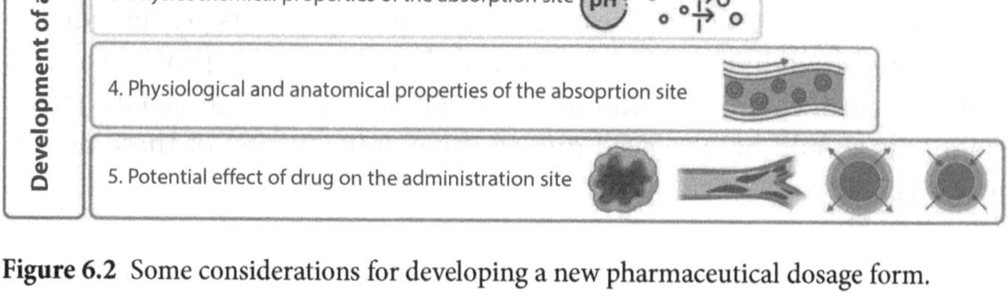

Figure 6.2 Some considerations for developing a new pharmaceutical dosage form.

subcutaneous tissue. Each layer differs considerably in thickness, depending on its anatomical site [19, 20].

Drug bioavailability is directly influenced by the route of administration, which, in turn, determines the onset and duration of action. Some considerations for developing a new dosage form are presented in Figure 6.2. For producing and evaluating biopharmaceuticals, these considerations are crucial; the design of dosage forms must take into account the physiology and pathology of the host body, which affect the absorption of drugs [5, 21, 22].

6.3 Parenteral Administration Routes

6.3.1 Intravenous

The intravenous (IV) route refers to the direct administration of drug formulation into the vein with a needle. It is the best route to administer drugs accurately and rapidly since the drug enters and directly reaches systemic circulation without the absorption lag-time. Therefore, it creates its therapeutic effect faster than all other administration routes. So, this pathway offers complete (100%) bioavailability of the active ingredient [21, 23, 24]. Some general considerations regarding the IV administration route are summarized in Table 6.1.

There are three basic IV modes for administering medications:

6.3.1.1 Fast IV Injection

Fast IV injection refers to drug administration directly into the blood circulation through a vein, often known as an "IV bolus," which is only suitable for small volumes (< 10 ml). Following IV bolus, the drug dilution occurs in the venous circulation. Then, the drug reaches the heart and subsequently is pumped throughout the body. Depending on the blood flow in the site of action, after 20 to 40 seconds, the therapeutic effect begins to appear, making it an admired method for emergencies (e.g., pain relief) [25, 26].

6.3.1.2 Slow IV Infusion

In many hospitalized occasions, medicines are injected into a vein over a long time, i.e., regularly more than 4 hours. The method is clinically called IV infusion. The infusion rate determines the amount of drug delivered

Table 6.1 IV administration route considerations.

IV route considerations	Description
Type of formulation	Only suitable for aqueous solutions, otherwise there is the risk of embolism and thrombophlebitis.
pH of formulation	The majority of drugs are poorly water soluble with the solubility highly dependent on pH. Accordingly, in many instances, there is no choice to use extreme pH ranges. For example, injectable phenytoin is prepared at a pH near 10–12. Hence, attention should be paid during IV administration to minimize extravasation of the solution and subsequent tissue injury, which can range from minor irritation to necrosis and ulceration.
Rate of injection	Even in aqueous solutions, precipitation of a poorly water-soluble drug might occur during injection, especially when cosolvents or surfactants are used to solubilize the drug.
Osmolarity of formulation	The mixing of drug formulation with circulation blood is not practically instantaneous, which sometimes may cause problems by local hyper- or hypo-osmolarity issues. Proper dilutions before use or using large veins or even a central IV route to obtain immediate access to the heart and, consequently, diluting the solution with a large volume of blood are the logical solutions.

within a definite period. The gravity or infusion pump drives the drug solution to flow through a plastic catheter into the vein. The most common veins for IV infusion are forearms and wrists veins, although other superficial veins can also be used [26].

6.3.1.3 Central IV Route

In some situations, it may be required to use a central IV route, which involves placing a catheter near the exit of the superior vena cava (right atrium). This is necessary when there is a special type of emergency, peripheral routes are unavailable, extended treatments are required, or hypertonic solutions must be supplied [27, 28].

6.3.2 Intramuscular

In intramuscular (IM) injection, the drug is injected into specific muscles. The most commonly used muscles for IM injection are the gluteus (buttocks) and deltoid (shoulder). This type of injection aims at using the interstitial fluid of the peripheral muscles as the site of drug administration, where drug absorption takes place via the endothelial walls of the capillaries into the systemic circulation.

6.3.2.1 Physiological Variables

Following IM administration, the drug is absorbed through diffusion from the muscular cells as well as their surrounding fluid to the circulation. The blood supply to different muscular tissues varies. For example, the deltoid has a greater blood flow than the gluteus muscle. The amount of blood flowing through the muscles rises when exercising or in case of fever. Low blood pressure, on the other hand, is associated with capillary constriction and insufficient muscle flow, which compromises drug absorption. The absorption process is, also, delayed by the presence of fatty tissue. It is, therefore, normal to see irregular absorption profiles following an IM injection in different patients. In specific populations (e.g., neonates and the elderly) and situations (e.g., pregnancy) the pharmacokinetic profile of IM-administered drugs may undergo significant alteration [29, 30].

6.3.2.2 Formulation

IM injections can be prepared intentionally to release the drug slowly or quickly at the site of administration via modifying the formulation. Since the free drug molecules are rapidly absorbed from the injection site, IM instant-release preparations are based on aqueous solutions with no limitation to drug dissolution. Lyophilized solids, for example, are dissolved in an aqueous medium just before administration. The drug effect takes normally around 30 minutes to start, the length is determined by the drug's biological half-life. Ionizable and weakly soluble compounds should always be prepared in buffer solutions with a pH near to physiological, since they may have precipitated at the site of injection otherwise. The solution of phenytoin, for instance, is commonly given intravenously not only due to the formulation pH (about 12) is seriously annoying to muscles, but also considering the pH value of the muscle tissue, which is less than the optimal pH, and the lack of dilution may lead the drug to crystallize, followed by a gradual redissolution, resulting in an erratic and subtherapeutic pharmacokinetic profile [31, 32].

For drugs with limited water solubility, other solvents such as propylene glycol or vegetable oils are good candidates as solvents or cosolvents. However, this raises a new concern that the drug may have a slow drug diffusion or partitioning into the bloodstream due to miscibility or viscosity issues. In this context, the drug must be partitioned between the formulation (or delivery system) and the physiologic extracellular environment before absorption which, in turn, leads to the prolonged release profiles observed with these carriers. Long-acting IM formulations that can be administered with a frequency of hours, days, or even weeks have been developed based on the aforementioned principle. These preparations are either oily solutions (e.g., estradiol and testosterone) or aqueous suspensions (e.g., penicillin G procaine and methylprednisolone). Furthermore, to obtain suitable pharmacokinetic profiles from each active ingredient, several colloidal dispersions containing new active ingredients (e.g., peptides and proteins) have been prepared using liposomes, nano- or microemulsions, and other nanovehicles as drug delivery carriers for IM administration [8, 10, 33]. Hypertonic products are generally contraindicated through the IM route and, therefore, must be administered IV with special precautions [27, 34].

6.3.3 Subcutaneous

In subcutaneous (SC) injection, the drug is injected at a 45- to 90-degree angle into the subcutaneous tissue, right beneath the dermis, into the deepest layers of the skin, hence it is also called "hypodermic" administration. SC injection is commonly done on the outside of the arm or thigh, or the front of the abdomen, and allows for smaller administration amounts than the intramuscular route [35, 36].

The injected fluids in the SC administration route must be neutral and isotonic; otherwise, they would irritate the skin and cause discomfort and necrosis. Oily solutions should not be injected via the SC route, unlike the IM route, because they can become clogged and induce a sterile abscess. The SC depot and continuous infusion, by which the medication is administered slowly and blood levels remain stable for a long time, are two of the most fascinating applications of this administration route [36].

Contraceptive implants (consisting of a thin bar of nonbiodegradable polymers containing hormones (e.g., etonogestrel and levonorgestrel) with modified drug release) are among the most well-known SC implants. These devices are surgically implanted in the inner part of the arm (under local anesthesia) and deliver the medicine for a set amount of time (usually 3–5 years), after which they must be removed [37].

Insulin pumps and morphine infusers are among the well-known examples of continuous SC infusion systems. These devices allow small amounts of solution to be injected into the subcutaneous tissue at a very slow, but programmed, rate. Insulin pumps come in a variety of shapes and sizes, but in general, they are mechanical, tiny, and portable electrically controlled devices that contain an insulin-containing reservoir connected to a cannula that is inserted into the SC tissue, and mostly in the abdomen. These devices release insulin over three or four days, decreasing the unpredictability associated with injections. Programmable devices are advance forms that allow for a basal infusion that can be personalized to the patient's lifestyle, and an insulin bolus on required to cover meal ingestion [37, 38].

Finally, elastomeric infusion systems, often known as infusers, are small devices that consist of a cylindrical plastic container with a balloon or elastomeric reservoir inside. The infused formulation is placed into this balloon, leading to swelling. The swollen balloon maintains a steady pressure and expels the contents through a tube connected to a catheter already implanted in the patient's SC tissue (at a usual speed of 0.5–2 mL/h). These systems are mostly used to provide analgesic medicines to chemotherapy patients who are undergoing palliative care or pain management, both ambulatory and during hospital stays [39, 40].

6.3.4 Specialized Parenteral Routes

There are cases where it is necessary to provide a medicine directly to a specific area, region, or organ to quickly obtain higher therapeutic concentrations at the site of action.

Intraspinal: Administration on or over the dura mater is referred to as epidural. To reach the neural roots, the medicine must pass through fat and veins, which delays the onset of the action. It allows for long-term catheter collation [41].

Intrathecal: This type of injection involves the direct injection into the subarachnoid space's cerebrospinal fluid (between the *pia mater* and arachnoid membranes) that allows the drug to directly impact the nervous system's roots. Intrathecal injection does not need any use of a permanent catheter. Local anesthesia (for example, by lidocaine or bupivacaine) and pain control (for example, an opioid analgesic) are the most common uses for this method [42, 43].

Intracerebroventricular: This method involves direct administration to the cerebral ventricles. When CNS-active medicines are unable to pass the blood–brain barrier and it is necessary to have high localized concentrations, the intracerebroventricular route is an alternative to the IV route (for example, antibiotics or chemotherapy of gliomas) [44].

Intra-arterial: Direct injection into an artery is usually used for effects on organs or tissues on a local level. Antineoplastic agents are sometimes injected into the tumor organ artery, i.e., in the vicinity of the tumor which is intended to decrease the systemic adverse effects. It can also be used to provide vasodilators or contrast media for arteriography in the case of arterial embolisms [45, 46].

Intracardiac: Due to the serious injuries that the needle might cause, direct injection into the heart is only utilized as an emergency route after a cardiac arrest (for example, injection of adrenaline into the heart chambers) [47].

Intradermal: The drug is injected into the dermis layer of the skin, usually in the ventral zone of the forearm. Intradermal dosing results in essentially no systemic absorption of the medication due to the extremely low blood supply of the dermis. It is commonly employed in the administration of vaccines and local anesthetics as well as hypersensitivity diagnostic tests [48, 49].

Intra-articular: Direct application of a substance to a joint, usually for local effects is a typical example of the intra-articular injection. Corticosteroids with anti-inflammatory properties, for example, are used to treat arthritis using this injection route [50].

Intralymphatic: Injection into a lymph node or a lymphatic vessel is another special way for injection. It is used to apply stem cells in the management of autoimmune diseases, antitumor treatment, and diagnostics (contrast agent injection) [51].

Intra-osseous: Injection into a long bone medulla allows for direct drug delivery within the medulla. Because the drug is promptly delivered to the human body through the vascular bed of these bones, it is also employed to induce systemic effects [52].

6.4 Classification

There are different classes of injectable drug delivery systems including conventional and novel formulations. A summary of these classes is shown in Figure 6.3.

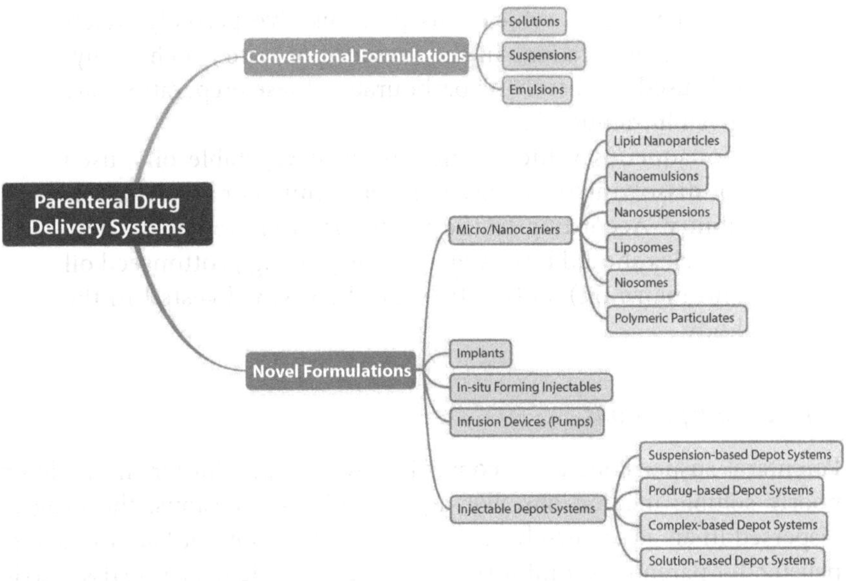

Figure 6.3 The classification of injectable drug delivery systems.

6.4.1 Conventional Parenteral Formulations

Solutions, suspensions, and emulsions are the most common parenteral formulations that have been used for a long time.

6.4.1.1 Solutions

An isotonic aqueous solution, with a pH near to that of the human blood and tissues (≈ 7.4), is the simplest and most practical way to present an injectable substance. Large-volume parenterals (LVPs), small-volume parenterals (SVPs), and irrigation solutions are among the typical examples of parenteral solutions. Aqueous solutions used in larger amounts than those generally given by intravenous injection are known as infusion fluids. Infusions are liquid preparations that are used for basic nourishment, electrolyte balance restoration, fluid replacement, and some other purposes. Vehicles and additional compounds are regularly included in the composition of solutions. There are three types of vehicles used for the preparation of injectable solutions [53, 54]:

- Aqueous vehicles, used in the preparation of injections made of water-soluble substances such as sodium chloride, Ringer's, and dextrose infusions.

- Water-miscible vehicles, used to dissolve partially water-soluble medications in the presence of water, such as glycols used for dissolving barbiturates. These preparations are given intramuscularly.
- Nonaqueous vehicles, such as fixed vegetable oils, used for dissolving the drugs with very limited or no water solubility. According to the regulatory requirements, these vehicles should be of vegetable origin (e.g., cottonseed oil and maize oil) so that they can be easily digested in the body.

6.4.1.2 *Suspensions*

Parenteral suspensions are a convenient way to administer insoluble or poorly soluble medications. For aqueous/oily suspensions, the drug is dispersed in an aqueous/oily vehicle. The most important factor of a suspension for parenteral administration is that it should not cause tissue irritation when injected. Because dispersed drugs have a high surface area, they have an acceptable aqueous solubility, which helps to guarantee the drug's availability for absorption. When compared to the injectable solutions, the parenteral suspensions show a longer drug release time at the injection site [55].

Suspensions are preferable over solutions because the medicine is present in solid form, which increases its resistance to hydrolysis and oxidation. Despite these advantages, suspensions have several drawbacks, including the difficulty of formulation, stabilization, and sterilization, as well as the possibility of dose nonuniformity at the time of administration. So, the development and manufacturing of suspension formulations is very challenging. Injectable suspensions may be prepared as ready-to-use or powders for reconstitution. Sterilization of suspensions through filtration is practically impossible, therefore they are either produced under aseptic situations or undergo terminal sterilizations (through ionizing radiation or heat) [56, 57].

There is a range of manufacturing complexity across these formulation approaches, depending on the number of parameters that control the release of the drug from the formulation. For example, particulate suspensions of insoluble drugs are largely dependent on particle size and dissolution for release control, and a fairly limited number of controlling parameters are important, making them somewhat easier to develop, scale up, and manufacture [58, 59].

6.4.1.3 *Emulsions*

When two immiscible liquids are combined to form a two-phase system, an emulsion is formed. One phase consists of droplets with a diameter larger than or equal to the colloidal particles (dispersed phase) and the other phase is spread throughout the whole system (continuous phase). The introduction of total parenteral nutrition (TPN) into the bloodstream necessitates the use of emulsions. Some patients, such as those in a coma, who cannot get food with a meal, need to be fed using TPN. Patients in surgical recovery also need this type of nutrition [17, 60]. Emulsions are generally divided into two categories: water-in-oil (w/o) and oil-in-water (o/w). Antioxidants, preservatives, pharmaceutical emulsifiers, and pharmaceutical oils (as described in nonaqueous vehicles in solutions) are ingredients commonly used in emulsion formulations. The type of drug and also the type of emulsion, determine the emulsion preparation steps and the internal phase best suited for the emulsion. To stabilize droplets formed during emulsification, pharmaceutical emulsifiers are added. In most cases, the drug/internal phase mixture is poured into the emulsifier/external phase mixture with homogenization at high pressure to break the drops of the internal phase. To stabilize the internal phase droplets, the entire surface is lined by the emulsifier. Thermodynamic instability is the main problem with emulsions which is caused by the increased surface free energy of the system which, in turn, depends on the increase in system surface area during emulsification, increasing interfacial tension, and the total surface area [61]. Parenteral emulsions are delivered in subcutaneous, intravenous, and intramuscular dosage forms. Liposin®, Lipofundin®, and Intralipid® are oily emulsions available commercially [62].

6.4.2 Novel Parenteral Formulations

Conventional parenteral formulations face different challenges, mainly including manufacturability, pharmacokinetic profile, stability (especially in biopharmaceuticals), and drug solubility both *in vitro* and *in vivo*, even though parenteral drugs have been used in clinical settings, e.g., in emergencies, for a long time. At present, parenteral drug delivery has made remarkable technological progress, leading to a variety of achievements in all aspects discussed. New parenteral formulations including parenteral formulations consisting of colloidal drug carriers (e.g., polymeric nanoparticles and liposomes), are among the cutting-edge research and development activities ongoing in many pharmaceutical companies.

6.4.2.1 Micro- and Nanocarriers

Microparticles and nanoparticles have become increasingly popular in pharma industries. Improved physicochemical properties, typically drug dissolution rate and solubility, drug stability, and release controllability, along with the enhanced bioavailability of the active ingredient have made these "tiny" carriers very attractive tools in the novel parenteral drug delivery field [63]. Table 6.2 presents some recent research about micro- and nanoparticles for parenteral drug delivery.

Table 6.2 Some micro- and nanocarriers investigated in research for parenteral drug delivery.

Carrier type	Drug	Major material(s)	Main Results	Ref.
Microparticles	Glatiramer acetate	Chitosan and PLGA	• Possibility of longer dose intervals • Increased safety due to fewer side effects • The removal of site injection reactions for GA-hydrogel-microparticles • Equal therapeutic efficacy, according to the animal and pathological tests	[64]
	Ivermectin	Hyaluronic acid and PLGA	• Getting repaired the insertion site's barrier property in 3 hours, in the rat *in vivo* study • Being proposed as a potential technique for controlled transdermal medication administration when compared to conventional hypodermic injection or the implantation of controlled-release devices	[65]

(*Continued*)

Table 6.2 Some micro- and nanocarriers investigated in research for parenteral drug delivery. (*Continued*)

Carrier type	Drug	Major material(s)	Main Results	Ref.
	Buprenorphine	PLGA	• As a lyophilized product, the potential to be kept for at least six months and used for at least 24 hours following reconstitution	[66]
	Ibuprofen	Castor oil and silica hybrid	• Gradual drug release over 15 days for subcutaneous conditions • excellent cellular viability, based on the cytocompatibility investigations on NIH-3T3 and Caco-2 cells • surface-limited hydrolysis of the particles, according to the biodegradability assays in simulated digestion medium suggested	[67]
	Tenoxicam	Sesame oil and poly-DL-lactide	• significantly improved oxidative stress biomarkers and reduced the inflammatory signs, according to the *in vivo* study revealed that microparticles	[68]
Nanoparticles	Paliperidone	Chitosan, Tween 80 , Pluronic F127, PVA and PCL	• High importance of the type of stabilizer and coating for the preparation of safe and effective parenteral nanoparticles	[69]

(*Continued*)

Table 6.2 Some micro- and nanocarriers investigated in research for parenteral drug delivery. (*Continued*)

Carrier type	Drug	Major material(s)	Main Results	Ref.
	α-mangostin	Silk fibroin	• A good control over particle charge, crystallinity, solubility, and dissolution • Reduction of drug hematotoxicity by 10-fold • High stability of nanoparticles after 6 months of storage	[70]
	Bedaquiline and Celecoxib	Glyceryl monostearate (GMS), Sodium deoxycholate, and Tween® 80	• Increased tmax, $t_{1/2}$, and mean residence time (MRT) after intramuscular and subcutaneous administration	[71]
	Melphalan	Tristearin, soya lecithin, and DSPE-m-PEG-2000	• Longer circulation, better residence time, and a lower elimination rate, when compared to pure melphalan solution	[72]

Lipid Nanoparticles

Parenteral lipid-based drug delivery systems conventionally include IM or SC injection of emulsions, suspensions, and oily solutions. These systems have a very simple preparation strategy of lipophilic drug incorporation in the lipid phase, with the subsequent release principally involving the separation of the oily vehicle from the drug at the injection site, where the environment is aqueous. It is, however, very difficult to introduce hydrophilic drugs into the lipid vehicles. The need for modified parenteral release of both hydrophilic and lipophilic drugs has led to the development of advanced lipid-based nano- and microcarriers. Intravenously injected nano-scale lipid particulate systems can be now added to the conventional IM and SC injections [73]. Lipid-drug conjugates (LDCs), nanostructured lipid carriers (NLCs), and solid lipid nanoparticles (SLNs) are nanocarriers prepared using different lipid materials. The typical formulations contain drugs, lipids, and stabilizers, mainly surfactants, as the essential ingredients. SLNs are nanoparticles composed of solid lipids (i.e., lipids that are solid at both body and room temperature) stabilized by a surfactant. The key features of SLNs that make them suitable for parenteral drug formulation are site-specific targeting, controlled drug release, protection of labile drugs against degradation, and good physical stability. Despite their benefits, SLNs have two main problems of drug expelling especially during storage as a result of polymorphic transitions, and the insufficient loading capacity [74].

Nanoemulsions

Nanoemulsions are w/o (water-in-oil) or o/w (oil-in-water) systems, also known as submicron emulsions or mini-emulsions [6]. Major problems associated with the stability of regular emulsions (such as sedimentation, coalescence, flocculation, and creaming) are not evident here in nanoemulsions. The energy involved in the preparation of nanoemulsion can be the basis for classifying their production methods. PIT (phase inversion temperature), solvent diffusion, and spontaneous emulsification are considered low-energy while ultra-sonification, microfluidization, and high-pressure homogenization are high-energy methods. PEG, vegetable oil, peanut oil, linseed oil, wheat germ oil, soybean oil, coconut oil, castor oil, and olive oil are some of the nanoemulsion system components. Other components include additives such as pH adjustment agents, preservatives, glycerol and sorbitol (as tonicity modifier), and tocopherol (as antioxidant). In nanoemulsion systems, oleylamine, steraylamine, polysorbates, poloxamers, and natural compounds like lecithins (from animal and plant sources) can be also used as emulsifiers/stabilizers. Pharmacokinetic and skin permeation studies have shown that the bioavailability of poorly soluble drugs may be improved using nanoemulsions [75, 76].

Nanosuspensions

Colloidal solid APIs at submicron dimensions stabilized by a polymer and/or surfactant and dispersed in an aqueous phase are called nanosuspensions. A nanosuspension may be administered parenterally. With a 200- to 600-nm average particle size, the particle size distribution of nanosuspension drugs is commonly not more than 1 micron. Nanosuspensions increase the bioavailability of poorly water-soluble drugs by increasing both the drug dissolution rate and saturation solubility. What makes nanosuspensions more stable over time is that, unlike coarse suspensions, they are less susceptible to Ostwald ripening. Many cost-effective and technically easy methods have been reported for the preparation of nanosuspensions, especially for hydrophobic formulations. The methods currently used for the preparation of nanosuspensions are media milling, high-pressure homogenization, and solvent diffusion [77].

Liposomes

Liposomes, discovered in the mid-1960s, are vesicles made of double layers of phospholipids. Amphiphilic, as well as fat-soluble drugs, are incorporated into the phospholipid double-layered membrane while water-soluble drugs are located in the inner aqueous core. The spontaneous accumulation of phospholipid molecules in an aqueous medium leads to liposome formation. Liposomes can be given intravenously; however, they may be uptaken mainly by the phagocytes of the liver and spleen mononuclear phagocytic systems (MPS). Liposomes are classified into giant liposomes (1000–10,000nm), large unilamellar vesicles (LUV, ≥ 60 nm), multilamellar large vesicles (MLV, 100–6,000 nm), and small unilamellar vesicles (SUV, 20–50 nm). Different methods of liposome preparation are used depending on the type of vesicle. The two liposome-based anticancer drugs containing doxorubicin, i.e., Myocet®, and Doxil®, are the result of research and the commercialization of liposomal drug delivery systems. Another example is the antifungal AmBisome®, an amphotericin B liposomal formulation used in systemic antifungal therapy. Liposomes have been attempted successfully in the treatment of cancer, infectious diseases, immune disorders, and eye diseases in addition to vaccine adjuvants and dermatology. Hormones, apoproteins, sugar residues, or antibodies have been used as ligands for liposome targeting [78–80].

Niosomes

Vesicular systems composed of nonionic surfactants are called Niosomes. These vesicles are obtained by hydrating synthetic nonionic surfactants in the presence or absence of cholesterol or other lipids. Niosomes are prepared using nonionic surfactants such as sorbitan esters (Span series) and polysorbates (Tween series). Niosomes are intrinsically stable and osmotically active. More interestingly, the drugs loaded in niosomes make niosomes even more stable. They can be administered parenterally. The surfactants used in niosomes' preparation should be nonimmunogenic, biocompatible, and biodegradable. Delaying circulatory clearance, limiting their effects on the target cells, and protecting the drug from destabilizing biological environments are the main reasons for the parenteral administration of niosomes. Niosomes and liposomes are produced using the same methods including the bubble method, reverse phase evaporation, microfluidization, ether injection, sonication, and film hydration.

Parenteral administration of several drugs has been attempted using Niosomes [81, 82].

Polymeric Particulates

Biodegradable polymer-based nanoparticles are among the promising novel parenteral delivery systems. Most sustained-release parenteral products are injectable polymeric microspheres that are delivered through intramuscular or subcutaneous injection for topical or systemic treatments. The method generally uses biocompatible and biodegradable polymers. Sophisticated technical equipment is usually required for the complicated nano- and micro-particle large-scale production processes. As organic solvents must be removed, the use of these solvents can further complicate the process and increase the cost [83].

6.4.2.2 Parenteral Implants

Implants are generally a polymeric reservoir or matrix-based solid material loaded by the drug of interest which are put inside the body, mostly subcutaneously, by a minor surgery and are intended to release their drug payload over a relatively long period of time. An effective formulation for maintaining concentrations over an extended length of time is critical for an implant. The foundational components utilized in implant preparation, therefore, must be biocompatible and safe. Polymers, both biodegradable and nonbiodegradable, are frequently employed as base materials for the fabrication of implants. The nonbiodegradable polymer has to be surgically removed after drug delivery, which makes the patient feel heavy and uncomfortable. On the other hand, biodegradable polymers have the benefit that no surgery is required because it is "absorbed" by the host body after the medication depletion. In particular, poly(lactic acid) and poly(lactide-co-glycolide) copolymer (so-called PLA and PLGA, respectively) are widely used as biocompatible as well as biodegradable polymers for the development of implants [37]. Drug release can be extended by the order of months or even years using implants made of these polymers/co-polymers. Another type of biodegradable polymers, namely polyanhydrides, demonstrate drug release for up to one year. As a biodegradable polymer, polyvinyl acetate is also used in this context. The skin, ear cochlea, heart, eye, and bone are typical examples of the various drug delivery targets that may be implanted [84, 85]. Osmotic drug delivery systems also can be used for implantation. In these systems, a semipermeable membrane surrounds a core containing osmogens and the active drug. Through the delivery ports, the drug solution is pushed out as the core absorbs water. Despite the pH and hydrodynamics of the dissolution medium, osmotic pumps release drugs at a constant rate. Semipermeable membrane (e.g., cellulose acetate), hydrophilic and hydrophobic polymers (e.g., carboxymethylcellulose and ethyl cellulose), wicking agents (e.g., colloidal silicon dioxide), solubilizing agents (e.g., Tween 20, 60, and 80), osmogens (e.g., potassium chloride), coating solvents (e.g., isopropyl alcohol and water), plasticizers (e.g., polyethylene glycol -600) and pore-forming agents (e.g., triethyl citrate) are used in the formulation of osmotic implantable drug delivery systems [86].

6.4.2.3 *In Situ Forming Systems*

One attractive alternative to drug delivery implants is the biodegradable injectable *in situ* depot-forming systems. The system consists of a solvent/cosolvent in which biodegradable carriers are dissolved or dispersed; basically, the drug is dispersed or dissolved in the system while it is in a liquid phase. The liquid dosage unit is administered through IM or SC injection followed by the gel transformation of the liquid at the injection site. There are many benefits to the controlled release of bioactive macromolecules through *in situ* forming systems, such as less intense production conditions for sensitive drugs, simpler formulation, lack of surgery, and ease of administration. Other benefits include lower dose and frequency of dosing, improved patient compliance, and, most importantly, "infusion-like" plasma level-time profiles. These systems are classified according to the depot formation mechanism into *in situ* hardening organogels, thermally induced gels, *in situ* cross-linked polymeric systems, and *in situ* polymeric precipitation thermoplastic pastes [87].

6.4.2.4 *Infusion Devices (Pumps)*

Several tools may be used for portable injections. These frameworks primarily consist of a speed-controlling module (e.g., a pump) and a drug reservoir. The primary gravitational pumps are based on the principle of gravity. An asynchronous motor is used in syringe pumps to propel the plunger up to one meter into the body to deliver medications. A particular volumetric peristaltic pump continuously transports drugs while driving a drug solution under external pressure [103, 104].

6.4.2.5 *Injectable Depot Systems*

Parenteral depot systems are long-lasting slow-release formulations that are designed to liberate their cargo over a certain period and allow less frequent administration [33, 88]. The rationale for designing depot systems is to overcome the disadvantages of conventional injections such as drug–plasma fluctuations and consequent adverse effects, low patient compliance, and lack of drug localization in the target site.

As previously mentioned, parental drug delivery is able to conquer several drawbacks associated with the oral administration of biomolecules such as degradation in the harsh environment of the gastrointestinal tract, inadequate absorption, and hepatic first-pass metabolism. For the variety of biomolecules (e.g. proteins), the injection is oftentimes the only feasible route of the administration for delivery to achieve appropriate therapeutic concentration. However, the invasive nature of injection makes it less-preferred by the patients. Depot injection is an effective approach to lower the dosing frequency and the required dose to maintain the plasma concentration of the drug at the therapeutic range [89, 90]. Depot systems are particularly beneficial for drugs with narrow therapeutic windows.

In some cases, particularly for the molecules with very short half-lives, a single injected dose may cause a fast increase in plasma drug concentration above the therapeutic range followed by a fast decline below the range thus creating a very short exposure window for the drug to exert its effect(s). The gold standard to precisely control the drug concentration in the plasma is continuous infusion, normally IV. However, continuous IV infusion is often not practical, especially for outpatients, due to the heightened risk of drug overdosing or underdosing as well as the need for an IV line connected to the patient within the daily life times as well as close medical supervision during drug administration courses. An alternative way, in this context, is continuous subcutaneous (SC) infusion, which is in use by about 10% of Type 1 diabetic patients to control insulin delivery. The first commercial system of this type was developed by MiniMed by Medtronic (Dublin, Ireland) and approved in the United States in 1983. Several companies have, then, developed and obtained approval for improved SC infusion pump devices [91]. Despite the advantages, pumps must be worn continuously by the patient causing discomfort and, therefore have not seen much market penetration except in cases where exquisite control is critical (e.g., Type 1 diabetes) or in postoperative settings.

Depot delivery systems offer the advantage of a more convenient and facile patient therapy which can be [92]:

- ✓ administered at home using patient-friendly devices or in-office procedures by a health care practitioner;
- ✓ designed to prevent/decrease the fluctuation in plasma concentration;

- ✓ developed to maintain the drug level in the therapeutic window by providing a continuous release;
- ✓ designed to minimize unwanted systemic effects.

The development of parental formulations with extended-release is practically challenging. So, several requirements should be carefully considered to designing a depot system. Generally, the drug molecules to be loaded in the depot system should be sufficiently potent in such a way that the total loads required to be released in a certain period of time are included in the formulation. Considering the intended duration of drug release, the volume of injection, and the drug loading capacity, it is possible to estimate the required potency. Disadvantageously, sometimes, extended-release formulation may provide reduced bioavailability compared to a solution injection, thus requiring higher doses.

An equally important factor is the ability to accurately predict the minimum efficacious continuous exposure that should be delivered from the depot formulation to the plasma. This concentration is pharmacology-dependent, but the ability to reach a steady-state exposure also depends on the drug's plasma half-life; the longer the half-life, the less drug is required to be continuously administered to achieve a given level. Reference to existing SC PK/PD data is preferable, and can often be used to predict the required input rates for a depot formulation since most modern extended-release products are being developed as SC depot formulations for self-administration by the patient. The pharmacokinetic data from this route of administration will support providing the absorption rate and extent into plasma, as well as possible local effects of the drug in tissues [93].

Another factor to be considered is the stability of drugs during the manufacturing process, storage shelf-life, and after administration (in the face of body temperature, enzymes, and so on). Moreover, the drug should be sufficiently soluble in the aqueous media as well as the organic solvent used in the manufacturing process [94]. Other criteria to be considered are PK/PD profile, absorption profile of the drug, and lack of irritation. Drug classes that require extended dosing periods (such as hormones, insulin, corticosteroids, and contraceptives) are appropriate candidates for depot formulations [95–98].

Suspension-Based Depot Systems

Dispersion of drugs in an aqueous or oily vehicle provides a suspension-based depot which is among the first long-acting injectable formulations. The majority of these systems rely on the *in vivo* dissolution properties of the suspension particles which, in turn, governs the drug release profile from the depot. An alternative strategy to suspension depot is to adsorb the therapeutic agent on a solid adsorbent which is common in vaccine formulations.

Prodrug-Based and Complex-Based Depot Systems

When the water solubility of a drug is low, suspension is considered a suitable formulation. However, if the aqueous solubility of a drug is too high, the solubility should be decreased. This could be realized by the prodrug strategy (e.g., esterification). An alternative strategy is the formation of complexes or salts with low water solubility.

Solutions (Oily Vehicles) as Depot Systems

In the cases where a drug has poor aqueous solubility/stability or the potential of irritation, being formulated as oily vehicles is an appropriate strategy. Compared to the suspensions, oily vehicles have a simple manufacturing process, low physical stability issues, and the capability of being sterilized through filtration. Drug candidates should have adequate stability and solubility in an oil vehicle. Ideal oils for depot formulation must be inert, low viscous, physically and chemically stable, nonirritating, and nonimmunogenic. A variety of fixed oils (e.g. corn oil, olive oil, almond oil, and castor oil) are commonly well-tolerated and hence, considered acceptable options as vehicles. However, they may cause allergic reactions in some patients. After injection, many oily vehicles are slowly dissolved in the body fluids or transported from the surface of the depot and consequently eliminated from the site of injection. The spread of oily depot following IM administration is not as extensive as the aqueous formulation, rather, it gets a pod-like shape. The surface area of the depot impacts the rate of drug release. Drug absorption from oily solutions follows first-order kinetics where the vehicle has a slower absorption than the drug. Here, the rate-limiting step of drug absorption from the depot is the diffusion of the drug through the aqueous layer surrounding the depot. The oil/water partition coefficient of the drug and the injection volume controls the rate constant. Conversely, drug absorption from oily suspensions follows zero-order kinetics because the suspension particles maintain the saturation solubility until they are completely dissolved [108–111].

IM injection is the most common route for the administration of depot formulations (and this may largely be due to the volume and dose requirements). However, it is possible to administer depot systems through SC, intradermal, and intra-articular routes [99, 100].

6.5 Formulation Considerations

In injectable suspensions, the drug is dispersed in an oily or aqueous vehicle. Coarse suspensions are the most common form of injectable suspensions with a particle size lower than 50 μm. The particle size significantly impacts the syringeability, physical stability, and release rate. Jet milling, spray drying, supercritical fluid, and media milling (for preparation of nanocrystal dispersions), are among the most commonly used techniques for particle size reduction [101]. In the submicron range, the homogeneity, syringeability, injectability (due to reduced viscosity), and physical stability (because of decreased settling rate) are improved [102]. Particle bridging may lead to clogging of the needle. To prevent this, it is recommended that the size of the largest particles is not more than $^{1}/_{4}$ to $^{1}/_{3}$ of the needle's inner diameter [33, 103]. The viscosity of the formulation should be adjusted to strike a balance between syringeability and the physical stability of the suspension. Thixotropy and shear thinning behaviors can be employed to achieve both goals. The effect of particle size of penicillin G procaine suspension on its release rate, stability, and syringeabiltiy has been extensively studied [104]. It has been shown that when the particle size of penicillin G procaine particles in aqueous IM suspensions was decreased from 60 to 100 mesh micronized size, its release rate got faster [105]. Generally, with a reduction in particle size, the absorption rate increases. When the particle size decreases to ~ 2 to 3μm or smaller, this effect is more obvious, probably due to the easy migration of smaller particles from the site of injection [106].

The particle surface of the drug may be modified with nonionic surfactants to get wet and sterically stable. Less commonly, povidone and lecithin may also be used [107]. To increase the viscosity of the continuous phase, different polymers like carboxymethylcellulose and high-molecular-weight PEGs (like PEG 3350) may be employed. Additional excipients for injectable formulations are buffers, preservatives, and electrolytes.

6.6 Pharmacokinetics

As an instance of a typical pharmacokinetic profile (plasma concentration of the drug over time), once the plasma concentration peak is obtained by IV bolus, immediately upon drug entry to the systemic circulation, drug concentration decreases exponentially and rapidly falls below the therapeutic level (i.e., below the minimum effective concentration, MEC).

Therefore, if a longer duration of action is required, a stable plasma drug level can be potentially achieved through slow IV infusions. In addition, dramatic plasma drug concentration fluctuations, like the ones that occur in recurrent IM or IV bolus administrations, are not observed in slow infusions. Considering the slow onset of the IV infusion therapeutic effect, administration of a loading dose for rapid therapeutic dose achievement, followed by adjustment of the infusion rate to maintain the therapeutic levels is a common practice. It is important to note that vascularized and perfused organs (e.g., kidneys, liver, lung, and heart) are exposed to high drug concentrations in a short period, as IV injections lead to high systemic concentrations very quickly, increasing the risk of unnecessary side effects. Thus, this type of administration should be controlled while monitoring for signs of possible toxic effects. For example, phenytoin solution should be injected intravenously at a speed slower than 50 mg/min to avoid central nervous system depression, cardiovascular collapse, or hypotension, i.e., complications of faster dosing [112–115]. Typical pharmacokinetic profiles of different types of parenteral administration are shown in Figure 6.4 as compared to the oral dose.

There is a delay in the onset of the therapeutic effect after IM injection when the drug is absorbed into the bloodstream. Figure 6.4 compares plasma concentration profiles upon oral, IM, and IV administration. As mentioned above, after IV injection, the drug almost instantaneously reaches its maximum plasma concentration, so the peak is often not noticed, but an exponential decline is observed [22, 116, 117].

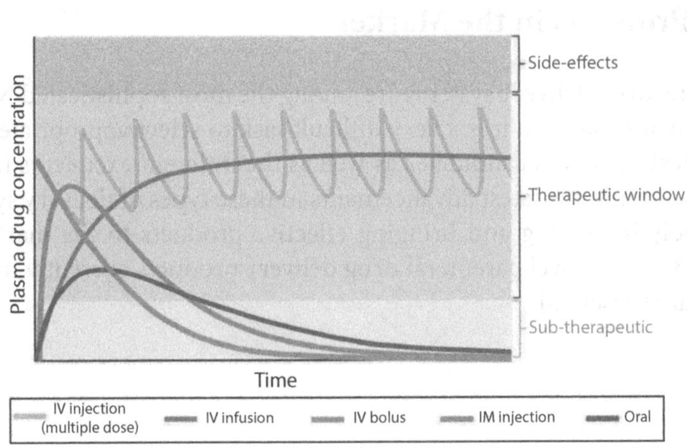

Figure 6.4 Various pharmacokinetic profiles upon parenteral administration as compared to the oral administration.

SC tissue is less vascularized than muscle; so, the SC pathway offers slower absorption kinetics compared to IM. However, it is highly affected by vasoconstriction and vasodilation with the percentage of variations usually being higher than the oral route. As the active transport mechanism is not associated with the SC route, drugs are absorbed mainly by passive diffusion with normal blood flow. Therefore, absorption limiting factors are those affecting diffusion namely diffusion distance (thickness of capillary), concentration gradient, drug diffusion coefficient, and absorbent capillary membrane area. Other factors including possible precipitation and/or aggregation of the drug, phagocytosis, cell uptake by endocytosis, injection site disintegration, or enzymatic sensitivity may limit the degree of interstitial space drug absorption [118–120].

Long-acting (depot) injectable drug delivery systems are ideally supposed to mimic the IV infusion PK profile to avoid any over- or under-dosing episodes during the parenteral therapy course. While this ideal case is still far away from being reached practically, significant improvements have been reported clinically upon using the controlled-release injectables thus providing many hopes by this way to continue the efforts. Almost all routes of parenteral administration have been attempted to reach the "steady state" of the plasma concentrations with a series of products with remarkable clinical success in the market, e.g., long-acting insulin formulations. The main focus of these efforts is on biopharmaceuticals evidently due to the more potent and potentially toxic nature of these products in addition to the more attractive market of these therapeutic agents.

6.7 Products in the Market

Parenteral drug delivery systems are among the most sophisticated NDDSs due to their design, so it is a very difficult task to select appropriate pharmaceutical actives as candidates as well as the different excipients in each case. As a result, the latest advancements in these types of delivery systems are rapidly increasing and bringing effective products to the market. In Table 6.3, some novel parenteral drug delivery products currently in clinical use are presented.

Table 6.3 List of some novel parenteral drug delivery systems in the market.

Product name	Route of administration	Active therapeutic agent	Drug delivery system	Indication(s)	Manufacturer	Approval date
Doxil®	IV	Doxorubicin	PEGylated nanoliposome	AIDS-related Kaposi's sarcoma Recurrent ovarian cancer Metastatic breast cancer Multiple myeloma	Baxter	1995
AmBisome®	IV	Amphptricin B	Liposome	Invasive fungal infections	Astellas	1997
Depocyt®	Intrathecal	Cytarabine	Liposome	Lymphomatous meningitis	DepoTech Corporation	1999
Viadur®	Implant	Leuprolide acetate	Osmotic pump Implant	Palliative treatment of advanced prostate cancer	ALZA Corporation	2000
Abraxane®	IV	Paclitaxel	Protein-Bound Particle (Albumin)	Breast Cancer Non-Small Cell Lung Cancer Pancreatic Cancer	American Biosciences	2005
Inflexal V®	IM	Inactivated hemagglutinin of influenza virus strains	Virosome	Influenza vaccine	Crucell	2008
Exparel®	Infiltration	Bupivacaine	Liposome	Long-acting non-opioid local analgesic for postsurgical local analgesia, and as a nerve block	Pacira Pharmaceuticals	2011

References

1. Abbas, M.N., Iqbal, W., Khan, S., Sterile Products, in: *Essentials of Industrial Pharmacy*, S.A. Khan (Ed.), Springer International Publishing, Cham, 177-201, 2022.
2. Brown, T.D., Whitehead, K.A., Mitragotri, S., Materials for Oral Delivery of Proteins and Peptides. *Nat. Rev. Mater.*, 5, 2, 127–48, 2020.
3. Madhav, N.V.S., Shakya, A.K., Shakya, P., Singh, K., Orotransmucosal Drug Delivery Systems: a Review. *J. Control. Release*, 140, 1, 2–11, 2009.
4. Lori, S.M., Ohadi, M., Estabragh, A.R.M., Afsharipour, S., Banat, M.I., Dehghannoudeh, G., Ph-Sensitive Polymer-Based Carriers as a Useful Approach for Oral Delivery of Therapeutic Protein: a Review. *Protein Pept. Lett.*, 28, 11, 1230–7, 2021.
5. Bhattacharjee, H., Loveless, V., Thoma, L.A., Parenteral Drug Administration: Routes of Administration and Devices, in: *Parenteral Medications*, CRC Press, Florida, United States, 2019.
6. Raeisi Estabragh, M.A., Bami, M.S., Ohadi, M., Banat, I.M., Dehghannoudeh, G., Carrier-Based Systems as Strategies for Oral Delivery of Therapeutic Peptides and Proteins: a Mini-Review. *Int. J. Pept. Res. Ther.*, 27, 2, 1589–96, 2021.
7. Dongare, S.D., Mali, S.S., Patrekar, P.V., Sterile Parenteral Products: a Narrative Approach. *J. Drug Deliv. Ther.*, 5, 1, 41–8, 2015.
8. Bose, S., Parenteral Drug Delivery for Older Patients, in: *Developing Drug Products in an Aging Society*, pp. 291–329, Springer, Berlin, Germany, 2016.
9. Chillistone, S. and Hardman, J.G., Factors Affecting Drug Absorption and Distribution. *Anaesth. Intensive Care Med.*, 18, 7, 335–9, 2017.
10. Gulati, N. and Gupta, H., Parenteral Drug Delivery: a Review. *Recent Pat. Drug Deliv. Formul.*, 5, 2, 133–45, 2011.
11. Fahr, A. and Liu, X., Drug Delivery Strategies for Poorly Water-Soluble Drugs. *Expert Opin. Drug Deliv.*, 4, 4, 403–16, 2007.
12. Jain, K.K., Current Status and Future Prospects of Drug Delivery Systems, in: *Drug Deliv. Syst.*, K.K. Jain (Ed.), pp. 1–56, Springer, New York, NY, 2014.
13. Shantha Kumar, T.R., Soppimath, K., Nachaegari, S.K., Novel Delivery Technologies for Protein and Peptide Therapeutics. *Curr. Pharm. Biotechnol.*, 7, 4, 261–76, 2006.
14. Ummadi, S., Shravani, B., Rao, N., Reddy, M.S., Sanjeev, B., Overview on Controlled Release Dosage Form. *System*, 7, 8, 51–60, 2013.
15. Kurnool, A.N. and Tavari, P., Long Acting Injectables-An Overview. *J. Pharm. Sci. Res.*, 11, 8, 2857–61, 2019.
16. Ratnaparkhi, M. and Gupta Jyoti, P., Sustained Release Oral Drug Delivery System-An Overview. *Terminology*, 3, 4, 10.22270, 2013.
17. Pittiruti, M., Hamilton, H., Biffi, R., Macfie, J., Pertkiewicz, M., ESPEN Guidelines on Parenteral Nutrition: Central Venous Catheters (Access, Care, Diagnosis and Therapy of Complications). *Clin. Nutr.*, 28, 4, 365–77, 2009.

18. Jewell, C., Weaver, M., Sgroi, C., Anderson, K., Sayeed, Z., Residential Addiction Treatment for Injection Drug Users Requiring Intravenous Antibiotics: a Cost-Reduction Strategy. *J. Addict. Med.*, 7, 4, 271–6, 2013.
19. Kolarsick, P.A., Kolarsick, M.A., Goodwin, C., Anatomy and Physiology of the Skin. *J. Dermatol. Nurses' Assoc.*, 3, 4, 203–13, 2011.
20. Gauglitz, G.G. and Schauber, J., Skin: Architecture and Function, in: *Dermal Replacements in General, Burn, and Plastic Surgery*, pp. 1–11, Springer, Berlin, Germany, 2013.
21. Ruiz, M.E. and Scioli Montoto, S., Routes of Drug Administration, in: *ADME Processes in Pharmaceutical Sciences*, pp. 97–133, Springer, Berlin, Germany, 2018.
22. Allam, A.N., El Gamal, S., Naggar, V., Bioavailability: a Pharmaceutical Review. *Int. J. Novel Drug Deliv. Tech.*, 1, 1, 77–93, 2011.
23. Nebendahl, K., Chapter 24 - Routes of Administration, in: *The Laboratory Rat*, G.J. Krinke (Ed.), pp. 463–83, Academic Press, London, 2000.
24. Jain, K.K., An Overview of Drug Delivery Systems, in: *Drug Deliv. Syst.*, K.K. Jain (Ed.), pp. 1–54, Springer, New York, NY, 2020.
25. Reinert, J.P., Maktabi, L., Branam, D., Snyder, M., Clinical Considerations for Rapid Administration of Undiluted or Minimally Diluted Levetiracetam Bolus Doses. *Expert Rev. Neurother.*, 22, 3, 231–6, 2022.
26. Gupta, D., Gupta, S.V., Yang, N., Understanding the Routes of Administration, in: *Handbook of Space Pharmaceuticals*, Y.V. Pathak, M.A. Dos Santos, L. Zea (Eds.), pp. 23–47, Springer International Publishing, Cham, 2022.
27. Yadav, B. and Vellody, R., Central Venous Access in Children – Recent Trends, in: *Anesthetic Management in Pediatric General Surgery: Evolving and Current Concepts*, S.T. Verghese and T.D. Kane (Eds.), pp. 21–43, Springer International Publishing, Cham, 2021.
28. Kolikof, J., Peterson, K., Baker, A.M., Central Venous Catheter, in: *Statpearls* [Internet], Statpearls Publishing, Florida, United States, 2022.
29. Kinnunen, M., Piirainen, P., Kokki, H., Lammi, P., Kokki, M., Updated Clinical Pharmacokinetics and Pharmacodynamics of Oxycodone. *Clin. Pharmacokinet.*, 58, 6, 705–25, 2019.
30. Tom-Revzon, C., Erratic Absorption of Intramuscular Antimicrobial Delivery in Infants and Children. *Expert Opin. Drug Metab. Toxicol.*, 3, 5, 733–40, 2007.
31. Strickley, R.G., Solubilizing Excipients in Oral and Injectable Formulations. *Pharm. Res.*, 21, 2, 201–30, 2004.
32. Fonte, P., Soares, S., Costa, A., Andrade, J.C., Seabra, V., Reis, S. *et al.*, Effect of Cryoprotectants on the Porosity and Stability of Insulin-Loaded PLGA Nanoparticles After Freeze-Drying. *Biomatter*, 2, 4, 329–39, 2012.
33. Cunningham, J.J., Kirchmeier, M.J., Mittal, S., Formulation of Depot Delivery Systems, in: *Pharmaceutical Dosage Forms-Parenteral Medications*, pp. 172–207, CRC Press, Florida, United States, 2016.

34. Tobias, J.D. and Ross, A.K., Intraosseous Infusions: a Review for the Anesthesiologist With a Focus on Pediatric Use. *Anesth. Analg.*, 110, 2, 391–401, 2010.
35. Aziz, A.-M., Subcutaneous Injections: Preventing Needlestick Injuries in the Community. *Br. J. Community Nurs.*, 17, 6, 258–64, 2012.
36. Sexson, K., Lindauer, A., Harvath, T.A., Administration of Subcutaneous Injections. *AJN Am. J. Nurs.*, 117, 5, 7–10, 2017.
37. Li, L., Gatto, G.J., Brand, R.M., Krovi, S.A., Cottrell, M.L., Norton, C. *et al.*, Long-Acting Biodegradable Implant for Sustained Delivery of Antiretrovirals (Arvs) and Hormones. *J. Control. Release*, 340, 188–99, 2021.
38. Press, M., Glucose Sensors and Insulin Pumps: Prospects for an Artificial Pancreas, in: *Artificial Organs*, N.S. Hakim (Ed.), pp. 77–91, Springer, London, 2009.
39. Calkins MD, Mu TS, Fitzgerald G, Gould M, Bentley TB. Infusion Pumps: Potential Application for Special Operations Environment. Walter reed army inst of research washington dc div of surgery. *Defense Technical Information Center*, ADA381172, 1–48, 2000, https://apps.dtic.mil/sti/citations/ADA381172.
40. Quinn, C., Intravenous Flow Control and Infusion Devices, in: *Intravenous Therapy in Nursing Practice*, pp. 197–224, 2008.
41. Cook, T., Combined Spinal–Epidural Techniques. *Anaesthesia*, 55, 1, 42–64, 2000.
42. Zhang, Z., Tachikawa, M., Uchida, Y., Terasaki, T., Drug Clearance From Cerebrospinal Fluid Mediated by Organic Anion Transporters 1 (Slc22a6) and 3 (Slc22a8) At Arachnoid Membrane of Rats. *Mol. Pharm.*, 15, 3, 911–22, 2018.
43. Shang, A.B. and Gan, T.J., Optimising Postoperative Pain Management in the Ambulatory Patient. *Drugs*, 63, 9, 855–67, 2003.
44. Cook, A.M., Mieure, K.D., Owen, R.D., Pesaturo, A.B., Hatton, J., Intracerebroventricular Administration of Drugs. *Pharmacother.: J. Hum. Pharmacol. Drug Ther.*, 29, 7, 832–45, 2009.
45. Joshi, S., Meyers, P.M., Ornstein, E., Intracarotid Delivery of Drugs: the Potential and the Pitfalls. *Anesthesiology*, 109, 3, 543–64, 2008.
46. Fernandes, B.F., Nikolitch, K., Coates, J., Novais, G., Odashiro, A., Odashiro, P.P. *et al.*, Local Chemotherapeutic Agents for the Treatment of Ocular Malignancies. *Surv. Ophthalmol.*, 59, 1, 97–114, 2014.
47. Vanden Hoek, T.L., Morrison, L.J., Shuster, M., Donnino, M., Sinz, E., Lavonas, E.J. *et al.*, Part 12: Cardiac Arrest in Special Situations: 2010 American Heart Association Guidelines for Cardiopulmonary Resuscitation and Emergency Cardiovascular Care. *Circulation*, 122, 18_Suppl_3, S829–S61, 2010.
48. Silberberg, A., Moeller-Bertram, T., Wallace, M.S., A Randomized, Double-Blind, Crossover Study to Evaluate the Depth Response Relationship of Intradermal Capsaicin-Induced Pain and Hyperalgesia in Healthy Adult Volunteers. *Pain Med.*, 16, 4, 745–52, 2015.

49. Strazar, A.R., Leynes, P.G., Lalonde, D.H., Minimizing the Pain of Local Anesthesia Injection. *Plast. Reconstr. Surg.*, 132, 3, 675–84, 2013.
50. Evans, C.H., Kraus, V.B., Setton, L.A., Progress in Intra-Articular Therapy. *Nat. Rev. Rheumatol.*, 10, 1, 11–22, 2014.
51. Patterson, A.M., Bonny, A.E., Shiels, W.E., Erwin, E.A., Three-Injection Intralymphatic Immunotherapy in Adolescents and Young Adults With Grass Pollen Rhinoconjunctivitis. *Ann. Allergy Asthma Immunol.*, 116, 2, 168–70, 2016.
52. Betzler, B.K., Chee, Y.-Y.J., Bin Abd Razak, H.R., Intraosseous Injections Are Safe and Effective in Knee Osteoarthritis: a Systematic Review. *Arthrosc. Sports Med. Rehabil.*, 3, 5, E1557–E67, 2021.
53. Eisenhauer, D.A., Schmidt, R., Martin, C., Schultz, S.G., 14 Processing of Small Volume Parenterals and Large Volume Parenterals, in: *Pharmaceutical Dosage Forms-Parenteral Medications: Volume 2: Facility Design, Sterilization and Processing*, vol. 2, p. 334, 2016.
54. Jain, D., Mahammad, S.S., Singh, P.P., Kodipyaka, R., A Review on Parenteral Delivery of Peptides and Proteins. *Drug Dev. Ind. Pharm.*, 45, 9, 1403–20, 2019.
55. Brazeau, G., Sauberan, S.L., Gatlin, L., Wisniecki, P., Shah, J., Effect of Particle Size of Parenteral Suspensions on *in Vitro* Muscle Damage. *Pharm. Dev. Technol.*, 16, 6, 591–8, 2011.
56. Townsend, M., Aseptic Processing of Protein Pharmaceuticals, in: *Development and Manufacture of Protein Pharmaceuticals*, S.L. Nail and M.J. Akers (Eds.), pp. 129–89, Springer US, Boston, MA, 2002.
57. Boquet, M.P. and Wagner, D.R., Injectable Formulations of Poorly Water-Soluble Drugs, in: *Formulating Poorly Water Soluble Drugs*, R.O. Williams III, A.B. Watts, D.A. Miller (Eds.), pp. 209–42, Springer, New York, NY, 2012.
58. Groves, M.J., *Parenteral Products: the Preparation and Quality Control of Products for Injection*, Elsevier, Netherlands, Amsterdam, 2014.
59. Merisko-Liversidge, E., Liversidge, G.G., Cooper, E.R., Nanosizing: a Formulation Approach for Poorly-Water-Soluble Compounds. *Eur. J. Pharm. Sci.*, 18, 2, 113–20, 2003.
60. Silberman, H. and Webster, J., Nutrition, Parenteral, in: *Encyclopedia of Medical Devices and Instrumentation*, vol. 5, pp. 124–34, 2006.
61. Khan, B.A., Akhtar, N., Khan, H.M.S., Waseem, K., Mahmood, T., Rasul, A. *et al.*, Basics of Pharmaceutical Emulsions: a Review. *Afr. J. Pharm. Pharmacol.*, 5, 25, 2715–25, 2011.
62. Xu, Z., Harvey, K.A., Pavlina, T.M., Zaloga, G.P., Siddiqui, R.A., Tocopherol and Tocotrienol Homologs in Parenteral Lipid Emulsions. *Eur. J. Lipid Sci. Technol.*, 117, 1, 15–22, 2015.
63. Sievens-Figueroa, L., Pandya, N., Bhakay, A., Keyvan, G., Michniak-Kohn, B., Bilgili, E. *et al.*, Using USP I and USP IV for Discriminating Dissolution Rates of Nano- and Microparticle-Loaded Pharmaceutical Strip-Films. *AAPS Pharm. Sci. Tech.*, 13, 4, 1473–82, 2012.

64. Hadidi, N. and Pazuki, G., Preparation, Characterization and *In-Vivo* Efficacy Study of Glatiramer Acetate (GA)-Hydrogel-Microparticles as Novel Drug Delivery System for GA in RRMS. *Sci. Rep.*, 12, 1, 22042, 2022.
65. Chen, B.Z., Yang, Y., Wang, B.B., Ashfaq, M., Guo, X.D., Self-Implanted Tiny Needles as Alternative to Traditional Parenteral Administrations for Controlled Transdermal Drug Delivery. *Int. J. Pharm.*, 556, 338–48, 2019.
66. Schreiner, V., Detampel, P., Jirkof, P., Puchkov, M., Huwyler, J., Buprenorphine Loaded PLGA Microparticles: Characterization of a Sustained-Release Formulation. *J. Drug Deliv. Sci. Technol.*, 63, 102558, 2021.
67. Doufène, K., Lapinte, V., Gaveau, P., Félix, G., Cacciaguerra, T., Chopineau, J. *et al.*, Tunable Vegetable Oil/Silica Hybrid Microparticles for Poorly Water-Soluble Drug Delivery. *Int. J. Pharm.*, 567, 118478, 2019.
68. Khattab, A., Abouhussein, D.M.N., Mohammad, F.E., Development of Injectable Tenoxicam *in Situ* Forming Microparticles Based on Sesame Oil and Poly-DL-Lactide: Characterization, Efficacy and Acute Toxicity. *J. Drug Deliv. Sci. Technol.*, 51, 682–94, 2019.
69. Elmowafy, M., Alruwaili, N.K., Shalaby, K., Alharbi, K.S., Altowayan, W.M., Ahmad, N. *et al.*, Long-Acting Paliperidone Parenteral Formulations Based on Polycaprolactone Nanoparticles; the Influence of Stabilizer and Chitosan on *in Vitro* Release, Protein Adsorption, and Cytotoxicity. *Pharmaceutics* [Internet], 12, 2, 160, 2020.
70. Pham, D.T., Saelim, N., Tiyaboonchai, W., Alpha Mangostin Loaded Crosslinked Silk Fibroin-Based Nanoparticles for Cancer Chemotherapy. *Colloids Surf., B*, 181, 705–13, 2019.
71. Elbrink, K., Van Hees, S., Chamanza, R., Roelant, D., Loomans, T., Holm, R. *et al.*, Application of Solid Lipid Nanoparticles as a Long-Term Drug Delivery Platform for Intramuscular and Subcutaneous Administration: *in Vitro* and *in Vivo* Evaluation. *Eur. J. Pharm. Biopharm.*, 163, 158–70, 2021.
72. Rudhrabatla, V.S.A.P., Sudhakar, B., Reddy, K.V.N.S., *in Vitro* and *in Vivo* Assessment of Designed Melphalan Loaded Stealth Solid Lipid Nanoparticles for Parenteral Delivery. *Bionanoscience*, 10, 1, 168–90, 2020.
73. Joshi, M.D. and Müller, R.H., Lipid Nanoparticles for Parenteral Delivery of Actives. *Eur. J. Pharm. Biopharm.*, 71, 2, 161–72, 2009.
74. Bayón-Cordero, L., Alkorta, I., Arana, L., Application of Solid Lipid Nanoparticles to Improve the Efficiency of Anticancer Drugs. *Nanomaterials*, 9, 3, 474, 2019.
75. Kumar, M., Bishnoi, R.S., Shukla, A.K., Jain, C.P., Techniques for Formulation of Nanoemulsion Drug Delivery System: a Review. *Prev. Nutr. Food Sci.*, 24, 3, 225–34, 2019.
76. Singh, Y., Meher, J.G., Raval, K., Khan, F.A., Chaurasia, M., Jain, N.K. *et al.*, Nanoemulsion: Concepts, Development and Applications in Drug Delivery. *J. Control. Release*, 252, 28–49, 2017.

77. Ahire, E., Thakkar, S., Darshanwad, M., Misra, M., Parenteral Nanosuspensions: a Brief Review From Solubility Enhancement to More Novel and Specific Applications. *Acta Pharm. Sin. B.*, 8, 5, 733–55, 2018.
78. Has, C. and Sunthar, P., A Comprehensive Review on Recent Preparation Techniques of Liposomes. *J. Liposome Res.*, 30, 4, 336–65, 2020.
79. Guimarães, D., Cavaco-Paulo, A., Nogueira, E., Design of Liposomes as Drug Delivery System for Therapeutic Applications. *Int. J. Pharm.*, 601, 120571, 2021.
80. Crommelin, D.J.A., Van Hoogevest, P., Storm, G., The Role of Liposomes in Clinical Nanomedicine Development. What Now? Now What? *J. Controlled Release*, 318, 256–63, 2020.
81. Raeisi Estabragh, M.A., Pardakhty, A., Ahmadzadeh, S., Dabiri, S., Malekpour Afshar, R., Farajli Abbasi, M., Successful Application of Alpha Lipoic Acid Niosomal Formulation in Cerebral Ischemic Reperfusion Injury in Rat Model. *Adv. Pharm. Bull.*, 12, 3, 541–9, 2022.
82. Gharbavi, M., Amani, J., Kheiri-Manjili, H., Danafar, H., Sharafi, A., Niosome: a Promising Nanocarrier for Natural Drug Delivery Through Blood-Brain Barrier. *Adv. Pharmacol. Sci.*, 2018, 6847971, 2018.
83. Manchanda, S., Das, N., Chandra, A., Bandyopadhyay, S., Chaurasia, S., Chapter 2 - Fabrication of Advanced Parenteral Drug-Delivery Systems, in: *Drug Deliv. Syst.*, R.K. Tekade (Ed.), pp. 47–84, Academic Press, Massachusetts, United States, 2020.
84. Davoodi, P., Lee, L.Y., Xu, Q., Sunil, V., Sun, Y., Soh, S. *et al.*, Drug Delivery Systems for Programmed and On-Demand Release. *Adv. Drug Deliv. Rev.*, 132, 104–38, 2018.
85. Hu, T., Yang, J., Cui, K., Rao, Q., Yin, T., Tan, L. *et al.*, Controlled Slow-Release Drug-Eluting Stents for the Prevention of Coronary Restenosis: Recent Progress and Future Prospects. *ACS Appl. Mater. Interfaces*, 7, 22, 11695–712, 2015.
86. Keraliya, R.A., Patel, C., Patel, P., Keraliya, V., Soni, T.G., Patel, R.C. *et al.*, Osmotic Drug Delivery System as a Part of Modified Release Dosage Form. *ISRN Pharm.*, 2012, 528079, 2012.
87. Cao, Z., Tang, X., Zhang, Y., Yin, T., Gou, J., Wang, Y. *et al.*, Novel Injectable Progesterone-Loaded Nanoparticles Embedded in SAIB-PLGA *in Situ* Depot System for Sustained Drug Release. *Int. J. Pharm.*, 607, 121021, 2021.
88. Anselmo, A.C. and Mitragotri, S., An Overview of Clinical and Commercial Impact of Drug Delivery Systems. *J. Control. Release*, 190, 15–28, 2014.
89. Patel, C.A. and Keraliya, R., a Review: Parenteral Depot Drug Delivery System. *J. Drug Deliv. Res.*, 3, 1, 1–10, 2014.
90. Neale, J., Tompkins, C.N.E., Mcdonald, R., Strang, J., Implants and Depot Injections for Treating Opioid Dependence: Qualitative Study of People Who Use or Have Used Heroin. *Drug Alcohol Depend.*, 189, 1–7, 2018.

91. Chilingerian, J., Origins of Drgs in the United States: a Technical, Political and Cultural Story, in: *The Globalization of Managerial Innovation in Health Care*, pp. 4–33, 2008.
92. Wolinsky, J.B., Colson, Y.L., Grinstaff, M.W., Local Drug Delivery Strategies for Cancer Treatment: Gels, Nanoparticles, Polymeric Films, Rods, and Wafers. *J. Control. Release*, 159, 1, 14–26, 2012.
93. Nanda, K., Amaral, E., Hays, M., Viscola, M.A.M., Mehta, N., Bahamondes, L., Pharmacokinetic Interactions Between Depot Medroxyprogesterone Acetate and Combination Antiretroviral Therapy. *Fertil. Steril.*, 90, 4, 965–71, 2008.
94. Schoenhammer, K., Petersen, H., Guethlein, F., Goepferich, A., Injectable *in Situ* Forming Depot Systems: PEG-DAE as Novel Solvent for Improved PLGA Storage Stability. *Int. J. Pharm.*, 371, 1, 33–9, 2009.
95. Wan, F. and Yang, M., Design of PLGA-Based Depot Delivery Systems for Biopharmaceuticals Prepared by Spray Drying. *Int. J. Pharm.*, 498, 1, 82–95, 2016.
96. Tomar, L., Tyagi, C., Kumar, M., Kumar, P., Singh, H., Choonara, Y.E. *et al.*, *in Vivo* Evaluation of a Conjugated Poly(Lactide-Ethylene Glycol) Nanoparticle Depot Formulation for Prolonged Insulin Delivery in the Diabetic Rabbit Model. *Int. J. Nanomed.*, 8, 505–20, 2013.
97. Machluf, M., Orsola, A., Atala, A., Controlled Release of Therapeutic Agents: Slow Delivery and Cell Encapsulation. *World J. Urol.*, 18, 1, 80–3, 2000.
98. Tang, Y., Oak, M., Mandke, R., Layek, B., Sharma, G., Singh, J., Thermosensitive Polymers for Controlled Delivery of Hormones, in: *Active Implants and Scaffolds for Tissue Regeneration*, M. Zilberman (Ed.), pp. 457–79, Springer, Berlin, Heidelberg, 2011.
99. Boyd, B.J., Whittaker, D.V., Khoo, S.-M., Davey, G., Lyotropic Liquid Crystalline Phases Formed From Glycerate Surfactants as Sustained Release Drug Delivery Systems. *Int. J. Pharm.*, 309, 1, 218–26, 2006.
100. Larsen, S.W., Thing, M.A., Larsen, C., Oily (Lipophilic) Solutions and Suspensions, in: *Long Acting Injections and Implants*, J.C. Wright and D.J. Burgess (Eds.), pp. 113–35, Springer US, Boston, MA, 2012.
101. Ding, Z., Wang, L., Xing, Y., Zhao, Y., Wang, Z., Han, J., Enhanced Oral Bioavailability of Celecoxib Nanocrystalline Solid Dispersion Based on Wet Media Milling Technique: Formulation, Optimization and *in Vitro/In Vivo* Evaluation. *Pharmaceutics*, 11, 7, 328, 2019.
102. Rahman, Z., Yang, Y., Korang-Yeboah, M., Siddiqui, A., Xu, X., Ashraf, M. *et al.*, Assessing Impact of Formulation and Process Variables on *In-Vitro* Performance of Directly Compressed Abuse Deterrent Formulations. *Int. J. Pharm.*, 502, 1, 138–50, 2016.
103. Webb, J.J., The Effect of Particle Size and Film Cooling on Nozzle Guide Vane Deposition, the Ohio State University. *J. Eng. Gas Turbine.*, 4, 1051–59, 2011.
104. Rhodes, C.A. and Malavia, N., Formulation of Depot Delivery Systems, in: *Parenteral Medications*, pp. 347–80, CRC Press, Florida, United States, 2019.

105. Lee, W.W. and Patterson, C.M., Biopharmaceutics of Injectable Formulations, in: *Biopharmaceutics*, pp. 253–63, 2022.
106. Miller, L.G. and Fincher, J.H., Influence of Drug Particle Size After Intramuscular Dosage of Phenobarbital to Dogs. *J. Pharm. Sci.*, 60, 11, 1733–6, 1971.
107. Rudd, N.D., Reibarkh, M., Fang, R., Mittal, S., Walsh, P.L., Brunskill, A.P.J. *et al.*, Interpreting *in Vitro* Release Performance From Long-Acting Parenteral Nanosuspensions Using USP-4 Dissolution and Spectroscopic Techniques. *Mol. Pharm.*, 17, 5, 1734–47, 2020.
108. Friedrich, I., Reichl, S., Müller-Goymann, C.C., Drug Release and Permeation Studies of Nanosuspensions Based on Solidified Reverse Micellar Solutions (SRMS). *Int. J. Pharm.*, 305, 1, 167–75, 2005.
109. Fredholt, K., Larsen, D.H., Larsen, C., Modification of *in Vitro* Drug Release Rate From Oily Parenteral Depots Using a Formulation Approach. *Eur. J. Pharm. Sci.*, 11, 3, 231–7, 2000.
110. Larsen, S.W., Østergaard, J., Friberg-Johansen, H., Jessen, M.N.B., Larsen, C., *In Vitro* Assessment of Drug Release Rates From Oil Depot Formulations Intended for Intra-Articular Administration. *Eur. J. Pharm. Sci.*, 29, 5, 348–54, 2006.
111. Larsen, S.W., Jessen, M.N.B., Østergaard, J., Larsen, C., Assessment of Drug Release From Oil Depot Formulations Using an *in Vitro* Model—Potential Applicability in Accelerated Release Testing. *Drug Dev. Ind. Pharm.*, 34, 3, 297–304, 2008.
112. Li, N., Feng, L., Tan, Y., Xiang, Y., Zhang, R., Yang, M., Preparation, Characterization, Pharmacokinetics and Biodistribution of Baicalin-Loaded Liposome on Cerebral Ischemia-Reperfusion After I.V. Administration in Rats. *Molecules*, 23, 7, 1747, 2018.
113. Onoue, S., Yamada, S., Chan, H.-K., Nanodrugs: Pharmacokinetics and Safety. *Int. J. Nanomed.*, 9, 1025–37, 2014.
114. Sysoev, Y.I., Whaley, A.K., Prikhodko, V.A., Semivelichenko, E.D., Eletskaya, E.I., Luzhanin, V.G. *et al.*, Pharmacokinetics of Intravenously and Intraperitoneally Administered Mafedine Sodium in Mice. *Pharm. Chem. J.*, 54, 12, 1193–7, 2021.
115. Terzi, E., Corum, O., Bilen, S., Kenanoglu, O.N., Atik, O., Uney, K., Pharmacokinetics of Danofloxacin in Rainbow Trout After Different Routes of Administration. *Aquaculture*, 520, 734984, 2020.
116. Jain, K.K., Drug Delivery Systems - an Overview, in: *Drug Deliv. Syst.*, K.K. Jain (Ed.), pp. 1–50, Humana Press, Totowa, NJ, 2008.
117. Harada, K., Ukai, Y., Kanakubo, K., Yamano, S., Lee, J., Kurosawa, T.A. *et al.*, Comparison of the Diuretic Effect of Furosemide by Different Methods of Administration in Healthy Dogs. *J. Vet. Emergency Crit. Care*, 25, 3, 364–71, 2015.
118. Kagan, L., Pharmacokinetic Modeling of the Subcutaneous Absorption of Therapeutic Proteins. *Drug Metab. Dispos.*, 42, 11, 1890–905, 2014.

119. Richter, W.F. and Jacobsen, B., Subcutaneous Absorption of Biotherapeutics: Knowns and Unknowns. *Drug Metab. Dispos.*, 42, 11, 1881–9, 2014.
120. Gradel, A.K.J., Porsgaard, T., Lykkesfeldt, J., Seested, T., Gram-Nielsen, S., Kristensen, N.R. *et al.*, Factors Affecting the Absorption of Subcutaneously Administered Insulin: Effect on Variability. *J. Diabetes Res.*, 2018, 1205121, 2018.

7

Implantable Drug Delivery Systems

Nasrin Zarei Chamgordani

Department of Pharmaceutics and Pharmaceutical Nanotechnology, School of Pharmacy, Shahid Beheshti University of Medical Sciences, Tehran, Iran

Abstract

Conventional routes of drug administration (such as oral and parental) encounter challenges such as degradation of active agents in the gastrointestinal tract, hepatic first-pass metabolism, compliance of the patient with treatment schedule especially in chronic conditions, and off-target effects. Precise delivery of therapeutics to the target site can overcome the challenges of conventional drug delivery systems. Implantable drug delivery systems (ImDDSs) are novel systems that can overcome the listed limitations and provide a practical means to achieve the therapeutic level of the drug during the duration of treatment and defeat the risk of drug concentration fluctuations. ImDDSs have also the potential to be on-demand, pulsatile, and chrono-controlled (i.e., release the drug according to the circadian rhythm). Rapid onset of action and the potential to be personalized are other advantages of them. ImDDSs have demonstrated an impressive example of the integration of engineering, medical, and drug delivery knowledge. To date, these systems have been successfully used in the field of contraception, pain management, and treatment of cancer, ocular, and cardiovascular diseases. ImDDSs are expected to grow rapidly in the future years. Therefore, knowledge about these systems helps for applications in disease treatment and prevention, as well as for the design and construction of new systems. This chapter aims to provide a brief overview of the different aspects of ImDDSs.

***Keywords*:** Implantable drug delivery systems, implants, intrathecal, subcutaneous, stent, osmotic pump, microchip

Email: n_zarei@sbmu.ac.ir

Mehrdad Hamidi and Mahsa Sayed Tabatabaei (eds.) Novel Drug Delivery Systems: Fundamentals and Applications, (235–274) © 2025 Scrivener Publishing LLC

7.1 Introduction

Implantable drug delivery systems (ImDDSs) are novel systems that are placed in the body, for example, subcutaneously through surgery, to release their payload for local or systemic effect without the need for repeated administration. Implantable systems are mostly addressed for a long time, controlled, and constant drug delivery. The concept of ImDDSs was introduced in 1938 for the long-term release of subcutaneously implanted pellets of a steroidal drug. Then, in 1960, the first real implant for drug release, composed of silicone polymer and drug was presented. Subsequently, several efforts were carried out to develop and commercialize ideal ImDDSs [1]. Today there are several FDA-approved ImDDSs for various indications which are presented later in this chapter.

ImDDSs overcome some of the limitations associated with conventional routes of administration. For instance, although oral administration is the most convenient route of administration, reduction in bioavailability as a consequence of drug degradation in the gastrointestinal tract, hepatic first-pass metabolism, inadequate absorption across gastrointestinal tract wall, and interaction of the administered drug with gastrointestinal contents, may render the oral route of administration ineffective. So, it demands multiple dosing or increasing the dose size leading to poor patient compliance, especially in chronic diseases. Furthermore, fluctuations in the plasma concentration of the drug following the multiple oral dosing may potentially lead to the risk of under- or over-dosing in some time intervals during the drug therapy course [2, 3].

To overcome the bioavailability issues associated with oral administration, intravenous administration has been suggested. The intravenous route doesn't have issues with bioavailability. Still, the short half-life of drugs in plasma requires patient compliance for frequent prescribing, especially when a long time and chronic treatment is needed [4].

Another shortcoming associated with the conventional routes of administration is the biological and physiological barriers against drug access to the target sites. For example, the blood-brain barrier and blood-spinalcord-barrier are the main barriers limiting drug delivery to the central nervous system (CNS). So, the conventional route of administration leads to a low level of drug concentration in the CNS. The intrathecal injection is an effective approach to bypass these barriers. However, in situations where there is a need for a long time and frequent administration of an analgesic (e.g., CNS cancer, chronic back and neck pain), intrathecal injection is inappropriate [5].

Transdermal and mucosal drug delivery are other routes of administration which, despite bypassing the hepatic first-pass effect, suffer from biological barriers such as stratum corneum in transdermal, cornea, precorneal, and conjunctival barriers in ocular, tight junction, and mucosal barriers in colon drug delivery which, in turn, may impair efficient drug delivery [6–8].

ImDDSs can overcome the limitation of conventional drug delivery systems, as described, for local or systemic drug delivery and provide a sensible means to achieve the therapeutic level of the drug through the entire duration of treatment and defeat the risk of drug concentration fluctuations.

In summary, the advantages of implantable systems are bypassing hepatic first-pass metabolism and biological barriers, preventing systemic side effects by releasing the drug close to the site of action, and providing the controlled, sustained, and long-term release of the drug. ImDDSs have also the potential to be designed for site-specific, targeted, on-demand, pulsatile, and chrono-controlled (i.e., based on circadian rhythm) drug release. Also, they can be stimuli-responsive, so release the cargo in response to different stimulations such as heat, pH changes, light, ultrasound, electric and magnetic fields, and so on. ImDDSs are ideal for treating chronic diseases (e.g., chronic pain, CNS disorders, eye diseases) and situations where a long-time treatment is needed (e.g., contraception) to improve patient adherence and compliance [4].

Despite the aforementioned favorable characteristics of ImDDSs, they have some limitations. One major drawback with ImDDSs is the requirement of surgery to implant the system in its site or expel it. Also, the termination of drug release is challenging when it is needed, e.g., in emergencies. The possibility of pain, irritation, or allergic reaction due to the high concentration of drug in the site of implantation are other potential limitations. Accordingly, this method of administration is currently limited to potent drugs when the cost-to-benefit ratio is logical [9].

This chapter aims to overview ImDDs in terms of the anatomy and physiology of some conventional implantation sites, classification, formulation, challenges, opportunities, and pharmacokinetics of these systems. Also, the currently marketed ImDDSs and those in advanced clinical phases will be emphasized.

7.2 Anatomy and Physiology

ImDDSs can be implanted in different parts of the body to provide local or systemic therapeutic effects (Figure 7.1). Depending on which area of

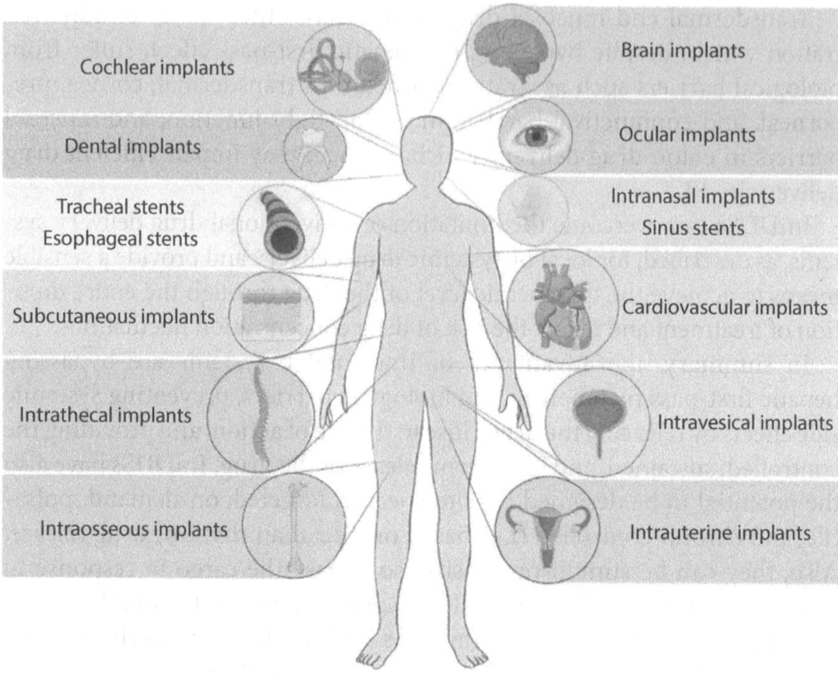

Figure 7.1 Different implantation sites of drug delivery systems (created with BioRender.com).

the body the different implantable systems are used in, their design may be different. An understanding of the anatomy and physiology of the implantation sites and the various physiological barriers influencing drug absorption is a prerequisite for the design and development of efficient ImDDSs. In the following, the anatomy and physiology of some of the common implantation sites and possible barriers to drug delivery are discussed.

7.2.1 Subcutaneous

Subcutaneous tissue, generally known as hypodermis, is the deepest skin layer that lies closest to the muscle and connects the skin to the underneath tissue (bones and muscles). The subcutaneous layer mainly consists of fat (adipose tissue) embedded as lobules in the connective tissue, blood vessels, and nerves. In addition to adipocytes, other cellular components of the subcutaneous layer are fibroblasts (which synthesize collagen) and macrophages. The interstitium also consists of fibrous collagen networks and extracellular components (including negatively charged glycosaminoglycans, salts, and plasma-originated proteins) that are synthesized from the cellular matrix. The extracellular matrices act as a barrier to drug delivery in subcutaneous administration. The passage of the drug through this matrix, the interaction of the drug with negatively charged components, and the possibility of catabolism of biopharmaceuticals (such as proteins) are factors that can affect the bioavailability of the subcutaneously administered materials [10, 11].

A network of blood and lymphatic vessels as well as nerves are also present in the subcutaneous layer. Upon subcutaneous administration, low molecular weight drugs and biomolecules (<16 kDa) are absorbed *via* capillary walls, and then introduced to the systemic circulation. This occurs following the transportation in the extracellular matrix of the subcutaneous tissue. Other (bio)materials with higher molecular weights are introduced to the lymphatic system and then enter the systemic circulation [12]. Subcutaneous implants are typically placed with the aid of special implantation devices, needles, or the use of surgery.

7.2.2 Intrathecal

Intrathecal drug delivery aims to sufficiently deliver the therapeutic agents directly to their site of action in the spinal cord. In this drug delivery system, the drug is stored in a reservoir (pump) that is implanted in the body (abdominal space) and is transported through a catheter to a suitable position (where the relevant receptors exist) in the spinal cord.

As can be seen in Figure 7.2 spinal cord includes the central canal (which carries the cerebrospinal fluid, CSF) which is surrounded by gray and white matter. This combination (central canal, gray and white matter)

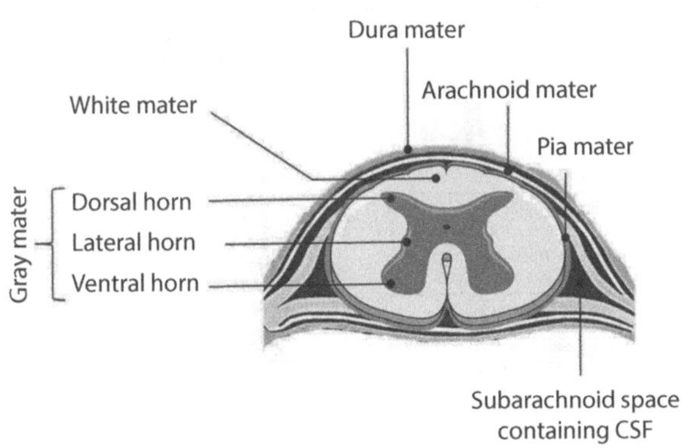

Figure 7.2 Cross-section of the spinal cord.

is covered by protective layers called meninges. Meninge includes three layers: (i) pia matter (inner layer), (ii) arachnoid, and (iii) dura matter. The space between the pia matter and arachnoid is called subarachnoid and contains CSF.

Once a drug is delivered to the intrathecal (subarachnoid) space containing CSF, it diffuses through pia matter-arachnoid and white matter to target the receptors in the superficial layers of the dorsal horn and ion channels (e.g. potassium channels and voltage-dependent calcium channels) [13].

The physicochemical characteristics of the intrathecally administered drug have a crucial effect on the penetration and distribution (so the bioavailability) of the drug through the white matter (more lipophile) and gray matter (relatively hydrophile), fat, and blood. CSF is a hydrophilic liquid, so the water-soluble drugs tend to distribute in the CSF. However, the penetration of the drug to the lipophilic layers of the spinal cord is challenging. In contrast, when lipid-soluble drugs enter the CSF tend to quickly diffuse to the epidural fat, thus, having a rapid distribution into the bloodstream, and showing less therapeutic effect on the site of action [14]. So more hydrophilic drugs are desired because of their higher bioavailability.

7.2.3 Vaginal

The fibromuscular tubular organ that connects different parts of the female reproductive system is called the vagina. As shown in Figure 7.3, the vaginal

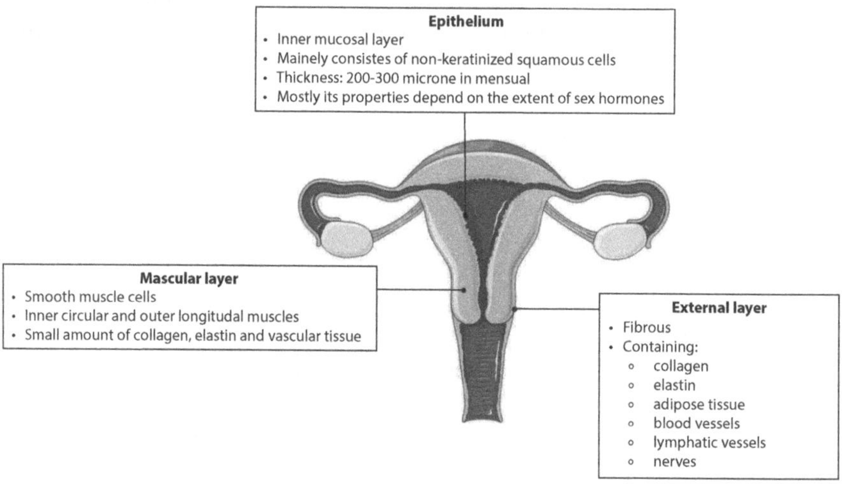

Figure 7.3 Vaginal wall constituents.

wall includes three layers: (i) the external layer, (ii) the muscular layer, and (iii) the epithelium [15, 16]. The thickness of vaginal epithelium is a major factor that affects drug absorption. An acidic vaginal fluid with a varied pH during the menstrual cycle is secreted from different parts of the vagina. This fluid mainly contains protein, lactic acid, electrolytes, carbohydrates, glycogen, leucocytes, enzymes, and pathogenic organisms. Vaginal fluid can affect drug absorption beneficially or detrimentally. For example, vaginal fluid may help drug dissolution and consequently increase the drug absorption. Vice versa, the components of the vaginal fluid may destabilize or degrade the active agent and reduce its bioavailability. Also, the absorption of ionizable drugs may be affected by the acidic environment of the vagina. On the other hand, the acidic pH of the vaginal is considered an opportunity to design pH-responsive systems [17].

After the dissolution or release from the inserted dosage form in the vaginal lumen, the drug is absorbed through the vaginal membrane. The absorption of drugs from the vagina can be transcellular (diffusion-controlled because of the concentration gradient), paracellular (through tight junctions), or receptor-mediated depending on the drug molecule's characteristics (such as molecular weight, hydrophilicity or lipophilicity, charge, and ionization state also may influence its absorption) [18].

7.2.4 Intravesicular

The bladder is a resilient muscular organ that stores and excretes urine. The bladder wall is comprised of several cellular layers: (i) the urothelium, (ii) lamina propria, (iii) muscularis mucosae detrusor muscle, (iv) adventitia, and (v) serous layers [19]. In the upper layer, urothelium, the cells are connected by tight junctions, so this layer acts as a barrier against drug penetration into underlying layers. However, this barrier may also be beneficial for local drug delivery as it prevents drugs from reaching blood circulation. The next barrier is the glycosaminoglycan (a mucin component) layer that covers the urothelium which prevents the adhesion of drug molecules or drug carriers (because of its hydrophilic nature) and thereby reduces the bioavailability [20]. Another bioavailability barrier in the intravesicular route is the limited residence time of the drug in the bladder, due to the constant filling of urine which dilutes the drug concentration and washes it out when emptying. Using intravesicular implants enhances the contact time of the drug and overcomes the challenge of drug dilution and washing out [21].

7.2.5 Ocular

The eye has three layers and three fluid-containing compartments. The outer layer includes the cornea, lamina cribrosa, and sclera. The middle layer contains a lot of blood vessels and includes the iris, ciliary body, and choroids. The inner layer forms the retina at the back. Eye compartments are the anterior chamber (between the cornea and iris), the posterior chamber (between the iris and lens), and the vitreous chamber (between the lens and retina) (Figure 7.4) [22].

Conventional ocular drug delivery systems encounter many barriers to effective drug delivery, namely tear film, corneal barrier, conjunctiva and sclera, and blood-ocular barriers [23]. These barriers have been discussed in detail in Chapter 11 of this book.

To overcome these barriers and enhance drug delivery to anterior and posterior sections, different strategies such as prodrugs, drug-cyclodextrin complexation, using penetration enhancers, colloidal systems, *in-situ* forming gels, contact lens, and ImDDSs have been studied [23, 24]. Most of the marketed ImDDSs are intravitreal, for drug delivery to the posterior section of the eye. Implants can also be intracameral, episcleral, and injected into the posterior chamber [6].

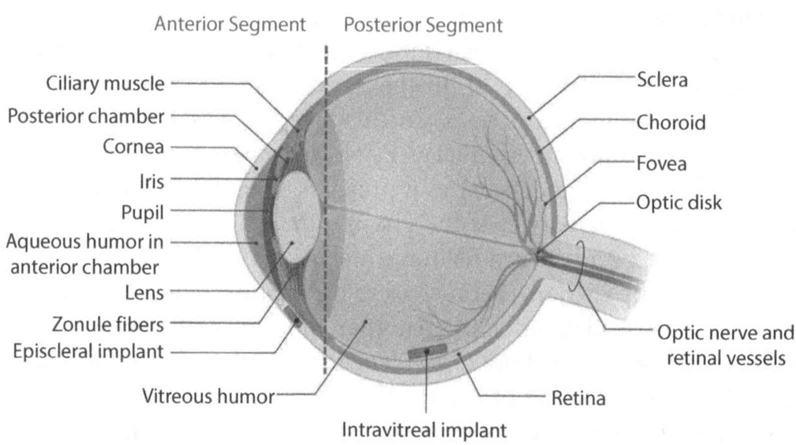

Figure 7.4 Anatomy of the eye (created with BioRender.com).

7.3 Classification and Formulation Considerations

Several approaches have been introduced in the literature to classify ImDDSs such as the mechanism of drug release, implantation site, and application [20, 21, 25]. In another common classification approach, ImDDSs are classified into active and passive forms. Passive implants do not need energy for drug release (e.g., polymeric matrix implants), while active (dynamic) systems are energy-dependent [26].

In this chapter, active systems are referred to as systems in which the release of drugs can be triggered, controlled, and regulated after implantation using different triggers (such as electrical, magnetical, mechanical, laser, or other actuation methods). In contrast, in passive implants, control of drug release after implantation is not possible and drug release is pre-determined by factors such as the construction method of implant, the material used for fabrication, and the formulation of the system. The release of drugs from a polymeric matrix *via* diffusion, degradation, or swelling are examples of passive release.

7.3.1 Passive-Release Implantable Systems

Passive ImDDSs are categorized into two main classes based on the mechanisms of drug release:

(i) Implants in which drug release is based on diffusion, degradation, or swelling (such as polymeric-, ceramic- and metal alloy-based implants)
(ii) Pump implants wherein the driving force for drug release is based on pumping and there is no controlling device such as valves or any electrical or mechanical circuits to control the release after implantation (e.g. osmotic and pressurized gas).

7.3.1.1 Polymeric Implantable Drug Delivery Systems

Polymeric ImDDSs are simple systems with non-movable parts that release drugs under the control of diffusion, swelling, erosion, or a combination of them, depending on the polymer nature, construction method, geometry, size of the device, site of implantation, and so on [2, 27]. Conventionally, polymeric drug delivery systems are constructed before implantation. However, some of them are formed *in situ* at the implantation site, so that drug-containing polymeric solutions or suspensions are injected into the body with a syringe and solidified *in situ* to form semi-solid or solid storage of the drug. *In situ* forming implants (ISFIs) are categorized as *in situ* cross-linkable polymer systems, organogels, and phase separation systems, according to the mechanism of implant gel formation [28]. In contrast to pre-manufactured implants, ISFIs do not need the surgery to implant on-site. The polymeric solution containing the drug is administered by a conventional syringe and needle at the site of the implantation. In addition, these systems do not require complicated and expensive manufacturing processes. However, the major disadvantage of these systems to be overcome is the burst release effect that occurs because of the time delay between injection and implant formation. Eligard®, Arestin®, and Zoladex® are some of the currently marketed products that use the ISF phenomenon for implant formation [29].

Polymeric ImDDSs are constructed from different materials in different designs, geometries, and sizes. All the materials used in implantable devices must be biocompatible, not irritative, chemically inert, and not stimulate allergic reactions [30]. The size of the device is preferred to be small enough that could be implanted without the need for general anesthesia and extensive surgery. Some of the formulation factors affecting drug release in polymeric ImDDSs are summarized in Figure 7.5.

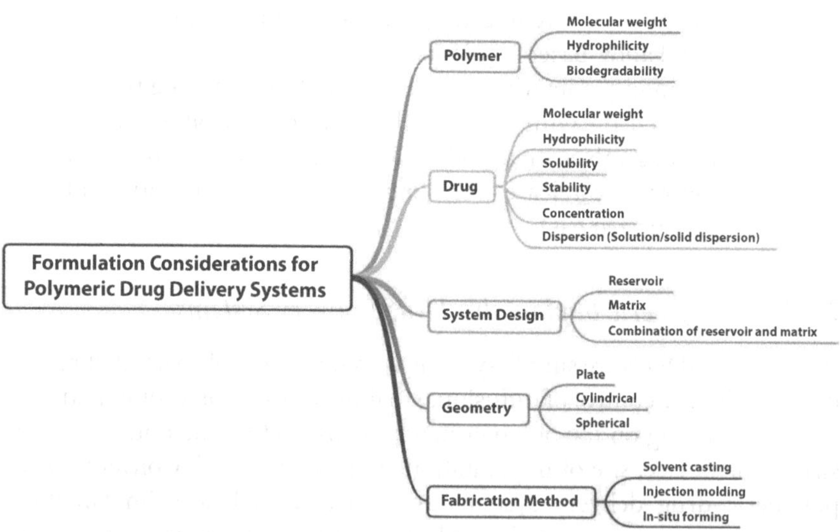

Figure 7.5 Formulation factors that affect drug release from polymeric systems.

Commonly used polymeric ImDDSs are reservoir-based, matrix (monolithic), or a combination of reservoir and matrix. Reservoir-based systems consist of a core containing the drug (reservoir) that is usually coated by a non-biodegradable (and less commonly, biodegradable) polymeric membrane. Diffusion of the drug through the reservoir membrane controls the release of the drug after its partitioning between core and membrane. The desired release profile will be achieved by varying membrane parameters such as molecular weight and composition of the polymer, the thickness of the membrane, the porosity of the membrane (nonporous, microporous), and the implant surface area [7]. The core in this type of implant can be a solid dispersion of a drug in liquid or a drug in solid form.

The possibility of near-zero-order release from reservoir-based systems is the most important advantage of these systems. Reservoir-based polymeric implants are commonly the first and the most convenient choice. However, the disadvantage of these systems is the complexity of production and the possibility of dose dumping due to defects in the membrane, which may lead to drug toxicity. A subcutaneous inserted contraceptive Norplant® is an example of a reservoir-based system. Vitrasert® (Durasert™ technology) is an intraocular reservoir-based system containing ganciclovir to treat cytomegalovirus (CMV) retinitis [31].

In matrix systems, the drug is dispersed or dissolved in a polymer matrix to form a monolithic system. Diffusion of the drug through the polymeric matrix or aqueous pores of the polymer controls the release of the drug from the matrix. In contrast to reservoir-based systems, matrix systems show an initial burst release effect. After a period of drug release because of the depletion of the active agent from the surface (thus increasing of diffusion pathway with passing time) the release rate decreases. Although it is difficult, with the appropriate design of the polymer matrix geometry, it is possible to compensate for drug depletion over time and achieve a constant release with an approximately zero-order mechanism. Plate, cylinder, and sphere are the main geometries for matrix systems. Some drug-eluting cardiac stents used to treat heart patients are based on polymeric matrix systems. Probuphine®, a subcutaneous device releasing buprenorphine, Xience®, an everolimus-eluting coronary stent, and PROPESS® a vaginal insertable dinoprostone-releasing system are examples of matrix-based non-biodegradable polymeric systems [32].

As already mentioned, polymeric systems can be made from biodegradable as well as non-biodegradable polymers. The most important advantage of biodegradable systems compared to non-biodegradable ones is that they do not need to be removed after the release of the drug is complete.

So far, many biodegradable polymers have been considered in studies for drug delivery, but a limited number of these polymers have been used in commercial products. Poly(lactide), poly(lactic-co-glycolic acid) (PLGA), and poly(anhydride) are among the most common biodegradable polymers used in commercial implantable systems. Gliadel® wafer is a poly(anhydride) carmustine-releasing system that is implanted in the brain after brain tumor surgery. Zoladex® is a subcutaneously implanted PLGA matrix system containing goserelin acetate for the palliative treatment of prostate and breast cancers in advanced carcinoma [31]. Eligard® is an *in situ* forming PLGA-based drug delivery system to deliver leuprolide acetate to treat prostate cancer, breast cancer, and endometriosis [33].

The non-degradable and biodegradable polymeric implantable systems can be constructed from different polymeric materials. Table 7.1 lists some of the commonly used polymers in commercial implantable products.

In addition to these polymers, other polymers or combinations of them have been considered in implantable systems. For instance, OncoGel® is an *in situ* forming thermosensitive sol-gel system made of PLGA-PEG-PLGA that delivers paclitaxel [29]. Surodex® is a rod-shaped ocular implant based on PLGA and hydroxy propyl methyl cellulose (HPMC) which is loaded with dexamethasone and releases the drug for 7-10 days [34, 35]. Poly(urethanes), poly(siloxanes), poly(caprolactone), poly(lactic acid), and natural polymers such as chitosan have been considered for implantable polymeric drug delivery in research and development stages [34, 33, 36].

Polymeric implants can be manufactured *via* different techniques such as compression, solvent casting, injection molding, hot melt extrusion, and 3D printing. These methods have been reviewed in detail by Stewart and coworkers [2].

7.3.1.2 Ceramic, Metal, and Metal Alloy Drug Delivery Systems

Ceramic, metal, and metal alloys are inorganic materials that have been investigated as ImDDSs. Ceramics are inorganic compounds such as bone cement, mesoporous silica, aluminum calcium phosphorous oxides, hydroxyapatite, tricalcium phosphate, and ceramic-metal hybrids. Metals and metal alloys such as titanium and titanium alloy, zirconium, stainless steel, and cobalt-chromium alloy, are the most common materials being used in the fabrication of implants. Some of the porous metals such as magnesium foam have also been investigated as a biodegradable matrix for drug delivery systems [4].

Ceramic and metallic compounds mostly can be used in orthopedic applications and bone disease, dental implants, and cardiovascular stents.

Table 7.1 Some commercialized polymeric implantable systems.

Implant type	Polymer	Product name	Prescription	Implantation site	Active ingredient	Design	Company	Duration of drug release
Non-degradable	Poly(ethylene-vinyl acetate) (PEVA)	Probuphine®	opioid addiction	Sub-cutaneous	buprenorphine	Rod/ Matrix	Titan Pharmaceuticals Inc.	6-12 month [37]
		Implanon® (discontinued)/ Nexplanon®	Contraception	Sub-cutaneous	Etonogestrel	Rod/ Reservoir-based Matchstick	N.V. Organon, Oss	3 years [37]
		Nuvaring®	Contraception	Intra-vaginal	Etonogestrel, Ethinyl estradiol	Ring/ thickness is about 4 mm, and diameter is 54 mm	Merck and Co., Kenilworth, NJ, USA	for 3 weeks, and a week interval between the subsequent applications [15]
	Poly(ethylene-vinyl acetate) (EVA), polyvinyl alcohol (PVA)	Vitrasert® (discountinued)	CMV retinitis in AIDS patients	Intra-Ocular / intravitreal	ganciclovir	2.5 mm diameter by 1 mm thick tablet continuously coated PVA and discontinuously coated EVA with a PVA suture tab	Bausch & Lomb, and Iluvien by Alimera sciences	over a 5 to 8 month [34]

(*Continued*)

Table 7.1 Some commercialized polymeric implantable systems. (*Continued*)

Implant type	Polymer	Product name	Prescription	Implantation site	Active ingredient	Design	Company	Duration of drug release
	Silicone	Norplant® (discountinued) and Jadelle®	Contraception	Sub-cutaneous	Levonorgestrel [38]	six rods (2.4 mm x 34 mm)/ e two-rod implants (2.5 mm x 43 mm)	Jadelle® manufactured by Bayer OY	Jadelle® provide contraception for five years
		Estring®	Menopausal symptoms	Intra-vaginal	Estradiol	Opaque ring with a whitish core/ reservoir Outer diameter - 55 mm; cross sectional diameter - 9 mm; core diameter - 2 mm	Pfizer (https://www.estring.com/)	3 months
	Silicone, PVA	Retisert®	chronic noninfectious uveitis	Intravitreal	fluocinolone acetonide [39]	Reservoir/ designed such as Vitrasert	Bausch & Lomb,	30 months
	Methacrylate based (2-hydroxyethyl methacrylate, 2-hydroxypropyl methacrylate, trimethylolpropane trimethacrylate) hydrogel (Hydron® implant technology)	Vantas®	Prostate Cancer	Sub-cutaneous	Histrelin Acetate [40]	Reservoir	Endo Pharmaceuticals Solutions Inc., Malvern, PA, USA	12 month
		Supprellin®LA [41]	Children with central precocious puberty	Subcutaneous Histrelin Acetate			Cranbury, NJ - Valera Pharmaceuticals, Inc	12 months

(*Continued*)

Table 7.1 Some commercialized polymeric implantable systems. *(Continued)*

Implant type	Polymer	Product name	Prescription	Implantation site	Active ingredient	Design	Company	Duration of drug release
	Polyimide, polyvinyl alcohol	ILUVIEN®	Diabetic macular edema	Intravitreal implant	Fluocinolone-acetonide	Rod/Reservoir [4]	Alimera Sciences, Inc.	36 months
Biodegradable	PLGA	Eligard® [42]	palliative treatment of advanced prostate cancer	Subcutaneous	leuprolide acetate	*In situ* forming	Tolmar Inc.	one-, three-, four-, or six-month depending on dose
		Sublocade® [42]	moderate to severe opioid use disorder	Sub-cutaneous	buprenorphine	*In situ* forming	Indivior Inc	over a one-month
		Ozurdex®	macular edema	Intravitreal	dexamethasone	length: 6.5 mm, diameter: 0.45 mm	Allergan Inc. Irvine, CA	2-6 month [43]
		Zoladex® [44]	Prostate cancer	Sub-cutaneous	Goserelin	*In situ* forming/monolithic	AstraZeneca	1 month or 3 months
		Prostap®SR [41]	Prostate cancer	Sub-cutaneous	leuprorelin acetate	Microsphere	Livzon pharmaceutical Inc	once every four weeks
		Risperdal consta® [45]	Schizophrenia	Intra-muscular	Risperidone	Microsphere	JANSSEN	2 weeks

(Continued)

Table 7.1 Some commercialized polymeric implantable systems. (*Continued*)

Implant type	Polymer	Product name	Prescription	Implantation site	Active ingredient	Design	Company	Duration of drug release
	Polyanhydride (1,3-bis-(*p*-carboxyphenoxy) propane and sebacic acid	Gliadel Wafers®	Primary malignant glioma	Intra-tumoral	Carmustine (BCNU)	1-20 μm microparticles compressed to form disk/1.45 cm in diameter and 1.0 mm thick	Eisai Co., Ltd.	-[46]
	PLGA+HPMC	Surodex® [35]	controle postoperative inflammation	anterior chamber	dexamethasone	rod-shaped 1.0×0.5 mm	Allergan, Inc., Irvine, CA, USA	7–10 days
	PLGA-PEG-PLGA [47]	Oncogel®	Oesophageal cancer	Intra-tumoral	paclitaxel	*In situ* forming	Clinical trial	4 to 6 weeks

Antimicrobial, anti-inflammatory, stenosis-preventive drugs and growth factors are examples of therapeutic agents that have been studied in orthopedic, dental, and cardiovascular implants. In these systems, the active agent (drug, biological compound, or any bioactive material) may be deposited directly on the surface of the device or attached to it by a covalent bond. However, the most popular way is the incorporation of the drug into a polymeric matrix (biodegradable or non-biodegradable) followed by coating the matrix on the implant surface. Alternatively, the drug can be embedded in the bulk of the device [48].

So far, several drug-releasing cardiac stents of this type are commercially available. For instance, TiOxNy-coated stents are marketed metal stents with a thin film of nitrogen-containing titanium oxide. In these systems, the TiOxNy film is deposited by magnetron sputtering. TiOxNy acts as a nitric oxide deposit to prevent restoration [49]. An FDA-approved zotarolimus-eluting stent, Resolute Onyx™ (Medtronic), is a core wire technology with a platinum-iridium core and cobalt alloy shell with BIOLINX polymer coating to control drug release and make the stent biocompatible [50]. Abbott's XIENCE is a fully bioresorbable everolimus eluting stent. In this system, the poly (vinylidene fluoride-co-hexafluoropropylene) (PVDF-HFP) and poly (n-butyl methacrylate) (PBMA) polymers are coated on the cobalt-chromium base [51].

7.3.1.3 Passive Pumps

Passive pumps, generally, are referred to the pumps that do not need an external electrical source for drug release. These pumps use chemical or mechanical energy to pump the drug. Osmotic pumps and propellant infusion pumps are two main subclasses of passive pumps.

Osmotic Pumps

Osmotic pumps are controlled-release passive pumps that can be implanted or administered orally. The osmotic pump's performance is based on the osmosis phenomenon. Although, osmotic pumps have different designs, most of them are composed of the same components and mechanisms. These pumps are commonly composed of an osmotic-driven agent (ionic components/inorganic salts such as sodium chloride and potassium chloride), a semi-permeable membrane (cellulose derivatives like acetate, diacetate, and butyrate), drug reservoir (e.g., polymeric elastomer, plasticizer, etc), a drug that can be in solid or liquid form (i.e., suspension or solution, respectively), flux regulator, and orifice. The semipermeable membrane controls the entrance and uptake of water by an osmotic agent, which leads to volume increase followed by pushing out the drug through the orifice of the drug reservoir. The characteristic of these systems is constant; so, the drug is continuously released with no control over initiation, turning off, and time setting up after implantation. The size of the orifice is critical in controlling drug release to achieve a zero-order profile. The size of the orifice should not be too large or too small. An orifice diameter of 0.075-0.275 mm is generally desired [52]. Osmogene concentration and membrane properties such as the composition and thickness are other factors affecting the drug release from the osmotic pump [3]. In this chapter, implantable osmotic pumps are considered and osmotic pumps which are used orally have been discussed in Chapter 4.

The osmotic pump systems could be designed as one-compartment, two-compartment, and multi-compartment systems [4, 53]. In one compartment osmotic pump, there is one drug reservoir and the drug itself acts as an osmotic agent (Figure 7.6). In this system, the key factor that affects the drug release rate from osmotic pumps is the water solubility of the drug.

High water-soluble drugs show a higher drug release rate. For poorly water-soluble drugs, some solubilizing agents should be added to regulate the drug release. LiRIS® (TARIS Biomedical®, Lexington, MA, USA) is an intravesical system, that delivers lidocaine to the bladder and is designed in one compartment format [53]. A special design of one-comportment osmotic pumps is a monolithic system. In this system, the drug is dispersed in a polymer matrix and there is not any membrane around the polymeric matrix. The diffusion of the water to the polymeric matrix and dilution of the drug leads to swelling, resulting in the drug leaching out. It should be

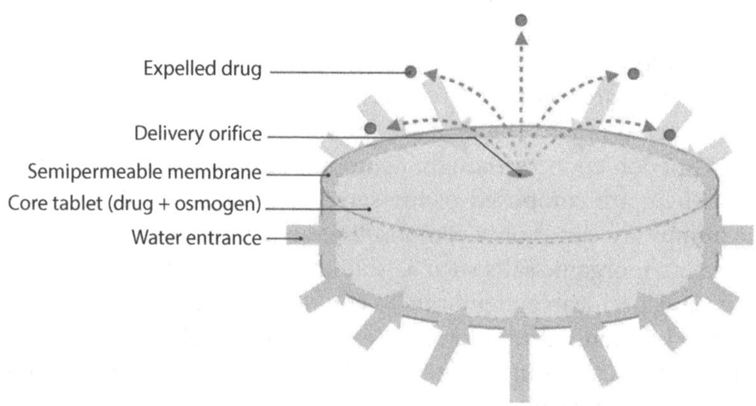

Figure 7.6 A schemeatic illustration of an osmotic pump.

considered that, in these systems, the drug release is not under the control of diffusion. Although monolithic or matrix systems are categorized as pump systems, these can also be included in the class of polymeric systems that release the drug under swelling.

In two-compartment osmotic pump systems, the drug and osmotic agent are placed in two separate compartments. The main advantage of the two-compartment systems is that, unlike single-compartment systems, the drug does not come into contact with water. Therefore, these drug delivery systems may be suitable for delivering water-sensitive drugs, especially those that are subject to hydrolysis by water, such as proteins and peptides [53].

Depending on the manufacturer, two-compartment systems can have different designs. ALZET® (DURECT Corp., Cupertino, CA, USA) is an osmotic pump that is commercialized and widely used in research studies. In this system, the drug is placed in a compressible reservoir. A layer of osmotic agent surrounds the reservoir and a semi-permeable coating encompasses the osmotic agent (Figure 7.7, left). The osmotic agent absorbs water which leads to an increase in the pressure in the drug reservoir, so the drug is forced out of the orifice. The amount of released drug can be manipulated by varying the concentration of the drug in a dispersion. Different volumes of ALZET pumps are available which determine the drug release rate and duration [53, 54].

DUROS® is another example of a two-compartment osmotic systems. This system is composed of a drug reservoir and an osmotic agent reservoir separated by a flexible piston. When the water penetrates through the semipermeable membrane, incorporated at the top of the osmotic agent reservoir, the osmotic compartment expands and moves the piston. The piston pushes the drug out of the orifice at the end of the drug reservoir (Figure 7.7, right) [53]. Similar to ALZET pumps, the amount

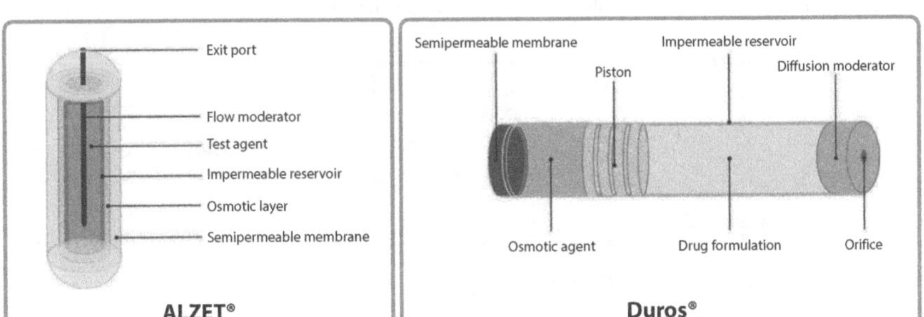

Figure 7.7 Examples of two-compartment osmotic pumps.

of drug release can be controlled by the concentration of the drug in the reservoir. FDA-approved Viadur® (Bayer AG, Leverkusen, Germany) was based on DUROS® technology that was used for the delivery of leuprolide acetate. However, because of the manufacturing costs, its production was discontinued.

Multi-compartment osmotic pumps are another type of osmotic pump that, in addition to the drug and the osmotic agent reservoirs, have also a solvent reservoir. Multi-compartment osmotic pumps that work like two-compartment osmotic pumps, based on the osmotic phenomenon. Rose-Nelson pump is a simple example of this type of pump (Figure 7.8) [52].

Osmotic pumps are almost constructed from non-biodegradable polymers, although, osmotic pumps with biodegradable polymers have also been investigated [55] and patented [56].

The advantage of osmotic pumps is that the drug delivery parameters are not affected by the drug release medium (e.g. pH) so the *in vitro* release behavior will be a good estimation of *in vivo* release in this regard.

Table 7.2 shows some of the studied and commercially available osmotic systems. More examples of other kinds of osmotic pumps and multi-compartment osmotic systems are completely reviewed by Herrlich *et al.* [53].

In addition to the different types of osmotic pumps discussed earlier, some other complicated designs of drug delivery systems that work based on the osmotic phenomenon, have recently been introduced. Acuros (available by Acuros GmbH, a spin-off company of the Humboldt University, Berlin, Germany) and hydrogel pumps are examples of these systems. The advantages of these systems are the possibility of self-administration, the potential of being on-demand, designing in large volumes, and fully

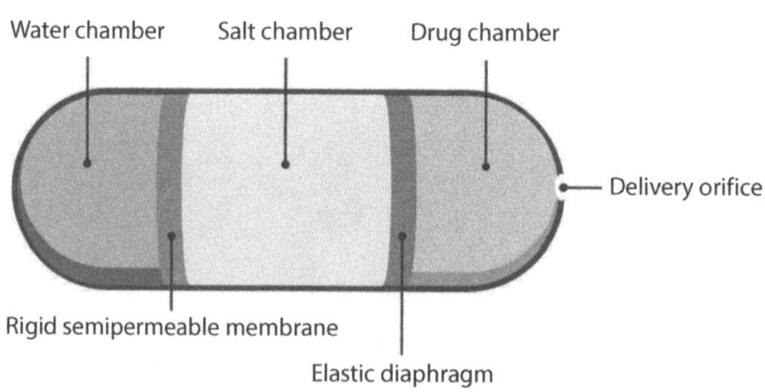

Figure 7.8 A schematic illustration of multi-compartment Rose-Nelson osmotic pump.

Table 7.2 Commercially available and under clinical trial osmotic implantable systems.

Product	Indication	Implantation site	Active ingredient	Design	Company	Duration of drug release
GemRIS	bladder cancer	Intra-vesical [57]	Gemcitabine/ solid	One compartment/semipermeable silicone	TARIS Biomedical	7 days
LiRIS®	painful bladder syndrome	Intravesical	Lidocaine	One compartment/silicone tube	TARIS Biomedical	two weeks [58]
ALZET®	for research purposes	Subcutaneous/ specific target site by catheter attachment	Any active agent	Two-compartment/drug reservoir made of thermoplastic hydrocarbon, the semipermeable coating made of cellulose ester	DURECT Corp	-
Viadur®	palliative treatment of advanced prostate cancer	subcutaneously in the inner aspect of the upper arm	leuprolide acetate	Two-compartment/DUROS® technology, the semi-permeable membrane, made of polyurethane polymer, a piston made from an elastomeric material, and cylindrical housing is made from titanium alloy	ALZA	12 months [59]
[59] Chronogesic™	Treatment of cancer pain and non-cancer pain	Subcutaneous Upper inner arm	sufentanil	Two-compartment/DUROS® technology	ENDO pharmaceuticals	3 months [60]

working without electronics. These systems are not implantable, so their details are not discussed here. The details of these systems are discussed in the reference [53].

Propellant Infusion Pump

These pumps are an alternative to the osmotic pump which uses a propellant (commonly fluorocarbon) to produce pressure instead of the osmosis. In these systems, drugs have a continuous and constant release with a predetermined rate and no control after implantation, similar to the osmotic pumps. Generally, propellant pumps are composed of a propellant chamber and a drug chamber separated by a metal bellow. The principle of drug flow in these pumps is based on the pressure produced by the change of propellant material from liquid to gas under the influence of body temperature, and as a result, it pushes the bellow to press the drug out. The principal components of propellant infusion pumps are summarized in Figure 7.9.

The body of these pumps as well as the accordion-like bellow are constructed from biocompatible titanium. The catheter is composed of flexible polymer. Chlorofluorocarbon is commonly used as a propellant, however, n-butanol is used also. The flow rate of drug release is controlled by a flow restrictor and the drug dosing is adjusted by the physician by varying the drug concentration when refilling the drug reservoir [61]. The pressure of the propellant, its flow rate, and the design of the device are the main factors affecting the drug release profile from propellant infusion pumps [3]. These pumps are commonly implanted in the abdomen and a catheter

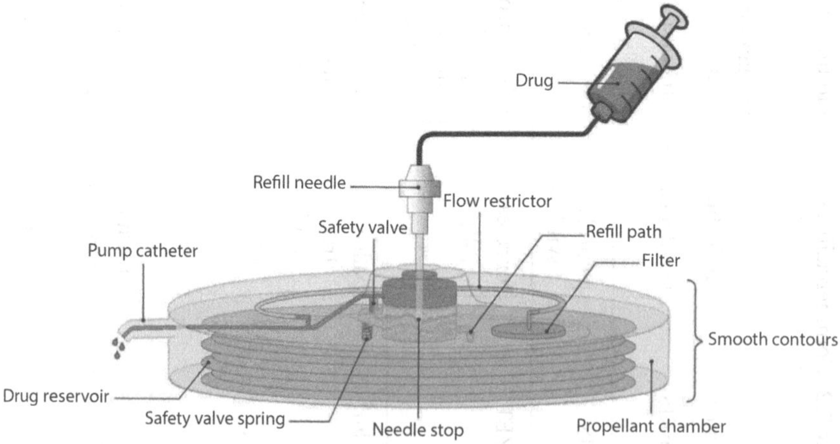

Figure 7.9 Infusion pump based on pressurized gas.

directly delivers the drug to the site of action. In this system, the dose of the administered drug is regulated by changing the concentration of the drug in the reservoir. The large reservoir of these fixed-rate infusion pumps makes them suitable for long-term drug delivery for chronic pain and cancer management. Codman® 3000 pump was an FDA-approved propellant infusion pump for intrathecal delivery of therapeutic agents for pain management or chemotherapy. However, the production of this pump was stopped in 2018, probably because of low demand and expenses [4].

As it is realized from the mechanism of operation of propellant infusion pumps, they work passively, however, by controlling the valves electrically or in other ways, the drug release can be controlled after the implantation of the system in the patient's body. These pumps are classified under the category of active pumps. Codman MedStream pump is a commercial programmable active system that pumps the drug based on the propellant infusion and works with programable valves. This pump works based on positive pressure force produced by gas (propellant) expansion for pumping (the pumping system is not electrically activated). In this system, a ceramic valve is regulated to control and program the drug flow [4].

7.3.2 Active (Programmable) Release Systems

Active programable systems have sophisticated control over the initiation, profile (timing), and turning off the drug release which makes it feasible for the patient or physician to control drug dosing. Active systems can deliver medicine at a constant rate, as a blouse, pulsatile, or in a complicated infusion program. These systems may have additional components such as a battery, pumping device, valves and channels, sensors, and a controlling system to set up precise and accurate dosing. Each part can be constructed from polymeric or metallic materials depending on the action, and usage. The pump (one of the most important components in an active system) and valves can be programmed to control and manage the flow. So far, various types of electrically active pumps, electromechanical pumps, with different pumping mechanisms have been studied and some of them have been used in commercial ImDDSs.

Peristaltic Pumps

Peristaltic pumps are composed of an electrically (battery) driven solenoid that drives on a flexible tube containing drug solution and pushes it through the tube to inject the drug into the application site (Figure 7.10).

The SynchroMed II pump is an example of a system that works based on peristaltic pumping and has been trialed for infusion of treprostinil. It has

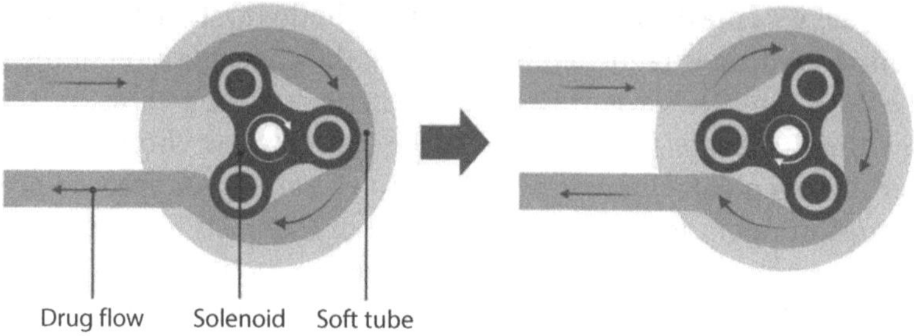

Figure 7.10 A schematic illustration of a peristaltic pump.

been approved for remodeling the treatment of pulmonary arterial hypertension [62, 63]. SynchroMed II also can be used for intrathecal delivery of opioids and ziconotide (for the management of chronic pain) and baclofen (for server spasms). In this system, the drug is driven through the pump tubing by the pressure produced by a gas-pressurized chamber sited under the drug chamber (Figure 7.11). SynchroMed II can be programmed to personalize the dosing of medication depending on the patient's needs [2].

The technology development in the field of micro equipment manufacturing (including micropumps, microvalves, micro-reservoirs, and other micro equipment) has led to the advancement in the construction of micro pumping systems. Generally, micro-pumping devices are categorized into mechanical and non-mechanical pumps [64]. In non-mechanical pumps, the force is directly transformed into the liquid and does not involve any movement part (diaphragm). In mechanical pumps, a stimulus causes deformation or bending in a flexible membrane (diaphragm), and the

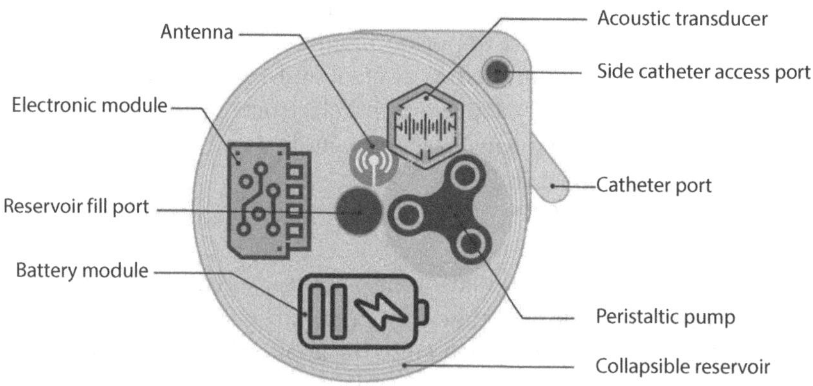

Figure 7.11 A schematic representation of SynchroMed II.

pressure created on the fluid as a result of bending in the membrane, draws the fluid into the pump chamber and then pumps it out. With the increase in the knowledge of microfabrication, the construction of microelectromechanical pumps grew increasingly. Microelectromechanical pumps have been considered in the field of ImDDSs to construct microelectromechanical systems (MEMS). MEMS are micro-scale devices including an electrical circuitry and movable parts that can accurately control the drug delivery. Nanoelectromechanical systems (NEMS) are also pump-based systems that overlap with microelectromechanical systems in terms of utilization and the method of action. The miniaturization of electronically-controlled pumping implants makes them more suitable for implementation in the body. On the other hand, low power usage of micropumps aids to compensate for the limited lifespan of the battery [65].

Some of the most popular mechanical pumping systems are electromagnetic, piezoelectric, and phase shape alloy types. Acoustic and electrochemical-based micropumps are categorized as non-mechanical micropumps [65].

Table 7.3 provides more detail about the mentioned mechanical and non-mechanical systems.

An interesting feature of electrochemical (non-mechanical) and magnetic-based (mechanical) micropumps is the potential of these systems to operate without a battery. Battery-less micropumps can be miniaturized for implantation purposes and thus the limitation on reducing the size of implantable battery-powered micropumps can be overcome.

Microchips

Microchips are an alternative group of micro-based systems that have been extensively investigated in the field of drug delivery and diagnosis. In microchips, the drug is stored in several reservoirs and can be released in a planned manner at the appropriate time and dose. Microchips are commonly composed of a substrate, in which the drug reservoirs are machined or etched, and a capping membrane that covers the reservoir. The membrane can be dissolved in response to an actuator based on electrochemical, electrothermal, and infrared radiation then the drug can be released (Figure 7.12). Microchips contain a circuit to control the process and a power supply [71]. These systems are fabricated based on microfabrication techniques like photolithography, soft lithography, film deposition, and etching. The substrate is commonly constructed from silicone, gold, or other materials, depending on the mechanism of stimuli to dissolve the capping membrane. Although microchips are generally classified as active systems, they can also work passively. In these systems, the capping agent is

Table 7.3 Classification and characterization of mechanical and non-mechanical micropumps.

Type of micropump	Physical driving forces	Principle	Reservoir	Stimuli	Example
Mechanical	Piezoelectric	Mechanical deformation of a diaphragm (lead zirconate titanate or piezoelectric ceramic) as the result of applied electric field	PDMS, polyethylene, parylene, SU-8 photoresist, hydrogel, among others, were considered for device building	Battery	Piezoelectric Silicon Micropump for Drug Delivery Applications [66]
	Magnetic	Deformation of the membrane (magnetic PDMS, iron oxide nanoparticles coated PDMS) due to applying a magnetic field	PDMS	External magnetic field	Intravitreal implantable magnetic micropump for on-demand VEGFR-targeted drug delivery [67]
	Phase change Thermally induce phase change and bend the alloy	Thermally induce phase change and bend the membrane (gallium or paraffin)	Silicon	metals micro heaters produce heat to make the phase change	A nanoliter resolution implantable micropump for murine inner ear drug delivery [68]

(*Continued*)

Table 7.3 Classification and characterization of mechanical and non-mechanical micropumps. (*Continued*)

Type of micropump	**Physical driving forces**	**Principle**	**Reservoir**	**Stimuli**	**Example**
Non-mechanical	Electrochemical	Electrochemical reactions (electrolysis) and production of gas (O_2 and H_2) and increasing the pressure. Titanium/platinum electrodes	PDMS	Battery/ triboelectric nanogenerator in which mechanical energy is converted to electricity.	A simple electrochemical micropump: Design and fabrication [68]
	Acoustic	Creation of flow pattern via microbubbles Oscillation or acoustically oscillating solid structures	PDMS	an external piezoelectric transducer creates an acoustic field [69]	A Digital Acoustofluidic Pump Powered by Localized Fluid-Substrate Interactions [70]

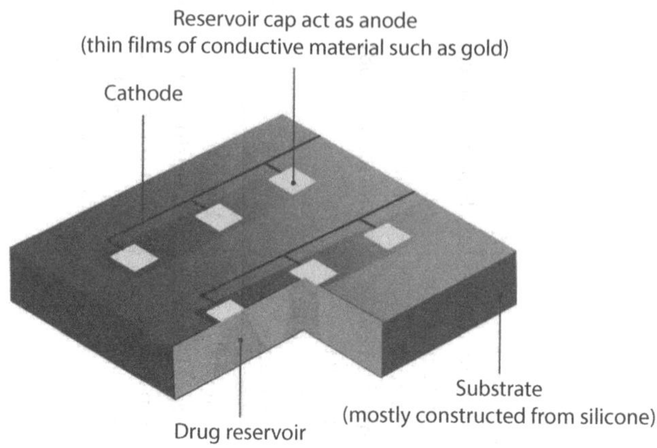

Figure 7.12 A schematic illustration of a microchip.

a degradable or permeable membrane and there is no real control on drug release after implantation.

7.4 Challenges and Opportunities

ImDDSs as a new type of drug delivery system face opportunities and challenges (Figure 7.13). Simultaneous delivery of different drugs, delivering the therapeutic agents in a pulsatile and staggered manner, being responsive or on demand, rapid delivery in ambulatory emergency care, and being controlled from the outside by the patient or the therapist are some of the advantages of these systems [26, 72, 73]. Most of the ImDDSs, regardless of the mechanism of action, may encounter challenges (Figure 7.13). In addition, ImDDSs have disadvantages such as the need for sterilization, the need to be bio- and cytocompatible and not-irritative, limitations in

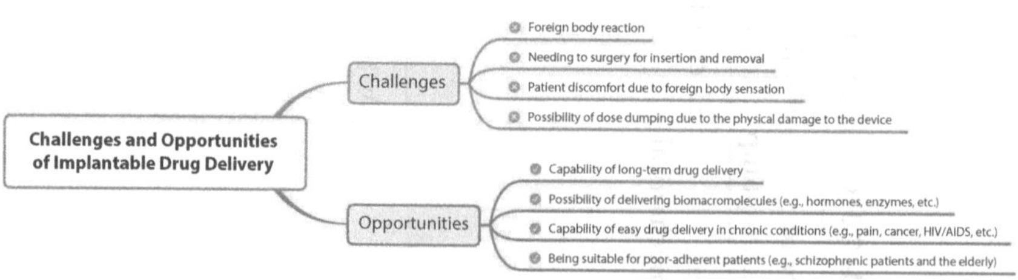

Figure 7.13 Challenges and opportunities of implantable drug delivery.

loading volume of the therapeutic agent and the need to refill the reservoir, and scale up and regulatory issues [74–76].

A specific disadvantage of passive release systems is the inability to control the amount, rate, onset, and ending of release after implantation in the body. However, these systems have a simple design and do not include electrical circuits. Active drug delivery systems overcome the limitations of passive systems in drug delivery. However, they have their limitations such as the limitation in the lifetime (capacity) and size of the battery and the need for an electrical circuit which is a challenge for long operation time and overall volume of the device. So, the development of active systems requires more studies to overcome the aforementioned limitations.

One of the main challenges with implantable systems that may affect the drug release amount, rate, and pharmacokinetics of dug, is the *in vivo* micro/macro environments. In some cases, it is observed that a collagen-based membrane (fibrous capsule) is formed around the system, which may alter the drug diffusion across the membrane and thereby, its release kinetics. The formation of this fibrous membrane is a result of inflammatory body reaction at the interface of the tissue and implanted system a long time after implantation [77]. The effect of this phenomenon would be more complicated realizing that each patient's body response to the foreign devices will be different and the thickness of this membrane around the system may be varied in patients. It is also demonstrated that the size of the instrument can affect the chronic inflammatory response of the body and therefore the fibrous capsule formation [78]. On the other hand, the larger the size of the system the larger site for the surgery. This greater trauma may lead to a greater response of the body. The implantation site is another factor that has a crucial effect on foreign body reactions because of the difference in the immune response of the host tissue. The body's response, also, is largely related to the physicochemical properties of the implant surface. As the hydrophilicity increases, the absorption of the protein on the surface is decreased. In this way, some of the hydrophilic polymers such as alginate, hyaluronan, and polyvinyl alcohol can be coated on the surface to mask the hydrophobic surface. However, this strategy cannot completely prevent fibrous tissue formation [79]. Since the formation of fibrous capsules may affect the long-term release of the drug from the implanted system, it seems necessary to study the release of the drug in the conditions in which the fibrous capsule is formed around the implant. In this regard, an *in vivo* study of the human parathyroid hormone fragment ([hPTH(1–34)]) on 7 patients, released from a fibrous encapsulated wirelessly controlled drug delivery microchip, demonstrated bioequivalence with the approved dose for injection. This study also demonstrated that the pharmacokinetic profile from fibrous-surrounded implants was comparable to multiple subcutaneous injections [80].

As most implantable systems are non-biodegradable, the foreign body response is unavoidable. Nevertheless, by miniaturization of the implantable system, using a biocompatible material, and choosing a suitable place for implantation, the foreign body response could be reduced to some extent.

7.5 Pharmacokinetics

ImDDSs overcome the bioavailability barriers and pharmacokinetic limitations associated with other routes of drug administration (e.g., blood-brain and cerebral barrier, the stratum corneum barrier, first past metabolism, intraocular barriers, gastrointestinal tract absorption barriers, etc.). In the development of ImDDSs, construction parameters of the designed system (e.g., the mechanism of drug release), physiological factors (e.g., physiological barriers of the implantation site), and body-system interaction (the reaction of the host body to the foreign system, e.g., fibrotic tissue formation around the system) should be carefully considered.

As discussed before, in polymeric drug delivery systems the drug release kinetic depends on factors such as the design and size of the system, the nature of the polymer, and drugs such as polymorphism, hydrophilicity/hydrophobicity, drug dispersion in the matrix (as solid or molecular dispersion) and its interaction with the polymer [2, 81]. Several mechanisms control the drug release from a drug-containing system. Generally, drug release from a polymeric system can be controlled by the dissolution of active agent, drug partitioning in the medium, diffusion or mass transfer of active molecules through the system and at the system/water interface, swelling of the system, erosion of the matrices, osmosis phenomenon, or a combination of these mechanisms [82].

Various mathematical models have been attempted to describe the drug release kinetic from polymeric matrices. First-order, zero-order, Higuchi, Hixson–Crowell, Korsmeyer–Peppas, Baker–Lonsdale, Weibull, Hopfenberg, and Gompertz are the most common mathematical models used so far [83]. Zero-order release is an ideal kinetic for passive drug delivery systems, especially for drugs with narrow therapeutic windows. Therefore, efforts have been made to manipulate various affecting parameters such as geometry, degradation behavior, drug chemistry, morphology, composition of the polymer, construction method, and many other factors to achieve zero-order release in polymeric implantable systems. One of the most attractive advantages of osmotic pumps (compared with diffusion-, swelling-, or erosion-controlled release systems) is the capability of providing ideal zero-order release. In these systems factors affecting drug releases such as orifice size, semipermeable membrane properties, and osmotic agent concentration and composition can be modified to reach a zero-order release. An alternative option is propellant infusion pumps which are considered among the best zero-order releasing implantable systems. In these systems, drugs can be released in a predetermined and constant flow for a long time. In programmable systems, the drug release is pulsatile and complex and can be tailored to the desired profile [4].

In addition to the factors mentioned above that control the drug release from the implantable systems, physiological factors can alter the drug release. The pH, enzymes, temperature, interstitial fluid volume, vascular system, and hydrophilicity of the implantation site are some of these parameters. Therefore, the *in vitro* conditions for studying the drug release must be a good simulation of the real *in vivo* situation. Once the drug is released from the system, it must be absorbed to show the appropriate therapeutic effect. The drug absorption is related to the physiological parameters of the implantation site like the boundary layers. All of the parameters that affect the release of the drug (related to the nature of the system or administration place) and absorption from the implantation site can affect the pharmacokinetics and bioavailability of the drug [84]. In the following, the pharmacokinetics of the drug in the drug-eluting stent, intrathecal, and subcutaneous drug delivery along with several studies reported in this field have been presented.

7.5.1 Drug-Eluting Stents

The influencing factors in local pharmacokinetics of drug-eluting stents relate to the physicochemical properties of the drug (e.g., molecular weight, charge, hydrophilicity/hydrophobicity) the endothelium barrier effect (barrier to transmural hydraulic flux and diffusion), arterial structure, geometry, and its physical status (thrombosis and atherosclerosis) and the design and placement of the stent [85].

The pharmacokinetic evaluation of everolimus-eluting stent (XIENCE V) was made to investigate the systemic exposure of everolimus. The total drug amount that the patients received was in the range of 53 to 588μg. This study showed in all doses that individual T_{max} (time to peak blood concentration) and C_{max} (peak blood concentration) were in the range of 0.13 to 2.17 h and 0.14 to 2.79 ng/ml, respectively, while the concentration was quantifiable only up to 720 h after the last implantation. These results indicated that a limited concentration of everolimus entered the systemic circulation. The mean whole concentration-time curve showed that the drug concentration reached its maximum level in 10 to 60 min after implant insertion and then decreased biphasically, indicating an initial fast distribution/elimination, and a second slow elimination [86].

7.5.2 Intrathecal

In intrathecal drug delivery, the physicochemical properties of drug molecules (such as ionization state, hydrophobicity, and size of the molecule) as well as the volume and rate of flow can affect the pharmacokinetics.

In a study to compare the effect of continuous and routine intrathecal administration of amphotericin B in rabbits, it was shown that when 0.15mg/kg/24h of the drug was administered continuously the C_{max} reached 7.01 μg/ml at 4h and then decreased to 1.0 to 1.34 μg/ml while in conventional administration of 0.015 mg/kg the C_{max} reached 3.41 μg/ml at 1 h and then decreased progressively [87].

The size (length) of the drug molecule is another factor that can affect the pharmacokinetics of intrathecal drug delivery. For example, the study on the pharmacokinetic behavior of oligodeoxynucleotides with different lengths revealed that the AUC (area under the concentration-time curve), clearance, and volumes of distribution of the drugs differ significantly depending on the size of the molecule [88].

7.5.3 Subcutaneous

Generally, the absorption of subcutaneously administered drugs can be affected by the physicochemical properties of drug molecules like charge, size, molecular weight, and hydrophilicity/hydrophobicity as well as the formulation characteristics such as ionic strength, concentration, volume, and pH [89]. The passage through tissues at the administration site to the vascular or lymphatic systems, penetration to the vascular endothelial cells, degradation by local enzymes, and elimination by endocytosis and phagocytosis are other factors that influence drug absorption through the subcutaneous route.

Pharmacokinetic analysis of 50 mg histrelin acetate subcutaneous implant showed that in most of the patients, the drug concentration was detectable in T_{max} equal to 12 h, it reached the peak concentration of 1.1ng/ml. The concentration then declined slowly over 52 weeks. The mean apparent clearance in patients who received this implant was 174 ± 56.5 ml/min which was comparable with the healthy volunteers, 179 ± 37.8 ml/

min, who received subcutaneous blouses [90]. Another pharmacokinetic study of biodegradable PLGA *in situ* forming risperidone implant (90 mg and 120 mg doses) showed two concentration peaks after subcutaneous injection of the system. The first peak appeared 4-8 hours after injection (initial burst release of the drug) and the second peak occurred at 10-14 days (the delayed release of the drug) [89].

7.6 Products in the Market

As mentioned in the previous sections, several ImDDSs are commercially available or under clinical trials. These systems have been used for various applications, including contraception, pain and spasm management,

Table 7.4 Some of FDA FDA-approved implantable drug delivery systems.

Implant	Implantation site	Active agent	Approval date	Indication	Drug release duration
Probuphine	subdermal	buprenorphine	2016	Treatment of opioid dependence	6 months
SCENESSE	subcutaneous	afamelantide	2019	increase pain-free light exposure in adult patients with a history of phototoxic reactions from erythropoietic protoporphyria	2 months
Synergy	stent	everolimus	2015	Coronary artery disease	3 months
Orsiro	Stent	Sirolimus	2019	Coronary artery disease	4 months
Xience apline	Stent	Everolimus	2014	Coronary artery disease	4 months
Prometrall	Intrathecal	Morphine, baclofen	2012	Severe spasticity, pain management	6 month to 1 year
YUTIQ	Intraocular	Fluocinolone acetonide	2018	Posterior segment uveitis	3 years
Durysta	Intraocular	Bimatoprost	2020	Glaucoma	6 months
Susvimo	Intraocular	Ranibizumab	2021	Neovascular age-related macular degeneration	6 months

and treatment of ocular, cardiovascular, and cancer diseases. Table 7.4 lists some of these implantable systems that have received FDA approval in the last ten years.

References

1. Kleiner, L.W., Wright, J.C., Wang, Y., Evolution of implantable and insertable drug delivery systems. *J. Control. Release*, 181, 1–10, 2014.
2. Stewart, S.A., Domínguez-Robles, J., Donnelly, R.F., Larrañeta, E., Implantable Polymeric Drug Delivery Devices: Classification, Manufacture, Materials, and Clinical Applications. *Polym. (Basel)*, 10, 12, 1379, 2018.
3. Laracuente, M.L., Yu, M.H., McHugh, K.J., Zero-order drug delivery: State of the art and future prospects. *J. Control. Release*, 327, 834–56, 2020.
4. Pons-Faudoa, F.P., Ballerini, A., Sakamoto, J., Grattoni, A., Advanced implantable drug delivery technologies: transforming the clinical landscape of therapeutics for chronic diseases. *Biomed. Microdevices*, 21, 1–22, 2019.
5. Nance, E., Pun, S., Asiga, R., Sellers, D., Drug delivery to the central nervous system. *Nat. Rev. Mater.*, 7, 314–31, 2022.
6. Mostafa, M., Al Fatease, A., Alany, R.G., Abdelkader, H., Recent Advances of Ocular Drug Delivery Systems: Prominence of Ocular Implants for Chronic Eye Diseases. *Pharmaceutics*, 15, 6, 1746, 2023.
7. Lemmens, G., Van Camp, A., Kourula, S., Vanuytsel, T., Augustijns, P., Drug Disposition in the Lower Gastrointestinal Tract: Targeting and Monitoring. *Pharmaceutics*, 13, 2, 161, 2021.
8. Gorzelanny, C., Mess, C., Schneider, S.W., Huck, V., Brandner, J.M., Skin Barriers in Dermal Drug Delivery: Which Barriers Have to Be Overcome and How Can We Measure Them? *Pharmaceutics*, 12, 7, 684, 2020.
9. Srivastava, V., Status of Implantable and Long-acting Injectable Technologies, in: Implantable Technologies: *Peptides and Small Molecules Drug Delivery*, Srivastava, V,. (Ed.), The Royal Society of Chemistry, London, p. 1–13, 2021.
10. Richter, W.F., Bhansali, S.G., Morris, M.E., Mechanistic determinants of biotherapeutics absorption following SC administration. *AAPS J.*, 14, 559–70, 2012.
11. McLennan, D.N., Porter, C.J., Charman, S.A., Subcutaneous drug delivery and the role of the lymphatics. *Drug Discov. Today Technol.*, 2, 1, 89–96, 2005.
12. Collins, D.S., Kourtis, L.C., Thyagarajapuram, N.R., Sirkar, R., Kapur, S., Harrison, M.W. *et al.*, Optimizing the Bioavailability of Subcutaneously Administered Biotherapeutics Through Mechanochemical Drivers. *Pharm. Res.*, 34, 10, 2000–11, 2017.
13. Kroin, J.S., Intrathecal drug administration: present use and future trends. *Clin. Pharmacokinet.*, 22, 5, 319–26, 1992.

14. Lynch, L., Intrathecal drug delivery systems. *Contin. Educ. Anaesth. Crit. Care Pain*, 14, 1, 27–31, 2014.
15. Osmalek, T., Froelich, A., Jadach, B., Tatarek, A., Gadziński, P., Falana, A. *et al.*, Recent advances in polymer-based vaginal drug delivery systems. *Pharmaceutics*, 13, 6, 884, 2021.
16. Musculature, PF, Contemporary views on female pelvic anatomy. *Cleve Clin. J. Med.*, 72, S3, 2005.
17. Vermani, K. and Garg, S., The scope and potential of vaginal drug delivery. *Pharm. Sci. Technol. Today*, 3, 10, 359–64, 2000.
18. Hussain, A. and Ahsan, F., The vagina as a route for systemic drug delivery. *J. Control. Release*, 103, 2, 301–13, 2005.
19. Joice, G.A., Bivalacqua, T.J., Kates, M., Optimizing pharmacokinetics of intravesical chemotherapy for bladder cancer. *Nat. Rev. Urol.*, 16, 10, 599–612, 2019.
20. Sarfraz, M., Qamar, S., Rehman, M.U., Tahir, M.A., Ijaz, M., Ahsan, A. *et al.*, Nano-Formulation Based Intravesical Drug Delivery Systems: An Overview of Versatile Approaches to Improve Urinary Bladder Diseases. *Pharmaceutics*, 14, 9, 1909, 2022.
21. Marchenko, I.V. and Trushina, D.B., Local Drug Delivery in Bladder Cancer: Advances of Nano/Micro/Macro-Scale Drug Delivery Systems. *Pharmaceutics*, 15, 12, 2724, 2023.
22. Kolb, H., Fernandez, E., Nelson, R. (Eds.), *Webvision: The Organization of the Retina and Visual System*, University of Utah Health Sciences Center, Salt Lake City (UT), 1995.
23. Kuno, N. and Fujii, S., Recent Advances in Ocular Drug Delivery Systems. *Polymers*, 3, 1, 193–221, 2011.
24. Achouri, D., Alhanout, K., Piccerelle, P., Andrieu, V., Recent advances in ocular drug delivery. *Drug Dev. Ind. Pharm.*, 39, 11, 1599–617, 2013.
25. Dash, A.K. and Cudworth, G.C., Therapeutic applications of implantable drug delivery systems. *J. Pharmacol. Toxicol. Methods*, 40, 1, 1–12, 1998.
26. Magill, E., Demartis, S., Gavini, E., Permana, A.D., Thakur, R.R.S., Adrianto, M.F. *et al.*, Solid implantable devices for sustained drug delivery. *Adv. Drug Deliv. Rev.*, 199, 114950, 2023.
27. Fung, S., Polymeric implants for cancer chemotherapy. *Adv. Drug Deliv. Rev.*, 26, 209–230, 1997.
28. Kempe, S. and Mäder, K., *In situ* forming implants—an attractive formulation principle for parenteral depot formulations. *J. Control. Release*, 161, 2, 668–79, 2012.
29. Sinha, V.R., Sharma, S., Silki, Kaur, M., Sarwal, A., 6 - Current Polyesteric Systems for Advanced Drug Delivery, in: *Nanoarchitectonics for Smart Delivery and Drug Targeting*, Holban, A.M., Grumezescu, A.M. (Eds), William Andrew Publishing, New York, p. 143–68, 2016.
30. Staples, M., Microchips and controlled-release drug reservoirs. *WIREs Nanomed. Nanobiotechnol.*, 2, 4, 400–17, 2010.

31. Yang, W.W. and Pierstorff, E., Reservoir-based polymer drug delivery systems. *J. Lab. Autom.*, 17, 1, 50–8, 2012.
32. Sanopoulou, M. and Papadokostaki, K.G., Controlled drug release systems: Mechanisms and kinetics, in: *Biomedical Membranes (Bio) Artificial Organs*, pp. 1–33, World Scientific, Singapore, 2017.
33. Gonella, A., Grizot, S., Liu, F., López Noriega, A., Richard, J., Long-acting injectable formulation technologies: challenges and opportunities for the delivery of fragile molecules. *Expert Opin. Drug Deliv.*, 19, 8, 927–44, 2022.
34. Haghjou, N., Soheilian, M., Abdekhodaie, M.J., Sustained release intraocular drug delivery devices for treatment of uveitis. *J. Ophthalmic Vis. Res.*, 6, 4, 317–29, 2011.
35. García-Estrada, P., García-Bon, M.A., López-Naranjo, E.J., Basaldúa-Pérez, D.N., Santos, A., Navarro-Partida, J., Polymeric Implants for the Treatment of Intraocular Eye Diseases: Trends in Biodegradable and Non-Biodegradable Materials. *Pharmaceutics*, 13, 5, 701, 2021.
36. Lowinger, M., Barrett, S., Zhang, F., Williams, R., Sustained Release Drug Delivery Applications of Polyurethanes. *Pharmaceutics*, 10, 2, 55, 2018.
37. Schneider, C., Langer, R., Loveday, D., Hair, D., Applications of ethylene vinyl acetate copolymers (EVA) in drug delivery systems. *J. Control. Release*, 262, 284–95, 2017.
38. Sey-Sawo, J., Jadelle implant acceptance amongst women in Western Health Region 1, The Gambia. *Int. J. Afr. Nurs. Sci.*, 9, 38–41, 2018.
39. Sanna, L., IIkka, I., Kaisu, K., Fluocinolone acetonide intravitreal implant (Retisert®) in the treatment of sight threatening macular oedema of juvenile idiopathic arthritis-related uveitis. *Acta Ophthalmol.*, 96, 6, 648–51, 2018.
40. Stewart, S.A., Domínguez-Robles, J., McIlorum, V.J., Mancuso, E., Lamprou, D.A., Donnelly, R.F. *et al.*, Development of a Biodegradable Subcutaneous Implant for Prolonged Drug Delivery Using 3D Printing. *Pharmaceutics*, 12, 2, 105, 2020.
41. Caló, E. and Khutoryanskiy, V.V., Biomedical applications of hydrogels: a review of patents and commercial products. *Eur. Polym. J.*, 65, 252–67, 2015.
42. Van Hemelryck, S., Wens, R., van Poppel, H., Luijks, M., Shahidi, K., Marczak, M. *et al.*, *In Vitro* Evaluation of Poly(lactide-co-glycolide) *In Situ* Forming Gels for Bedaquiline Fumarate Salt and Pharmacokinetics Following Subcutaneous Injection in Rats. *Pharmaceutics*, 13, 8, 1231, 2021.
43. Lee, D. and Chan, C., Modified insertion technique for a sustained-release dexamethasone intravitreal implant (Ozurdex®). *Am. J. Ophthalmol. Cas. Rep.*, 30, 19, 100725, 2020.
44. Sequeira, J.A.D., Santos, A.C., Serra, J., Veiga, F., Ribeiro, A.J., Chapter 10 - Poly(lactic-co-glycolic acid) (PLGA) matrix implants, in: *Nanostructures for the Engineering of Cells*, Tissues and Organs, A.M. Grumezescu (Ed.), pp. 375–402, William Andrew Publishing, New York, 2018.

45. Sah, H. and Lee, B., Development of New Microencapsulation Techniques Useful for the Preparation of PLGA Microspheres. *Macromol. Rapid Commun.*, 24, 21, 1845–51, 2006.
46. Wolinsky, J.B., Colson, Y.L., Grinstaff, M.W., Local drug delivery strategies for cancer treatment: gels, nanoparticles, polymeric films, rods, and wafers. *J. Control. Release*, 159, 1, 14–26, 2012.
47. Cho, H., Gao, J., Kwon, G.S., PEG-b-PLA Micelles and PLGA-b-PEG-b-PLGA Sol-Gels for Drug Delivery. *J. Control. Release*, 240, 191–201, 2016.
48. Lyndon, J.A., Boyd, B.J., Birbilis, N., Metallic implant drug/device combinations for controlled drug release in orthopaedic applications. *J. Control. Release*, 179, 63–75, 2014.
49. Sagib, M., Kuzumin, O., Kraśkiewicz, H., Wasyluk, Ł, Cuniberti, G., Ficai, A. *et al.*, Evaluation of *in Vitro* Corrosion Behavior of Titanium Oxynitride Coated Stainless Steel Stents. *IEEE J. Mag., IEEE Xplore., IEEEE Access.*, 9, 59766–82, 2021.
50. Blum, M., Cao, D., Mehran, R., Device profile of the Resolute Onyx Zotarolimus eluting coronary stent system for the treatment of coronary artery disease: overview of its safety and efficacy. *Expert Rev. Med. Devices*, 17, 4, 257–65, 2020.
51. Kamberi, M., Pinson, D., Pacetti, S., Perkins, L.E.L., Hossainy, S., Mori, H. *et al.*, Evaluation of chemical stability of polymers of XIENCE everolimus-eluting coronary stents *in vivo* by pyrolysis-gas chromatography/mass spectrometry. *J. Biomed. Mater. Res. B Appl. Biomater.*, 106, 5, 1721–9, 2018.
52. Keraliya, R.A., Patel, C., Patel, P., Keraliya, V., Soni, T.G., Patel, R.C. *et al.*, Osmotic drug delivery system as a part of modified release dosage form. *ISRN Pharm.*, 2012, 528079, 2012.
53. Herrlich, S., Spieth, S., Messner, S., Zengerle, R., Osmotic micropumps for drug delivery. *Adv. Drug Deliv. Rev.*, 64, 14, 1617–27, 2012.
54. Alzet Home, ALZET® Osmotic Pumps, Available from: https://www.alzet.com/.
55. Ryu, W., Hung, Z., Prinz, F., Goodman, S., Fasching, R., Biodegradable micro-osmotic pump for long-term and controlled release of basic fibroblast growth factor. *J. Control. Release*, 124, 1–2, 98–105, 2007.
56. P. Blaskovich, R. Ohri, S. Bennett, Biodegradable Osmotic Pump Implant For Drug Delivery. US20120156289A1, 2012.
57. Grimberg, D., Shah, A., Inman, B., Overview of Taris GemRIS, a Novel Drug Delivery System for Bladder Cancer. *Eur. Urol. Focus*, 6, 4, 620–2, 2020.
58. Watanabe, T., Sadahira, T., Watanabe, M., Maruyama, Y., Wada, K., Sekito, T. *et al.*, Intravesical indwelling lidocainereleasing devices for IC/BPS (Review). *World Acad. Sci. J.*, 4, 4, 1–7, 2022.
59. Moul, J.W. and Civitelli, K., Managing advanced prostate cancer with Viadur (leuprolide acetate implant). *Urol. Nurs.*, 21, 6, 385–8, 393–4, quiz 395–6, 2001.

60. Stevenson, C.L., Santini, J.T., Langer, R., ChronogesicTM drug delivery systems utilizing microtechnology. *Adv. Drug Deliv. Rev.*, 64, 14, 1590–602, 2012.
61. Jannsen, H., Klein, S., Nestler, B., An adjustable flow restrictor for implantable infusion pumps based on porous ceramics. *Biomed. Tech. (Berl.)*, 60, 4, 359–64, 2015.
62. Gomberg-Maitland, M., Bourge, R.C., Shapiro, S.M., Tarver, J.H., Zwicke, D.L., Feldman, J.P. *et al.*, Long-term results of the DelIVery for Pulmonary Arterial Hypertension trial. *Pulm. Circ.*, 9, 4, 1–9, 2019.
63. Chappel, E., Chapter 7 - Implantable drug delivery devices, in: *Drug Delivery Devices and Therapeutic Systems*, E. Chappel (Ed.), pp. 129–56, Academic Press, Cambridge, 2021.
64. Nisar, A., Afzulpurkar, N., Mahaisavariya, B., Tuantranont, A., MEMS-based micropumps in drug delivery and biomedical applications. *Sens. Actuators B Chem.*, 130, 2, 917–42, 2008.
65. Mendoza, L., Scilletta, N., Bellino, M., Desimone, M., Catalano, P., Recent Advances in Micro-Electro-Mechanical Devices for Controlled Drug Release Applications. *Front. Bioeng Biotechnol.*, 8, 1–28, 2020, Frontiers.
66. Bußmann, A., Leistner, H., Zhou, D., Wackerle, M., Congar, Y., Rchter, M. *et al.*, Piezoelectric Silicon Micropump for Drug Delivery Applications. *Appl. Sci.*, 11, 8008, 2021.
67. Wang, C., Seo, S.J., Kim, J.S., Lee, S.H., Jeon, J.K., Kim, J.W. *et al.*, Intravitreal implantable magnetic micropump for on-demand VEGFR-targeted drug delivery. *J. Control. Release*, 283, 105–12, 2018.
68. Uvarov, I.V., Lemekhov, S.S., Melenev, A.E., Naumov, V.V., Koroleva, O.M., Izyumov, M.O. *et al.*, A simple electrochemical micropump: Design and fabrication. *J. Phys. Conf. Ser.*, 741, 1–6, 2016.
69. Gao, Y., Wu, M., Lin, Y., Zhao, W., Xu, J., Acoustic bubble-based bidirectional micropump. *Microfluid. Nanofluid.*, 24, 4, 29, 2020.
70. Wu, Z., Cai, H., Nunez, A., Liu, H., Bondesson, M., Guo, S. *et al.*, A Digital Acoustofluidic Pump Powered by Localized Fluid-Substrate Interactions. *Anal. Chem.*, 91, 11, 7097–103, 2019.
71. Sumi, C. and Sutradhar, K., Full article: Implantable microchip: the futuristic controlled drug delivery system. *Drug Deliv.*, 23, 1, 1–11, 2014.
72. Choy, J.H., Cho, Y.C., Lee, S., Implantable multireservoir device with stimulus-responsive membrane for on-demand and pulsatile delivery of growth hormone. *PNAS*, 116, 24, 11664–72, 2019.
73. Mazidi, Z., Javanmardi, S., Naghib, S.M., Mohammadpour, Z., Smart stimuli-responsive implantable drug delivery systems for programmed and on-demand cancer treatment: An overview on the emerging materials. *Chem. Eng. J.*, 433, 134569, 2022.
74. Tng, H., Hu, R., Roy, P., Yong, K.T., Micromachines | Free Full-Text | Approaches and Challenges of Engineering Implantable Microelectromechanical Systems

(MEMS) Drug Delivery Systems for *in Vitro* and *in Vivo* Applications. *Micromachines*, 3, 4, 615–31, 2012.
75. Bazaka, K. and Jacob, M., Implantable Devices: Issues and Challenges. *Electronics*, 2, 1–34, 2012.
76. Al-Jawadi, S., Capasso, P., Sharma, M., The road to market implantable drug delivery systems: a review on US FDA's regulatory framework and quality control requirements. *Pharm. Dev. Technol.*, 23, 10, 953–63, 2018.
77. Rujitanaroj, P.O., Jao, B., Yang, J., Wang, F., Anderson, J.M., Wang, J. *et al.*, Controlling fibrous capsule formation through long-term down-regulation of collagen type I (COL1A1) expression by nanofiber-mediated siRNA gene silencing. *Acta Biomater.*, 9, 1, 4513–24, 2013.
78. Wang, Y., Vaddiraju, S., Gu, B., Papadimitrakopoulos, F., Burgess, D.J., Foreign Body Reaction to Implantable Biosensors: Effects of Tissue Trauma and Implant Size. *J. Diabetes Sci. Technol.*, 9, 5, 966–77, 2015.
79. Kastellorizios, M., Tipnis, N., Burgess, D., Foreign Body Reaction to Subcutaneous Implants, in: *Immune Responses to Biosurfaces*, Lambris, J., Ekdahl, K., Ricklin, D., Nilsson, B. (Eds), pp. 93–108, Springer, Cham.
80. Farra, R., Sheppard, N.F., McCabe, L., Neer, R.M., Anderson, J.M., Santini, J.T. *et al.*, First-in-human testing of a wirelessly controlled drug delivery microchip. *Sci. Transl. Med.*, 4, 122, 122ra21, 2012.
81. Borandeh, S., van Bochove, B., Teotia, A., Seppälä, J., Polymeric drug delivery systems by additive manufacturing. *Adv. Drug Deliv. Rev.*, 173, 349–73, 2021.
82. Bruschi, M., Main mechanisms to control the drug release, in: *Strategies to Modify the Drug Release from Pharmaceutical Systems*, pp. 15–28, Elsevier, 2015.
83. Dash, S., Murthy, P.N., Nath, L., Chowdhury, P., Kinetic modeling on drug release from controlled drug delivery systems. *Acta Pol. Pharm.*, 6, 217–33, 2010.
84. Liechty, W.B., Kryscio, D.R., Slaughter, B.V., Peppas, N.A., Polymers for Drug Delivery Systems. *Annu. Rev. Chem. Biomol. Eng.*, 1, 149–73, 2010.
85. Yang, C. and Burt, H.M., Drug-eluting stents: factors governing local pharmacokinetics. *Adv. Drug Deliv. Rev.*, 58, 3, 402–11, 2006.
86. Wiemer, M., Seth, A., Chandra, P., Neuzner, J., Richardt, G., Piek, J. *et al.*, Systemic exposure of everolimus after stent implantation: a pharmacokinetic study. *Am. Heart J.*, 154, 4, 751, 2008.
87. Fang, M., Lü, T.M., Ma, A.D., Wang, L., Li, G.L., Qiu, A.Z. *et al.*, Comparative Pharmacokinetics of Continuous and Conventional Intrathecal Amphotericin B in Rabbits. *Antimicrob. Agents Chemother.*, 56, 10, 5253–7, 2012.
88. Krupp, J.L. and Bernards, C.M., Pharmacokinetics of intrathecal oligodeoxynucleotides. *Anesthesiology*, 100, 2, 315–22, 2004.
89. Tchobaniouk, L.V., McAllister, E.E., Bishop, D.L., Carpentier, R.M., Heins, K.R., Haight, R.J. *et al.*, Once-Monthly Subcutaneously Administered

Risperidone in the Treatment of Schizophrenia: Patient Considerations. *Patient Prefer. Adherence*, 13, 2233–41, 2019.

90. Schlegel, P., A review of the pharmacokinetic and pharmacological properties of a once-yearly administered histrelin acetate implant in the treatment of prostate cancer. *BJU Int.*, 103, Suppl 2, 7–13, 2009.

8

Inhalable Novel Drug Delivery Systems

Mahdieh Farhangi[1]* and Fariba Hajifathaliha[2]

[1]Varian Pharmed Research Center, Varian Pharmed Pharmaceutical Co., Tehran, Iran
[2]Department of Pharmaceutics and Pharmaceutical Nanotechnology, School of Pharmacy, Shahid Beheshti University of Medical Sciences, Tehran, Iran

Abstract

The respiratory system has been a route of drug administration for a long time. The anatomy and special characteristics of the lungs facilitate local and systemic drug delivery through this organ. Efficient pulmonary drug delivery is greatly dependent on the formulation alongside the design of the device. The introduction of novel drug delivery systems and especially various nanocarriers to the pharmaceutical industry, has made noticeable improvements in drug efficacy and patient compliance. Recently, various studies have been conducted for the preparation and characterization of novel inhalable drug delivery systems. The goal of all these studies is to design an efficient novel pulmonary drug delivery system that can overcome the challenges facing this route of drug administration. This chapter aims to describe the anatomy and physiology of the respiratory system and also shed light on the influential parameters such as physicochemical characteristics of the drug and its formulation, pharmaceutical considerations, and physiopathology of the respiratory system in the development and production of novel inhalable drug delivery systems. Challenges facing inhalable drug delivery systems, whether biological barriers or pharmaceutical limitations, and different novel approaches to overcome these limitations, will be reviewed. Additionally, pharmacokinetics and the fate of novel inhalable drug delivery systems within the body will also be discussed. Finally, those products that have reached the market after a long journey from laboratories to the clinic will be introduced.

***Keywords*:** Respiratory system, pulmonary drug delivery, lung drug delivery, nanocarriers, pharmacokinetics

**Corresponding author*: Farhangi.Mahdieh@varianpharmed.com

Mehrdad Hamidi and Mahsa Sayed Tabatabaei (eds.) Novel Drug Delivery Systems: Fundamentals and Applications, (275–306) © 2025 Scrivener Publishing LLC

8.1 Introduction

The history of drug delivery to the lungs goes back to ancient Egypt more than 3500 years ago when patients or physicians used to prepare both the aerosols and the custom-made delivery systems themselves [1]. Since then, many efforts have been made to improve the inhalable drug delivery system (DDS); but introducing the first pressurized metered-dose inhaler (pMDI) was a true revolution in this area. The first medicines that were administered at that time were adrenaline and isoprenaline, which were later superseded by albuterol and the steroid beclomethasone dipropionate [2].

Several factors make the respiratory tract a desirable route of drug administration, whether as a local or a systemic target organ. However, enumeration of the advantages of this route would be more accurate if we first define the aim of drug delivery, i.e., its delivery of medicines to or through the respiratory tract.

Drug delivery to the lungs as local delivery of drugs to manage different disease situations like chronic obstructive pulmonary disease (COPD), cystic fibrosis, asthma, lung cancers, and pulmonary arterial hypertension (PAH) is the main objective of inhalable DDSs. Like any other targeted delivery approach, administration of drugs to the lungs has several advantages such as using relatively low doses of medicines, facing less incidence of systemic adverse effects, therefore fewer expenses of therapy, and for some specific drugs, reaching a rapid onset of action. On the other hand, large surface area (>100 m^2), thin epithelial layer (<1μm), bypassing the first-pass metabolism, and as a result, higher bioavailability, are some of the advantages reported for drug delivery through the lungs, which, collectively, make it an interesting approach for systemic delivery of medicines [3]. In addition, inhalable DDSs can not only be as effective as parenteral systems but also have the advantage of being non-invasive, which could increase patient compliance [4].

Regardless of the numerous advantages of the respiratory tract as a desirable target for drug delivery, thereby attracting substantial investment in the development of new inhalable formulations, the number of inhalable DDSs in the market is relatively low. This shows that the design and production of a successful inhalable DDS require novel strategies and formulations that consider drug, formulation, device, and so on at the same time. Novel DDSs are designed to achieve higher bioavailability, controlled *in-vivo* distribution, and enhanced physicochemical and deposition properties [5–7]. These novel systems include liposomes, dendrimers, micellar formulations, polymeric nanoparticles, microparticles, and lipid-polymer hybrid nanoparticles.

8.2 Anatomy and Physiology

The primary responsibility of the respiratory system is supplying oxygen from the atmosphere for cellular respiration and the removal of carbon dioxide (CO_2) from cells back into the air. To perform this vital role, the lungs need a large portion of the total blood flow in the body. This need is met so that the lungs receive about 10% of total blood flow in a normal healthy person [8]. The pulmonary vascular system consists of two divided circulations; systemic (bronchial) circulation and pulmonary circulation. In pulmonary circulation, deoxygenated blood containing CO_2 reaches the lungs through the right ventricle *via* pulmonary arteries. After the gas exchange in the alveolar-capillary network, the oxygenated blood returns to the heart to be pumped throughout the tissues, including the lungs. Through systemic circulation, the lungs, like other organs in the body, are provided oxygenated blood and send the blood containing CO_2 to the heart [9].

Besides gas exchange, the lungs also have a significant effect on fluid homeostasis and immune defense. The two latter are carried out by the lymphatic system. Lymphatic circulation drains excess and leaked fluid into the vascular system. It also provides a platform for dendritic cells and lymphocytes as parts of the immune system to travel throughout the body [10].

The respiratory system could be anatomically and functionally divided into different parts. Anatomically, it is divided into the upper and lower respiratory systems (Figure 8.1). The upper respiratory tract includes the nose, nasal cavities, pharynx, and larynx. Besides recognizing smells and speech ability, the upper respiratory system is responsible for making the air warm, moist, and filtered. The air is filtered by the tiny hairs located in the nasal cavity, and its large particles are removed. In addition, sneezing is the other mechanism for the removal of dust particles and pollens, which are triggered by sensitive receptors in the upper airway [11]. The trachea, primary bronchi, and the entire lungs are the constituents of the lower respiratory system. The remaining particles in inhaled air are trapped in the trachea and pushed upward towards the pharynx, which later would be swallowed or expectorated.

The lungs are the main organ in the respiratory system. They are surrounded by two protective membranes called pleura, in which there is a

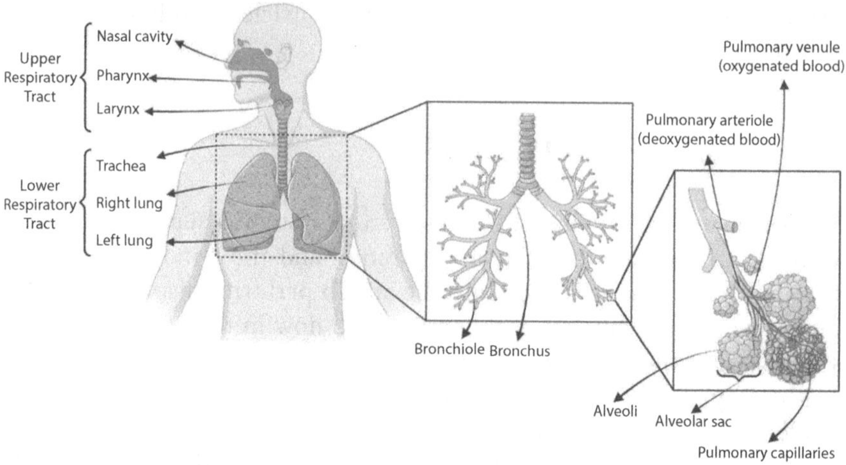

Figure 8.1 Anatomy of the respiratory system (Created with BioRender.com).

lubricant fluid to decrease friction and to make respiration more comfortable. Although it is a paired organ, its two parts are not the same size. The left lung has two lobes and is slightly smaller than the right lung with three lobes [12].

In our airways, after the trachea, some bronchi make the bronchial tree and form numerous bifurcations, that is primary bronchi divide into secondary bronchi, tertiary bronchi, and further into bronchioles. The bronchioles finally make alveolar ducts, and tiny air sacs called alveoli [11]. It is estimated that there are approximately 480 million alveoli in humans (ranging from 274 to 790 million); the larger the lungs, the more the number of alveoli [13]. Alveolar sacs are the primary site of gas exchange in the pulmonary system. Oxygen passes through the upper respiratory parts and ultimately enters the bloodstream *via* a capillary network around the alveolar sacs. Carbon dioxide, along with other gaseous metabolic byproducts, is also exited from the body by exhalation. Through inhalation, the diaphragm contracts and is pulled down while external intercostal muscles expand toward the outside/up direction; consequently, the lungs expand as well. Expansion of the lungs during inhalation decreases the intrapulmonary pressure, which allows the air to enter the lungs. In contrast, relaxation of external intercostal muscles and diaphragm during exhalation increases intrapulmonary pressure, which causes air to be expelled from the lungs. In a deep and vigorous breath, the other muscles like the abdominal muscles, internal intercostal, and pectoralis are also involved. Although the rate and depth of breathing are under control, the whole breathing process is known as an involuntary activity [11, 12, 14, 15].

The respiratory system is innervated by different kinds of sympathetic, parasympathetic, and sensory nerve fibers. Noradrenergic innervation originates from the sympathetic ganglia and is provided for bronchial blood vessels and mucus glands through sympathetic innervation while parasympathetic innervation is supplied by the vagus nerve. Although different sources have been known for pulmonary sensory nerve fibers, they are also provided mainly by the vagus nerve [12].

As mentioned earlier, functionally, the respiratory system is also divided into different parts; the conducting zone and respiratory zone (Figure 8.2). The conducting zone includes the upper respiratory system plus the trachea, bronchi, upper parts of bronchioles, while the respiratory bronchioles, alveolar ducts, and alveolar sacs construct the respiratory zone. Ciliated epithelial cells and mucus-producing goblet cells are two major epithelial cell types in the conducting zone, which together form the mucociliary escalator that, by sweeping out the entrapped particles in mucus, plays a crucial role in the elimination of inhaled particles. The mucus is about 8μm thick and composed of proteins, glycoproteins (mucin), inorganic salts, lipids, and water. It should be kept in mind that if drug particles deposit in airways before reaching their target site or alveolar region, they would be sent to the trachea under the influence of the mucociliary escalator without creating any effect. The conducting zone also includes dendritic cells, which act as a part of the immune system in the airways. These cells are found almost everywhere in the respiratory tract. They are

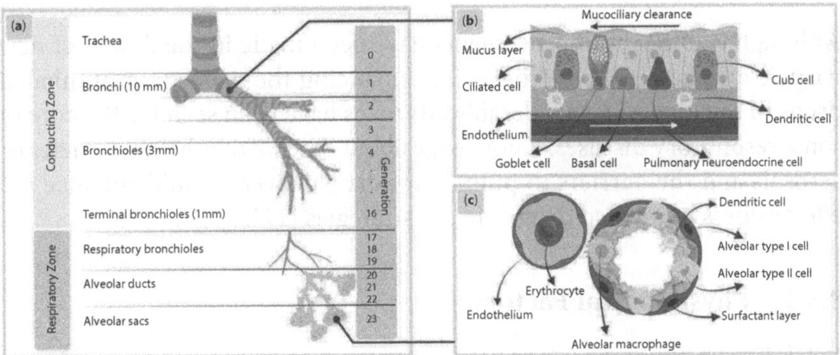

Figure 8.2 (a) Functional division of respiratory system, (b) anatomy of the conducting zone, (c) anatomy of the respiratory zone (created with BioRender.com).

antigen-presenting cells (APC) that introduce the particles and pathogens to the T cells in the lymphoid tissues.

The respiratory zone comprises alveolar epithelial type I (AT-I) and alveolar epithelial type II (AT-II) cells. The majority of alveolar surface area (90%) is covered by AT-I, which is responsible for the vital function of gas exchange in alveoli. AT-II cells are smaller than AT-I cells. These cells synthesize the pulmonary surfactant, which covers the alveolar epithelium. Different lipids such as dipalmitoylphosphatidylcholine (DPPC), cholesterol, acidic phospholipids, phosphatidylglycerol (PG), and to a lesser extent, the hydrophilic and lipophilic proteins, are the components of the pulmonary surfactant (10% proteins compared to 90% lipids). Different constituents of pulmonary surfactant and mucus would affect drug solubility, diffusion, and the interaction between drug and receptors [8, 16].

In addition to the dendritic cells mentioned earlier, the mononuclear phagocytic system (MPS), especially macrophages in the alveolar region, is the other member of the immune system in the airways. They serve as the front line of defense against pathogens and particles that phagocytes and transfer them into the phagosome, where they finally degrade. In the design of inhalable novel delivery systems, the alveolar phagocytic activity can be considered both as a challenge that should be avoided and as an opportunity that can be targeted, especially in certain infectious diseases. Although the readers are aware that there are also other types of cells in respiratory systems, such as Clara cells (non-ciliated bronchiolar secretory cells) and serous cells (Periciliary fluid secretory cells), here only the most dominant types, have been discussed.

8.3 Factors Affecting Pulmonary Drug Delivery

Although significant improvements have been made in the design of new particle and device technologies for enhancing the deposition of inhaled drugs in the lung, no considerable advances have been seen for the cure of some respiratory diseases. Knowledge about the fate of inhaled particles or aerosols and the barriers to pulmonary drug delivery would enhance the effectiveness of inhaled drug delivery strategies [17].

8.3.1 Physiological Factors

8.3.1.1 Mucociliary Clearance

Mucociliary clearance is the most important mechanism of clearance in the upper respiratory tract. Both non-cellular (the mucus blanket) and cellular (ciliated epithelial cells, secretory cells) elements are involved in this fast and unspecific clearance mechanism. As the inhaled particles are trapped in the mucus, the escalation movement of cilia would result in the swallowing of the mucus and elimination of the administered drug. About 80–90% of an inhaled drug would be excreted by this mechanism within 24 hours [16, 17].

8.3.1.2 Alveolar Macrophages

Phagocytosis of the inhaled drug is another mechanism of clearance in the lung. Alveolar macrophages are the most important immune cells in the lower respiratory tract, which can phagocyte the deposited particles and degrade them in phagosomes [16].

8.3.1.3 Pulmonary Surfactant Layer

The pulmonary surfactant layer, as a second non-cellular barrier, has important effects on the fate of inhaled particles. The surface properties of particles affect the interactions with the surfactant layer and subsequent events. Opsonization of inhaled particles is the result of the absorption of surfactant proteins, especially SP-A and SP-D, on the particles' surface. By contrast, interactions with surfactant lipids decrease the uptake of particles by alveolar macrophages [17].

8.3.1.4 Enzymatic Degradation

Although the enzymatic degradation of drugs in the lung is much lower than liver, enzymes such as antitrypsin, proteases, trypsin, and cytochrome P450 families can cause clearance of inhaled particles from the pulmonary system. The surface properties of inhaled particles have great effects on their enzymatic degradation in the lung. For example, coating polylysine dendrimers with high molecular weight polyethylene glycol (PEG) decreased the clearance of nanoparticles by the degradation mechanism [18, 19].

8.3.2 Pathological Characteristics

The most commonly occurring lung diseases include COPD, cystic fibrosis, asthma, pneumonia, pulmonary edema, lung carcinoma, pulmonary embolus, and pulmonary arterial hypertension (PAH), which may alter the anatomic structure of respiratory tracts, the viscosity of mucus, and the ventilatory conditions and breathing pattern of patients. Also, pH changes in the respiratory tract from neutral to acidic situations would occur in chronic bacterial infections. These alterations have significant effects on the deposition, absorption, and clearance of inhaled particles. Obstructive lung diseases reduce the deposition of inhaled particles in the alveoli. Chronic airway inflammations are related to enhanced epithelium permeability which would increase the systemic absorption of drugs. The higher viscosity of mucus in COPD, cystic fibrosis, asthma, and pneumonia would result in a lower rate of mucociliary clearance. For COPD and severe asthma, the phagocytic activity of alveolar macrophages decreases [20].

8.3.3 Physicochemical Parameters

8.3.3.1 Particle Size

The particle size of pulmonary drug delivery formulations has a great influence on the efficacy of these systems. The size of particles in these formulations is expressed as mass median aerodynamic diameter (MMAD). Aerodynamic diameter is the diameter of a spherical particle of unit density (1 g/cm^3) that possesses the same gravitational settling velocity as the particle under examination. This diameter could be calculated by the following equation:

$$d_a = d_p \times \sqrt{(\rho/\rho_0)} \tag{8.1}$$

where d_a and d_p represent the aerodynamic diameter and the observed diameter, respectively; ρ is the density of the particle under examination, and ρ_0 equals 1 g/cm^3. Particles with MMAD sizes in the range of 1–5 μm could reach the lower respiratory tract [16].

Penetration of particles into the mucus layer is a size-dependent process. The mucus layer consists of a crosslinked network from mucin fibers which controls the filtration of particles according to their sizes. The diffusion of nanoparticles that are larger than the cut-off mucus size is sterically restricted. By contrast, smaller nanoparticles and pathogens can easily diffuse through the mucus layer, and their retention time increases [21].

Nanoparticles would be detected and captured by phagocytes to a lesser extent than the microparticles (1–5 μm) [16]. Extremely small or large particles have fewer contacts with the macrophages' membrane in comparison to intermediate-size particles. Particles in the range of 2–3 μm have maximal attachments with the macrophages' membrane and, subsequently, are captured more efficiently by phagocytes [22]. Macrophage uptake of inhalable particles is also affected by their shape (aspect ratio). Long or thin particles and those that have pointy spikes would be captured less [23].

8.3.3.2 Surface Charge

The surface charge of nanoparticles has a great effect on their mucociliary clearance. The effective electric charge on the nanoparticle's surface is measured by the zeta potential. The potential difference between the dispersion medium and the stationary layer of fluid attached to the particle is called zeta potential [24]. The negatively charged groups of mucus electrostatically interact with positively charged nanoparticles and this mucoadhesion effect increases the retention time of the drug in the lung. Nanoparticles that have a high negative surface charge cannot diffuse through the mucus layer due to the repulsive interactions. The neutral particles show the best mucus permeability properties. Coating with neutral polymers such as PEG would balance the surface properties of nanoparticles which results in higher penetration of these DDS into the mucus layer [16].

The effect of surface charge on the alveolar macrophage phagocytosis of particles is an area of conflict. Some studies have reported that particles with a positive surface charge would be caught more efficiently due to the negatively charged sialic acid on the surface of macrophages [16, 25]. Other studies have indicated that liposomes with negative surface charges are phagocytosed more easily [26]. It has been indicated that particles with higher surface charge (either positive or negative) would be uptaken to a higher extent by macrophages [27].

8.3.3.3 Surface Hydrophobicity

Mucin consists of so-called PTS domains and non-PTS regions. PTS domains are rich in proline, threonine, and serine which are heavily glycosylated. The presence of non-PTS regions in mucin structure enhances the adhesion of hydrophobic nanoparticles to the mucus layer and their clearance. Hydrophilic nanoparticles present better mucus penetration and a longer retention time. PEG is the polymer commonly used for the preparation of hydrophilic particles. The molecular weight of PEG is an important factor. While higher molecular weight PEGs would interact with mucin chains and enhance mucociliary clearance, lower molecular weight PEGs (2–5 kDa) increase nanoparticle hydrophilicity as well as their mucus diffusion. The development of alternative hydrophilic polymers is necessary since the anti-PEG antibodies enhance the entrapment of PEG-based particles in mucus secretions [16].

Although hydrophilic particles are less phagocytosed by macrophages, the small size of nanoparticles facilitates their escape from alveolar macrophage uptake, and there is no need for coating inhalable nanoparticles with hydrophilic polymers [16].

8.3.4 Pharmaceutical Parameters

The therapeutic goals determine the formulation strategies. The pharmacokinetic parameters of a pulmonary DDS such as the highest concentration of a drug in the blood (C_{max}), the time it takes to reach the maximum concentration (T_{max}), and the elimination half-life ($t_{1/2}$) are strongly dependent on the drug formulation. When a rapid action of the drug is needed, highly soluble and permeable active pharmaceutical ingredients or particulate carriers which could be dissolved quickly and transported to the bloodstream through pulmonary epithelial cells are preferable. Also, small hydrophobic molecules such as nicotine are absorbed within seconds to minutes. For targeting lung tissue in the treatment of pulmonary diseases, prolonged release of a drug would be beneficial. The design of sustained-release pulmonary DDS would lead to a longer duration of action, reduced dosing frequencies, and increased patient compliance [17].

Excipients play a significant role in the design of effective pulmonary DDSs. Few excipients have approval for use in these products. The most commonly used excipients in pulmonary formulations are lactose, mannitol, glucose, and phospholipids like DPPC [28]. Phospholipids are biocompatible and can enhance the solubility of active pharmaceutical ingredients due to their amphiphilic properties. However, the fusion of phospholipids with endogenous pulmonary surfactants can result in the formation of some vesicles, which entrap the active pharmaceutical ingredient and slow its dissolution and absorption [17]. Bile salts such as cholates, glycocholates, or taurocholates are another class of excipients that could be used as absorption enhancers. The absorption enhancement of these compounds is due to the alteration of mucus properties, interaction with cellular membranes, and opening of intercellular tight junctions that can cause toxicity [17]. Cyclodextrin is another absorption enhancer that shows fewer toxic effects. However, all excipients with absorption enhancement properties have toxicity, and it would be better to use a mixture of these compounds in formulations [28, 29]. Furthermore, with the addition of suitable excipients to the formulation, some physical properties of these products, such as pH and osmolality, should be adjusted. The optimum pH and osmolality of inhalable formulations are around neutral and 150-1200 mOsm/kg, respectively. Any deviations from these criteria could lead to coughing, bronchospasm, and tissue irritation [30].

During formulation development and stability studies, the physicochemical properties of inhalable DDSs including zeta potential, size, polydispersity index (PDI), the percentage of drug that is successfully entrapped into the nanoparticle (encapsulation efficiency; EE%), pH, osmolality, drug release, and viscosity should be considered [31]. The zeta potential value of stable formulations is in the range of ± 30 mV [32, 33]. As discussed earlier, the size and PDI of particles have great effects on the deposition pattern of a drug in the respiratory tract. The viscosity of formulations is also affected by the size of particles. Larger particles increase the viscosity of formulations and decrease the output of nebulizers [34].

8.3.5 Particles' Deposition Mechanisms

Undoubtedly, one of the most important factors to be considered in the development of a successful inhalable DDS is the deposition of drug particles in different parts of the respiratory tract. Three main mechanisms of particle deposition are impaction by inertial force, sedimentation due to gravity, and Brownian diffusion (Figure 8.3). In addition, there are also some relatively insignificant mechanisms such as turbulent flows, interception, and electrostatic precipitation. The influential factors determining the deposition of particles could be generally divided into size, shape, and

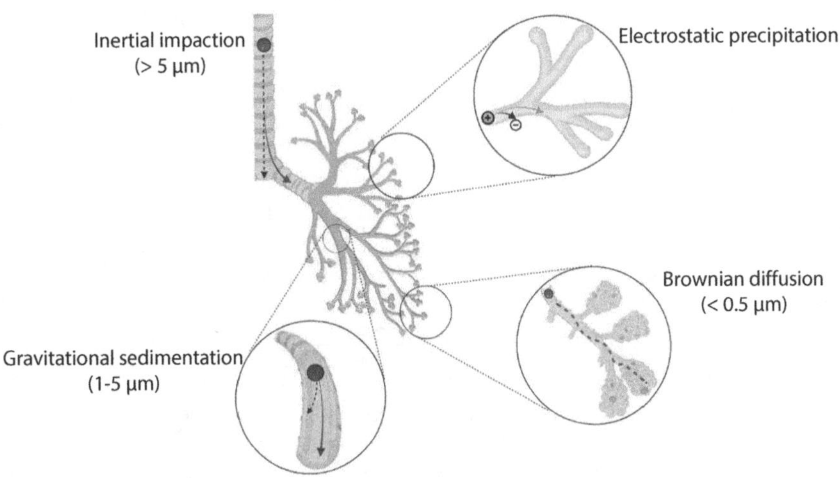

Figure 8.3 Mechanisms of particles deposition in the lung (created with BioRender.com).

density of particles, breathing rate and residence time of particles in the airway, and anatomical characteristics of the respiratory system.

Impaction occurs due to the inertia of the inhaled particles when reaching a bifurcation in the respiratory tract and their inability to change their direction. This mechanism involves mainly the deposition of large particles in the upper parts of the respiratory tract, i.e. the oropharyngeal region and larynx. It is stated that inertial impaction is responsible for filtering the particles above 5 μm out in the upper part of the airway [35]. Gravitational sedimentation, as its name implies, leads to particle deposition due to gravity and is known as the mechanism of deposition for particles in the approximate size range of 1-5 μm (as stated earlier, larger particles are usually deposited by inertial impaction). This mechanism of deposition is seen in small airways and alveolar regions wherein the particles are close to the walls. The chance of sedimentation will increase for particles with a larger aerodynamic diameter and also with a longer residence time of particles in the respiratory tract [36]. Unlike sedimentation and impaction mechanisms, the decrease in the size of particles increases Brownian diffusion. It is the main mechanism of deposition for particles smaller than 0.5 μm [34].

8.4 Challenges and Opportunities

As mentioned earlier, pulmonary drug delivery has various pros and cons. The opportunities associated with this route of administration along with

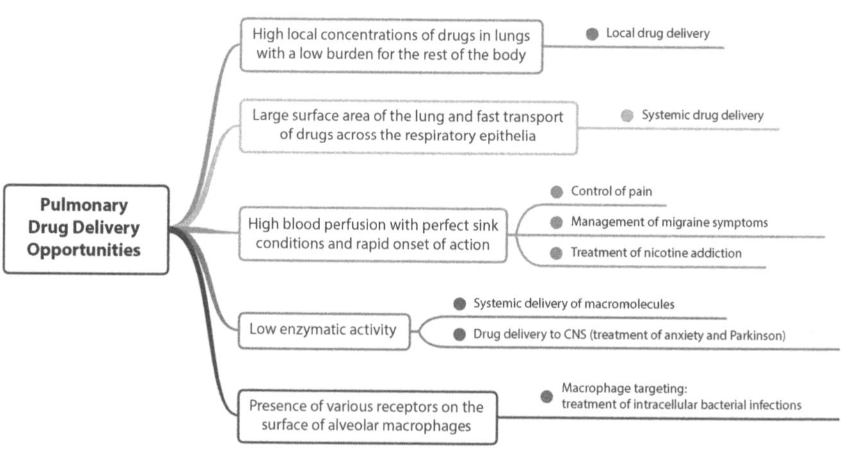

Figure 8.4 Opportunities of the respiratory tract in the development of inhalable DDS.

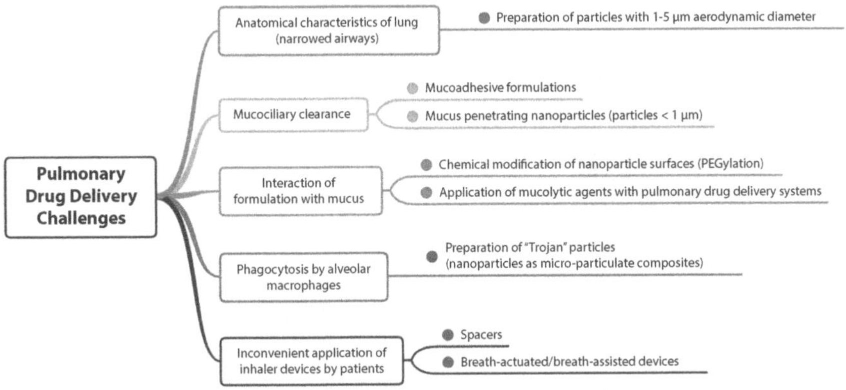

Figure 8.5 Pulmonary drug delivery challenges and strategies to overcome.

the challenges and strategies to overcome them are summarized in Figure 8.4 and Figure 8.5, respectively [16, 17, 37–40].

8.5 Classification and Formulation Considerations

Inhalation drug delivery has been known and used since ancient times. The preparation of modern inhalers such as pMDIs, dry powder inhalers (DPIs), and nebulizers has had a great effect on the improvement of pulmonary drug delivery (Figure 8.6).

It has been predicted that this market will reach 41.5 billion dollars by 2026. The efficacy of these systems is affected by the formulation, device, and patient's clinical condition. Unfortunately, the limitations in FDA/EMA-approved excipients for inhalation drug delivery have complicated the process of development [41].

Nebulizers were the first devices that introduced into the market. Nebulizers can be categorized as jet, mesh, and ultrasonic types according to their mechanism of aerosol production. However, nebulizers have some disadvantages that have limited their application including low efficiency and reproducibility, time-consuming administration, and large immovable devices. Nebulizer formulations are simple aqueous, sterile, and isotonic solutions that are packaged as unit doses. The stability of active pharmaceutical ingredients (APIs) in an aqueous medium is another concern [28].

pMDIs have been employed as the main inhalable DDS for the treatment of asthma since the 1950s. These systems may be formulated as either suspensions or solutions of APIs in a liquefied propellant. The most important

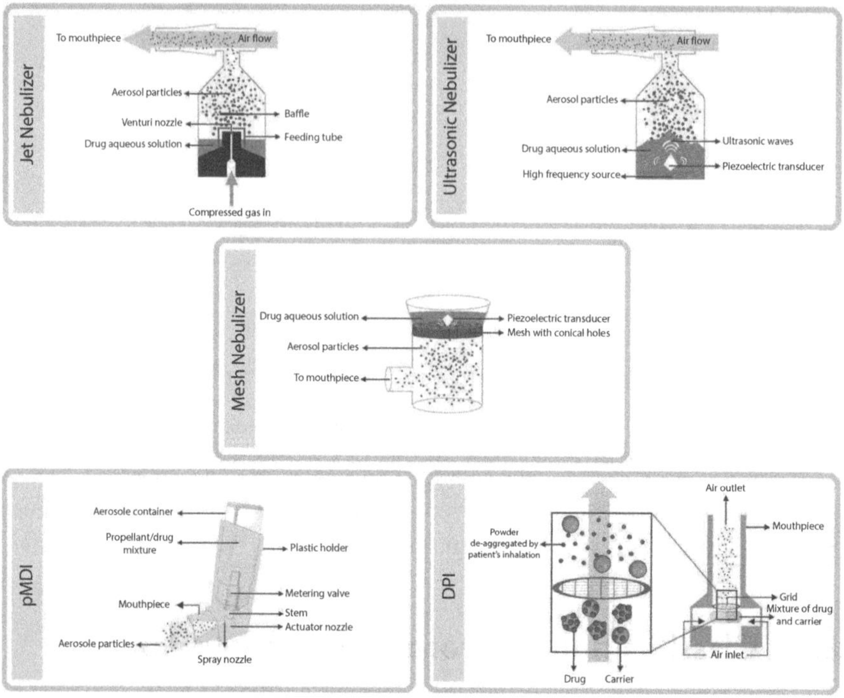

Figure 8.6 Schematic diagram of different devices for inhalation drug delivery.

drawbacks of these devices are limited delivery of emitted dose to the deep lung and the need for inhalation/actuation coordination. Unfortunately, pMDIs are not environmentally friendly due to the presence of propellants in their formulation and damage to the ozone layer. The actuation/inhalation coordination problem has been addressed by using spacers with pMDIs or the development of Breath-actuated inhalers. Autohaler® (3M Pharmaceuticals, U.S.A.) and Easybreathe® (Norton Healthcare, U.K.) are breath-actuated inhalers that just release the drug after a patient's inhalation [28].

Nowadays, DPIs are among the widespread inhalable DDSs. Solid powder blend formulations that are physically stable and propellant-free are preferable for inhalation delivery. These devices are categorized into three different groups: (i) unit-dose devices with a drug in hard gelatin capsules, (ii) multi-dose devices with a drug in foil blisters, and (iii) multi-dose devices with drug preloaded in an inhaler. Powder flow has a critical effect on the drug delivery efficacy of DPIs. Ordered mixing is a suitable strategy for the production of freely flowing formulations. Fine drug particles are adsorbed on the surface of larger inert carriers (usually lactose) that would be deaggregated by turbulent airflow generated within the inhalation device. The larger carrier particles deposit in the throat while fine drug particles reach the deep lung. Spray drying, spray freeze-drying, and supercritical fluids are three effective techniques for the production of particles with appropriate and controlled sizes [28].

Pulmonary drug delivery strategies and formulations have been improved continuously during the past decades. The following sections introduce these recent improvements. In addition, suitable devices for the administration of each novel pulmonary DDS have also been discussed.

8.5.1 Nanocarriers

Controlled release, the increased water solubility of lipophilic drugs, higher retention time in the lung, lower dosing frequency, and targeted drug delivery are some benefits of nanocarriers that have attracted scientists to the development of inhalable nanoparticles [16].

8.5.1.1 Polymeric Nanoparticles

Various natural (e.g., albumin, hyaluronic acid, carrageenan, gelatin, chitosan) and synthetic (e.g., poly (lactic acid) (PLA), poly (lactic-co-glycolide acid) (PLGA), poly (vinyl alcohol) (PVA), acrylic acid derivatives) polymers have been used for the preparation of inhalable polymeric nanoparticles. Biocompatibility and biodegradability are two critical properties that should be considered. Chitosan, hyaluronic acid, and PLGA are the commonly used polymers for the preparation of inhalable nanoparticles [19].

Chitosan is a natural polymer that is made by alkaline treatment of chitin. It can improve drug absorption and controlled release kinetic due to its mucoadhesive properties and the ability to open tight junctions [42]. In addition, it has been shown that chitosan can enhance the dispersion of particles in DPI formulations [43]. Chitosan is insoluble in the neutral pH of physiological environments due to its pKa value (5.5–6.5). Synthesis of chitosan derivatives such as glycol chitosan would improve polymer solubility. Moreover, thiolated derivatives have higher mucoadhesive properties in comparison to non-thiolated ones [44].

Hyaluronic acid is an endogenous polymer in the lung that has been used for the controlled delivery of drugs such as insulin to the respiratory tract [45]. Hyaluronic acid can hinder phagocytosis by agglutination of alveolar macrophages. In addition, high molecular weight hyaluronic acid has bioadhesive properties which make it suitable for efficient pulmonary drug delivery [46].

PLGA is the most preferred polymer for the preparation of inhalable polymeric nanoparticles. Alongside its biocompatibility, the biodegradation and drug release rate of nanoparticles could be adjusted by altering polymer molecular weight, chemical structure, and the lactide: glycolide ratio [19].

8.5.1.2 Liposomes

Liposomes are widely studied for pulmonary drug delivery and are the only FDA-approved nanocarriers for inhalation [16]. Liposomes are prepared from phospholipids that are biocompatible to the lung due to their intrinsic existence. Solubilization of lipophilic APIs, controlled drug release, and the ability to prepare the formulations with various particle sizes and physicochemical characteristics are liposome advantages in this context [19, 47]. However, liposomes' poor stability and high leakage rate of drugs in solution and biological fluids have limited their application. These drawbacks have adverse effects on liposome storage, nebulization, and spray drying [48, 49]. Coating liposomes with polymers such as chitosan would improve formulation stability, aerodynamic properties, and drug bioavailability [49, 50]. Although nebulization of liposomes is a widespread and commercialized route of administration, scientists have put efforts into the preparation of liposome-based DPI formulations using lyophilization, spray drying, spray freeze drying, or supercritical fluid technology [51].

8.5.1.3 Solid Lipid Nanoparticles and Nanostructured Lipid Carriers

Solid lipid nanoparticles (SLNs) are lipid-based nanocarriers that can be administered to the lungs. These formulations are prepared using lipids that are solid at room/body temperature, such as glyceryl dibehenate or glyceryl tristearate. The addition of some safe surfactants such as tween 80 to the formulation would decrease instabilities due to particle aggregation [52]. Emulsification and low-temperature solidification and hot high-shear homogenization are the main preparation methods of SLNs [52, 53]. The most important advantage of SLNs in comparison to liposomes is their stability in a liquid phase and during nebulization [54]. The absence of organic solvents in SLN formulation as well as ease of scale-up and sterilization are other benefits of these nanocarriers in comparison to liposomes. However, poor drug loading, high drug release rate, and gelation due to the lipids polymorphism have decreased SLN application [52].

Nanostructured Lipid Carriers (NLCs) are the second generation of solid lipid nanocarriers that have overcome the drawbacks of SLNs [55]. These nanocarriers have higher encapsulation efficiency and lower drug leakage during the shelf-life period. The lipophilic nature and bioadhesion of these nanoparticles improve their therapeutic effects [56].

8.5.1.4 Dendrimers

Dendrimers are biocompatible, stable, and monodisperse nanocarriers in the range of 2-20 nm. Various molecules, including drugs, proteins, targeting ligands, and solubility-enhancing agents (e.g., PEGs) could be conjugated on the polyfunctional surface of dendrimers. In addition, dendrimers can encapsulate APIs in their structure. Enzymatic processes are responsible for the degradation of poly amino acid dendrimers such as poly-lysine to low molecular weight products that could be eliminated through kidneys. These low immune-reactive nanocarriers could be administered as controlled-release pulmonary DDSs [57–59].

8.5.1.5 Polymeric Micelles

Amphiphilic polymers can self-assemble as core/shell-like structures in the aqueous media, which are, then, called polymeric micelles. These biocompatible and highly stable nanocarriers can encapsulate lipophilic APIs. Chitosan-fatty acid grafted conjugates and Pluronic® (poloxamer micelles) have been used as DDS in the lung [60, 61]. In addition to nebulization, these nanocarriers could be administered as nano-aggregate DPI formulations [62].

8.5.1.6 Inorganic Nanoparticles

Nanomagnetosols (i.e., aerosol droplets containing superparamagnetic iron oxide nanoparticles) are inorganic nanoparticles that have been used for pulmonary drug delivery. These nanoparticles could be directed and deposited in a specific lung region by an external magnetic field. Nanomagnetosols containing anticancer drugs have been administered for the treatment of bronchial cancer or mid-to-lower lung regions. In addition, locally targeted hyperthermia has a beneficial effect on the specific death of cancerous cells. It has been indicated that DPI formulation of iron oxide nanoparticles could be prepared by spray drying process with mannitol [63, 64]. Mesoporous silica nanoparticles are another group of inorganic nanoparticles that have been used for pulmonary drug delivery. These nanoparticles can encapsulate a large amount of drug due to their porous structure. However, the safety concern is the main reason for their limited application [65, 66].

8.5.1.7 Polymer–Drug Conjugates

Polymer–drug conjugates are covalently linked conjugations of API to a hydrophilic polymer, especially PEG. These formulations have a sustained drug release rate and hinder the “burst release” common for other nanoparticles [16]. This strategy could be used for the pulmonary delivery of anticancer drugs and proteins. The increased retention time of the drug in the lung and lower local toxicity are beneficial [67, 68].

8.5.2 Formulation Considerations

Similar to the conventional formulations, drug-loaded nanoparticles could be administered using nebulizers, pMDIs, and DPIs. Nebulizers and pMDIs, deliver the suspension form of nanoparticles, while dry solid particles are needed for DPIs [69].

Nebulizers (especially jet or ultrasonic nebulizers) are widely used devices for the delivery of nanoparticles. Nanoparticle aggregation is the most important problem during nebulization. Therefore, the surface properties and concentration of nanoparticles should be considered. The addition of stabilizers and solubilizing agents such as surfactants to the formulation is beneficial for the hindrance of agglomeration. In addition, ultrasonic nebulizers can reduce nanoparticle aggregation more efficiently [69].

pMDIs deliver nanoparticles’ suspensions in propellant as inhalable droplets. Although some studies have indicated that pMDIs could be used for the transformation of nanocarriers’ suspension into inhalable droplets with high fine particle fractions, the application of these devices is limited. As discussed earlier, this limitation is due to the significant drawbacks of pMDIs including the limited delivery of emitted dose to the deep lung and the need for inhalation/actuation coordination [70, 71].

DPIs are stable formulations that reduce the physicochemical instabilities like drug leakage/degradation and aggregation of nanoparticles in aqueous media. Inhalable dry microparticles with MMAD in the range of 1–5 μm could be prepared using freeze drying, spray freeze drying, and spray drying processes [71]. Trojan and strawberry particles are two formulation strategies for nanoparticle delivery by DPIs (Figure 8.7) [69]. Trojan particles are hollow microparticles formed by nanoparticles' aggregation in the shell layer [72]. These particles could be prepared by spray drying or spray freeze drying of suspensions containing nanoparticles and a carrier (e.g., lactose, mannitol, trehalose, leucine, and phenylalanine) [19]. The type and number of carriers have significant effects on the size and surface properties of microparticles. It has been shown that microparticles with corrugated surfaces have a higher fine particle fraction and a lower tendency to be adsorbed to the device parts than the smooth ones [62]. Strawberry particles are formed by the physical adsorption of nanoparticles on the surface of micron-sized carriers. Freeze-drying of a mixture containing nanoparticles and lactose is a method of preparation [69]. After inhalation of these DPI formulations, the carrier dissolves in the respiratory tract, and microparticles convert to the starting nanoparticles. Effervescent technology has been used for the preparation of rapidly dissolving microparticles.

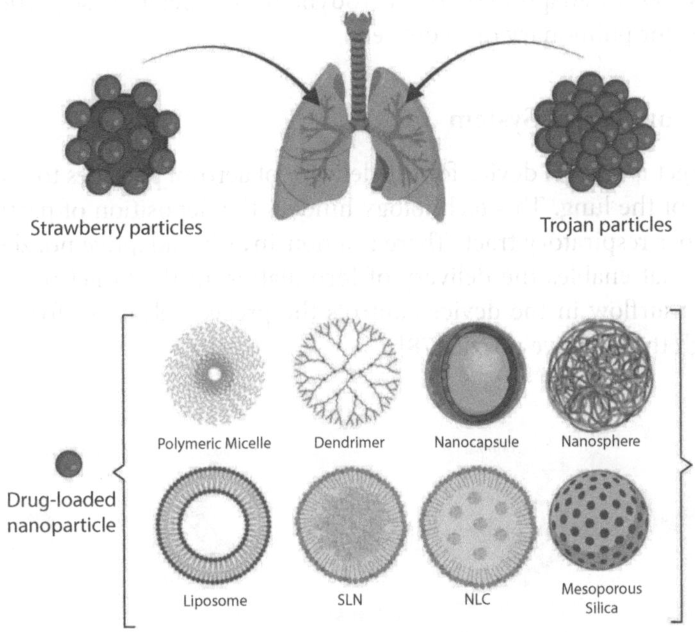

Figure 8.7 DPI formulations (including trojan particles and strawberry particles) for drug delivery of nanoparticles to the lung (created with BioRender.com).

These formulations could be prepared by spray-drying nanoparticles in the presence of effervescent excipients (e.g., a combination of citric acid with ammonia and sodium bicarbonate) [73].

8.5.3 Template-Assisted Particles

Template-assisted approaches are recently introduced methods for preparing reproducible particles capable of pulmonary drug delivery. Particle replication in non-wetting templates (PRINT) technology enables scientists to produce monodisperse particles with desirable aerodynamic diameters [74, 75]. Wilson *et al.* [76] have used this top-down manufacturing method for the production of a protein-containing DPI formulation. The process could be divided into two separate steps. The first step is particle fabrication, in which the formulation (including protein, trehalose, and glycerol) is applied to a mold (patterned with 1 μm cylinders). Heating is necessary for the filling of the mold. Then, solidified particles are collected after the cooling of the mold (harvesting step) (Figure 8.8). Also, Kohler *et al.* [77] have prepared uniform nanometer fibers using template-assisted approaches. By adaption of their aerodynamic diameter, these particles are suitable for pulmonary drug delivery.

8.5.4 on-Target System

on-Target is a novel device for the delivery of aerosol particles to a specific region of the lung. This technology hinders the deposition of particles in the upper respiratory tract. There is a non-invasive adaptive nozzle in the device that enables the delivery of formulation to the target region. The laminar airflow in the device controls the precise release of formulation through the adaptive nozzle [78].

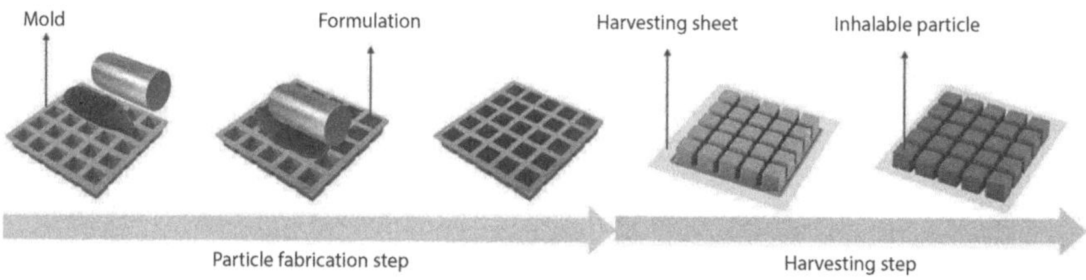

Figure 8.8 A schematic representation of PRINT technology to prepare particles for pulmonary drug delivery.

8.6 Pharmacokinetics

8.6.1 A Summary of Pulmonary Pharmacokinetics

The fate of inhaled particles depends on their physicochemical properties such as size and hydrophobicity. There are three main mechanisms for the clearance of particles from the lung, including mucociliary clearance, phagocytosis, and systemic uptake. Hydrophobic particles with MMAD higher than 5 μm are usually deposited in the mucus layer of the upper respiratory tract. The escalation movement of cilia would result in the clearance of particles by swallowing or coughing out. Detection of inhaled drugs in the feces is due to the swallowing of eliminated particles. Phagocytosis is the primary mechanism of clearance in the lower respiratory tract. As mentioned earlier, particles in the range of 1–5 μm would be phagocytosed more. The systemic uptake of low molecular weight lipophilic molecules is higher than other inhaled compounds [71]. Also, it has been indicated that lipophilicity is more important than size [79]. There are three different groups of lung epithelial membrane transporters, including ATP-binding cassette (ABC) efflux transporters, solute carrier (SLC) transporters, and organic anion transporting polypeptides (OATPs). These transporters have dominating effects on the fate of some inhaled drugs. For example, ciprofloxacin diffuses rapidly through the lung epithelial cells, and its plasma concentration rapidly equilibrates with lung concentration [80]. Considering the basic principles of pharmacokinetics in the respiratory system, we will discuss how a novel DDS alters the pharmacokinetic characteristics in the following section.

8.6.2 Pharmacokinetics of Nanoparticles

The knowledge about the pharmacokinetics of the novel inhalable formulations is not yet complete. This issue is known as one of the essential reasons for the rare occurrence of these formulations in the market. In addition, the majority of studies in this area only investigate the fate of the loaded drugs without considering their carriers. Last but not least, all the information about the pharmacokinetics of novel inhalable DDSs is based on pre-clinical studies. Despite all these pitfalls, we can find some rare studies investigating the pharmacokinetics of drug carriers, such as liposomes and SLNs [81]. In addition, recent studies showed that the conformation or size of nanoparticles has great effects on their fate compared to the solutions of starting materials. For example, the elimination half-life of albumin nanoparticles from the lung is longer than the albumin solution.

Endogenous mechanisms for protein clearance from the lungs are responsible for eliminating inhaled albumin solution, but, by contrast, are not effective for the clearance of albumin nanoparticles [82].

Haque *et al.* have compared the pharmacokinetics of SLNs and liposomes as two pulmonary drug carriers. The authors, in agreement with the previous studies, have introduced mucociliary clearance as the main mechanism of removal of the carriers such as SLNs, liposomes, biodegradable macromolecules, and non-biodegradable nanomaterials from the lungs [59, 81, 83]. The surface properties of nanoparticles have dominating effects on their mucoadhesion and fate. It has been shown that mucociliary clearance has a direct relationship with the mucoadhesion of the carriers. Hydrophilic surfaces of carriers (e.g., PEG-based nanocarriers) have a decreasing effect on mucoadhesion [84]. It has been indicated that the recovery of methotrexate-loaded dendrimers in feces was twice as high as that of PEGylated ones [85]. PLGA nanoparticles that were coated by chitosan, were deposited in more significant quantities in the upper respiratory tract in comparison to PVA-coated ones [86]. Although mucoadhesion would increase the clearance of nanoparticles due to the mucociliary escalation, this property could enhance the lung retention time of nanoparticles that are incapable of diffusion across the mucus layer [87].

However, for that portion of the formulation reached the alveolar region, the main method of clearance is the alveolar macrophages. It has also been shown that using novel DDSs (such as SLNs, liposomes, etc.) increases the residence time of the particles in the lungs [81]. As mentioned earlier, nanoparticles are phagocytosed in less extent than microparticles by alveolar macrophages. To solve the problem of low deposition of nanoparticles in the deep lung, the preparation of Trojan particles is a good strategy [16, 40, 88]. Also, PEGylation of nanocarriers would reduce their phagocytosis *via* inhibition of opsonization by surfactant proteins [89]. Mannosylated nanoparticles, however, are prepared for targeting alveolar macrophages in the treatment of intracellular lung infections [90]. Mannose receptors cover the surface of alveolar macrophages and administration of mannose-functionalized nanoparticles containing antibacterial agents such as isoniazid and rifampicin is a preferable strategy in the treatment of tuberculosis [38, 91].

Although systemic absorption of nanoparticles from the lung does not play a big part in the fate of these formulations, small amounts of nanoparticles could be transported to the blood circulation and lymphatic system through cellular or non-cellular pathways. The absorption of nanoparticles in the lymphatic system is more than blood circulation. The lymphatic absorption is through cellular (lung APCs) or non-cellular (lymphatic drainage from interstitial space) pathways. As a general rule, nanoparticles with smaller sizes are more easily transported to the blood circulation or lymph system [88, 92]. Conjugating high molecular weight PEG chains to the nanoparticles would decrease their systemic absorption rate [71].

Encapsulation of APIs in the nanoparticles reduces systemic absorption of the drug through the lung epithelium. It has been reported that the inhalation of a liposomal formulation of ciprofloxacin increases the concentration of the drug in lung epithelial lining fluid and alveolar macrophages in comparison to the free drug [93]. Nanoparticles with negative or neutral surface charge easily reach blood circulation [92]. The lymphatic absorption of nanoparticles through the cellular pathway would not be affected by the surface charge of nanoparticles [94]. The neutral nanoparticles with hydrophilic coatings are transported to the lymph nodes more rapidly through the non-cellular pathway than positively charged ones [95].

Controlled drug release is a result of inhibiting pre-mature elimination or degradation of nanoparticles. Generally, lipid nanoparticles are subject to degradation due to the surfactant lining layer of the lung. Nanocarriers with an optimum size conjugated to external ligands can escape the alveolar macrophages as well as systemic uptake through the lung epithelium [71]. Moreover, the preparation of active targeting nanoparticles can improve their pharmacokinetics and pharmacodynamic properties. In addition to macrophage-targeted nanocarriers, this strategy could be used for the treatment of lung cancers. Transferrin or folic acid-conjugated liposomes have been used for the active targeting of A549 cancer cells [96–98].

As discussed earlier in Section 3.2., pathological conditions could affect the pharmacokinetic profile of inhalable DDSs. For example, clearance pathways and distribution profiles of inhaled nanoparticles are affected by inflammatory lung diseases [99, 100]. Inflammation causes an increase in alveolar macrophage number as well as mucus viscosity and also changes mucus composition [101]. Overall, the elimination rate of nanoparticles decreases in these conditions. In addition, the higher viscosity of mucus impairs alveolar macrophage mobility and, therefore, their eliminating function [102].

8.7 Products in the Market

As stated earlier, despite decades of working on the development of inhalable DDSs, very few formulations have already found their way to the market. However, scientists and companies keep working on novel strategies and formulations for inhalable formulations to pass the preclinical and clinical phases and hit the market. Liang W *et al.* [103] have listed the novel formulations in clinical stages. It is worth mentioning that through literature screening, we found several formulations that were introduced as novel inhalable products in the market, such as Apulmig (by Aradigm), Cisplatin-based DPI, and Paclitaxel-based DPI (by InhaTarget); while both FDA and EMA have rejected the first one, according to the manufacturer, the last two, are still at preclinical stages of toxicity and efficacy studies in animal models [104, 105]. Another controversial issue is that products such as Dornase alfa (by Genentech), Survanta (by Abbvie), Alveofact (by Boehringer-Ingelheim Pharma), and Curosurf (by Dey LP and Chiesi Farmaceutici SpA) are sometimes listed as novel inhalable DDS. These products, however, consist of surfactants,

Table 8.1 Novel inhalable DDS in the market.

Product	Description	Active therapeutic agent	Indication	Manufacturer	Approval date
ARIKAYCE	Liposomal suspension for inhalation, administered only using a Lamira Nebuliser System.	Amikacin	Non-tuberculous mycobacterial (NTM) lung infections caused by Mycobacterium avium Complex (MAC)	Insmed Incorporated	09/28/2018
TOBI Podhaler	Using Pulmosphere™ technology (emulsion-based spray drying process) to produce light porous particles, dry-powder formulations	Tobramycin	Cystic fibrosis with Pseudomonas aeruginosa	Novartis	3/22/2013
Afrezza	A drug-device consisting of Technosphere® insulin inhalation powder pre-metered into single-use dose cartridges for administration with an AFREZZA™ inhaler	Insulin	Diabetes mellitus	MannKind Corporation	06/27/2014

phospholipids, and enzymes, which could not be categorized into any classes of novel inhalable DDSs such as liposomes. To the best of our knowledge, the current existing approved novel inhalable DDS in the market can be summarized in Table 8.1.

References

1. Stein, S.W. and Thiel, C.G., The history of therapeutic aerosols: a chronological review. *J. Aerosol Med. Pulm. Drug Deliv.*, 30, 1, 20–41, 2017.
2. Crompton, G., A brief history of inhaled asthma therapy over the last fifty years. *Prim. Care Respir. J.*, 15, 6, 326–31, 2006.
3. Newman, S.P., Drug delivery to the lungs: challenges and opportunities. *Ther. Deliv.*, 8, 8, 647–61, 2017.
4. Khan, I., Lau, K., Bnyan, R., Houacine, C., Roberts, M., Isreb, A. *et al.*, A facile and novel approach to manufacture paclitaxel-loaded proliposome tablet formulations of micro or nano vesicles for nebulization. *Pharm. Res.*, 37, 1–19, 2020.
5. Mehta, P.P., Ghoshal, D., Pawar, A.P., Kadam, S.S., Dhapte-Pawar, V.S., Recent advances in inhalable liposomes for treatment of pulmonary diseases: concept to clinical stance. *J. Drug Deliv. Sci. Technol.*, 56, 101509, 2020.
6. Huang, Z., Kłodzińska, S.N., Wan, F., Nielsen, H.M., Nanoparticle-mediated pulmonary drug delivery: state of the art towards efficient treatment of recalcitrant respiratory tract bacterial infections. *Drug Deliv. Transl. Res.*, 11, 1634–1654, 2021.
7. Xu, Y., Liu, H., Song, L., Novel drug delivery systems targeting oxidative stress in chronic obstructive pulmonary disease: a review. *J. Nanobiotechnol.*, 18, 1, 1–25, 2020.
8. Mortensen, N.P. and Hickey, A.J., Targeting inhaled therapy beyond the lungs. *Respiration*, 88, 5, 353–64, 2014.
9. Stoilova, T. and Ruhrberg, C., Lung blood and lymphatic vascular development, in: *Lung Stem Cells in Development, Health and Disease (ERS Monograph) Sheffield*, pp. 31–43, European Respiratory Society, Sheffield, UK, 2021.
10. El-Chemaly, S., Levine, S.J., Moss, J., *Lymphatics in lung disease*, vol. 1131, p. 195, Annals of the New York Academy of Sciences, New York, 2008.
11. Peate, I., Anatomy and physiology, 10. The respiratory system. *Br. J. Healthcare Assistants*, 12, 4, 178–81, 2018.
12. Aung, H., Sivakumar, A., Gholami, S., Venkateswaran, S., Gorain, B., An Overview of the Anatomy and Physiology of the Lung, in: *Nanotechnology-Based Targeted Drug Delivery Systems for Lung Cancer*, pp. 1–20, 2019.
13. Ochs, M., Nyengaard, J.R., Jung, A., Knudsen, L., Voigt, M., Wahlers, T. *et al.*, The number of alveoli in the human lung. *Am. J. Respir. Crit. Care Med.*, 169, 1, 120–4, 2004.

14. De Virgiliis, F. and Di Giovanni, S., Lung innervation in the eye of a cytokine storm: neuroimmune interactions and COVID-19. *Nat. Rev. Neurol.*, 16, 11, 645–52, 2020.
15. Hsia, C.C., Hyde, D.M., Weibel, E.R., Lung structure and the intrinsic challenges of gas exchange. *Compr. Physiol.*, 6, 2, 827–95, 2011.
16. Liu, Q., Guan, J., Qin, L., Zhang, X., Mao, S., Physicochemical properties affecting the fate of nanoparticles in pulmonary drug delivery. *Drug Discovery Today*, 25, 1, 150–9, 2020.
17. Ruge, C.A., Kirch, J., Lehr, C.-M., Pulmonary drug delivery: from generating aerosols to overcoming biological barriers—therapeutic possibilities and technological challenges. *Lancet Respir. Med.*, 1, 5, 402–13, 2013.
18. Haque, S., Whittaker, M.R., McIntosh, M.P., Pouton, C.W., Kaminskas, L.M., Disposition and safety of inhaled biodegradable nanomedicines: Opportunities and challenges. *Nanomed. Nanotechnol. Biol. Med.*, 12, 6, 1703–24, 2016.
19. Liang, Z., Ni, R., Zhou, J., Mao, S., Recent advances in controlled pulmonary drug delivery. *Drug Discovery Today*, 20, 3, 380–9, 2015.
20. Wang, Y.-B., Watts, A.B., Peters, J.I., Williams III, R.O., The impact of pulmonary diseases on the fate of inhaled medicines—A review. *Int. J. Pharm.*, 461, 1-2, 112–28, 2014.
21. Huckaby, J.T. and Lai, S.K., PEGylation for enhancing nanoparticle diffusion in mucus. *Adv. Drug Deliv. Rev.*, 124, 125–39, 2018.
22. Champion, J.A., Walker, A., Mitragotri, S., Role of particle size in phagocytosis of polymeric microspheres. *Pharm. Res.*, 25, 8, 1815–21, 2008.
23. Champion, J.A. and Mitragotri, S., Shape induced inhibition of phagocytosis of polymer particles. *Pharm. Res.*, 26, 1, 244–9, 2009.
24. Barba, A.A., Bochicchio, S., Dalmoro, A., Caccavo, D., Cascone, S., Lamberti, G., Chapter 10 - Polymeric and lipid-based systems for controlled drug release: an engineering point of view, in: *Nanomaterials for Drug Delivery and Therapy*, A.M. Grumezescu (Ed.), pp. 267–304, William Andrew Publishing, NY, USA, 2019.
25. Seydoux, E., Rodriguez-Lorenzo, L., Blom, R.A., Stumbles, P.A., Petri-Fink, A., Rothen-Rutishauser, B.M. *et al.*, Pulmonary delivery of cationic gold nanoparticles boost antigen-specific CD4+ T cell proliferation. *Nanomed. Nanotechnol. Biol. Med.*, 12, 7, 1815–26, 2016.
26. Maretti, E., Rossi, T., Bondi, M., Croce, M.A., Hanuskova, M., Leo, E. *et al.*, Inhaled Solid Lipid Microparticles to target alveolar macrophages for tuberculosis. *Int. J. Pharm.*, 462, 1-2, 74–82, 2014.
27. Patel, B., Gupta, N., Ahsan, F., Particle engineering to enhance or lessen particle uptake by alveolar macrophages and to influence the therapeutic outcome. *Eur. J. Pharm. Biopharm.*, 89, 163–74, 2015.
28. Pilcer, G. and Amighi, K., Formulation strategy and use of excipients in pulmonary drug delivery. *Int. J. Pharm.*, 392, 1-2, 1–19, 2010.

29. Morales, J.O., Peters, J.I., Williams, R.O., Surfactants: their critical role in enhancing drug delivery to the lungs. *Ther. Deliv.*, 2, 5, 623–41, 2011.
30. Chen, Y., Du, S., Zhang, Z., He, W., Lu, E., Wang, R. *et al.*, Compatible Stability and Aerosol Characteristics of Atrovent®(Ipratropium Bromide) Mixed with Salbutamol Sulfate, Terbutaline Sulfate, Budesonide, and Acetylcysteine. *Pharmaceutics*, 12, 8, 776, 2020.
31. Elbatanony, R.S., Parvathaneni, V., Kulkarni, N.S., Shukla, S.K., Chauhan, G., Kunda, N.K. *et al.*, Afatinib-loaded inhalable PLGA nanoparticles for localized therapy of non-small cell lung cancer (NSCLC)—development and *in-vitro* efficacy. *Drug Deliv. Transl. Res.*, 11, 3, 927–43, 2021.
32. Satari, N., Taymouri, S., Varshosaz, J., Rostami, M., Mirian, M., Preparation and evaluation of inhalable dry powder containing glucosamine-conjugated gefitinib SLNs for lung cancer therapy. *Drug Dev. Ind. Pharm.*, 46, 8, 1265–77, 2020.
33. Varsou, D.D., Afantitis, A., Tsoumanis, A., Papadiamantis, A., Valsami-Jones, E., Lynch, I. *et al.*, Zeta-Potential Read-Across Model Utilizing Nanodescriptors Extracted via the NanoXtract Image Analysis Tool Available on the Enalos Nanoinformatics Cloud Platform. *Small*, 16, 21, 1906588, 2020.
34. Mishra, B. and Singh, J., Novel drug delivery systems and significance in respiratory diseases, in: *Targeting Chronic Inflammatory Lung Diseases Using Advanced Drug Delivery Systems*, pp. 57–95, Elsevier, Amsterdam, Netherlands, 2020.
35. Caimmi, D., Martocq, N., Trioleyre, D., Guinet, C., Godreuil, S., Daniel, T. (Eds.), Positive effect of liposomal amikacin for inhalation on Mycobacterium abcessus in cystic fibrosis patients, in: *Open forum infectious diseases*, Oxford University Press, US, 2018.
36. Darquenne, C., Deposition mechanisms. *J. Aerosol Med. Pulm. Drug Deliv.*, 33, 4, 181–5, 2020.
37. Islam, N. and Rahman, S., Improved treatment of nicotine addiction and emerging pulmonary drug delivery. *Drug Discov. Ther.*, 6, 3, 123–32, 2012.
38. Costa, A., Sarmento, B., Seabra, V., Mannose-functionalized solid lipid nanoparticles are effective in targeting alveolar macrophages. *Eur. J. Pharm. Sci.*, 114, 103–13, 2018.
39. Pei, Y. and Yeo, Y., Drug delivery to macrophages: challenges and opportunities. *J. Controlled Release*, 240, 202–11, 2016.
40. Beck-Broichsitter, M., Merkel, O.M., Kissel, T., Controlled pulmonary drug and gene delivery using polymeric nano-carriers. *J. Controlled Release*, 161, 2, 214–24, 2012.
41. Yıldız-Peköz, A. and Ehrhardt, C., *Advances in Pulmonary Drug Delivery*, Multidisciplinary Digital Publishing Institute, Basel, Switzerland, 2020.
42. Murata, M., Nakano, K., Tahara, K., Tozuka, Y., Takeuchi, H., Pulmonary delivery of elcatonin using surface-modified liposomes to improve systemic

absorption: polyvinyl alcohol with a hydrophobic anchor and chitosan oligosaccharide as effective surface modifiers. *Eur. J. Pharm. Biopharm.*, 80, 2, 340–6, 2012.

43. Li, H.-Y. and Birchall, J., Chitosan-modified dry powder formulations for pulmonary gene delivery. *Pharm. Res.*, 23, 5, 941–50, 2006.
44. Makhlof, A., Werle, M., Tozuka, Y., Takeuchi, H., Nanoparticles of glycol chitosan and its thiolated derivative significantly improved the pulmonary delivery of calcitonin. *Int. J. Pharm.*, 397, 1-2, 92–5, 2010.
45. Surendrakumar, K., Martyn, G., Hodgers, E., Jansen, M., Blair, J., Sustained release of insulin from sodium hyaluronate based dry powder formulations after pulmonary delivery to beagle dogs. *J. Controlled Release*, 91, 3, 385–94, 2003.
46. Sheth, P. and Myrdal, P.B., Polymers for pulmonary drug delivery, in: *Controlled pulmonary drug delivery*, pp. 265–82, Springer, Salmon Tower Building, New York City, 2011.
47. Swaminathan, J. and Ehrhardt, C., Liposomes for pulmonary drug delivery, in: *Controlled pulmonary drug delivery*, pp. 313–34, Springer, Salmon Tower Building, New York City, 2011.
48. Swaminathan, J. and Ehrhardt, C., Liposomal delivery of proteins and peptides. *Expert Opin. Drug Deliv.*, 9, 12, 1489–503, 2012.
49. Manca, M.L., Valenti, D., Sales, O.D., Nacher, A., Fadda, A.M., Manconi, M., Fabrication of polyelectrolyte multilayered vesicles as inhalable dry powder for lung administration of rifampicin. *Int. J. Pharm.*, 472, 1-2, 102–9, 2014.
50. Manca, M.L., Manconi, M., Valenti, D., Lai, F., Loy, G., Matricardi, P. *et al.*, Liposomes coated with chitosan–xanthan gum (chitosomes) as potential carriers for pulmonary delivery of rifampicin. *J. Pharm. Sci.*, 101, 2, 566–75, 2012.
51. Misra, A., Jinturkar, K., Patel, D., Lalani, J., Chougule, M., Recent advances in liposomal dry powder formulations: preparation and evaluation. *Expert Opin. Drug Deliv.*, 6, 1, 71–89, 2009.
52. Gaspar, D.P., Faria, V., Gonçalves, L.M., Taboada, P., Remuñán-López, C., Almeida, A.J., Rifabutin-loaded solid lipid nanoparticles for inhaled antitubercular therapy: physicochemical and *in vitro* studies. *Int. J. Pharm.*, 497, 1-2, 199–209, 2016.
53. Ji, P., Yu, T., Liu, Y., Jiang, J., Xu, J., Zhao, Y. *et al.*, Naringenin-loaded solid lipid nanoparticles: preparation, controlled delivery, cellular uptake, and pulmonary pharmacokinetics. *Drug Des. Dev. Ther.*, 10, 911, 2016.
54. Jaafar-Maalej, C., Elaissari, A., Fessi, H., Lipid-based carriers: manufacturing and applications for pulmonary route. *Expert Opin. Drug Deliv.*, 9, 9, 1111–27, 2012.
55. Patil-Gadhe, A. and Pokharkar, V., Pulmonary targeting potential of rosuvastatin loaded nanostructured lipid carrier: optimization by factorial design. *Int. J. Pharm.*, 501, 1-2, 199–210, 2016.

56. Patlolla, R.R., Chougule, M., Patel, A.R., Jackson, T., Tata, P.N., Singh, M., Formulation, characterization and pulmonary deposition of nebulized celecoxib encapsulated nanostructured lipid carriers. *J. Controlled Release*, 144, 2, 233–41, 2010.
57. Kaminskas, L.M., Kelly, B.D., McLeod, V.M., Sberna, G., Owen, D.J., Boyd, B.J. *et al.*, Characterisation and tumour targeting of PEGylated polylysine dendrimers bearing doxorubicin via a pH labile linker. *J. Controlled Release*, 152, 2, 241–8, 2011.
58. Kaminskas, L.M., McLeod, V.M., Ryan, G.M., Kelly, B.D., Haynes, J.M., Williamson, M. *et al.*, Pulmonary administration of a doxorubicin-conjugated dendrimer enhances drug exposure to lung metastases and improves cancer therapy. *J. Controlled Release*, 183, 18–26, 2014.
59. Ryan, G.M., Kaminskas, L.M., Kelly, B.D., Owen, D.J., McIntosh, M.P., Porter, C.J., Pulmonary administration of PEGylated polylysine dendrimers: absorption from the lung versus retention within the lung is highly size-dependent. *Mol. Pharm.*, 10, 8, 2986–95, 2013.
60. Farhangi, M., Kobarfard, F., Mahboubi, A., Vatanara, A., Mortazavi, S.A., Preparation of an optimized ciprofloxacin-loaded chitosan nanomicelle with enhanced antibacterial activity. *Drug Dev. Ind. Pharm.*, 44, 8, 1273–84, 2018.
61. Pellosi, D.S., d'Angelo, I., Maiolino, S., Mitidieri, E., di Villa Bianca, R.dE., , Sorrentino, R. *et al.*, *In vitro/in vivo* investigation on the potential of Pluronic® mixed micelles for pulmonary drug delivery. *Eur. J. Pharm. Biopharm.*, 130, 30–8, 2018.
62. Farhangi, M., Mahboubi, A., Kobarfard, F., Vatanara, A., Mortazavi, S.A., Optimization of a dry powder inhaler of ciprofloxacin-loaded polymeric nanomicelles by spray drying process. *Pharm. Dev. Technol.*, 24, 5, 584–92, 2019.
63. Dames, P., Gleich, B., Flemmer, A., Hajek, K., Seidl, N., Wiekhorst, F. *et al.*, Targeted delivery of magnetic aerosol droplets to the lung. *Nat. Nanotechnol.*, 2, 8, 495–9, 2007.
64. Stocke, N.A., Meenach, S.A., Arnold, S.M., Mansour, H.M., Hilt, J.Z., Formulation and characterization of inhalable magnetic nanocomposite microparticles (MnMs) for targeted pulmonary delivery via spray drying. *Int. J. Pharm.*, 479, 2, 320–8, 2015.
65. Mohseni, M., Gilani, K., Mortazavi, S.A., Preparation and characterization of rifampin loaded mesoporous silica nanoparticles as a potential system for pulmonary drug delivery. *Iran J. Pharm. Res. IJPR*, 14, 1, 27, 2015.
66. Taratula, O., Garbuzenko, O.B., Chen, A.M., Minko, T., Innovative strategy for treatment of lung cancer: targeted nanotechnology-based inhalation co-delivery of anticancer drugs and siRNA. *J. Drug Targeting*, 19, 10, 900–14, 2011.
67. Luo, T., Loira-Pastoriza, C., Patil, H.P., Ucakar, B., Muccioli, G.G., Bosquillon, C. *et al.*, PEGylation of paclitaxel largely improves its safety and anti-tumor

efficacy following pulmonary delivery in a mouse model of lung carcinoma. *J. Controlled Release*, 239, 62–71, 2016.
68. Koussoroplis, S.J., Paulissen, G., Tyteca, D., Goldansaz, H., Todoroff, J., Barilly, C. *et al.*, PEGylation of antibody fragments greatly increases their local residence time following delivery to the respiratory tract. *J. Controlled Release*, 187, 91–100, 2014.
69. Rogueda, P.G. and Traini, D., The nanoscale in pulmonary delivery. Part 2: formulation platforms. *Expert Opin. Drug Deliv.*, 4, 6, 607–20, 2007.
70. Conti, D.S., Brewer, D., Grashik, J., Avasarala, S., da Rocha, S.R., Poly (amidoamine) dendrimer nanocarriers and their aerosol formulations for siRNA delivery to the lung epithelium. *Mol. Pharm.*, 11, 6, 1808–22, 2014.
71. Shen, A.M. and Minko, T., Pharmacokinetics of inhaled nanotherapeutics for pulmonary delivery. *J. Controlled Release*, 326, 222–44, 2020.
72. Al-Hallak, M.K., Sarfraz, M.K., Azarmi, S., Roa, W.H., Finlay, W.H., Löbenberg, R., Pulmonary delivery of inhalable nanoparticles: dry powder inhalers. *Ther. Deliv.*, 2, 10, 1313–24, 2011.
73. Ely, L., Roa, W., Finlay, W.H., Löbenberg, R., Effervescent dry powder for respiratory drug delivery. *Eur. J. Pharm. Biopharm.*, 65, 3, 346–53, 2007.
74. Perry, J.L., Herlihy, K.P., Napier, M.E., DeSimone, J.M., PRINT: a novel platform toward shape and size specific nanoparticle theranostics. *Acc. Chem. Res.*, 44, 10, 990–8, 2011.
75. Garcia, A., Mack, P., Williams, S., Fromen, C., Shen, T., Tully, J. *et al.*, Microfabricated engineered particle systems for respiratory drug delivery and other pharmaceutical applications. *J. Drug Deliv.*, 2012, 10 pp., 2012.
76. Wilson, E.M., Luft, J.C., DeSimone, J.M., Formulation of high-performance dry powder aerosols for pulmonary protein delivery. *Pharm. Res.*, 35, 10, 1–11, 2018.
77. Kohler, D., Schneider, M., Krüger, M., Lehr, C.M., Möhwald, H., Wang, D., Template-Assisted Polyelectrolyte Encapsulation of Nanoparticles into Dispersible, Hierarchically Nanostructured Microfibers. *Adv. Mater.*, 23, 11, 1376–9, 2011.
78. Kleinstreuer, C. and Seelecke, S., Inhaler system for targeted maximum drug-aerosol delivery, Google Patents, US7900625B2, 2011.
79. Otterson, G.A., Villalona-Calero, M.A., Hicks, W., Pan, X., Ellerton, J.A., Gettinger, S.N. *et al.*, Phase I/II Study of inhaled doxorubicin combined with platinum-based therapy for advanced non–small cell lung cancer. *Clin. Cancer Res.*, 16, 8, 2466–73, 2010.
80. Ong, H.X., Traini, D., Bebawy, M., Young, P.M., Ciprofloxacin is actively transported across bronchial lung epithelial cells using a Calu-3 air interface cell model. *Antimicrob. Agents Chemother.*, 57, 6, 2535–40, 2013.
81. Haque, S., Whittaker, M., McIntosh, M.P., Pouton, C.W., Phipps, S., Kaminskas, L.M., A comparison of the lung clearance kinetics of solid lipid

nanoparticles and liposomes by following the 3H-labelled structural lipids after pulmonary delivery in rats. *Eur. J. Pharm. Biopharm.*, 125, 1–12, 2018.

82. Woods, A., Patel, A., Spina, D., Riffo-Vasquez, Y., Babin-Morgan, A., De Rosales, R. *et al.*, *In vivo* biocompatibility, clearance, and biodistribution of albumin vehicles for pulmonary drug delivery. *J. Controlled Release*, 210, 1–9, 2015.
83. Geiser, M., Update on macrophage clearance of inhaled micro-and nanoparticles. *J. Aerosol Med. Pulm. Drug Deliv.*, 23, 4, 207–17, 2010.
84. Lai, S.K., Wang, Y.-Y., Hanes, J., Mucus-penetrating nanoparticles for drug and gene delivery to mucosal tissues. *Adv. Drug Deliv. Rev.*, 61, 2, 158–71, 2009.
85. Haque, S., McLeod, V.M., Jones, S., Fung, S., Whittaker, M., McIntosh, M. *et al.*, Effect of increased surface hydrophobicity via drug conjugation on the clearance of inhaled PEGylated polylysine dendrimers. *Eur. J. Pharm. Biopharm.*, 119, 408–18, 2017.
86. Ungaro, F., d'Angelo, I., Miro, A., La Rotonda, M.I., Quaglia, F., Engineered PLGA nano-and micro-carriers for pulmonary delivery: challenges and promises. *J. Pharm. Pharmacol.*, 64, 9, 1217–35, 2012.
87. Ensign, L.M., Schneider, C., Suk, J.S., Cone, R., Hanes, J., Mucus penetrating nanoparticles: biophysical tool and method of drug and gene delivery. *Adv. Mater.*, 24, 28, 3887–94, 2012.
88. Smyth, H.D. and Hickey, A.J., *Controlled pulmonary drug delivery*, Springer, Salmon Tower Building, New York City, 2011.
89. Chono, S., Suzuki, H., Togami, K., Morimoto, K., Efficient drug delivery to lung epithelial lining fluid by aerosolization of ciprofloxacin incorporated into PEGylated liposomes for treatment of respiratory infections. *Drug Dev. Ind. Pharm.*, 37, 4, 367–72, 2011.
90. Chono, S., Tanino, T., Seki, T., Morimoto, K., Efficient drug targeting to rat alveolar macrophages by pulmonary administration of ciprofloxacin incorporated into mannosylated liposomes for treatment of respiratory intracellular parasitic infections. *J. Controlled Release*, 127, 1, 50–8, 2008.
91. Maretti, E., Costantino, L., Rustichelli, C., Leo, E., Croce, M.A., Buttini, F. *et al.*, Surface engineering of Solid Lipid Nanoparticle assemblies by methyl α-d-mannopyranoside for the active targeting to macrophages in anti-tuberculosis inhalation therapy. *Int. J. Pharm.*, 528, 1-2, 440–51, 2017.
92. Zhong, Q., Merkel, O.M., Reineke, J.J., da Rocha, S.R., Effect of the route of administration and PEGylation of poly (amidoamine) dendrimers on their systemic and lung cellular biodistribution. *Mol. Pharm.*, 13, 6, 1866–78, 2016.
93. Chono, S., Tanino, T., Seki, T., Morimoto, K., Efficient drug delivery to alveolar macrophages and lung epithelial lining fluid following pulmonary administration of liposomal ciprofloxacin in rats with pneumonia and estimation of its antibacterial effects. *Drug Dev. Ind. Pharm.*, 34, 10, 1090–6, 2008.

94. Fromen, C.A., Rahhal, T.B., Robbins, G.R., Kai, M.P., Shen, T.W., Luft, J.C. *et al.*, Nanoparticle surface charge impacts distribution, uptake and lymph node trafficking by pulmonary antigen-presenting cells. *Nanomed.: Nanotechnol. Biol. Med.*, 12, 3, 677–87, 2016.
95. Choi, H.S., Ashitate, Y., Lee, J.H., Kim, S.H., Matsui, A., Insin, N. *et al.*, Rapid translocation of nanoparticles from the lung airspaces to the body. *Nat. Biotechnol.*, 28, 12, 1300–3, 2010.
96. Anabousi, S., Bakowsky, U., Schneider, M., Huwer, H., Lehr, C.-M., Ehrhardt, C., *In vitro* assessment of transferrin-conjugated liposomes as drug delivery systems for inhalation therapy of lung cancer. *Eur. J. Pharm. Sci.*, 29, 5, 367–74, 2006.
97. Yuan, Z., Chen, D., Zhang, S., Zheng, Z., Preparation, characterization and evaluation of docetaxel-loaded, folate-conjugated PEG-liposomes. *Yakugaku Zasshi*, 130, 10, 1353–9, 2010.
98. Zhu, X., Kong, Y., Liu, Q., Lu, Y., Xing, H., Lu, X. *et al.*, Inhalable dry powder prepared from folic acid-conjugated docetaxel liposomes alters pharmacodynamic and pharmacokinetic properties relevant to lung cancer chemotherapy. *Pulm. Pharmacol. Ther.*, 55, 50–61, 2019.
99. Barlow, P.G., Brown, D.M., Donaldson, K., MacCallum, J., Stone, V., Reduced alveolar macrophage migration induced by acute ambient particle (PM 10) exposure. *Cell Biol. Toxicol.*, 24, 3, 243–52, 2008.
100. Johnson, L.N. and Koval, M., Cross-talk between pulmonary injury, oxidant stress, and gap junctional communication. *Antioxid. Redox Signaling*, 11, 2, 355–67, 2009.
101. Lai, S.K., Wang, Y.-Y., Wirtz, D., Hanes, J., Micro-and macrorheology of mucus. *Adv. Drug Deliv. Rev.*, 61, 2, 86–100, 2009.
102. Haque, S., Feeney, O., Meeusen, E., Boyd, B.J., McIntosh, M.P., Pouton, C.W. *et al.*, Local inflammation alters the lung disposition of a drug loaded pegylated liposome after pulmonary dosing to rats. *J. Controlled Release*, 307, 32–43, 2019.
103. Liang, W., Pan, H.W., Vllasaliu, D., Lam, J.K., Pulmonary delivery of biological drugs. *Pharmaceutics*, 12, 11, 1025, 2020.
104. https://bronchiectasisnewstoday.com/bronchiectasis-experimental-treatments/apulmiq-ciprofloxacin/
105. https://www.inhatarget.com/clinical-trial.

9

Intranasal Novel Drug Delivery Systems

Mehrnoosh Nikpour

Department of Pharmaceutics, School of Pharmacy, Zanjan University of Medical Sciences, Zanjan, Iran

Abstract

Intranasal drug delivery (INDD) is an attractive route of drug administration, due to its great capabilities for absorption of a wide range of drug molecules including small molecules and biopharmaceuticals, both for systemic and topical purposes. Among the most important of these capabilities, one can mention the large surface area with high blood supply and porous endothelium as well as the circumvented hepatic first-pass metabolism. These allow rapid drug absorption and improved bioavailability while a low amount of drug dose is needed to create a therapeutic effect. Moreover, the intranasal cavity provides a special pathway for drug delivery to the central nervous system (CNS) while bypassing the blood–brain barrier (BBB). Furthermore, intranasal is a preferred way for vaccine administration. Nevertheless, INDD has its challenges as well, namely limited space volume of the nasal cavity, rapid clearance, and enzymatic metabolism. Such challenges, although limiting the widespread use of INDD, have recently led to the development of novel technologies and systems to conquer them to an acceptable level (e.g., mucoadhesion and nanocarriers). In this chapter, after a brief review of the anatomy and physiology of the nasal cavity, the factors influencing INDD will be discussed. Then, the challenges and opportunities of INDD will be investigated and the ways to overcome or take advantage of each one will be explained, respectively. Also, available intranasal pharmaceutical preparations will be classified according to their specific properties. A general pharmacokinetic overview of INDD will be given, and it will be discussed how novel INDD systems may improve pharmacokinetic profiles of intranasally administered drugs. Finally, a summarized list of available commercialized intranasal preparations will be given.

***Keywords*:** Intranasal drug delivery, nose-to-brain drug delivery, mucoadhesive, nanotechnology, pharmacokinetics

Email: Mehrnoosh.nik@gmail.com

Mehrdad Hamidi and Mahsa Sayed Tabatabaei (eds.) Novel Drug Delivery Systems: Fundamentals and Applications, (307–348) © 2025 Scrivener Publishing LLC

9.1 Introduction

Drug delivery systems are of high importance as they determine to a high extent if an administered therapeutic agent reaches its target site and imposes its pharmacological impact or not [1]. Intranasal drug delivery (INDD) is considered an appropriate route of administration either for topical or systemic purposes and can be applied to diverse types of medicines including biopharmaceuticals, proteins, peptides, and non-peptides [2]. INDD has been used for a long time for many topical and systemic treatment purposes, because of its ease of access, rapid onset of action, and high bioavailability compared to some other parenteral routes of administration. The intranasal airway provides a large and easily accessible surface area with high blood flow and porous endothelium [3]. This large surface area permits drugs to be freely in contact with the porous endothelium and results in high systemic absorption [4]. Hence, lower drug doses would be needed and a faster onset of therapeutic action would occur [5]. The intranasally administered drugs can directly access the systemic blood circulation and even the central nervous system (CNS) while bypassing the liver first-pass metabolism as well as the blood-brain barrier (BBB), respectively [6, 7]. So, INDD can be extremely advantageous for the administration of drugs that undergo significant chemical decomposition when exposed to the gastrointestinal (GI) acidic and enzymatic environment, especially for those molecules that are highly metabolized by the liver first-pass effect. Thus intranasal route can increase the bioavailability of such therapeutic agents [6, 8]. Furthermore, INDD makes the self-treatment facile which enhances patient compliance with little risk of getting over-dosed [7]. Additionally, INDD is considered as an alternative route for topical administration of some drugs such as corticosteroids that have undesirable side effects when administered systemically. Topical intranasal corticosteroids can be used in the treatment of nasal disorders like allergic rhinitis, rhino-sinusitis, and polyps [9, 10].

Despite these advantages, INDD has its limitations such as the small volume that can be used, rapid clearance, and enzymatic metabolism. Also, thought the INDD is recently regarded as a serious easily accessible route for brain drug delivery without the challenge of bypassing BBB, its efficiency is low due to nasal anatomic, physiological, and histological properties [11, 12]. Such challenges necessitate the use of novel approaches and technologies. Novel drug delivery systems including nano-based systems and mucoadhesive preparations are good strategies that can overcome these obstacles to an acceptable level. This chapter aims to provide a brief but comprehensive review of the anatomy and physiology of the nasal cavity, the challenges and opportunities associated with INDD, and the classification of novel INDD systems. Then, the pharmacokinetic aspects of INDD along with the commercialized novel INDD systems will be reviewed.

9.2 Anatomy and Physiology

The upper part of the respiratory tract plays an important role in making the inhaled air warm (37°C), humidified (100% relative humidity), and particle-free [13]. Furthermore, it captures the majority of viruses and bacteria and makes the inhaled air healthier [14]. The nose is the principal part of this system which conditions the inspired air. Also, millions of the neurons responsible for the sense of smell, called olfactory nerves, are located in the upper section of the nasal cavity [15]. The total structure of the nose can be divided into two separate parts internal and external. The external structure consists of cartilage and bones that protect the internal section which is consisted of cavities and mucosa. The upper third section of the nose is comprised of a pair of bones that are laterally attached to the maxilla. This structure forms the bony opening of the nasal airway on the face which is called the pyriform aperture (Figure 9.1a). The size and shape of the pyriform aperture can be different for each person. In general, it is narrower in females. The rest two-third structure of the nose is made up of cartilage. At the lowest part, they comprise the naris, the two horseshoe-shaped inlets surrounded by lateral boundaries known as *ala*. (Figure 9.1b). The whole exterior nose is covered by muscular fibers and skin [16–18]. It is structurally similar to a three-sided pyramid with the top on the forehead between two bony arches that slide down slightly to its base above the upper lip. The junction area where the nasal bone attaches to the frontal bone is named radix. The dorsum is the ridge that links the radix to the nasal tip which is the most frontward part of the nose [16]. Two nostrils are located at the base of the pyramid and are partitioned by a blade named membranous columella (Figure 9.1c). This septum has a bony-cartilaginous structure. The internal surface area of the nose is almost 150 cm^2 excluding microvilli and about 96 m^2 if the microvilli are considered with a volume of about 15 ml [19, 20]. In Figure 9.2, the bony and cartilage structure of the nose along with its wall and valves is depicted.

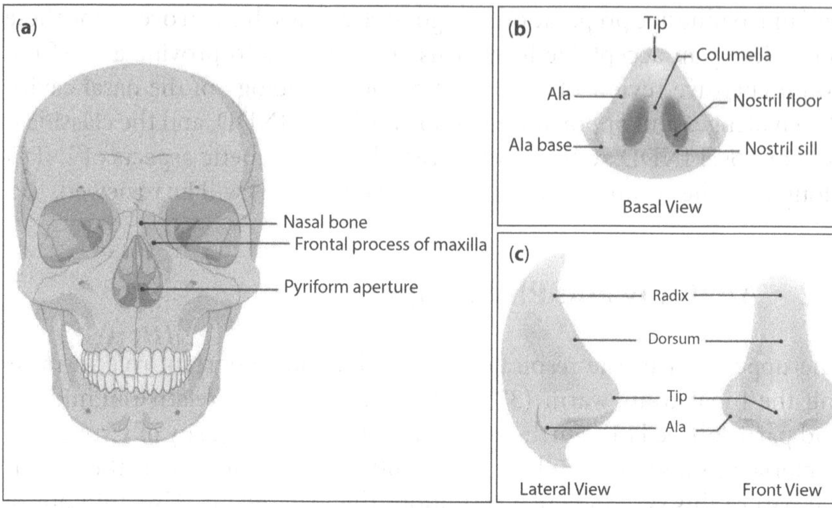

Figure 9.1 Anatomy of the nose; (a) The nose bony framework and pyriform aperture; (b) External anatomy of the nasal lower part from the basal view, and (c) Surface anatomy of the nose from the lateral and front view (Created with BioRender.com).

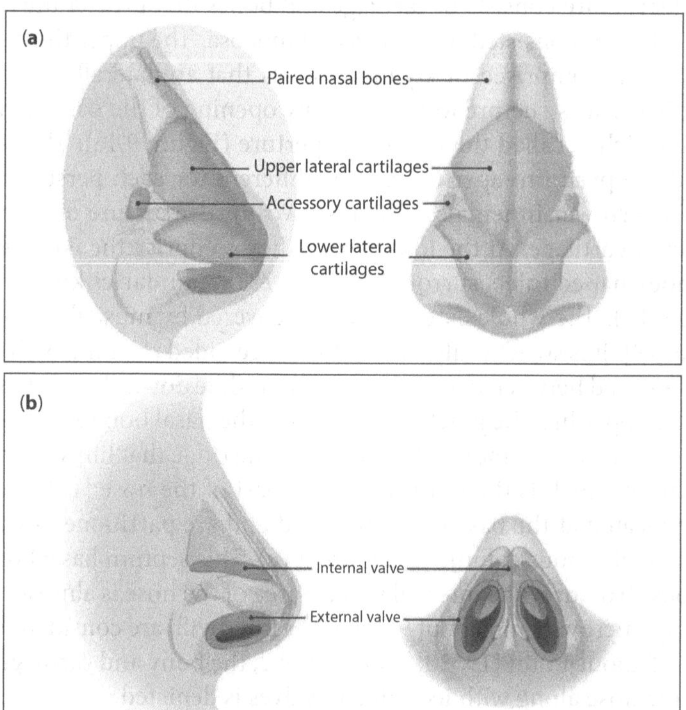

Figure 9.2 (a) Bony and cartilage structure of the nose; (b) Lateral view of the nasal wall and valves (Created with BioRender.com).

The nasal epithelium can be generally divided into two types [21]:

(i) Olfactory epithelium: it is of the pseudostratified columnar type and includes olfactory cells, supporting cells, serous glands, and mucous glands;
(ii) Non-olfactory epithelium: it is a highly vascularized tissue overlaid by a ciliated pseudostratified columnar epithelium.

There are various absorption mechanisms in the nose such as passive diffusion, carrier-mediated transport, and transcytosis. However, there are also hindering mechanisms that efflux the drug molecule out of the cells such as glycoprotein transporter [22, 23]. Low molecular weight lipophilic drugs can be rapidly absorbed by passive diffusion through the mucosa. In such cases, the bioavailability of the administered drug by the intranasal route is comparable to that in the intravenous route [22]. Because of the lower acidic pH of nasal mucus (5.5–6.5) and mild enzymatic functions compared to the GI tract, INDD provides high bioavailability for the administered drugs [24, 25].

Each person breathes about 12 to 24 times per minute and inhales about 10,000 liters of air during the day [26]. The inspired air enters the nostrils at the speed of 2 to 3 m/s. It turns at a 60° angle and becomes straight after passing the posterior vestibular segment. The speed of the inhaled air reaches its maximum (12–18 m/s) at the narrowest part of the nasal airway where the vestibule joins the nasal cavity which is called the intranasal valve (INV). The speed descends again to 2-3 m/s when reaching the turbinates [18]. Turbinates are special curved shelve-like structures, made of bones, vessels, and soft tissues located along the lateral walls of the nose, near the nasal septum. They provide appropriate levels of moisture and heat for the passing through air by their alternative bulge and shrinkage which controls the air passage [27]. In addition to the previously mentioned roles of the nose, it has other vital functions and in summary, we can categorize the physiological nasal assignments as taking part in warming, humidifying, and filtering the inhaled air, olfaction, trigeminal sensation, immunology, and airflow dynamics. All these functions have key roles in protecting the lower parts of the respiratory system [19, 26]. Each of these functions is briefly described below.

9.2.1 Olfaction

Human beings can sense more than 10,000 distinct odors and are capable of distinguishing about 5,000 of them. This susceptibility owes to the millions of neurons located in the olfactory epithelium and is implemented *via* odorant-binding proteins (OBPs). OBPs are a group of so-called lipocalin proteins. The lipocalin superfamily includes small secreted proteins with diverse structural and functional properties. These proteins bind to other small molecules, especially the hydrophobic ones, and ferry them to their corresponding receptors. OBPs solubilize hydrophobic pheromones and carry them from the environment to the olfactory receptors. This process leads to a nearly 10,000 times increase in the concentration of pheromone molecules in comparison with the external environment and causes the sensation of a diverse range of smells [26, 28–30].

9.2.2 Trigeminal Sensation

Numerous nasal trigeminal neurons are responsible for sensing pain, itching, burning, and irritation. These nerves have pivotal life-protecting roles and may cause some defensive reflexes such as sneezing, weeping, and nasal dripping after being stimulated [26].

9.2.3 Immunology

The immunological functions of the nose can be generally classified into two physical and chemical groups. The physical immunology is provided by the mucociliary clearance and the nasopharynx-associated lymphoid tissue. In a normal situation, almost 20-40 ml of nasal mucus is produced per day. Nasal mucus operates as a barrier that lines the mucosal epithelium of the internal nose and prevents the penetration of exogenous particles and pathogens into the epithelium. It acts as a defensive mechanism against allergens, irritants, or any other hazardous substances that may enter the nostrils [19, 31, 32]. The mucus is composed of 95% water and 5% glycoproteins and ions [33]. It is constantly produced and secreted which lubricates the airway and takes part in the exchange of nutrients and gases *via* hydrating the epithelia [34]. Mucus can trap bout 80% of large suspended particles in the inhaled air with an aerodynamic diameter of more than 12.5 μm and prevent them from reaching the pharynx. The proper function of cilia transport the mucus and its contents through the nose to the pharynx where they may be either expectorated or swallowed [35]. At the chemical level, there are numerous immunological substances in the nasal secretion including IgG, IgA, IgE, and IgM immunoglobulins, some enzymes like lysozymes and lactoferrin, lymphocytes, neutrophils, and also protective proteins such as complements [26]. The nose-associated lymphoid tissue (NALT) provides the possibility of mucosal vaccination for a variety of bacterial and viral infections *via* B and T lymphocytes [31]. Nasal vaccination is needle-free and provides both local and systemic immunization. Providing local immunization generates more efficient protection against infectious agents [36, 37].

9.2.4 Airflow Dynamics

The air flow and functionality of the upper respiratory system vary according to gender, race, age, and vestibule shape in different people [14, 15, 38]. The nasal cycle is considered the spontaneous alternative congestion and decongestion of each side of the nose. Almost 80% of healthy people experience the nasal cycle in 3-7 hour intervals. Although the mechanism of this phenomenon is not clear, it provides about 70% of respiratory resistance. Some non-physiological factors can influence the nasal cycle. For instance, doing exercises will decrease airway resistance while smoking and drinking alcohol can increase it [26].

9.3 Factors Affecting Intranasal Drug Delivery

Having a good and clear knowledge of the factors that influence INDD is essential for designing a proper formulation and also the device [39]. These factors can be related to the anatomical parameters, physiological conditions, environmental situations, physicochemical properties of the drug molecule,

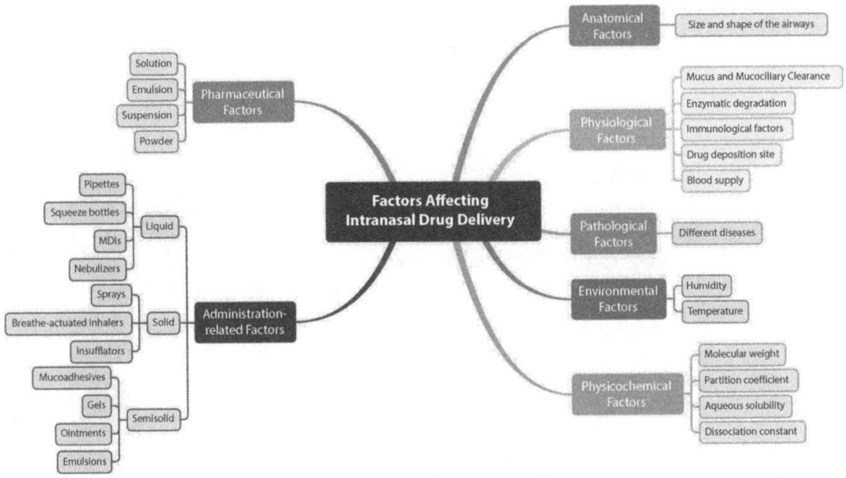

Figure 9.3 Factors affecting intranasal drug delivery.

the drug formulation, and the drug delivery device (Figure 9.3) [7, 39]. In the following, how these factors affect INDD is discussed.

9.3.1 Physiological Factors

The nasal passage length is approximately 12 cm from the entrance of the nostrils to the nasopharynx. There are superior, middle, and inferior turbinates in this pathway that make the main nasal passage with a length of about 6 cm. it is full of vasculatures and is covered by ciliated epithelium [39–41]. At the end of the nostrils where the ostium is located, the airway gets tighter which causes an increase in the air velocity. Afterward, in the preturbinate region, known as the atrium, the speed of the inspired air declines, its direction changes, and the air enters into the turbinates where the resistance gets more and the turbulence increases (Figure 9.4). According to the nasal anatomical properties, there are two main deposition parts in the nasal airway [41]:

(i) Next to the ostium;
(ii) In the frontal section of the middle turbinate.

Other factors influence the site of particle deposition such as the patient's age, the shape and size of the nasal cavity, and the person's body weight and height. The vestibule, atrium, and the origin of the turbinates have non-ciliated epithelium where the majority of the drug absorption takes place. Physiological factors are described below separately [40]:

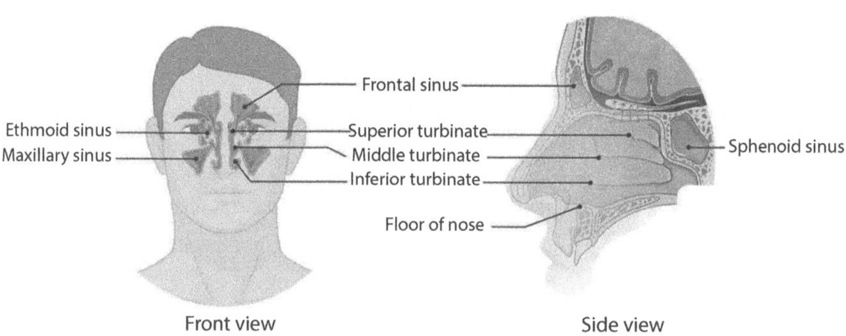

Figure 9.4 General anatomy of the internal nose from front and sided view (Created with BioRender.com).

9.3.1.1 *Mucus and Mucociliary Clearance (MCC)*

There are about 100 to 250 cilia on each ciliated cell with a dimension of almost 0.3 μm in width and 5 μm in length. These cilia have beating movements with a frequency of about 1000 times per minute [42]. However, this functionality varies in different parts of the nasal cavity [43]. The mucociliary clearance causes the drug deposition in a secondary location. This phenomenon leads to the non-metabolic clearance of the intra-nasally administered drugs since it usually transfers the deposited compounds to a second region within 30 minutes [44] which, in turn, influences the pharmacodynamics and pharmacokinetics of the drug and must be seriously noted [40].

9.3.1.2 *Enzymatic Cleavages*

Nasal mucosa enzymatic activity causes the metabolic clearance of the inhaled substances which can be considered the nasal first-pass effect. These enzymes are classified as shown in Table 9.1.

The noteworthy point is that the existence of the esterase enzymes helps the formulators design prodrugs for INDD purposes while avoiding the nasal first-pass effect. In this regard, the ester prodrug is easily absorbed through the mucosa and metabolized by the esterase enzymes. Afterward, the parent compound would be distributed in the systemic blood circulation. This strategy prevents the decrease in bioavailability [45].

9.3.1.3 *Immunological Parameters*

Nasal mucosa, also called nasopharynx-associated lymphoid tissue, plays an essential immunological role against xenobiotics and allergens [32] by providing a variety of immunological agents that are listed in Table 9.2 [32, 46, 47].

Table 9.1 Nasal mucosa enzymes.

Oxidative phase I enzymes	Conjugative phase II enzymes
Cytochrome P-450 Flavin-containing monooxygenase Aldehyde dehydrogenase Epoxide hydroxylase Carboxylesterase Carbonic anhydrase	Glucuronyl and sulfate transferases Glutathione transferase (GST)

Table 9.2 Nasal immunological agents.

Nasal immunological agents
IgA (subclasses of IgA1 and IgA2)
IgM
IgG
Memory B lymphocytes
Helper T-lymphocytes (TH1 and TH2)
Cytotoxic T lymphocytes (CTLs)
Plasma cells (PCs)
Dendritic cells (DCs)
Microfold (M) cells
Macrophages

9.3.1.4 Blood Supply

Having sufficient blood flow is necessary for proper drug absorption. The nasal blood flow is supplied by 6 arterial branches derived from external and internal carotid arteries. Intranasal drug absorption is facilitated owing to highly vascularized and ciliated epithelial tissue, rich in lymphatic network, and having porous endothelia which has made the main nasal passage a suitable candidate for drug delivery purposes [20, 40].

9.3.1.5 Drug Deposition Site

The key factor that determines the deposition area of the inhaled drug is the aerodynamic diameter of the particles. The main mechanism of deposition in the nose and the oropharynx for particles in the size range of 0.5–1 μm is inertial impaction. In this size range, normally local effects arise. Particles smaller than 0.5 μm are mostly exhaled, particles in the range of 1-5 μm reach the conducting airways where they assert their systemic clinical effects, and particles larger than 5 μm settle in the mouth and esophageal site [48].

9.3.2 Pathological Situations

Pathological status such as asthma, chronic sinusitis, infections, nasal polyposis, allergy, nasal septum deviation, congestion, rhinitis, Kartagener's syndrome, Sjogren's syndrome, and the common cold can change the mucus condition and the functionality of the ciliated cells which influences drug delivery efficiency and its bioavailability. If the pathological condition causes the reduction of MCC, it leads to an increase in contact time between the drug and the mucosa which eventuates in the increased drug absorption and *vice versa*. A runny nose dilutes the administered drug and reduces its concentration which declines the absorption extent. If the pathological situation changes the nasal pH, the ionization of the drug will change which impacts the net drug absorption as well [7].

9.3.3 Environmental Conditions

Temperature and humidity can also affect the INDD by influencing drug dissolution and absorption in the nasal cavity [49].

9.3.4 Physicochemical Factors

9.3.4.1 Molecular Weight

In INDD, there is an inverse correlation between the amount of drug absorption and its molecular weight [50]. Drugs with molecular weight (MW) < 300 Da are easily absorbed through membrane aqueous channels [22]. The amount of absorption in drugs with MW > 1000 Da is seriously reduced and usually, special strategies are needed to enhance their bioavailability [51].

9.3.4.2 Partition Coefficient and Lipophilicity

The nasal cell membrane has a lipophilic structure, hence, the rate of intranasal drug absorption increases as its lipophilicity increases. It has been proven that this increase follows a linear relationship [25, 52].

9.3.4.3 Aqueous Solubility

Each molecule should be first dissolved in nasal secretions to be absorbed afterward [53]. Normally the volume of administration in INDD is low in comparison to the oral route; so, it is necessary that the administered drug be sufficiently water-soluble and potent to hinder the need for a large quantity of administration. Otherwise, drug delivery will face significant challenges [54].

9.3.4.4 *Dissociation Constant*

The pH of the nasal mucosa lies in the range of 5.0-6.5. The greatest absorption extent arises for non-ionized molecules while for weak electrolytes it depends on their pKa and their ionization degree [55].

9.3.5 Pharmaceutical (Formulation-Related) Factors

Formulation type, concentration of the API(s), pH, osmolarity, density (especially for powders), viscosity, and volume of the administered formulation may influence the efficacy of INDDs [56]. The pH of the formulation should be within the 4.5-6.5 range to hinder any mucosal irritation. This pH range also prevents any bacterial growth in the dosage form and provides good permeation conditions [37].

There are four main basic formulations generally used for INDD purposes: solution, emulsion, suspension, and powder. The formulation type is a critical parameter in choosing the type of drug delivery device. The device should generate proper droplets or solid particles and its material must be compatible with the formulation contents [40]. An efficient intranasal therapy method requires an accurate and repeatable drug delivery pattern which depends on the proper design and function of the intended device. Inhaled particle deposition in the nasal airway occurs via five mechanisms:

(i) Inertial impaction
(ii) Gravitational sedimentation
(iii) Brownian diffusion
(iv) Electrostatic precipitation
(v) Interception

Inertial impaction, gravitational sedimentation, and Brownian diffusion are the most important mechanisms among which inertial impaction is considered the most significant one that arises due to a variation in the orientation of the inhaled air [40, 57]. The possibility of the inertial impaction can be described by the below equation:

$$\textit{The probability of impaction in a bent airway} = Ud^2 \sin \theta / R \qquad (9.1)$$

Where θ is the angle of the bend, U is the velocity of the air, d is the aerodynamic diameter of the particle, and R represents the radius of the airway.

As it is shown in Eq. 9.1, the inertial impaction depends on the shape of the airway, the size of the particles, and the inhaled air flow [40].

9.3.6 Administration-Related Factors

INDD could be aided by different devices for the delivery of liquids (including pipettes, squeeze bottles, meter-dose inhalers, and nebulizers) and solids (powder devices including sprays, breath-actuated inhalers, and insufflators) as well as bioadhesive formulations.

9.3.6.1 Pipettes

Pipettes are usually applied for intranasal delivery of drops. Drops are one of the oldest dosage forms for intranasal drug administration. This device is inexpensive and operates by a "blow-fill-seal" mechanism. Drops are more recommended for single-dose and over-the-counter (OTC) products like decongestants and saline. Their compliance, however, is limited due to the extreme bent position of the patient's head and neck which is required for gravity-dependent administration of the drug [58, 59].

9.3.6.2 Squeeze Bottles

This kind of device is generally used for delivering drugs for topical treatments such as decongestants. It consists of an air-filled plastic that delivers the formulation into the nasal cavity by getting squeezed. The volume of administration varies depending on the applied pressure and is not proper for precise drug delivery. This method is not that hygiene because nasal secretions and microorganisms can be sucked into the device after applying the working force [40, 60].

9.3.6.3 Metered-Dose Inhalers (MDIs)

Metered dose inhalers (MDIs) are one of the most applicable forms of devices intended for INDD which is generally used for the treatment of asthma and chronic obstructive pulmonary disease (COPD) [61]. They can pump 25-250 µl of dosage per puff into the nasal cavity and operate highly reproducible [40]. The great benefit of this device is that preservatives are not needed for drugs delivered by MDIs due to their special design and structure. The device consists of a vial (made from aluminum or plasticized glass), an actuator, a metering valve, and a dose counter in more recent generations. It is equipped with an ambulant piston and a collapsible bag that compensate for the volume of the emitted dose and give the device the advantage of being used upside down without the risk of getting polluted by sucking the nasal contents [62]. The size of emitted particles depends on the pump specifications, formulation properties, the orifice of the actuator, and, sometimes, the applied force [60]. The formulation is composed of API, propellant, and necessary excipients. The amount of emitted particles with an aerodynamic diameter of less than 5 μm is a key factor in determining the efficacy of INDD by MDIs [61].

9.3.6.4 Nebulizers/Atomizers

These devices convert solution, emulsion, or suspension formulations into small aerosol particles using compressed neutral gasses, and ultrasonic or mechanical power. Patients can inhale the particles *via* their mouth or nose. The low speed and fine size of the particles help their sedimentation in the upper and narrower parts of the nose in comparison with pMDIs [63]. A new generation of nebulizers has been marketed which produces aerosols by the vibration of a perforated membrane called "Mesh nebulizer" [64, 65].

9.3.6.5 Powder Devices

Powders are advantageous dosage forms because of their high stability and the potential to be formulated without necessarily using preservatives. The efficacy of powder INDDs depends on a variety of parameters including their hygroscopicity, water-solubility, particle size and shape, and velocity properties which will influence their deposition site and absorption [40]. Powder devices are generally placed in three groups:

- (i) Sprays, which release a plume of the drug after the applied pressure is released;
- (ii) Breath-actuated inhalers, the mechanism of which is patient-dependent and the user must inhale the drug into their nostril;
- (iii) Insufflators, which are comprised of a mouthpiece and a nosepiece connected; The mechanism of these devices is also patient-dependent. The user exhales into the

mouthpiece which pushes the drug into the nasal cavity through the nosepiece.

9.3.6.6 *Bioadhesive Formulations*

The bioadhesion is considered as the adhesion of a drug carrier to either an epithelial cell membrane or a mucus layer. In the latter case, it is called mucoadhesion. Mucoadhesive formulations are good candidates for both topical and systemic INDD purposes. They provide a longer retention time for the drug at the contact site which improves its bioavailability. Most of the time, the drug carrier is comprised of synthetic or natural polymers [66, 67]. A variety of theories are applied to describe the mechanism of mucoadhesion [66], including wetting theory [68], electrostatic theory [69], diffusion theory [70], adsorption theory [71], and fracture theory [72].

Various natural or synthetic polymers have mucoadhesive properties. Their common characteristic is that they all have plenty of hydrophilic groups in their chemical structures such as amide, sulfate, hydroxyl, and carboxyl groups. These groups interact with mucus by different mechanisms including hydrogen bonding or binding to specific mucosal receptors. An ideal mucoadhesive polymer should have the following characteristics [66]:

- ✓ Being safe and nonirritant
- ✓ Being site-specific
- ✓ Having the capability of quick adherence to mucus
- ✓ Having fast drug-releasing property
- ✓ Having a good shelf life
- ✓ Being cost-effective.

Chitosan is a popular cationic polymer that interacts with sialic acid groups of mucus. It absorbs high amounts of water and swells in its interaction site. This phenomenon opens the nasal epithelial tight junctions which leads to enhanced drug absorption and permeability [73]. Other examples of mucoadhesive polymers include carbopol, pluronic, polycarbophil, and cellulose derivatives [74]. Lectins and thiolated polymers operate site specifically. Lectins bind to carbohydrate groups of mucus and thiolated polymers form a strong covalent bond with cysteines of mucosal glycoproteins *via* free thiol groups in their chemical structure [75–77].

9.4 Challenges and Opportunities

As one of the oldest routes of drug administration, INDD has a variety of opportunities including easy accessibility, large mucosal surface area, the possibility to achieve rapid absorption, and improved bioavailability [11, 78]. The main challenges of INDD are related to the nose-to-brain and also systemic delivery purposes. The olfactory neuron cell bodies are located in the peripheral epithelium in the upper third part of the nasal cavity and influence drug delivery to the CNS [78]. Olfactory neurons are considered ideal candidates for brain drug delivery which bypasses BBB [79, 80]. Thus, the region where the drug is deposited in the nasal cavity strictly influences the target site where the drug will be delivered and determines the outcome of the treatment including topical, systemic, immunological, or CNS-related (Table 9.3) [37]. This limitation insists on the necessity of having a region-specific device or method that can precisely deliver the drug to the desired site and, indeed, is considered a critical drawback of the intranasal route of administration [81, 82]. The opportunities, challenges, and strategies to overcome the challenges of INDD are summarized in Figure 9.5.

One of the most promising strategies that can properly provide INDD purposes is using nanotechnology. The drug amount that may ensure attaining the therapeutic level, however, relies on the ability of the nanocarrier to conquer the biological barriers [79, 80, 83]. Nanotechnology can optimize INDD by providing [83, 84]:

1) Increased drug solubility
2) Enhanced drug absorption

Table 9.3 Different target sites and purposes pursued by intranasal drug delivery.

Target site	Purpose	Example of disease
Mucosa	Local Drug Delivery	Rhinitis, Decongestion, Polyposis
Mucosa	Systemic	Pain, Diabetes, Enuresis
Mucosa/NALT	Immunotherapy	Herpes, Influenza. Tetanus
Olfactory mucosa	CNS	Alzheimer's disease, Brain Tumors, Depression

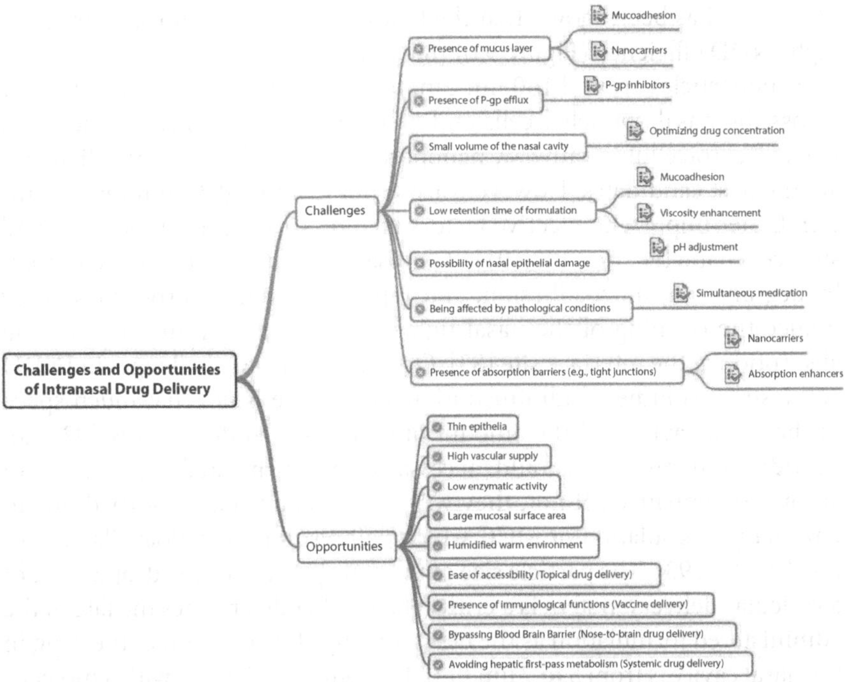

Figure 9.5 Opportunities, challenges and their corresponding solutions for intranasal drug delivery strategies.

3) Higher bioavailability
4) Enzymatic cleavage inhibition
5) Mucoadhesion
6) Prolonged intranasal retention time
7) Optimized drug release kinetics
8) Less adverse effects
9) Better nose-to-brain delivery
10) More efficient systemic delivery
11) More efficacious mucosal vaccination.

Among all kinds of nanoparticles, nanocapsules with an oily reservoir core encapsulated in a polymeric membrane are the most promising type especially ideal for the delivery of hydrophobic molecules. These nanoparticles enhance drug penetration to the brain by promoting encapsulated drug transport through neuroepithelium *via* transcellular or paracellular pathways. Nanocapsules also protect the drug molecule from enzymatic degradation and increase their nasal retention time by lowering MCC

[83, 85]. It has been shown that the following types of nanocarriers assert high INDD efficiency (Figure 9.6) [86–89].

Nanoparticles around 100 nm can, to a high extent, pass intracellularly across the nasal epithelial cells and olfactory neurons to reach the CNS. For the extracellular pathway, nanoparticles with about 20 nm diameter are the best candidates. However, size is not the only determining parameter. Other important effective factors include structural composition and surface characteristics [90, 91]. Chitosan-coated nanoparticles present biodegradation, mucoadhesion, and rapid drug-release properties. They induce the opening of the nasal tight junctions and prompt faster drug absorption at the interface site [83]. One of the serious challenges in INDD is the small volume of administration since there is not that much space in the nasal cavity and the amount of administered dose is low [92, 93]. Peptides and proteins should necessarily be formulated along with an absorption enhancer. If not, they will not be sufficiently absorbed due to enzymatic degradation or MCC which will result in low bioavailability of the drug [37, 93]. The particle size of the drug powder or the droplet size of a vesicular delivery system is a critical factor that determines the fate of the administered formulation as the best size range for retention of the drug in the nasal cavity is from 5 to 7 μm [92]. Lipophilic molecules with a molecular weight of less than 1 KDa can be freely absorbed from nasal epithelium, but polar molecules do not easily pass the epithelial cell membrane [94].

Till now, very few adjuvants have been introduced for being used in nasal vaccination mainly because the clinical trial of this strategy is complicated and takes a long time to be completed. Furthermore, a good and potent adjuvant must be able to trigger a proper immune response regarding the aim of vaccination [95]. Due to MCC, liquid dosage forms do not stay in the nasal cavity for a long time. Therefore, repeating the administration

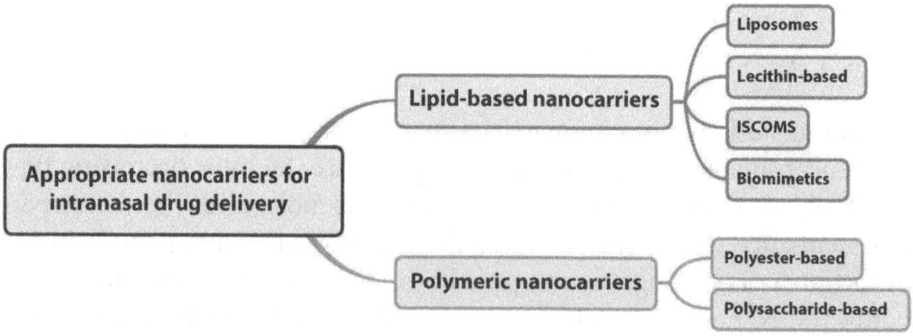

Figure 9.6 Appropriate nanocarriers for intranasal drug delivery.

would be required which may lead to epithelial cell damage or loss and also decrease the patient's compliance [96]. To augment the retention time and reduce drug clearance, gelling agents can be used in the formulation of the drug to increase its viscosity [97]. For enhancing nasal drug permeation, films and some nanocarriers such as nanoemulsions, microspheres, liposomes, and niosomes have been suggested as good candidates [37]. Nasal drug absorption can be strictly affected by the efflux mechanism of P-glycoproteins (P-gps) [98]. A way to prevail against this phenomenon is the simultaneous administration of a P-gp inhibitor but, it must be noted that P-gps play a key role in preventing toxic agents to pass the epithelium. So, further considerations should be made before incorporating a P-gp inhibitor in the formulation [54].

9.5 Classification and Formulation Considerations

Pharmaceutical dosage forms used for INDD can be categorized into three main groups solids, semisolids, and liquids. Each group is described in the following.

9.5.1 Solids

Solid preparations have the best physicochemical and microbial stability status among all formulation types [99]. Powders are solid dosage forms widely used for INDD purposes. They are especially suitable for formulating poorly water-soluble APIs. They do not usually need a complicated composition of excipients and are generally simply prepared. A high dose of API can be delivered by powders. They provide better drug diffusion, absorption, nasal retention time, and bioavailability in comparison to liquid forms. Powders are, however, less common in the market than liquids [100]. The only commercialized nasal powder that is used for systemic treatment is Onzetra Xsail® (Avanir Pharmaceuticals Inc., Aliso Viejo, CA, USA), which contains sumatriptan and is administered for palliating migraine symptoms. In January 2016, this product was approved by the US Food and Drug Administration (FDA) [101]. Powders have the potential to be comprised of just the API but, most of the time it is not practical because the physicochemical properties of the raw material are not always appropriate to be used alone. For instance, the delivery of pure drug particles does not usually occur in a repeatable manner due to the effect of particle shape and size on its flow properties. In addition, pure API does not dissolve in the mucosa to an acceptable extent so it cannot efficiently emerge its pharmacological effect. Making prodrugs is a suitable manner to dominate poor solubility of solid particles in nasal mucosa. The important issue to be noted is that the proposed pharmacological effect and the safety of the drug have to be strictly monitored after chemical modifications [100, 102]. There are also two main techniques to overcome these obstacles including lyophilization and spray drying. Lyophilization generates porous and fast-dissolving powders, but the high expense and the need for some specific conditions have limited its application. Today, the spray drying technique is used more prevalently [103]. If the described methods to defeat the API imperfections could not help, normally utilization of excipients including fillers, absorption enhancers, and mucoadhesive agents (if necessary) would be inevitable [100, 104]. In addition to lyophilization and spray drying, supercritical-fluid-assisted spray drying, spray freeze drying and agglomeration of micronized powders are other common techniques to produce nasal powders [100].

9.5.2 Liquids

As the most common types of INDD formulations, liquids have some critical drawbacks such as the need for a large volume of administration, high MCC, and hence low nasal retention time. Using excipients such as preservatives and other stabilizing agents is often indispensable. Proteins and peptides are difficult to formulate as liquids due to the need for special additives such as absorption enhancers. A majority of drugs in solution form cannot easily pass through epithelial cell membranes which leads to decreased bioavailability [105]. Liquids are formulated as solutions or suspensions for INDD purposes. Liquid forms induce much less mucosal irritation than solid forms, but it must be noted that liquids are isotonic, otherwise, serious irritations may arise [54, 60].

9.5.3 Semisolids

Gels are the most common type of semisolids used in INDD. They are more viscous than liquids providing more retention time and less MCC. Gels reduce the post-nasal diffusion of the drug and also decline its frontal escape and cause an increase in bioavailability [37]. Mucoadhesive polymers such as sodium carboxymethylcellulose (NaCMC) are mainly used as a gelling agent. In this situation, the bioavailability reaches much higher levels. Viscous emulsions are also placed in the semisolid category. Creams as the main type of topical emulsions as well as ointments are other semi-solid dosage forms attempted for topical intranasal treatment purposes [106].

9.6 Pharmacokinetics

Lipophilic drugs are usually absorbed rapidly when administered intranasally with a pharmacokinetic profile similar to that of intravenous (IV) injection with a bioavailability of nearly 100%. The absorption of hydrophilic polar drugs is however very low, which causes their bioavailability to be about 10% upon intranasal administration [107]. A critical factor causing low absorption of nasally administered drugs, whether they are liquids or solids, is their short residence time in the nasal cavity (about 15–20 minutes) which occurs due to rapid nasal MCC. Another key factor that influences intranasal absorption extent is the volume of the administration. The optimum range is 25–200 μL (ideally 100 μL) per nostril [52]. Nasal enzymatic metabolism is the other factor that decreases the absorption extent. Enzymatic activities of CYP450 (oxidative phase 1 enzymes), conjugative phase 2 enzymes, lactate dehydrogenase, oxidoreductases, hydrolases, acid phosphatase, and esterases are the main source of this phenomenon [6, 108, 109].

It has been shown that for a large number of drugs, the plasma pharmacokinetic profile obtained after intranasal drug administration is very similar to that of intravenous (IV) bolus injection [110]. INDD provides the possibility of better controlling the drug plasma concentration levels and decreases the probability of many adverse effects incidence [110, 111]. The absorption of peptides and proteins is, however, challenging through intranasal administration (IN) due to a variety of reasons including MCC, enzymatic metabolisms, and low epithelial permeation. This phenomenon causes very low blood concentration levels of proteins and peptides. Accordingly using absorption enhancers in their formulations is inevitable. For this reason, only a few peptides or protein-based dosage forms are available on the market as INDD systems [6, 110, 112].

Nose-to-brain drug delivery is very attractive due to its prominence in bypassing BBB. A majority of drug molecules can be directly delivered to the brain via olfactory and trigeminal pathways. So, INDD is considered a promising route for the treatment of plenty of neurodegenerative diseases and psychiatric disorders. Many lipophilic molecules can cross BBB by a passive diffusion mechanism. The more a molecule is lipophilic the easier it bypasses the barrier. In comparison to other parenteral routes, intranasal administration provides faster CNS drug delivery and higher CNS-availability. Nevertheless, the extent of drug absorption through nasal mucosa is negligible in many cases, and also the volume of administration is a serious limiting factor. Therefore, nose-to-brain drug delivery is problematic, although it is still attractive for the pharma industry. A good strategy to overcome INDD challenges is using nanosystems including liposomes, nanoemulsions, polymeric nanocarriers, inorganic nanocarriers, and lipid-based nanocarriers [113, 114]. Solid lipid nanoparticles (SLNs) and nanostructured lipid carriers (NLCs) are two main instances of such promising nanosystems. They increase drug bioavailability by enhancing drug solubility and absorption and diminishing enzymatic degradation [114].

To evaluate and also compare the PK profile of various nose-to-brain administered formulations, different PK parameters can be studied such as AUC_{brain}, $C_{max(brain)}$, $T_{max(brain)}$, drug targeting efficiency (DTE), drug transport percentage (DTP), the ratio of measured drug concentration in the brain to that in blood at each time point, and the brain drug concentration in terminal time point [115–117]. DTE% defines the amount of accumulated drug in the brain after intranasal administration relative to that after IV administration and is calculated by Eq. 9.2 as follows:

$$\boldsymbol{DTE\%} = \frac{(\boldsymbol{AUC(Brain, IN)/AUC(Blood, IN)})}{(\boldsymbol{AUC(Brain, IV)/AUC(Blood, IV)})} \times \mathbf{100} \qquad (9.2)$$

DTE% values can vary between 0 and ∞ which determines the efficiency of drug delivery to the brain [118). In Table 9.4, the relation between DTE% and brain targeting efficiency is demonstrated.

DTP% specifies the amount of drug that enters the brain just *via* direct pathways (not from blood to the brain through crossing BBB which is considered an indirect pathway) including olfactory and trigeminal routes. DTP% can be calculated using the Eq. 9.3 as follows:

$$DTP = 1 - \left(\frac{100}{DTE\%}\right) \times 100 \qquad (9.3)$$

DTP% value lies in the range of -∞ to 100 [118]. Each value range can be described as shown in Table 9.4.

Table 9.4 Description of different pharmacokinetic parameters for nose-to-brain drug delivery.

Parameter	**Range**	**Value**	**Description**
DTE%	0 to ∞	< 100	Non-efficiency
		> 100	Efficiency
DTP%	-∞ to 100	< 0	DTE% <100
		0	No direct drug absorption
		> 0	Direct drug absorption
		100	The inability of drugs to cross BBB
B% $_{\text{(IN/IV)}}$	NA*	< 100	Brain drug accumulation after IV administration is greater than that after IN administration
		>100	Brain drug accumulation after IN administration is greater than that after IV administration

* Not applicable

Although DTE% and DTP% are valuable parameters that determine the amount of drug delivered to the brain after IN administration, they are not precise enough because in the case of very low amounts of $AUC_{(Blood,\ IN)}$, DTE% and DTP% present high values. Hence, another parameter, so-called $B_{IN/IV}$ is used which shows the ratio of the accumulated drug in the brain after INDD to that after IV administration and is calculated using Eq. 9.4 as below:

$$B\%_{\ (IN/IV)} = \frac{AUC(Brain, IN)}{AUC(Brain, IV)} \times 100 \tag{9.4}$$

Different value ranges of $B\%_{\ (IN/IV)}$ demonstrate the amount of accumulated drug in the brain as shown in Table 9.4.

Mucoadhesive polymers can be used to coat nanoparticles. In the case of using chitosan, as described previously, the surface charge of nanoparticles will get positive which promotes electrostatic interactions between nanoparticles and the mucosal membrane. Furthermore, the drug penetration will be increased due to the opening of nasal tight junctions so, the drug bioavailability will get higher [119, 120]. Another strategy to increase the efficiency of nanocarriers is to formulate them as gelling systems which can increase their residence in the nasal cavity and thus increase brain bioavailability [114].

Different models can be applied to carry out intranasal pharmacokinetic studies. Normally, *in vitro* models describe the diffusion and permeation of drugs, *in vivo* models allow absorption studies in a real pharmacokinetic context, and *ex vivo* models include nasal perfusion experiments [121].

9.6.1 *In Vivo* Models

To date, a variety of animal species have been used to perform *in vivo* experiments for nasal delivery purposes, including rats, mice, rabbits, dogs, guinea pigs, monkeys, and sheep. It is important to investigate the anatomical properties of the aimed animal before doing the study [122]. In general, rat and mouse models are used for studying drug absorption *via* nose-to-brain pathway (CNS delivery) while sheep, dog, and rabbit models are suitable for systemic PK studies. However, it must be noted that although animal models are helpful, they are not representative of the drug delivery situation in human beings due to the great anatomical and physiological differences between these species [123]. The volume of administration is about 10 μl in mice and 45 μl in rats. Greater volumes can be applied to bigger animals [122].

Direct absorption of the administered dose from nose to brain bypasses BBB and inhibits drug dilution due to protein binding. This phenomenon helps prepare the nasal formulations at very lower concentrations than oral dosage forms (0.01–1% of the oral concentration).

Drug solubility and potency are two serious limiting factors. The drug must become rapidly dissolved in a few microliters of nasal liquid before being eliminated by nasal clearance mechanisms. Furthermore, the drug should be freely absorbed and must be potent enough to exert its pharmacological effects [122].

9.6.2 *In Vitro* Models

Although *in vivo* models are fundamental for investigating drug absorption and permeation through the nasal cavity, from the mechanistic point of view, *in vitro* models describe the way molecules absorb or permeate. RPMI 2650 and CaCo-2 are the two main cell lines used for this purpose [122].

9.6.3 *Ex Vivo* Models

To evaluate the toxicity of the excipients and also the drugs' transmucosal transport, *ex vivo* models are applicable using excised animal tissues. For INDD studies, nasal mucosa is obtained from laboratory animals and even sometimes from humans if necessary. *Ex vivo* studies permit gathering information about metabolism and efflux phenomena as well. The major drawback of *ex vivo* experiments is that their outcomes cannot be precisely extrapolated to *in vivo* conditions [122, 123]. The most common *ex vivo* model for studying nasal permeability is the Ussing chamber. Working with this model is very easy and provides the possibility of investigating the quantity of passive diffusion, active transport, efflux, and carrier-mediated routes of transport [124].

9.7 Products in the Market

Despite the numerous advantages of intranasal drug administration, limited numbers of INDD systems have been commercialized due to their challenges. A summary of the marketed formulations is demonstrated in Table 9.5 [100, 125–127]. In addition to the listed products, numerous preparations containing sodium chloride are present in the market used as lubricants.

Table 9.5 A summarized list of commercialized products for intranasal administration.

Product brand name	API	Device/ dosage form	Main indication	Manufacturer
Rhinocort®	Budesonide (100 µg)	Breath-actuated inhaler (Multi-dose)	Seasonal and allergic rhinitis management	AstraZeneca
Teijin Rhinocort®	Beclomethasone dipropionate (50 µg)	Spray, powder (Capsule-based, single dose)	Asthma maintenance treatment Seasonal and allergic rhinitis management	Teijin Pharma

(*Continued*)

Table 9.5 A summarized list of commercialized products for intranasal administration. (*Continued*)

Product brand name	API	Device/ dosage form	Main indication	Manufacturer
Erizas®	Dexamethasone cipecilate	400 µg: capsule-based breath-actuated 200 µg: multi-dose nasal spray	Allergic rhinitis	Nippon Shinyaku Co., Ltd.
Onzetra Xsail®	Sumatriptan succinate (11 mg)	Breath-actuated inhaler	Treatment of the migraine headache symptoms	Avanir Pharmaceuticals
Allergo-Comod	Cromolyn sodium	Spray, solution	Allergic rhinitis management	Ursapharm
Atronase	Ipratropium bromide	Spray, solution	Bronchospasm treatment	Boehringer Ingelheim, Belg
Imigran®	Sumatriptan	Spray, solution	Treatment of the migraine headache symptoms	GlaxoSmithKline
Miacalcic®	Calcitonin	Spray, solution	Post-menopausal osteoporosis	Novartis Pharmaceutical Ltd
Miacalcin®	Salmon calcitonin	Spray, solution	Osteoporosis	Novartis Pharma
Migranal®	Dihydroergotamine mesylate	Spray, solution	Treatment of the migraine headache symptoms	Novartis Pharma
Desmospray	Desmopressin acetate	Spray, solution	Nocturnal Enuresis	Ferring Pharmaceuticals, United Kingdom

(*Continued*)

Table 9.5 A summarized list of commercialized products for intranasal administration. (*Continued*)

Product brand name	API	Device/ dosage form	Main indication	Manufacturer
Minrin®	Desmopressin acetate	Spray, solution	Nocturnal enuresis, Diabetes insipidus management, Hemophilia A, Von Willebrand's disease (type 1)	Ferring, Mexico
Octostim®	Desmopressin acetate	Spray, solution	Nocturnal enuresis, Diabetes insipidus management, Hemophilia A, Von Willebrand's disease (type 1)	Ferring, Mexico
Nascobal®	Cyanocobalamine	Spray, solution	Vit-B12 deficiency	Endo Pharmaceuticals
Suprecur®	Buserelin (acetate)	Spray, solution	Prostate carcinoma, endometriosis	Sanofi-Aventis
Profact®	Buserelin (acetate)	Spray, solution	Prostate carcinoma, endometriosis	Sanofi-Aventis
Suprefact®	Buserelin (acetate)	Spray, solution	Prostate carcinoma, endometriosis	Sanofi-Aventis
Onzetra®	Sumatriptan	Nasal powder (Breath-powered)	Treatment of the migraine headache symptoms	Avanir Pharmaceuticals, Inc.

(*Continued*)

Table 9.5 A summarized list of commercialized products for intranasal administration. (*Continued*)

Product brand name	API	Device/ dosage form	Main indication	Manufacturer
Xsail®	Sumatriptan	Nasal powder (Breath-powered)	Treatment of the migraine headache symptoms	Avanir Pharmaceuticals, Inc.
Synarel®	Nafarelin acetate	Spray, solution	Endometriosis symptoms	Pfizer Inc.
Syntocinon®	Oxytocin	Spray, solution	Lactation induction	Novartis Pharma
Zomig®	Zolmitriptan	Spray, solution	Treatment of the migraine headache symptoms	Cipla
Dymista	Azelastine and fluticasone	Spray, solution	Controlling seasonal allergy symptoms	Meda Pharmaceuticals Inc
Afrin	Oxymetazoline	Spray, solution	Temporary relief of nasal congestion	Bayer
Vicks Sinex®	Phenylephrine	Spray, solution	Temporary relief of nasal congestion	Procter & Gamble
Astelin®	Azelastine	Spray, solution	Treat nasal symptoms of allergies	Meda AB
Astepro®	Azelastine	Spray, solution	Treatment of nasal allergy symptoms	Meda Pharmaceuticals Inc
Nostrilla	Oxymetazoline	Spray, solution	Temporary relief of nasal congestion	Insight Pharmaceuticals LLC

(*Continued*)

Table 9.5 A summarized list of commercialized products for intranasal administration. (*Continued*)

Product brand name	API	Device/ dosage form	Main indication	Manufacturer
Privine	Naphazoline	Spray, solution	Temporary relief of nasal congestion	Insight Pharmaceuticals Llc
Patanase	Olopatadine	Spray, solution	Treat seasonal allergy symptoms	Novartis Pharmaceuticals
Sinex Long-Acting	Oxymetazoline	Spray, solution	Temporary relief of nasal congestion	Procter & Gamble
Zicam	Oxymetazoline	Spray, solution	Temporary relief of nasal congestion	Mattrix Initiatives
Nasalcrom	Cromolyn sodium	Spray, solution	Treatment of asthma, sinus infections, or the common cold symptoms	Prestige Consumer Healthcare, Inc.
Atrovent	Ipratropium	Spray, solution	Treatment of runny nose	Boehringer Ingelheim
Neo-Synephrine	Phenylephrine	Spray, solution	Temporary relief of nasal congestion	B.F. Ascher & Company, Inc.
Tyzine	Tetrahydrozoline	Drop, solution	Treatment of nose stuffiness	YOSHINDO
Sudafed	Oxymetazoline	Spray, solution	Temporary relief of nasal congestion	McNeil Consumer Healthcare
Sinarest	Oxymetazoline	Spray, solution	Temporary relief of nasal congestion	Centaur Pharmaceuticals Pvt. Ltd

(*Continued*)

Table 9.5 A summarized list of commercialized products for intranasal administration. (*Continued*)

Product brand name	API	Device/ dosage form	Main indication	Manufacturer
Rhinall	Phenylephrine	Spray, solution	Temporary relief of nasal congestion	Product Quest Manufacturing LLC
Benzedrex	Propylhexedrine	Nasal inhaler	Temporary relief of nasal congestion	B.F. Ascher & Co.
Flonase	Fluticasone	Spray, solution	Treatment of nasal congestion, sneezing, runny nose	GlaxoSmithKline
Nasonex	Mometasone	Spray, solution	Anti-Inflammatory Treatment and prevention of allergic rhinitis symptoms	Merck
Nasacort	Triamcinolone	Spray, solution	Treatment of nasal congestion, sneezing, runny nose	Chattem, Inc.
QNASL	Beclomethasone	Spray, solution	Treatment of nasal congestion, sneezing, runny nose	Teva Pharmaceutical Industries Ltd.

(*Continued*)

Table 9.5 A summarized list of commercialized products for intranasal administration. (*Continued*)

Product brand name	API	Device/ dosage form	Main indication	Manufacturer
Veramyst	Fluticasone	Spray, solution	Treatment of nasal congestion, sneezing, runny nose, and allergic itchy or watery eyes	GlaxoSmithKline
Omnaris	Ciclesonide	Spray, solution	Treatment of nasal congestion, sneezing, runny nose	Covis Pharma.
Xhance	Fluticasone	Spray, solution	Treatment of nasal congestion, sneezing, runny nose	Optinose
Nasacort	Triamcinolone	Spray, solution	Treatment of nasal congestion, sneezing, runny nose	Chattem, Inc.
Flonase	Fluticasone	Spray, solution	Ease allergy signs	GlaxoSmithKline
Rhinocort	Budesonide	Spray, solution	Treatment and prevention of nasal congestion, sneezing, runny nose, and inflammation	AstraZeneca

(*Continued*)

Table 9.5 A summarized list of commercialized products for intranasal administration. (*Continued*)

Product brand name	API	Device/ dosage form	Main indication	Manufacturer
Nasacort	Triamcinolone	Spray, solution	Treatment and prevention of nasal congestion, sneezing, runny nose, and inflammation	Chattem, Inc.
Beconase	Beclomethasone	Spray, solution	Treatment and prevention of nasal congestion, sneezing, runny nose, and inflammation	Valois S.A. Le Prieure BPG,
Sinuva	Mometasone	Spray, solution	Treatment and prevention of nasal congestion, sneezing, runny nose, and inflammation	Intersect ENT, Inc.
Ticanase	Fluticasone propionate	Spray, solution	Management of the nasal symptoms of perennial nonallergic rhinitis	Apotex Inc.

(*Continued*)

Table 9.5 A summarized list of commercialized products for intranasal administration. (*Continued*)

Product brand name	API	Device/ dosage form	Main indication	Manufacturer
Nasarel	Flunisolide	Spray, solution	Control of allergic rhinitis (hay fever) symptoms	Teva branded pharm.
Zetonna	Ciclesonide	Spray, solution	Treatment of nasal congestion, sneezing, runny nose	Covis Pharma
Vancenase	Beclomethasone	Spray, solution	Treatment and prevention of nasal congestion, sneezing, runny nose, and inflammation	SCHERING
Propel	Mometasone	Spray, solution	Treatment of nasal congestion, sneezing, runny nose	Intersect ENT
AllerNaze	Triamcinolone	Spray, solution	Treatment of nasal symptoms of both seasonal and perennial allergic rhinitis	Collegium Pharmaceutical, Inc

References

1. Zhang, Y., Chan, H.F., Leong, K.W., Advanced materials and processing for drug delivery: the past and the future. *Adv. Drug Deliv. Rev.*, 65, 1, 104–20, 2013.
2. Singh, A.K., Singh, A., Madhv, N.S., Nasal cavity, a promising transmucosal platform for drug delivery and research approaches from nasal to brain targetting. *J. Drug Deliv. Ther.*, 2, 3, 22–33, 2012.
3. Gupta, S. and Kumar, S., An Overview on Intranasal Drug Delivery System: Recent Technique and Its Contribution in Therapeutic Management. *Curr. Res. Pharm. Sci.*, 09, 02, 17–23, 2019.
4. Wiedmann, T.S., *Pharmaceutical Inhalation Aerosol Technology*, A.J. Hickey (Ed.), p. 603, Marcel Dekker, New York, Elsevier, 2004, 2005.
5. Kumar, S., Dang, S., Nigam, K., Ali, J., Baboota, S., Selegiline nanoformulation in attenuation of oxidative stress and upregulation of dopamine in the brain for the treatment of Parkinson's disease. *Rejuvenation Res.*, 21, 5, 464–76, 2018.
6. Illum, L., Nasal drug delivery—possibilities, problems and solutions. *J. Controlled Release*, 87, 1-3, 187–98, 2003.
7. Ugwoke, M.I., Verbeke, N., Kinget, R., The biopharmaceutical aspects of nasal mucoadhesive drug delivery. *J. Pharm. Pharmacol.*, 53, 1, 3–22, 2001.
8. Appasaheb, P.S., Manohar, S.D., Bhanudas, S.R., Anjaneri, N., A review on intranasal drug delivery system. *J. Adv. Pharm. Educ. Res.*, 3, 4, 333–346, Oct-Dec. 2013.
9. Keeler, J.A., Patki, A., Woodard, C.R., Frank-Ito, D.O., A computational study of nasal spray deposition pattern in four ethnic groups. *J. Aerosol Med. Pulm. Drug Deliv.*, 29, 2, 153–66, 2016.
10. Kalish, L., Snidvongs, K., Sivasubramaniam, R., Cope, D., Harvey, R.J., Topical steroids for nasal polyps. *Cochrane Database Syst. Rev.*, 12, 1–96, 2012.
11. Fortuna, A., Schindowski, K., Sonvico, F., Intranasal Drug Delivery: Challenges and Opportunities. *Front. Pharmacol.*, 13, 1–3, 2022.
12. Keller, L.-A., Merkel, O., Popp, A., Intranasal drug delivery: Opportunities and toxicologic challenges during drug development. *Drug Deliv. Transl. Res.*, 12, 4, 735–57, 2022.
13. Sahin-Yilmaz, A. and Naclerio, R.M., Anatomy and physiology of the upper airway. *Proc. Am. Thorac. Soc.*, 8, 1, 31–9, 2011.
14. Zhu, J.H., Lee, H.P., Lim, K.M., Lee, S.J., Evaluation and comparison of nasal airway flow patterns among three subjects from Caucasian, Chinese and Indian ethnic groups using computational fluid dynamics simulation. *Respir. Physiol. Neurobiol.*, 175, 1, 62–9, 2011.
15. Hazeri, M., Farshidfar, Z., Faramarzi, M., Sadrizadeh, S., Abouali, O., Details of the physiology of the aerodynamic and heat and moisture transfer in the normal nasal cavity. *Respir. Physiol. Neurobiol.*, 280, 103480, 2020.

16. Suh, M.K., Surgical anatomy and physiology of the nose, in: *Atlas of Asian Rhinoplasty*, pp. 1–65, Springer, The PLUS Plastic Surgery Clinic, Seoul, Korea, 2018.
17. Sesenna, E., Leporati, M., Brevi, B., Oretti, G., Ferri, A., Congenital nasal pyriform aperture stenosis: diagnosis and management. *Ital. J. Pediatr.*, 38, 1, 1–5, 2012.
18. Watelet, J.-B. and Cauwenberge, P.V., Applied anatomy and physiology of the nose and paranasal sinuses. *Allergy*, 54, 14–25, 1999.
19. Van Cauwenberge, P., Sys, L., De Belder, T., Watelet, J.-B., Anatomy and physiology of the nose and the paranasal sinuses. *Immunol. Allergy Clin.*, 24, 1, 1–17, 2004.
20. Gizurarson, S., Anatomical and histological factors affecting intranasal drug and vaccine delivery. *Curr. Drug Deliv.*, 9, 6, 566–82, 2012.
21. Uchenna Agu, R., Jorissen, M., Willems, T., Van den Mooter, G., Kinget, R., Augustijns, P., Effects of pharmaceutical compounds on ciliary beating in human nasal epithelial cells: a comparative study of cell culture models. *Pharm. Res.*, 16, 9, 1380–5, 1999.
22. Hussain, A.A., Dakkuri, A., Itoh, S., Nasal absorption of ondansetron in rats: an alternative route of drug delivery. *Cancer Chemother. Pharmacol.*, 45, 5, 432–4, 2000.
23. Crowe, T.P., Greenlee, M.H.W., Kanthasamy, A.G., Hsu, W.H., Mechanism of intranasal drug delivery directly to the brain. *Life Sci.*, 195, 44–52, 2018.
24. England, R., Homer, J., Knight, L., Ell, S., Nasal pH measurement: a reliable and repeatable parameter. *Clin. Otolaryngol. Allied Sci.*, 24, 1, 67–8, 1999.
25. Costantino, H.R., Illum, L., Brandt, G., Johnson, P.H., Quay, S.C., Intranasal delivery: physicochemical and therapeutic aspects. *Int. J. Pharm.*, 337, 1-2, 1–24, 2007.
26. Jones, N., The nose and paranasal sinuses physiology and anatomy. *Adv. Drug Deliv. Rev.*, 51, 1-3, 5–19, 2001.
27. Georgakopoulos, B. and Le, P.H., *Anatomy, head and neck, nasal concha*, StatPearls Publishing, Treasure Island (FL), Florida, The United States of America, 2019.
28. Pelosi, P. and Knoll, W., Odorant-binding proteins of mammals. *Biol. Rev.*, 97, 1, 20–44, 2022.
29. Jones, N. and Rog, D., Olfaction: a review. *J. Laryngol. Oto.*, 112, 1, 11–24, 1998.
30. Flower, D.R., The lipocalin protein family: structure and function. *Biochem. J.*, 318, 1, 1–14, 1996.
31. Pabst, R., Mucosal vaccination by the intranasal route. Nose-associated lymphoid tissue (NALT)—Structure, function and species differences. *Vaccine*, 33, 36, 4406–13, 2015.
32. Gallo, O., Locatello, L.G., Mazzoni, A., Novelli, L., Annunziato, F., The central role of the nasal microenvironment in the transmission, modulation,

and clinical progression of SARS-CoV-2 infection. *Mucosal Immunol.*, 14, 2, 305–16, 2021.
33. Quraishi, M., Jones, N., Mason, J., The rheology of nasal mucus: a review. *Clin. Otolaryngol. Allied Sci.*, 23, 5, 403–13, 1998.
34. Leal, J., Smyth, H.D., Ghosh, D., Physicochemical properties of mucus and their impact on transmucosal drug delivery. *Int. J. Pharm.*, 532, 1, 555–72, 2017.
35. Bustamante-Marin, X.M. and Ostrowski, L.E., Cilia and mucociliary clearance. *Cold Spring Harbor Perspect. Biol.*, 9, 4, a028241, 2017.
36. Xu, Y., Yuen, P.-W., Lam, J.K.-W., Intranasal DNA vaccine for protection against respiratory infectious diseases: the delivery perspectives. *Pharmaceutics*, 6, 3, 378–415, 2014.
37. Jaiswal, N., Sharma, P.K., Kumar, A., Scope of nasal drug delivery: current challenges. *World J. Med. Sci.*, 14, 4, 97–112, 2017.
38. Borojeni, A.A., Noga, M.L., Vehring, R., Finlay, W.H., Measurements of total aerosol deposition in intrathoracic conducting airway replicas of children. *J. Aerosol Sci.*, 73, 39–47, 2014.
39. Gizurarson, S., The relevance of nasal physiology to the design of drug absorption studies. *Adv. Drug Deliv. Rev.*, 11, 3, 329–47, 1993.
40. Kublik, H. and Vidgren, M., Nasal delivery systems and their effect on deposition and absorption. *Adv. Drug Deliv. Rev.*, 29, 1-2, 157–77, 1998.
41. Hilding, A., Phagocytosis, mucous flow, and ciliary action. *Arch. Environ. Health*, 6, 1, 61–73, 1963.
42. Gizurarson, S., The effect of cilia and the mucociliary clearance on successful drug delivery. *Biol. Pharm. Bull.*, b14-00398, 38, 497–506, 2015.
43. App, E., Zayas, J., King, M., Rheology of mucus and transepithelial potential difference: small airways versus trachea. *Eur. Respir. J.*, 6, 1, 67–75, 1993.
44. Lee, V., *Peptide and protein drug delivery*, CRC Press, University of Southern California, School of Pharmacy, Los Angeles, California, 1990.
45. Sarkar, M.A., Drug metabolism in the nasal mucosa. *Pharm. Res.*, 9, 1, 1–9, 1992.
46. Kirkeby, L., Rasmussen, T.T., Reinholdt, J., Kilian, M., Immunoglobulins in nasal secretions of healthy humans: structural integrity of secretory immunoglobulin A1 (IgA1) and occurrence of neutralizing antibodies to IgA1 proteases of nasal bacteria. *Clin. Diagn. Lab. Immunol.*, 7, 1, 31–9, 2000.
47. Porzia, A., Cavaliere, C., Begvarfaj, E., Masieri, S., Mainiero, F., Human nasal immune system: a special site for immune response establishment. *J. Biol. Regul. Homeost. Agents*, 32, 1 Suppl. 1, 3–8, 2018.
48. Haughney, J., Price, D., Barnes, N.C., Virchow, J.C., Roche, N., Chrystyn, H., Choosing inhaler devices for people with asthma: current knowledge and outstanding research needs. *Respir. Med. CME*, 3, 3, 125–31, 2010.
49. Farage, M.A. and Maibach, H.I., Morphology and physiological changes of genital skin and mucosa, in: *Topical Applications and the Mucosa*, vol. 40, pp. 9–19, 2011.

50. Henkin, R., Intranasal insulin: from nose-to-brain, in: *Nutrition*, vol. 26, pp. 624–33, 2010.
51. Illum, L., Hinchcliffe, M., Davis, S., The effect of blood sampling site and physicochemical characteristics of drugs on bioavailability after nasal administration in the sheep model. *Pharm. Res.*, 20, 9, 1474–84, 2003.
52. Romeo, V., Sileno, A., Pimplaskar, H., Behl, C., Effects of physicochemical properties and other factors on systemic nasal drug delivery. *Adv. Drug Deliv. Rev.*, 29, 1-2, 89–116, 1998.
53. Kumar, M., Misra, A., Mishra, A., Mishra, P., Pathak, K., Mucoadhesive nanoemulsion-based intranasal drug delivery system of olanzapine for brain targeting. *J. Drug Targeting*, 16, 10, 806–14, 2008.
54. Bhise, S.B., Yadav, A.V., Avachat, A.M., Malayandi, R., Bioavailability of intranasal drug delivery system. *Asian J. Pharm. (AJP)*, 2, 4, 1–5, 2008.
55. Yu, H. and Kim, K., Direct nose-to-brain transfer of a growth hormone releasing neuropeptide, hexarelin after intranasal administration to rabbits. *Int. J. Pharm.*, 378, 1-2, 73–9, 2009.
56. Bahadur, S. and Pathak, K., Physicochemical and physiological considerations for efficient nose-to-brain targeting. *Expert Opin. Drug Deliv.*, 9, 1, 19–31, 2012.
57. Gonda, I. and Gipps, E., Model of disposition of drugs administered into the human nasal cavity. *Pharm. Res.*, 7, 1, 69–75, 1990.
58. Aggarwal, R., Cardozo, A., Homer, J., The assessment of topical nasal drug distribution. *Clin. Otolaryngol. Allied Sci.*, 29, 3, 201–5, 2004.
59. Merkus, P., Ebbens, F.A., Muller, B., Fokkens, W.J., Influence of anatomy and head position on intranasal drug deposition. *Eur. Arch. Oto-Rhino-Laryngol. Head Neck*, 263, 9, 827–32, 2006.
60. Djupesland, P.G., Nasal drug delivery devices: characteristics and performance in a clinical perspective—a review. *Drug Deliv. Transl. Res.*, 3, 1, 42–62, 2013.
61. Myrdal, P.B., Sheth, P., Stein, S.W., Advances in metered dose inhaler technology: formulation development. *AAPS PharmSciTech*, 15, 2, 434–55, 2014.
62. Marple, B., Roland, P., Benninger, M., Safety review of benzalkonium chloride used as a preservative in intranasal solutions: an overview of conflicting data and opinions. *Otolaryngology-Head Neck Surgery.*, 130, 1, 131–41, 2004.
63. Laube, B.L., Devices for aerosol delivery to treat sinusitis. *J. Aerosol Med.*, 20, s1, S5–S18, 2007.
64. Möller, W., Schuschnig, U., Khadem Saba, G., Münzing, W., Canis, M., Häussinger, K. *et al.*, Nasally inhaled pulsating aerosols: assessing lung and nose deposition, in: *C37 Exercise, Aerosols and Airway Dynamics*, pp. A4432–A, American Thoracic Society, PARI Pharma GmbH, Germany, 2011.
65. Pritchard, J.N., Hatley, R.H., Denyer, J., Hollen Dv. Mesh nebulizers have become the first choice for new nebulized pharmaceutical drug developments. *Ther. Deliv.*, 9, 2, 121–36, 2018.

66. Shaikh, R., Singh, T.R.R., Garland, M.J., Woolfson, A.D., Donnelly, R.F., Mucoadhesive drug delivery systems. *J. Pharm. Bioallied Sci.*, 3, 1, 89, 2011.
67. Leung, S.-H.S. and Robinson, J.R., The contribution of anionic polymer structural features to mucoadhesion. *J. Controlled Release*, 5, 3, 223–31, 1987.
68. McBain, J. and Hopkins, D., On adhesives and adhesive action. *J. Phys. Chem.*, 29, 2, 188–204, 2002.
69. Derjaguin, B. and Smilga, V., *Adhesion: fundamentals and practice*, pp. 54–60, McLaren, London, 1969.
70. Jiménez-castellanos, M.R., Zia, H., Rhodes, C., Mucoadhesive drug delivery systems. *Drug Dev. Ind. Pharm.*, 19, 1-2, 143–94, 1993.
71. Ahuja, A., Khar, R.K., Ali, J., Mucoadhesive drug delivery systems. *Drug Dev. Ind. Pharm.*, 23, 5, 489–515, 1997.
72. Gu, J., Robinson, J., Leung, S., Binding of acrylic polymers to mucin/epithelial surfaces: structure-property relationships. *Crit. Rev. Ther. Drug Carrier Syst.*, 5, 1, 21–67, 1988.
73. Hasegawa, H., Ichinohe, T., Strong, P., Watanabe, I., Ito, S., Tamura, S.I. *et al.*, Protection against influenza virus infection by intranasal administration of hemagglutinin vaccine with chitin microparticles as an adjuvant. *J. Med. Virol.*, 75, 1, 130–6, 2005.
74. Park, H. and Robinson, J.R., Mechanisms of mucoadhesion of poly (acrylic acid) hydrogels. *Pharm. Res.*, 4, 6, 457–64, 1987.
75. Nicholls, T.J., Green, K.L., Rogers, D.J., Cook, J.D., Wolowacz, S., Smart, J.D., Lectins in ocular drug delivery: an investigation of lectin binding sites on the corneal and conjunctival surfaces. *Int. J. Pharm.*, 138, 2, 175–83, 1996.
76. Hornof, M., Weyenberg, W., Ludwig, A., Bernkop-Schnürch, A., Mucoadhesive ocular insert based on thiolated poly (acrylic acid): development and *in vivo* evaluation in humans. *J. Controlled Release*, 89, 3, 419–28, 2003.
77. Albrecht, K., Zirm, E., Palmberger, T., Schlocker, W., Bernkop-Schnürch, A., Preparation of thiomer microparticles and *in vitro* evaluation of parameters influencing their mucoadhesive properties. *Drug Dev. Ind. Pharm.*, 32, 10, 1149–57, 2006.
78. Gänger, S. and Schindowski, K., Tailoring formulations for intranasal nose-to-brain delivery: a review on architecture, physico-chemical characteristics and mucociliary clearance of the nasal olfactory mucosa. *Pharmaceutics*, 10, 3, 116, 2018.
79. Spindler, L.M., Feuerhake, A., Ladel, S., Günday, C., Flamm, J., Günday-Türeli, N. *et al.*, Nano-in-Micro-Particles Consisting of PLGA Nanoparticles Embedded in Chitosan Microparticles via Spray-Drying Enhances Their Uptake in the Olfactory Mucosa. *Front. Pharmacol.*, 2282, 1–19, 2021.
80. Garg, Y., Kapoor, D.N., Sharma, A.K., Bhatia, A., Drug Delivery Systems and Strategies to Overcome the Barriers of Brain. *Curr. Pharm. Des.*, 28, 8, 619–641, 2022.

81. Maigler, F., Ladel, S., Flamm, J., Gänger, S., Kurpiers, B., Kiderlen, S. *et al.*, Selective CNS Targeting and Distribution with a Refined Region-Specific Intranasal Delivery Technique via the Olfactory Mucosa. *Pharmaceutics*, 13, 11, 1904, 2021.
82. Ladel, S., Flamm, J., Zadeh, A.S., Filzwieser, D., Walter, J.-C., Schlossbauer, P. *et al.*, Allogenic Fc domain-facilitated uptake of IgG in nasal lamina propria: Friend or foe for intranasal CNS delivery? *Pharmaceutics*, 10, 3, 107, 2018.
83. Clementino, A.R., Pellegrini, G., Banella, S., Colombo, G., Sonvico, F. *et al.*, Structure and fate of nanoparticles designed for the nasal delivery of poorly soluble drugs. *Mol. Pharm.*, 18, 8, 3132–46, 2021.
84. Flamm, J., Hartung, S., Gänger, S., Maigler, F., Pitzer, C., Schindowski, K., Establishment of an Olfactory Region-specific Intranasal Delivery Technique in Mice to Target the Central Nervous System. *Front. Pharmacol.*, 12, 789780–, 2021.
85. Erdoğar, N., Akkın, S., Bilensoy, E., Nanocapsules for drug delivery: an updated review of the last decade. *Recent Pat. Drug Deliv. Formul.*, 12, 4, 252–66, 2018.
86. Barbieri, S., Sonvico, F., Como, C., Colombo, G., Zani, F., Buttini, F. *et al.*, Lecithin/chitosan controlled release nanopreparations of tamoxifen citrate: loading, enzyme-trigger release and cell uptake. *J. Controlled Release*, 167, 3, 276–83, 2013.
87. Clementino, A., Batger, M., Garrastazu, G., Pozzoli, M., Del Favero, E., Rondelli, V. *et al.*, The nasal delivery of nanoencapsulated statins– an approach for brain delivery. *Int. J. Nanomed.*, 11, 6575, 2016.
88. Pohlmann, A.R., Fonseca, F.N., Paese, K., Detoni, C.B., Coradini, K., Beck, R.C. *et al.*, Poly (ϵ-caprolactone) microcapsules and nanocapsules in drug delivery. *Expert Opin. Drug Deliv.*, 10, 5, 623–38, 2013.
89. Frank, L.A., Chaves, P.S., D'Amore, C.M., Contri, R.V., Frank, A.G., Beck, R.C. *et al.*, The use of chitosan as cationic coating or gel vehicle for polymeric nanocapsules: Increasing penetration and adhesion of imiquimod in vaginal tissue. *Eur. J. Pharm. Biopharm.*, 114, 202–12, 2017.
90. Morrison, E.E. and Costanzo, R.M., Morphology of the human olfactory epithelium. *J. Comp. Neurol.*, 297, 1, 1–13, 1990.
91. Mistry, A., Stolnik, S., Illum, L., Nose-to-brain delivery: investigation of the transport of nanoparticles with different surface characteristics and sizes in excised porcine olfactory epithelium. *Mol. Pharm.*, 12, 8, 2755–66, 2015.
92. Donovan, M.D. and Huang, Y., Large molecule and particulate uptake in the nasal cavity: the effect of size on nasal absorption. *Adv. Drug Deliv. Rev.*, 29, 1-2, 147–55, 1998.
93. Ishikawa, F., Murano, M., Hiraishi, M., Yamaguchi, T., Tamai, I., Tsuji, A., Insoluble powder formulation as an effective nasal drug delivery system. *Pharm. Res.*, 19, 8, 1097–104, 2002.
94. Rahisuddin, S.P., Garg, G., Salim, M., Review on nasal drug delivery system with recent advancement. *Int. J. Pharm. Pharm. Sci.*, 3, 2, 1–5, 2011.

95. Peek, L.J., Middaugh, C.R., Berkland, C., Nanotechnology in vaccine delivery. *Adv. Drug Deliv. Rev.*, 60, 8, 915–28, 2008.
96. Chatterjee, B., Nose-to-brain drug delivery: A recent update. *J. Formul. Sci. Bioavailab.*, 1, 2, 1–2, 2017.
97. Arora, P., Sharma, S., Garg, S., Permeability issues in nasal drug delivery. *Drug Discovery Today*, 7, 18, 967–75, 2002.
98. Potschka, H., Fedrowitz, M., Löscher, W., P-glycoprotein and multidrug resistance-associated protein are involved in the regulation of extracellular levels of the major antiepileptic drug carbamazepine in the brain. *NeuroReport*, 12, 16, 3557–60, 2001.
99. Tanaka, A., Furubayashi, T., Enomura, Y., Hori, T., Shimomura, R., Maeda, C. *et al.*, Nasal drug absorption from powder formulations: effect of fluid volume changes on the mucosal surface. *Biol. Pharm. Bull.*, 40, 2, 212–9, 2017.
100. Fasiolo, L.T., Manniello, M.D., Tratta, E., Buttini, F., Rossi, A., Sonvico, F. *et al.*, Opportunity and challenges of nasal powders: Drug formulation and delivery. *Eur. J. Pharm. Sci.*, 113, 2–17, 2018.
101. Silberstein, S., Winner, P.K., McAllister, P.J., Tepper, S.J., Halker, R., Mahmoud, R.A. *et al.*, Early Onset of Efficacy and Consistency of Response Across Multiple Migraine Attacks From the Randomized COMPASS Study: AVP-825 Breath Powered® Exhalation Delivery System (Sumatriptan Nasal Powder) vs Oral Sumatriptan. Headache. *J. Head Face Pain*, 57, 6, 862–76, 2017.
102. Lee, Y.H., Kim, K.H., Yoon, I.K., Lee, K.E., Chun, I.K., Rhie, J.Y. *et al.*, Pharmacokinetic evaluation of formulated levodopa methyl ester nasal delivery systems. *Eur. J. Drug Metab. Pharmacokinet.*, 39, 4, 237–42, 2014.
103. Rassu, G., Soddu, E., Cossu, M., Brundu, A., Cerri, G., Marchetti, N. *et al.*, Solid microparticles based on chitosan or methyl-β-cyclodextrin: A first formulative approach to increase the nose-to-brain transport of deferoxamine mesylate. *J. Controlled Release*, 201, 68–77, 2015.
104. Colombo, G., Bortolotti, F., Chiapponi, V., Buttini, F., Sonvico, F., Invernizzi, R. *et al.*, Nasal powders of thalidomide for local treatment of nose bleeding in persons affected by hereditary hemorrhagic telangiectasia. *Int. J. Pharm.*, 514, 1, 229–37, 2016.
105. Illum, L., Nasal drug delivery: new developments and strategies. *Drug Discovery Today*, 7, 23, 1184–9, 2002.
106. Bitter, C., Suter-Zimmermann, K., Surber, C., Nasal drug delivery in humans, in: *Topical Applications and the Mucosa*, vol. 40, pp. 20–35, 2011.
107. Cramptom, H.J., A comparison of the relative clinical efficacy of a single dose of ketotifen fumarate 0.025% ophthalmic solution versus placebo in inhibiting the signs and symptoms of allergic rhinoconjunctivitis as induced by the conjunctival allergen challenge model. *Clin. Ther.*, 24, 11, 1800–8, 2002.
108. Benedetti, M.S., Whomsley, R., Poggesi, I., Cawello, W., Mathy, F.-X., Delporte, M.-L. *et al.*, Drug metabolism and pharmacokinetics. *Drug Metab. Rev.*, 41, 3, 344–90, 2009.

109. Lazo, J.S., Parker, K.L., Bruton, L., *Goodman & Gilman's the pharmacological basis of therapeutics*, McGraw-Hill Publishing, Minion Pro by Knowledge Works Global Ltd., The United States of America, 2005.
110. Kumar, A., Pandey, A.N., Jain, S.K., Nasal-nanotechnology: revolution for efficient therapeutics delivery. *Drug Delivery*, 23, 3, 671–83, 2016.
111. Van der Kuy, P., Lohman J, M., Hooymans, P., Ter Berg, J., Merkus, F., Bioavailability of intranasal formulations of dihydroergotamine. *Eur. J. Clin. Pharmacol.*, 55, 9, 677–80, 1999.
112. Pisal, P.B., Patil, S.S., Pokharkar, V.B., Rheological investigation and its correlation with permeability coefficient of drug loaded carbopol gel: influence of absorption enhancers. *Drug Dev. Ind. Pharm.*, 39, 4, 593–9, 2013.
113. Trapani, A., De Giglio, E., Cometa, S., Bonifacio, M.A., Dazzi, L., Di Gioia, S. *et al.*, Dopamine-loaded lipid based nanocarriers for intranasal administration of the neurotransmitter: A comparative study. *Eur. J. Pharm. Biopharm.*, 167, 189–200, 2021.
114. Nguyen, T-T-L and Maeng, H.-J., Pharmacokinetics and Pharmacodynamics of Intranasal Solid Lipid Nanoparticles and Nanostructured Lipid Carriers for Nose-to-Brain Delivery. *Pharmaceutics*, 14, 3, 572, 2022.
115. Pokharkar, V., Patil-Gadhe, A., Palla, P., Efavirenz loaded nanostructured lipid carrier engineered for brain targeting through intranasal route: *In-vivo* pharmacokinetic and toxicity study. *Biomed. Pharmacother.*, 94, 150–64, 2017.
116. Salem, L.H., El-Feky, G.S., Fahmy, R.H., El Gazayerly, O.N., Abdelbary, A., Coated lipidic nanoparticles as a new strategy for enhancing nose-to-brain delivery of a hydrophilic drug molecule. *J. Pharm. Sci.*, 109, 7, 2237–51, 2020.
117. Bhatt, P.C., Srivastava, P., Pandey, P., Khan, W., Panda, B.P., Nose-to-brain delivery of astaxanthin-loaded solid lipid nanoparticles: fabrication, radio labeling, optimization and biological studies. *RSC Adv.*, 6, 12, 10001–10, 2016.
118. Pires, P.C. and Santos, A.O., Nanosystems in nose-to-brain drug delivery: A review of non-clinical brain targeting studies. *J. Controlled Release*, 270, 89–100, 2018.
119. Vieira, A.C., Chaves, L.L., Pinheiro, S., Pinto, S., Pinheiro, M., Lima, S.C. *et al.*, Mucoadhesive chitosan-coated solid lipid nanoparticles for better management of tuberculosis. *Int. J. Pharm.*, 536, 1, 478–85, 2018.
120. Trotta, V., Pavan, B., Ferraro, L., Beggiato, S., Traini, D., Des Reis, L.G. *et al.*, Brain targeting of resveratrol by nasal administration of chitosan-coated lipid microparticles. *Eur. J. Pharm. Biopharm.*, 127, 250–9, 2018.
121. Chhajed, S., Sangale, S., Barhate, S., Advantageous nasal drug delivery system: a review. *Int. J. Pharm. Sci. Res.*, 2, 6, 1322, 2011.
122. Erdő, F., Bors, L.A., Farkas, D., Bajza, Á., Gizurarson, S., Evaluation of intranasal delivery route of drug administration for brain targeting. *Brain Res. Bull.*, 143, 155–70, 2018.

123. Cho, H.-J., Termsarasab, U., Kim, J.-S., Kim, D.-D., *In vitro* nasal cell culture systems for drug transport studies. *J. Pharm. Invest.*, 40, 6, 321–32, 2010.
124. Li, H., Sheppard, D.N., Hug, M.J., Transepithelial electrical measurements with the Ussing chamber. *J. Cyst. Fibros.*, 3, 123–6, 2004.
125. Pozzoli, M., Rogueda, P., Zhu, B., Smith, T., Young, P.M., Traini, D. *et al.*, Dry powder nasal drug delivery: challenges, opportunities and a study of the commercial Teijin Puvlizer Rhinocort device and formulation. *Drug Dev. Ind. Pharm.*, 42, 10, 1660–8, 2016.
126. Gotoh, M., Okubo, K., Okuda, M., Hamada, C., Clinical pharmacology study of the corticosteroid nasal spray dexamethasone cipecilate (NS-126): examination of the durability of efficacy in the nasal induction test. *Expert Opin. Invest. Drugs*, 19, 12, 1475–86, 2010.
127. Drugs.com, Nasal antihistamines and decongestants, 2022, Available from: https://www.drugs.com/drug-class/nasal-antihistamines-and-decongestants.html.

10

Dermal and Transdermal Novel Drug Delivery Systems

Shiva Hashemi[1†], Saeedeh Abedin[1†], Azadeh Vaezi Moghaddam[1] and Mahsa Sayed Tabatabaei[2*]

[1]*Department of Pharmaceutics and Pharmaceutical Nanotechnology, School of Pharmacy, Shahid Beheshti University of Medical Sciences, Tehran, Iran*
[2]*HIDA Pharmaceutical Technologies Development Center, HIDA Pharmaceuticals, Tehran, Iran*

Abstract

The skin, due to its large surface area and ease of accessibility, is an attractive pathway for both local (dermal) and systemic (transdermal) drug delivery. Drug administration through the skin provides several advantages such as high patient compliance, consistent blood levels of the drug, avoiding drug degradation in the gastrointestinal tract as well as hepatic first-pass metabolism, and reduction of systemic side effects. However, the skin (and particularly, the *stratum corneum*) acts as an impervious barrier against drug permeation. Practically, the number of drugs with optimal physicochemical properties to permeate through the skin is very low. Therefore, novel approaches and systems have been developed to conquer this limitation. This chapter first gives a brief introduction to the skin and its barrier properties, as well as the opportunities and challenges ahead, and factors affecting percutaneous drug delivery. Then, we delve into different scenarios for overcoming the skin barrier, including temporal disruption of the *stratum corneum*, bypassing it, molecular modification of drugs to acquire optimal features to permeate through the skin, and using advanced carriers, formulations, or devices to improve drug delivery. In the following, a variety of novel strategies and systems including chemical and physical permeation enhancement methods, nanocarriers, microneedles, patches, transdermal sprays, and advanced transdermal formulations that have been developed to facilitate the delivery of therapeutics to the skin will be reviewed. Next, the fate of the drug after entering the skin (i.e., dermal

**Corresponding author*: tabatabaei@hidapharma.com
†Authors contributed to the manuscript equally.

Mehrdad Hamidi and Mahsa Sayed Tabatabaei (eds.) Novel Drug Delivery Systems: Fundamentals and Applications, (349–392) © 2025 Scrivener Publishing LLC

and transdermal pharmacokinetics) will be overviewed; and finally, the novel skin products available in the market will be highlighted.

Keywords: Skin, *Stratum Corneum*, percutaneous drug delivery, permeation enhancer, nanocarrier, dermal pharmacokinetics

10.1 Introduction

The skin is a complex multilayered structure with a large surface area (1.5-2 m^2) and easy access, which makes it of great interest as a drug delivery route both for systemic (transdermal) and local (dermal) purposes [1]. Dermal drug administration refers to localized delivery of therapeutics in order to treat skin disorders. So, the systemic exposure, and consequently, systemic adverse effects are minimized. Transdermal drug delivery (TDD), however, is considered for systemic drug delivery through the skin. Recent efforts have focused on the design of new approaches, technologies, devices, and systems to facilitate the delivery of therapeutics and cosmetics to the skin. In this regard, the appropriate permeability of the drug to reach the action site at therapeutic levels and sufficient retention time is of crucial importance. Molecular weight (MW) and oil/water partition coefficient are the main factors in predicting skin permeation of drugs. The ideal MW for a drug to pass through the skin barrier is < 500 Da [2]. This makes it difficult for macromolecules, to pass through the *stratum corneum* as the outermost layer of the skin with major barrier function, at therapeutically acceptable rates. To conquer such a formidable barrier, applying one strategy alone to improve drug permeability may not be enough. So, a combination of different methods (for instance, physical and chemical approaches) may be used to create a synergistic effect and improve the skin penetration of macromolecules. Nanocarriers are oftentimes considered excellent auxiliary options to improve skin drug delivery.

One of the most important advantages of TDD is improving patient compliance, especially when oral administration is not possible (for example, for unconscious patients or those with swallowing problems). Also, TDD is known as a pain-free and safe administration route of drug delivery with the potential to reduce the frequency of drug application. Additionally, TDD creates more consistent blood levels of the drug, prevents degradation of the drug in the gastrointestinal (GI) tract, avoids hepatic first-pass effect, and also reduces systemic side effects.

On the other side, the main drawback of TDD systems is inadequate permeation through the skin, which limits drug penetration. In addition, skin irritation (e.g. itching and erythema) or allergic reactions may occur by some drugs and/or excipients used in skin formulations [3].

10.2 Anatomy and Physiology

Skin is the largest body organ with an area of about 1.7 m^2, which constitutes 16% of the total body mass of a normal person. Primarily, the skin is a protective layer that separates the body from the external environment. So, it is considered a formidable barrier against xenobiotics (e.g., microorganisms), ultraviolet (UV) radiation, and other external threats. Also, the skin prevents the water loss [4]. Anatomically, as shown in Figure 10.1, the human skin structure is categorized into three main layers from the outside to the inside, namely epidermis, dermis, and hypodermis.

10.2.1 Epidermis

The upper layer of the skin is the epidermis (composed of the *stratum corneum* and viable epidermis). This layer provides a barrier against pathogens, UV light, chemicals, and mechanical injuries. Also, it controls the release of water from the body into the atmosphere (transepidermal water loss, TEWL), and consequently, keeps skin hydration. Also, this layer produces new skin cells and determines the skin color [5]. The functions of the epidermis are performed by the *stratum corneum* and viable epidermis.

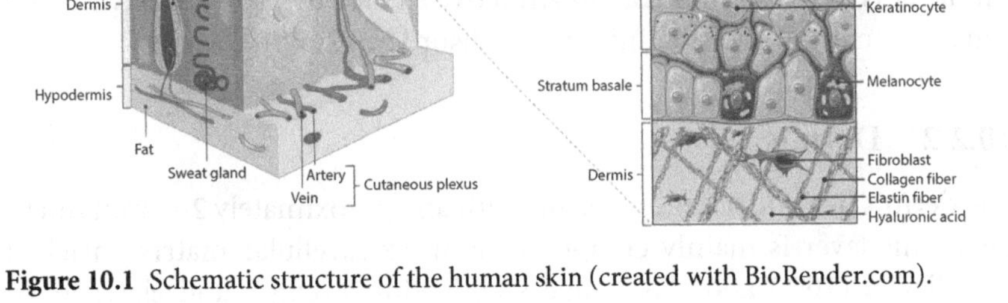

Figure 10.1 Schematic structure of the human skin (created with BioRender.com).

10.2.1.1 Stratum Corneum

The *stratum corneum* is considered a major barrier to the external environment, due to its unique structure, great density (about 1.4 g/cm^3 in the dry state), as well as poor hydration (15–20%). It has approximately 10–20 μm thickness and consists of 10–15 layers of corneocytes mainly composed of keratins (70%) and lipids (20%), enclosed by a lipid-enriched matrix that consists of ceramides, free fatty acids, and cholesterol. Ceramides of the lipid matrix undertake the most crucial role in the barrier function of the skin. In the case of some skin disorders closely related to the abnormality of the barrier function (e.g., psoriasis and atopic dermatitis), the amount of ceramides is significantly decreased. Oxidative interactions have also been detected in the *stratum corneum*. The presence of α-tocopherol in the *stratum corneum* is essential since its antioxidant effect can help stabilize the lipid layers and resist the lipid peroxidation induced by environmental factors such as UV radiation of air pollution [2, 6].

10.2.1.2 Viable Epidermis

The viable epidermis, located underneath the *stratum corneum*, is which consists of *stratum basale* (the deepest part of the epidermis), *stratum spinosum*, *stratum granulosum*, and *stratum lucidum* (see Figure 10.1). This layer with a 50–100 μm thickness is devoid of vessels and nerves. Numerous cells are metabolically active in the *stratum basale* such as keratinocytes, melanocytes, Merkel cells, and Langerhans cells. Several skin diseases (e.g., psoriasis, melanoma, and dermatomycoses) may influence the viable epidermis. For instance, it is thought that melanocytes in the basal epidermis are in the pathogenesis of melanoma which accounts for the majority of global skin cancer deaths. This necessitates drug delivery to the epidermis so as not to be transferred into the dermis. Furthermore, the viable epidermis is considered the first immunologic defense of the skin with the contribution of the skin immune system cells (e.g., Langerhans cells). So, this layer is in the spotlight as an attractive site for the delivery of vaccines, gene vaccination, and treating genetic disorders [2, 7].

10.2.2 Dermis

The dermis lies under the epidermis with an approximately 2–3 mm thickness. This layer is mainly composed of an extracellular matrix enriched in collagen (70%) and elastic fibers endowing strength and elasticity to the skin. Hair follicles, nerves, blood and lymphatic vessels, and a variety of immune cells (e.g., dendritic cells, mast cells, and macrophages) are housed in the dermis [8]. The blood vessels supply the nutrients for the dermis and epidermis. The immune cells play crucial roles in the case of dermal inflammation, infections, tumor progression, angiogenesis, wound healing, and tissue remodeling. Consequently, for both preventive and therapeutic vaccination and immunological treatments, novel approaches have been adopted for targeted delivery into the dermis [9]. As an intercellular penetration route, transfollicular drug delivery has recently garnered considerable attention. The optimum particle size for transfollicular drug delivery is about 400-700 nm [10]. In addition, hair follicle stem cells are considered a target for gene delivery [11].

10.2.3 Hypodermis

The innermost layer of the skin is called hypodermis, which consists of fat cells, fascia, as well as a large network of lymphatic and blood vessels. This layer is located between the dermis and the underlying fascia of muscles and bones. Accordingly, the hypodermis undertakes a major role in supporting, the conduction of the skin's vascular and nerve signals, protecting against physical shock and heat insulation [12].

10.3 Barriers to Transdermal Drug Delivery

As mentioned earlier, the skin offers several advantages as a route of drug delivery, such as high surface area, great accessibility, ease of application, circumvented systemic adverse effects, the possibility of drug accumulation at the local site, bypassed first-pass metabolism, and the possibility of both dermal and transdermal drug delivery [13]. However, it is not as simple as it sounds; since for effective transdermal drug delivery, the drug molecule should first cross or bypass the *stratum corneum* as the major barrier against any xenobiotic, including drugs. Such an impervious resistance against foreign molecules, as already mentioned, is due to the complexity of the skin barrier composed of serial lipophilic (the *stratum corneum*) and hydrophilic (the viable epidermis and dermis) compartments. Generally, there are two different ways ahead: (i) transepidermal and (ii) transappendageal (shunt) routes. The transepidermal pathway is divided into intracellular (also called transcellular) and intercellular routes. In the former, drug molecules should pass through the *stratum corneum*. Here, the lipophilic nature of the multi-layered matrix between the cells circumvents the passage of hydrophilic molecules. In the latter, the drugs only need

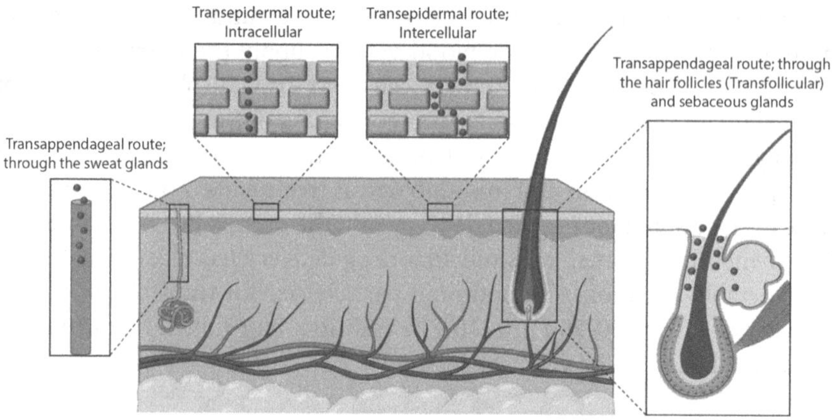

Figure 10.2 Schematic illustration of available pathways for a drug to cross through the skin including (i) Transappendageal route (through the hair follicles, sebaceous glands and sweat glands) and (ii) Transepidermal route (including intracellular and intercellular pathways) (created with Biorender.com).

to pass through the lipid matrix. The transappendageal pathway includes drug transport through the hair follicles, sebaceous glands, or sweat glands. This is the route of choice for high-molecular-weight, bulky, and charged molecules [14]. These pathways are schematically illustrated in Figure 10.2.

Although appendages take up only a small area of the skin, they play an important role in skin permeation. For instance, hair follicles are of growing interest for drug delivery into deeper sites of the skin as well as prolonged residence time; particularly for particle-based formulations and hydrophilic, high-molecular-weight drugs. Other appendages, however, have been less noticed, presumably because of difficult sampling for the drug quantification [2, 15].

It should be noted that, however, the mere existence of such pathways is not adequate to achieve effective concentrations of many drugs in their sites of action. Therefore, different strategies have been proposed so far to improve skin absorption. These strategies are explained in detail in the following sections.

10.4 Factors Affecting Dermal/Transdermal Drug Delivery

Several factors including biological, pathological, demographic, physicochemical, and pharmaceutical parameters have the potential to affect drug permeation through the skin. Here, how these factors impact transdermal drug delivery has been briefly discussed.

10.4.1 Biological Factors

10.4.1.1 Regional Skin Site

There is a high structural skin variation, not only among the individuals (i.e., intersubject variation) but also in different anatomical parts of the skin of a person's body. For instance, the epidermis thickness of the foot sole and hand palm are up to 13 times higher than that of lips or eyelids. So, it seems that drug permeation should be less from areas of higher thickness, but interestingly this is not always the case; since similar permeation profiles have been obtained in skin regions with different thicknesses of *stratum corneum*, and conversely, different permeation profiles in areas with the same thickness. Nevertheless, a general order for permeation through different skin sites can be considered as follows: epigenital region > head and neck > trunk region > arms > legs. For transdermal studies, the formulation is often applied to the trunk or arm, due to the lower sensitization probability as well as better adhesion and ease of removal (due to the lower hair follicle density) [16].

10.4.1.2 Skin Metabolism

Although the enzymatic activity of the skin is considerably lower than the liver (about 10%), it is considered a metabolically active organ. So, when drugs permeate through the skin, they are exposed to different enzymes located in the skin (mainly in the epidermis, appendages, and even the *stratum corneum*). These enzymes, collectively, are responsible for first-pass skin metabolism which can decrease the bioavailability of the applied drug. Both Phase I metabolic processes, which convert lipophilic molecules to more polar ones (e.g., reduction, oxidation, or hydrolysis), and Phase II metabolic reactions, during which hydrophilic groups are conjugated to the original molecule (e.g., methylation, glucuronidation), occur in the skin. Also, some microorganisms (e.g., *Staphylococcus epidermidis*) in the superficial skin layers may contribute to the metabolism of some drugs [16, 17].

10.4.1.3 Skin Hydration

Water, as a penetration enhancer, could significantly impact the *stratum corneum* and improve the permeation of different drugs [18]. In the physiological state, water only makes up about 10-20% of the *stratum corneum*'s weight. However, if shortly soaked in water, the *stratum corneum* could absorb water up to 20 times more than usual. Several disease states (e.g., ichthyoses and eczema) could also impact the degree of hydration [16].

10.4.1.4 Skin Temperature

Although the internal temperature of the body is maintained at about 37°C, the skin surface temperature may change significantly under the influence of several factors such as environmental parameters (e.g., ambient temperature, humidity, air circulation, and sunlight) and emotional and psychological conditions. As a general rule, it has been shown that there is a direct correlation between skin temperature and drug permeation. So, the higher the temperature, the more fluidized lipid domains in the *stratum corneum*, and thereby, a higher drug flux through the skin. Occlusion is a simple strategy to prevent temperature fluctuations, although it can raise slightly the local temperature [16].

10.4.2 Barrier Function-Related Factors

Several factors can change the barrier nature of the skin and, thereby, alter its permeability, generally including [17, 19]:

(i) Chemical factors, like solvents and detergents
(ii) Physical factors, such as sunlight and weather
(iii) Chronological factors, e.g., seasonal variations and circadian rhythm
(iv) Psychological parameters such as stress
(v) Mechanical factors, like injury
(vi) Pathological conditions, such as atopic dermatitis, psoriasis, ichthyoses, impetigo, eczema), and
(vii) Lifestyle habits, e.g., diet.

In addition, some of these damages (e.g., caused by chemical, thermal, mechanical, and microbial factors) may initiate the inflammatory process which, in turn, can further reduce the barrier properties of the skin [16].

10.4.3 Demographic Factors

10.4.3.1 Gender

Among the most important factors that differ between the characteristics of the female and male skin are (i) the size of keratinocytes (larger in male skin than female), (ii) the size of skin pores including sebaceous and sweat glands (again, larger in male skin than female), (iii) the activity of the sebaceous glands (higher activity in male skin than female) and pH (more acidic in male skin than female). Although these differences seem to affect the skin permeation of drugs, most of the evidence has demonstrated that drug permeation through the skin of males is virtually no different from that of females [16].

10.4.3.2 Age

With age, many skin factors change, including the thickness of the skin, the density of pores/hair, blood flow, pH, the content of water and skin lipids, and so on [20]. However, there are conflicting data on whether aging affects the skin absorption of drugs. Some studies show that, despite major changes occurring in different aspects of the skin, the *stratum corneum* (and thereby, its barrier function) remains intact and unchanging throughout human life. Conversely, some studies state that, while aging does not significantly impact the permeation of hydrophobic drugs, the permeation of hydrophilic compounds decreases with age. This is, probably, due to lower water content of the *stratum corneum* in the elderly [16].

10.4.4 Physicochemical Factors

For optimal percutaneous drug delivery, several physicochemical properties of drug molecules (e.g., molecular size and shape, lipophilicity, and water solubility) should be carefully considered. These factors along with their optimum values are shown in the Figure 10.3.

10.4.5 Pharmaceutical Factors

One of the most important formulation-related factors to be adjusted is the concentration of the drug in the formulation. To ensure the maximum drug permeation through the skin, the formulation should be saturated, or

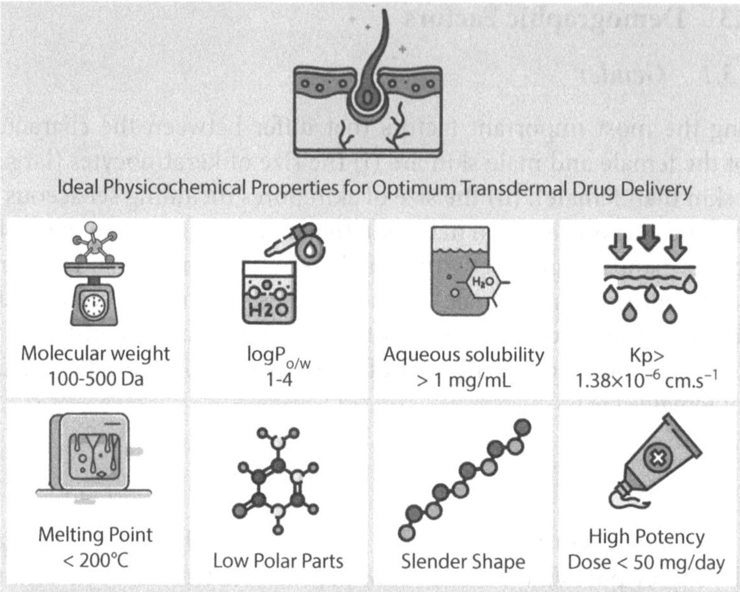

Figure 10.3 Ideal physicochemical characteristics of a drug molecule to pass transepidermal (data obtained from Ref. [21]).

more precisely, supersaturated. This is because, the thermodynamic activity of the drug in the supersaturated system is > 1, which drives the drug out of the formulation and improves its skin permeation [21, 22]. Despite this major benefit, supersaturated formulations usually suffer from instability issues. This could be overcome using different strategies such as adding suitable polymers [21].

10.5 Challenges and Opportunities

The strategic location of the skin as the first defense barrier of the body creates several barriers to the entrance of external molecules, and the drug is no exception to this rule. So, percutaneous drug delivery faces several challenges. Nevertheless, the skin provides an interesting pathway for drug delivery mainly due to its accessible and large surface. Figure 10.4 lists the main challenges and opportunities of transcutaneous drug delivery.

Figure 10.4 Challenges and opportunities of (trans)dermal drug delivery.

10.6 Classification and Formulation Considerations

Considering the barrier function of the *stratum corneum*, several approaches could be considered to improve transdermal drug delivery:

(i) Temporary disruption of the *stratum corneum* using physical or chemical permeability enhancement methods
(ii) Bypassing the *stratum corneum* using physical devices (e.g., microneedles) or appendageal routes of drug delivery
(iii) Modifying the permeability of drug molecules using structural modification methods
(iv) Using novel carriers/formulations to help the permeation of drug molecules with different mechanisms
(v) Extending the drug residence time on the applied site through bioadhesion
(vi) Combinational approaches to achieve synergistic effect

Being so broad and specialized, the explanation of all these approaches and systems can be the subject of one or more independent books. Therefore, in this chapter, we will review the most common ones (Figure 10.5).

10.6.1 Chemical Enhancement Methods

Chemical enhancement strategies are broadly categorized into two approaches. First, the use of chemical permeation enhancers to temporarily disrupt the arrangement of the *stratum corneum*; and second, modifying the structure of the drug to permeate better through the skin.

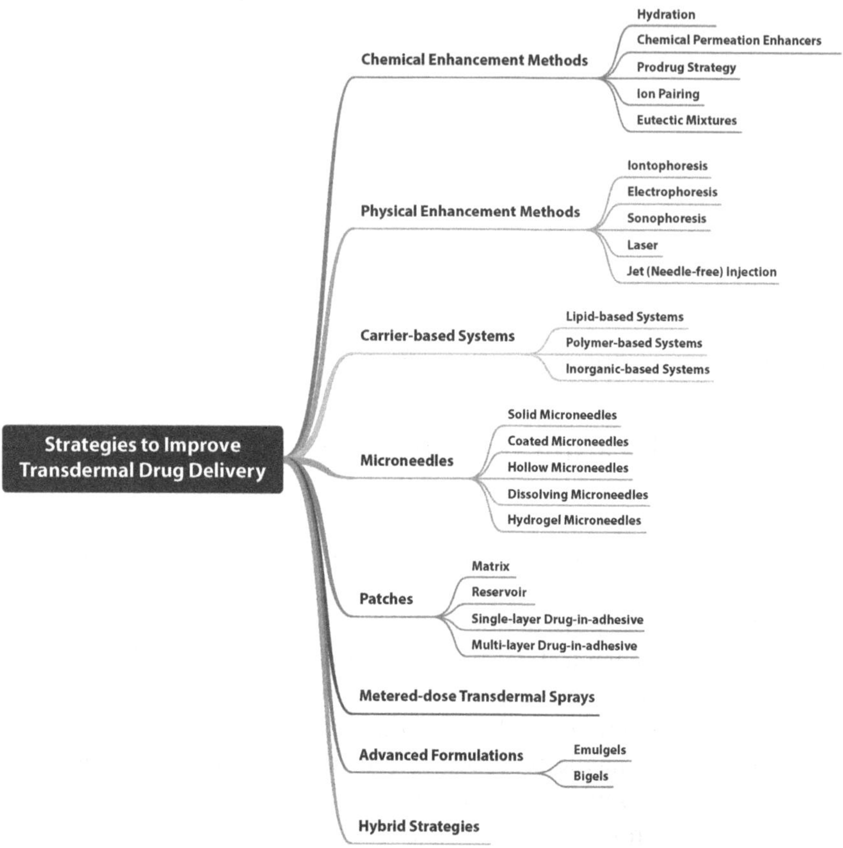

Figure 10.5 Different approaches, strategies and systems to improve drug delivery via the skin at a glance.

10.6.1.1 Using Chemical Permeation Enhancers

To increase skin permeation, it is possible to modify the *stratum corneum* through hydration and using different chemical enhancers. If an occlusive dressing is applied or water loss is prevented (for instance, through the addition of lipid excipients), water content can be maintained which leads to the disruption of the lipid packaging in the *stratum corneum* and increased permeation. Besides, chemical permeation enhancers can also be effective in enhancing skin permeation by changing the integrity of the lipid bilayer and solvent nature of the *stratum corneum*, denaturation the proteins of the skin, and modifying the partition coefficient of drugs. However, safety concerns should be considered when using such components. They should be nontoxic or nonirritant at the concentration used. Examples of chemical permeation enhancers that have been applied successfully are sulphoxides (e.g., DMSO), azones, fatty acids, alcohols (e.g., ethanol, isopropyl alcohol, and propylene glycol), surfactants (e.g., sodium lauryl sulfate (SLS) and Tweens), urea, and also terpenes (such *as* limonene, menthol, cineol and bisabolol) [23, 24].

10.6.1.2 Drug Modification

To improve drug penetration through the *stratum corneum*, the physicochemical properties of the parent drug like oil-water partition coefficient (log*P*) and water solubility can be modified through conjugation or derivatization (e.g., prodrug strategy) [25, 26]. The prodrug passes more efficiently through the skin than the parent drug and is then metabolized and converted to its parent form in the skin layers. Ion-pairing is another permeation enhancement strategy wherein a pair of two oppositely charged ions non-covalently bonded with Coulombic forces; the neutralized electricity provides a modified partition coefficient which allows a more effective skin penetration. Then, the paired molecules are dissociated and the parent drug is released. Another method is the preparation of eutectic systems. They are supramolecular compounds with noncovalent and nonionic bonds interaction of active drug molecules and cocrystal forms. Generally, an eutectic mixture composition contains a hydrogen bond acceptor such as zinc chloride, choline chloride, and tetrabutylammonium bromide as well as a hydrogen bond donor with common chemicals such as urea, citric acid, and oxalic acid, leading to charge delocalization and then reduction in melting point [27, 28].

10.6.2 Physical Enhancement Methods

10.6.2.1 Iontophoresis

Iontophoresis is a physical method in which small electrical currents (usually 0.5-20 mA) are employed to enhance drug permeation through the skin. Anionic drugs are located under the cathode and neutral/cationic drugs are placed underneath the anode. When a low electric current is applied, ions permeate through the skin. So that the cations are repelled from the anode and mobilized to the cathode, and *vice versa* for the anions (Figure 10.6). This strategy is more common for ionizable molecules. However, weak-charged or neutral drugs could be delivered through *electroosmosis*. Several factors influence the effectiveness of drug delivery by iontophoresis including formulation parameters, physicochemical properties of the drug,

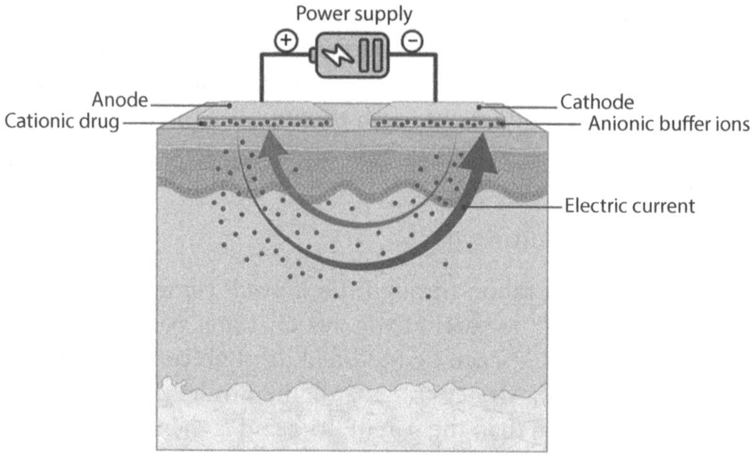

Figure 10.6 Mechanism of iontophoresis to improve skin absorption of drugs (created with BioRender.com).

physiology of the application site, instrumental parameters, and duration of the process. To date, iontophoresis has been explored for improving the delivery of several small-molecule drugs (e.g., NSAIDs, almotriptan, granisetron, and donepezil) as well as insulin. Although it is a powerful physical enhancement method, the commercial use of iontophoresis devices is restricted by some limitations, notable among which is the probability of local irritation and skin damage caused by current, requiring device setup, unwieldy use for patients, and a high cost compared to the other transdermal approaches. Besides, there is a molecular weight cut-off (< 10,000 Da) for the delivery of drugs. Moreover, the electrodes may be corroded in contact with the aqueous gel of the adhesive. Furthermore, it is difficult to predict drug bioavailability [29].

10.6.2.2 Electroporation

Electroporation is another physical permeation enhancement technique, in which a high voltage (normally 10–1000 V) is applied for a very short time (i.e., less than several hundred milliseconds), in a non-thermal process. The pulse waves induce the formation of aqueous micropores in the *stratum corneum,* through which drug molecules can penetrate the deeper layers of skin (Figure 10.7). This method is particularly beneficial for the delivery of macromolecules. Drug delivery using electroporation is influenced by several instrumental parameters (e.g., voltage, number of pulses,

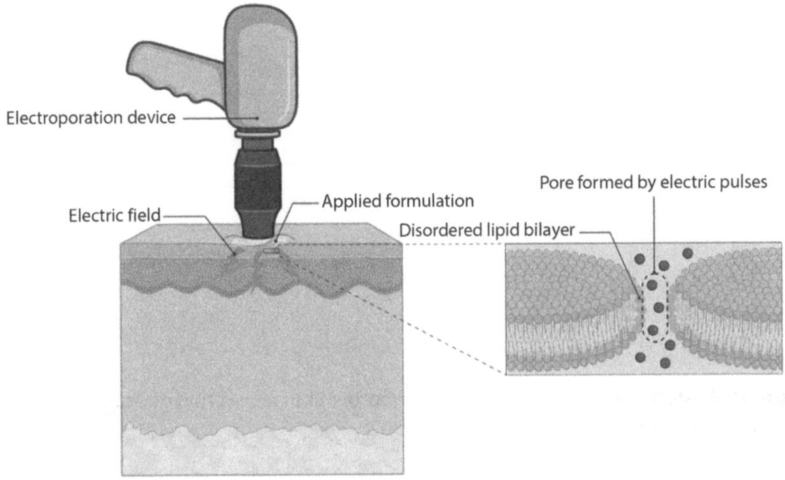

Figure 10.7 Mechanism of electroporation to improve skin absorption of drugs (created with BioRender.com).

and the length of pulse) as well as formulation and drug-related factors. However, similar to the previous case, electroporation is a complicated, expensive, and time-consuming system. Also, there is the possibility of cell damage. Moreover, since the aqueous pores are created in a limited area, the throughput of this technique is low [29].

10.6.2.3 Sonophoresis (Phonophoresis)

Sonophoresis is a physical enhancement method, wherein ultrasound waves are used to alter the arrangement of the lipid bilayer of the *stratum corneum* and enhance the drug permeation. The use of this technique to improve drug delivery across the skin was first proposed in 1954. The drug solution is located on the skin under the probing equipment and, subsequently, a specified frequency of ultrasound is applied. There are two main mechanisms by which ultrasound improves the permeability of drugs including thermal effect and cavitation. When ultrasound is applied, the temperature of the skin increases leading to a high diffusivity. Additionally, cavities and bubbles are formed in the *stratum corneum* (Figure 10.8). Cavitation is assumed the main mechanism involved in drug absorption. This technique is convenient, economical, and relatively non-invasive. The frequencies that are applied in sonophoresis are between 20 KHz to 16 MHz and intensities up to 14 W/cm^2. The permeability of various drugs including hydrophilic and large molecular weight drug molecules could

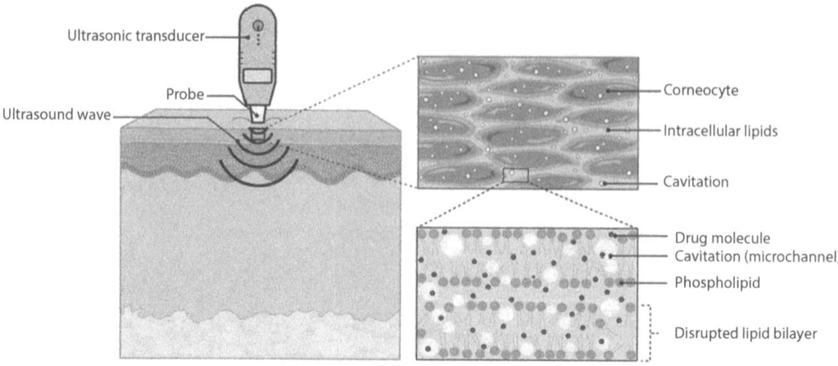

Figure 10.8 Mechanism of sonophoresis to improve skin absorption of drugs (created with BioRender.com).

be increased by this method, especially under low-frequency sonophoresis. However, sonophoresis has some drawbacks, such as being time-consuming, the need for complicated instruments, and the necessity for the skin to be in a healthy situation during the procedure [30].

10.6.2.4 Laser Technology

The ablative laser is an effective physical ablation method to improve drug permeation through the skin. The term laser is the abbreviated form of "light amplification by stimulated emission of radiation". A laser is an intense beam of coherent monochromatic radiation that stimulates photon emission from excited molecules or atoms. Increasing drug penetration by laser allows more effective treatment compared to passive as a result of the shorter duration of the therapy, needing to lower drug dose and lower side effects. In contrast to other techniques, the control of the etched depth is possible by fine-tuning the energy. Laser-activated drug penetration includes two techniques full ablation and fractional ablation. In addition, non-ablative lasers have been shown to be effective in improving drug penetration. In ablative lasers, laser light energy is absorbed by water and skin pigments which is then converted into thermal energy and leads to water excitation and explosive evaporation from the epidermis [31]. Figure 10.9 schematically illustrates the mechanism of laser in improving drug absorption through the skin.

Er:YAG and CO2 lasers are among the most popular laser modalities for drug delivery. They are both fractional lasers that produce vertical microchannel arrays as drug permeation pathways that act as local drug

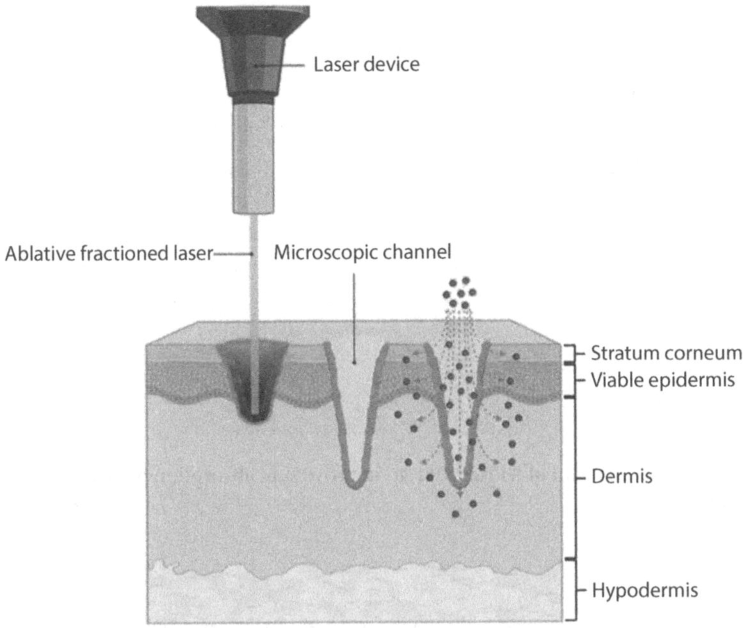

Figure 10.9 Mechanism of fractional ablative laser technology to improve skin absorption of drugs (created with BioRender.com).

reservoirs. Using laser thermal ablation, especially in the case of Er:YAG, the penetration of drugs can be increased more than 100 times. This could be beneficial for the delivery of a variety of drugs, particularly biomacromolecules (e.g., proteins, peptides, DNAs, and vaccines). Compared with traditional laser resurfacing, fractional ablative lasers are considered new devices with enhanced safety and recovery duration. The technical specifications such as laser energy, spot diameter, spot density, pulse duration, and ablation area can be tuned according to the specific needs [32].

10.6.2.5 *Jet Injection (Needle-Free Injection)*

An interesting alternative to conventional injection is jet injection with a mechanical needle-free device that generates high-speed fluid jets of about 120 to 200 m/s. There is a liquid inside the injector which is compressed and then pushed through a tiny orifice (100-300 μm). Subsequently, a liquid flow with a high speed is forced out (Figure 10.10). This force is capable of penetrating the liquid across the *stratum corneum* and into the underneath tissues. Recently, the jet injection has been investigated as a device for vaccination as well as the transdermal delivery of

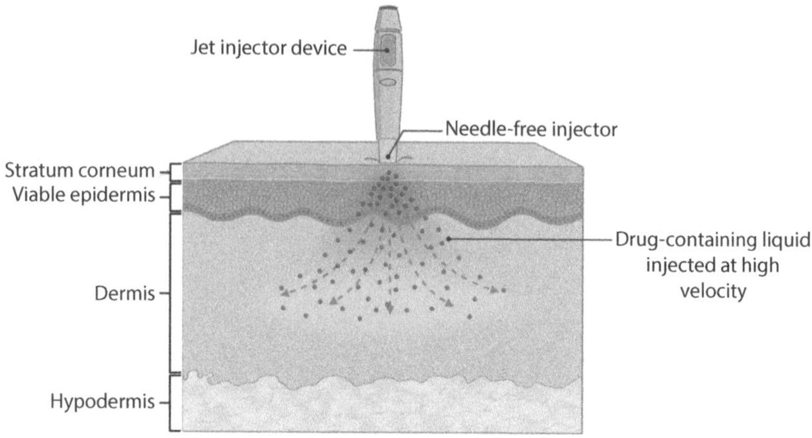

Figure 10.10 Mechanism of jet injection to improve skin absorption of drugs (created with BioRender.com).

macromolecules, such as monoclonal antibodies. Furthermore, jet injection has been recently considered to deliver intact phospholipid vesicles and exosomes. Besides several advantages of jet injectors like the capability for the daily hormone, there are some disadvantages such as cross-contamination due to splash-back on the injection nozzles, especially in mass vaccinations. This can be overcome to some extent using single-use nozzles. However, jet injection faces some drawbacks, such as the probability of bleeding/bruising at the injection site because of the high pressure of the spray. There are two kinds of jet injectors; (i) the traditional ones that deliver a large volume of drug deeper into the skin, and (ii) laser-based injectors that are suitable for superficial targets and deliver smaller volumes [33, 34].

10.6.3 Carrier-Based Systems

Nanocarriers are considered a promising approach for improving transdermal drug delivery with great potential for clinical applications. To date, several types of nanocarriers have been in the spotlight for skin drug delivery purposes. The most common types of these systems are explained below.

10.6.3.1 Lipid-Based Nanocarriers

Lipid-based nanocarriers include vesicular lipid-based systems (e.g., liposomes, ethosomes, transfersomes, and niosomes), lipid nanoparticles (e.g., SLNs and NLCs), and nano/microemulsions.

Liposomes

Liposomes are small bilayer vesicles made up of cholesterol and phospholipids. They consist of an aqueous core surrounded by one or more lipid bilayers. Their size may differ from 25 nm to 2.5 μm. Liposomes may have one or more bilayer membranes, with both size and number of bilayers affecting drug encapsulation efficiency. Based on these two factors, liposomes can be classified into two main categories (see Figure 10.11):

(i) Unilamellar vesicles include small (SUVs), large (LUVs), and giant unilamellar vesicles (GUVs),
(ii) Multilamellar vesicles (MLVs) with onion structure.

Liposomes have been successfully used as an efficient drug delivery system and several formulations have been accepted in the market for the treatment of some diseases. The advantages of liposomes as drug delivery systems include increased efficacy and therapeutic index of the drug as well as drug stability *via* encapsulation, being non-immunogenic for systemic and local administration, and being capable of decorating by ligands

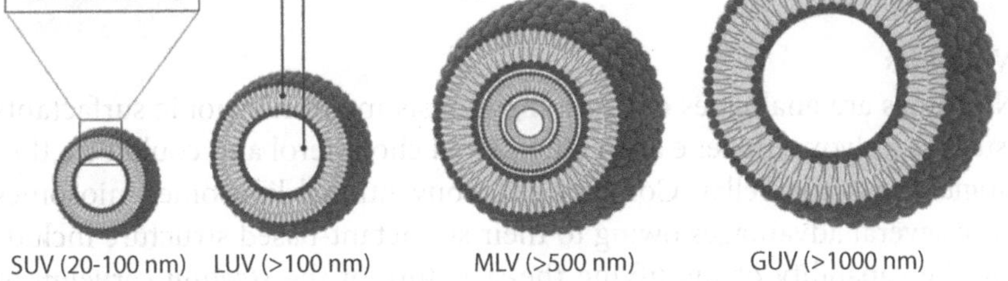

Figure 10.11 Schematic structure of different types of liposomes (created with BioRender.com).

for active targeting. Nevertheless, short half-life and low stability are critical challenges of liposome formulations [35, 36].

Transfersomes

Conventional liposomes cannot penetrate deeply into the skin layers. To deal with this issue, new vesicular carriers have been formulated by introducing new additives known as edge activators to create ultradeformable elastic liposomes named transfersomes. These new systems are the first generation of ultra-deformable vesicles and as one of the most successful carriers for skin delivery. Insertion of an edge activator (such as sodium cholate, sodium deoxycholate, Tweens, and Spans) between the phospholipids, in combination with a small amount (≤ 10%) of ethanol, increases the carrier deformability [37].

Ethosomes

Ethosomes, as the second generation of flexible vesicular nanocarriers, were introduced in the 1990s. They possess a high concentration of ethanol (20–50%) in their structure. Thermal analysis revealed that the presence of ethanol reduces the transition temperature of the phospholipid membranes. So, ethosomes have a higher elasticity and deformability compared to conventional liposomes. Some advantages of ethosomes include enhancement of stability (up to 1 year), high entrapment efficiency, and compatibility with the skin [38].

Transethosomes

To improve drug delivery into deeper skin layers, a novel ultradeformable hybrid system of ethosome and trasfersome (called transethosomes) has been developed which consists of phospholipid, ethanol, and an edge activator. Oleic acid and propylene glycol are proper candidates as permeation enhancers, the combination of which with ethanol and propylene provides a dual penetration enhancement strategy [39, 40].

Niosomes

Niosomes are analogues of liposomes consisting of non-ionic surfactants (such as polyoxyethylene alkyl ethers) and cholesterol and could be either single or multilamellar. Compared to conventional liposomes, niosomes have several advantages owing to their surfactant-based structure including the capability of modifying the structure of the *stratum corneum* to increase permeability, superior stability, and the economical feature of manufacturing. The handshaking method, ether injection, and reverse-phase evaporation are among the most common methods for niosome

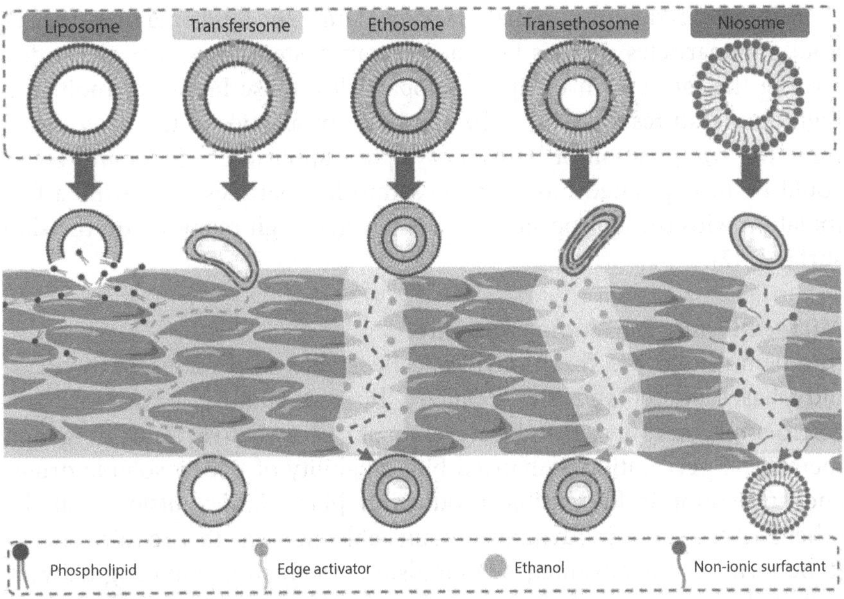

Figure 10.12 Comparative permeation of liposomes, ethosomes, transfersomes, transethosomes and niosomes through the skin (created with BioRender.com).

preparation [39]. Comparative permeation of the above-discussed lipid vesicles through the skin is demonstrated in Figure 10.12.

SLNs and NLCs

Solid lipid nanoparticles (SLNs), developed in the early 1990s, are promising alternatives to liposomes, emulsions, and polymeric nanoparticles. They are composed of lipids that are solid at ambient temperature. The second generation of lipid nanoparticles are nanostructured lipid carriers (NLCs) that contain a mixture of solid and liquid lipids, which collectively remain solid at room and body temperatures. The mean particle size of SLNs and NLCs is about 40 to 1000 nm, based on the composition of the lipid matrix and the production method [41].

Despite several advantages of SLN over the nanocarriers, such as low toxicity, chemical versatility, and the possibility of industrial-scale production, it faces some serious drawbacks such as low encapsulation efficiency and the possibility of drug leakage during storage. Lipids' crystalline structure of SLN limits the incorporation of drugs. To overcome this problem, the liquid lipids are incorporated into the NLC structure which creates significant defects in the crystal lattice, thereby, improving drug loading capacity and reducing the drug leakage. Different lipids may be used in lipid nanoparticles such as fatty acids, waxes, steroids, and triglycerides. During the fabrication of lipid nanoparticles, these lipids are molten at high temperatures (> 80°C). The solubility properties of the drug determine the composition of lipids. For topical application, SLNs and NLCs could be incorporated into suitable semisolid matrices to provide a formulation with the appropriate consistency and high retention on the skin surface [42].

Microemulsions and Nanoemulsions

Microemulsions are thermodynamically stable dispersions of two liquid phases (water and oil) stabilized by surfactant and co-surfactant. Microemulsions offer several features for skin drug delivery including high membrane permeation, enhanced bioavailability of poorly soluble drugs, and an economically feasible production [44]. Nanoemulsions, on the other hand, are thermodynamically unstable systems, that require energy to be formed. The advantages of emulsion systems for skin drug delivery include low toxicity, non-irritancy, controlled/extended drug release, and smooth skin feeling. Multiple emulsions, mainly water-in-oil-in-water (w/o/w), offer some additional advantages, such as the possibility of modified drug release, permeation through the skin by zero-order kinetic, and prolonged deposition in the target area [44]. The schematic structure of the SLNs, NLCs, and nanoemulsions is demonstrated in Figure 10.13.

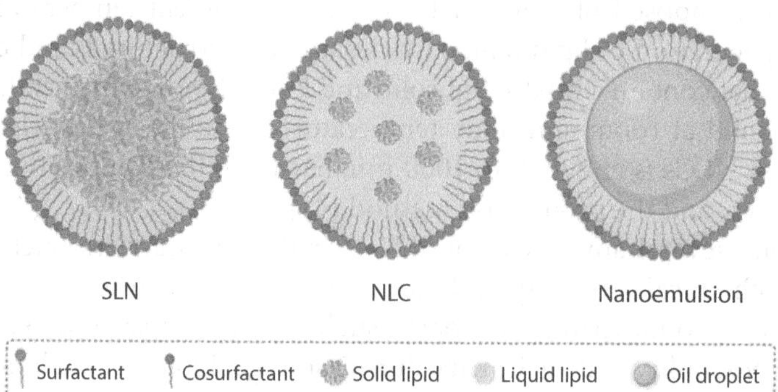

Figure 10.13 Schematic structure of SLN, NLC, and nanoemulsion (created with BioRender.com).

10.6.3.2 Polymer-Based Nanocarriers

Polymeric nanoparticles are potential topical nanocarriers to improve skin penetration. They can mask the undesired physicochemical properties of drugs and protect them from denaturation and degradation. Also, they can provide controlled drug release and reduce side effects. Some of the main subgroups of polymeric nanocarriers used for skin drug delivery are nanocapsules (with a core-shell structure), nanospheres (containing a polymeric matrix), and polymeric micelles (Figure 10.14). Different natural, semi-synthetic, and synthetic polymers can be used in nanoparticle preparation [45]. The physicochemical properties of polymers could be improved using interpenetrating polymeric networks (IPNs). Polymeric nanoparticles can also be dispersed in hydrophilic gels can improve their skin penetration because of the enhanced contact and retention time of drugs on the skin surface [46].

There are several methods to prepare polymeric nanoparticles such as nanoprecipitation (the most common technique), *in-situ* polymerization, emulsification-diffusion-solvent extraction, and salting-out. Poly (Ɛ-caprolactone) as a biocompatible and biodegradable polymer is one of the most commonly used polymers for this purpose. One of the synthetic polymeric structures with a core-shell assembly and spherical branched arrangement is dendrimers, which could be used to enhance topical delivery since they can interact with the skin lipids and denature the keratin proteins [47].

One of the attractive applications of polymeric nanoparticles is their use in sunscreen products since they can be designed in such a way as to provide extended release while minimizing skin penetration [48]. The potential of follicular accumulation is another advantage of polymeric nanoparticles which is seen, for instance, with non-biodegradable polystyrene nanoparticles [49].

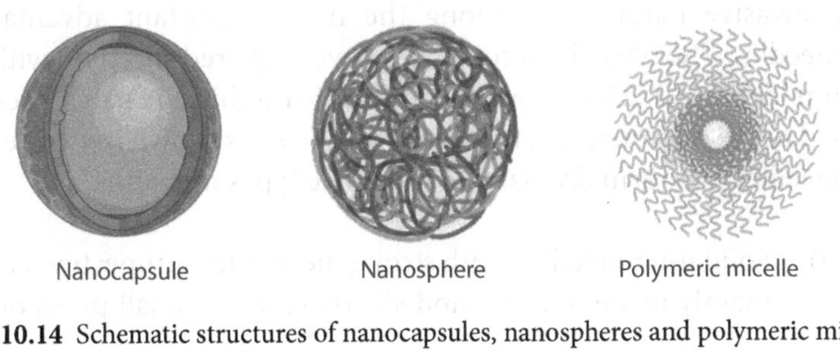

Figure 10.14 Schematic structures of nanocapsules, nanospheres and polymeric micelles (created with BioRender.com).

10.6.3.3 Inorganic Nanocarriers

Carbon-based nanomaterials, especially carbon nanotubes (CNTs), are promising drug delivery systems owing to their unique chemical and physical features. CNTs offer numerous advantages for drug delivery purposes, including high loading capacity, great aspect ratio, and high stability. These outstanding physicochemical properties, along with the exceptional needle shape architecture and very small size (about 10-20 nm) of CNTs, make them an excellent option for improving transdermal drug delivery [50, 51].

Metal and metal oxides are another subclass of inorganic nanomaterials. They are rigid particles composed of metals or metal oxides (e.g., gold, silver, titanium oxide, and silica) that are widely used for different dermal and transdermal applications. In contrast to organic nanomaterials, the inorganic compounds are bio-persistent; namely, they are not disintegrated in the body. So, their safety aspects should be carefully considered. Their advantages such as adjustable physicochemical characteristics, reproducible synthesis procedure, and ease of scale make them appropriate candidates to be used in the pharmaceutical industry. Also, they are suitable candidate for cosmetic application, since they mainly accumulate in the superficial layers of the skin. Inorganic nanoparticles can be used as carriers, rheology modifiers, and even active compounds (e.g., antimicrobials and UV filters) in the formulations. For instance, silver nanoparticles (AgNPs) have recently garnered great attention for transdermal applications due to their antimicrobial and anti-inflammatory properties. It has been proposed that inorganic nanoparticles be integrated with other strategies (e.g., microneedles) to improve their efficiency [52, 53].

10.6.4 Microneedles

Microneedle is a unique platform for transdermal drug delivery that bypasses the *stratum corneum.* Ease of use, capability of self-administration, providing improved bioavailability and rapid onset of action, and minimally invasive nature, are among the most important advantages of microneedles. The typical microneedles have a tapered sharp tip with 150–1500 μm length, 50–250 μm width, and 1–25 μm thick. A patch backing is attached to the microneedle array for ease of use. As shown in Figure 10.15, microneedles are normally classified into five types including:

(i) Solid microneedles, with strong mechanical properties, are mainly made of metal and silicon creating small pores on

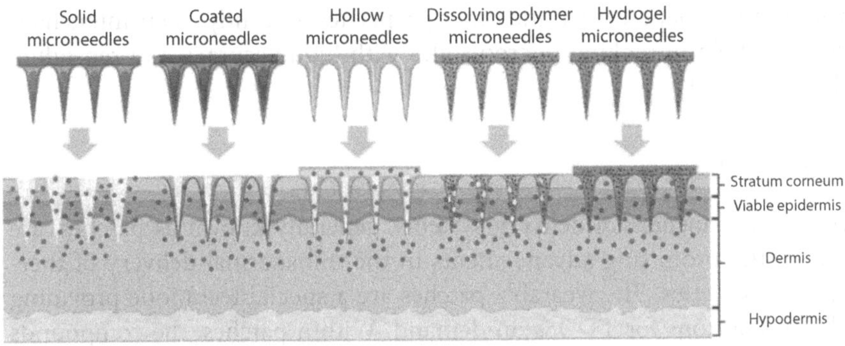

Figure 10.15 Different types of microneedles (created with BioRender.com).

the skin surface but do not contain drugs and excipients. After microneedle insertion, the drug should be applied.

(ii) Coated microneedles contain the drug in a water-soluble matrix coating, so it is delivered simultaneously with the application. However, this system has dose limitations and is not suitable for a large dose of the drug.

(iii) Hollow microneedles which contain an empty region inside, through which the drug can directly enter the skin.

(iv) Dissolving microneedles which are fabricated from water-soluble or biodegradable materials blended with the drug. They rapidly dissolve or disintegrate upon contact with the skin and no considerable waste is produced after application. However, like coated microneedles, this system has limitations in the delivery of large doses.

(v) Hydrogel microneedles consist of swellable polymers that allow drugs to be delivered slowly. In this type of microneedle, the drug can exist in the entire area such as the tip and base of the microneedle as well as the patch backing. This system is appropriate for the delivery of large doses; however, its drawback is the slow rate of delivery which makes the patch-wearing time long.

Microneedles can be made up of various materials like silicon, metals, synthetic polymers, glass, ceramics, and more recently, carbohydrates and proteins. The selection of material for microneedle fabrication is based on the type of microneedles, according to solubility for coated and dissolving microneedles and mechanical strength for non-dissolving ones [56]. Various manufacturing techniques are employed to produce microneedles with unique shapes, sizes, and other properties. Recently, 3D printers have come to help in making microneedles with high accuracy and reproducibility [54, 55].

10.6.5 Patches

Adhesive transdermal delivery systems, generally known as patches, are one of the promising advancements in the transdermal delivery of therapeutic moieties. The wearable patches are a special technique providing ability solutions for TDDSs on demand. Within patches, the compounds are either dissolved in a gel or liquid-based reservoir or blended with a solid polymer matrix [57].

Being patient-friendly, non-invasive, and self-administrable with no risk of infection are among the merits of these formulations. However, the application of patches is associated with the potential risk of skin sensitization or irritation for some patients, the possibility of imperfect skin adhesion, and a higher cost compared to conventional topical forms [58, 74].

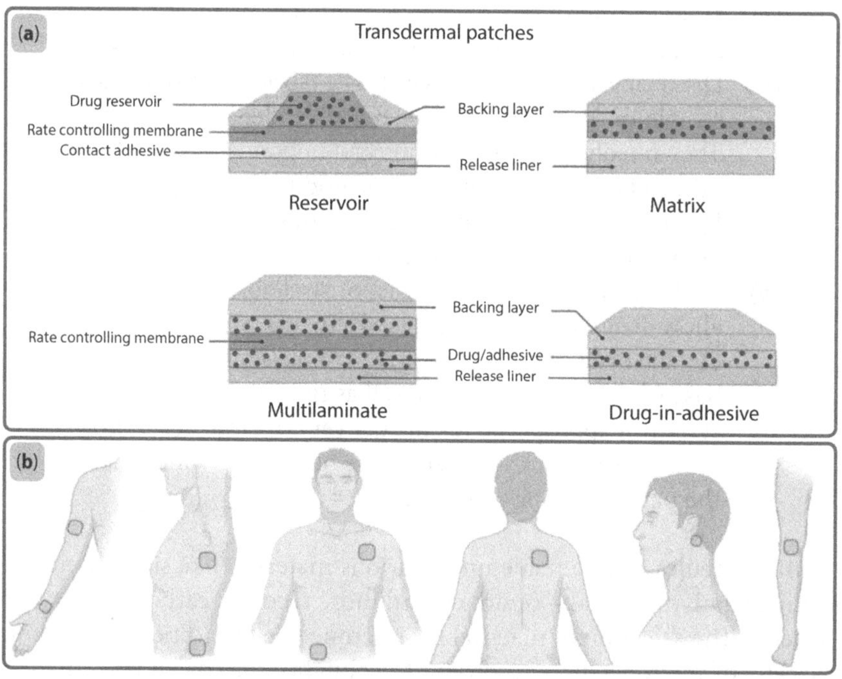

Figure 10.16 (a) Different types of transdermal patches; (b) examples of transdermal patch application sites (created with BioRender.com).

Different types of transdermal patch formulations include single-layer drug-in-adhesive, multi-layer drug-in-adhesive, reservoir, and matrix (Figure 10.16). The drug release rate from the patch is controlled by the surface area of the patch in contact with the skin, the dose size, and the usage frequency [59].

10.6.6 Metered-Dose Transdermal Sprays (MDTSs)

Metered-dose transdermal spray (MTDS) is a powerful alternative to patches. It is a single-phase aerosol composed of polymer, permeation enhancer, and therapeutic agent. Following spray application, a fast-drying non-occlusive film is formed. This system is user-friendly and self-administrable [60].

10.6.7 Advanced Formulations

10.6.7.1 Emulgels

Emulgels (also called gelled emulsions or creamed gels) are an advanced form of topical formulations, consisting of a dispersed liquid phase in a structured network, which combine the advantages of gels and emulsions. So, they can encapsulate both lipophilic and hydrophilic drugs and conquer inadequate solubility, stability, and absorption issues. Also, they can provide dual-controlled release. Emulgels have a non-greasy and easily spreadable nature with a transparent appearance. Compared to other types of transdermal dosage forms (e.g., solutions, emulsions, ointments, and patches), emulgels are considered superior formulations [61].

10.6.7.2 Bigels

Since the dispersed (internal) and continuous (external) phases of the emulgels are inconsistent, they may face instability challenges. To overcome this, a novel solid-like system, known as bigels, has been developed, both phase of which is in the gel state. In fact, bigels are the combination of hydrogels and organogels [62].

10.7 Pharmacokinetics

Dermal pharmacokinetic research is vital in the selection of drug candidates for further assessment, evaluation of formulation efficiency, and bioequivalence studies. The pharmacokinetics of topically applied drugs depend on not only their absorption through the skin but also the time course of administration [63]. As previously stated, despite the widely available surface area, relatively low enzymatic degradation, and the possibility of long-term use of the formulation on the skin, the dermal permeability of drugs is severely impeded by the *stratum corneum* [64]. Since topical formulations deliver their drugs exactly on or close to the intended site of action, investigation of drug uptake into and elimination from the *stratum corneum* may provide valuable pharmacokinetic data. Even in cases where the *stratum corneum* is not the target site of pharmacological effect, the drug must still cross this barrier. So, it is of crucial importance to estimate the rate and extent of drugs permeating into/across the *stratum corneum* following the application of topical products [65]. Moreover, it is worth noting that, when the *stratum corneum* concentration-time curves of two formulations are comparable, they are presumably bioequivalent. So, assessment of drug content in different layers of skin can provide supportive information on topical products' bioequivalence [63, 65]. There are various methods to evaluate penetration/permeation of drugs through

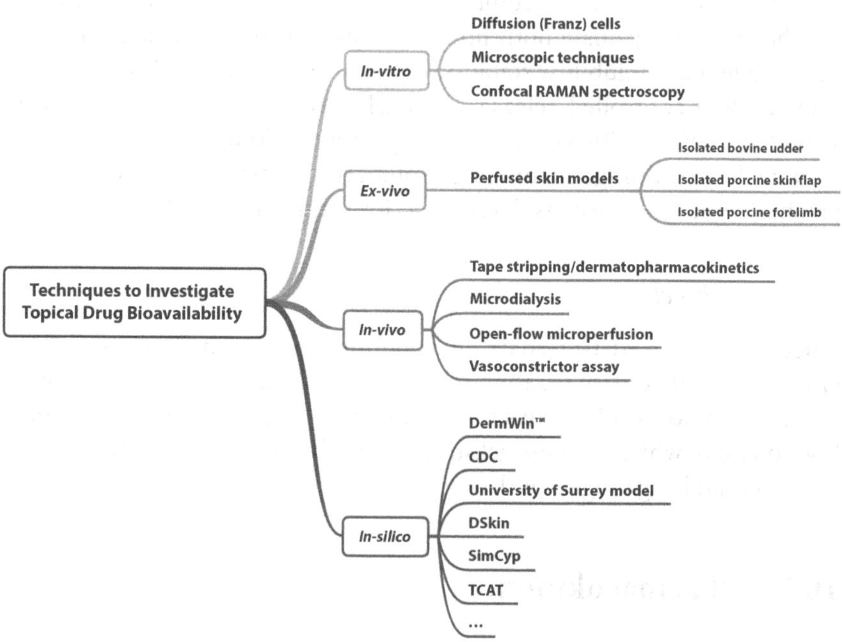

Figure 10.17 Techniques to assess topical bioavailability of drugs (date obtained from [66, 67]).

the skin including *in vitro*, *in vivo*, and *in silico* approaches (Figure 10.17). Here, we focus on common *in vivo* techniques (applied to humans) to evaluate dermal pharmacokinetics.

10.7.1 Dermatopharmacokinetic (Tape Stripping) Approach

The concentration of drugs within the skin layers is estimated by the dermatopharmacokinetic (DPK) method -akin to plasma/urine pharmacokinetic studies when drugs are administered systematically or orally [65]. Accordingly, it is assumed that (i) the *stratum corneum* is the rate-limiting barrier for drug absorption, (ii) there is a direct relationship between the drug concentration at the *stratum corneum* and diffusion of the drug into the dermis, and (iii) the total drug concentration in the skin layers (i.e., in *stratum corneum*, epidermis, and dermis) explains the dermatological efficacy of the drug. Dermatopharmacokinetics parameters include C_{max} (the highest concentration of drug in the skin layers), T_{max} (the time at which C_{max} is achieved), and AUC (area under the curve of concentration-time) [68]. How to calculate these parameters is explained in detail elsewhere [65].

Although dermatopharmacokinetic is a general term that encompasses any measurement of the concentration of a drug in the skin, in many articles, it is limited to describing the tape-stripping technique [46]. Tape stripping, as a straightforward and minimally invasive procedure, is the most often used technique to quantify the drug in the *stratum corneum*, which is applicable for both *in vitro* and *in vivo* permeation experiments. First, an adhesive tape strip is uniformly stuck on the skin surface by a roller, followed by peeling off. In this way, the layers are removed one after the other. To extract the drug from adhesives, the collected tape strips are subjected to a solvent, in which the drug is highly soluble. It is followed by centrifugation and quantification of the drug with a validated analytical method (e.g., HPLC) [68]. Figure 10.18 shows a schematic illustration of the tape-stripping technique.

There are several limitations associated with this technique such as the interference of adhesive, inter-individual and site-related variability of the strip application times required to remove the layer, the different number of *stratum corneum* layers in various species, and differences between healthy and diseased skins. So, the tape stripping procedure should be thoroughly validated before the main experiments. Besides, the accumulation of the drug in the skin furrows and hair follicles may mislead the analyst. So, the simultaneous use of more than one technique is usually preferred [68].

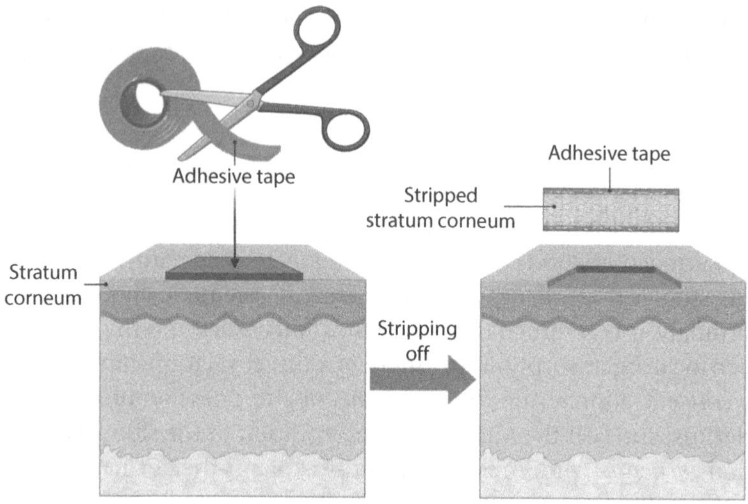

Figure 10.18 Schematic illustration of tape stripping technique (see the text for more information) (created with BioRender.com).

Using emerging techniques that measure the drug that has passed through the *stratum corneum* can be helpful.

10.7.2 Microdialysis

Microdialysis is a probe-based sampling method to directly determine the pharmacokinetics of drugs in the skin. In this technique, an ultrathin semipermeable hollow fiber (called the microdialysis probe) is embedded within the dermis. One end of the probe is connected to a syringe that pumps a tissue-compatible sterile buffer (perfusate) at a very low rate and the samples are collected from the other end (Figure 10.19). The probe mimics the role of an artificial vessel and exchanges small molecules between the buffer and the surrounding tissue based on simple diffusion laws. The concentration of unbound pharmacologically active drugs in the dermis is reflected by the quantification of the recovered sample. Similar to the tape stripping approach, it is possible to test both test and reference samples at the same time on the same individuals. Due to providing valuable chronological pharmacokinetic data, this technique is rapidly growing in the field of drug penetration studies. Also, this technique could be employed for sampling from perturbed or diseased skin. Nevertheless,

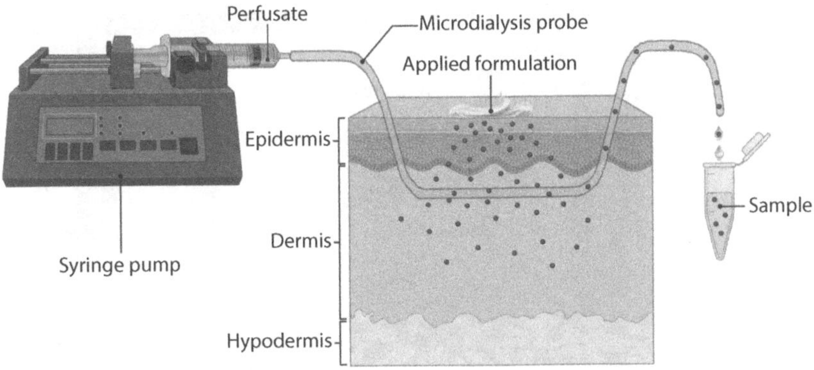

Figure 10.19 Schematic illustration of skin microdialysis technique (created with BioRender.com).

there are some challenges when this technique is used for the sampling of highly lipophilic drugs and those with high protein-bound fraction (i.e., lipophilicity and size limit). Moreover, microdialysis is more invasive than tape stripping [69].

10.7.3 Dermal Open-Flow Microperfusion (dOFM)

Dermal open-flow microperfusion is another probe-based sampling method that circumvents the restrictions of the microdialysis technique. The dOFM device (developed by HEALTH—Joanneum Research GmbH, Graz, Austria) is mainly composed of a sterile membrane-free (braided) probe and a wearable pump. Being fenestrated, this type of probe expands the range of analytes to be entered (i.e., all molecules in the interstitial fluid including highly lipophilic and large particles can also permeate). Thanks to the presence of a wearable multi-channel pump, it is possible to parallel sampling over an extended period ($\geq$ 24 hours). When the probe is inserted intradermally, the sterile perfusate is continuously flowed by the pump (push). To control the sampling rate, the outlet leads to a sampling vial which is evacuated by the pump (pull). Such a push-pull system provides a volume-controlled sampling [70, 71]. Figure 10.20 schematically illustrates this technique.

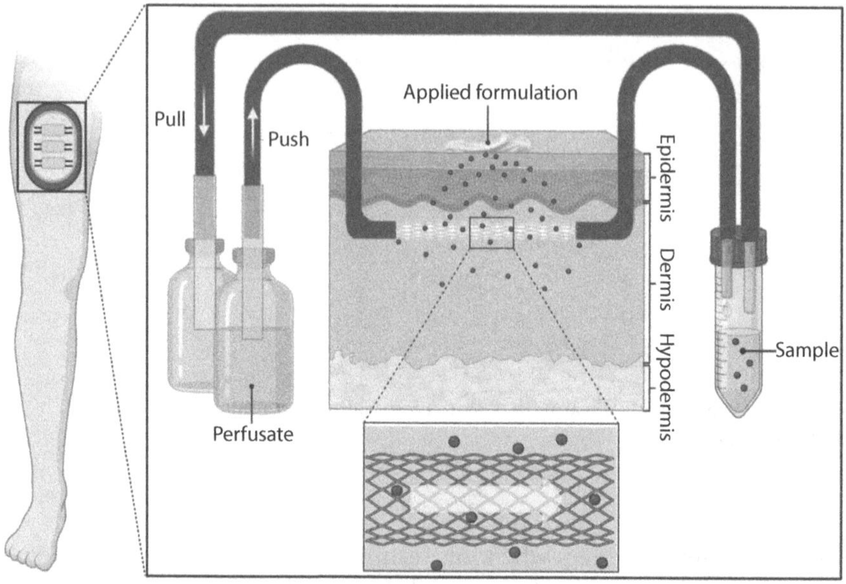

Figure 10.20 Schematic illustration of dermal open-flow microperfusion (dOFM) technique (created with BioRender.com).

10.7.4 Vasoconstrictor (Skin Blanching) Assay

Vasoconstrictor assay is a corticosteroid-specific technique determining the bioavailability based on the skin blanching effect (as a pharmacodynamic endpoint). Once topical corticosteroids are applied, local vasoconstriction occurs which can be measured by different methods including visual inspection, chromametry, and digital image analysis. Visual inspection includes an examination by the naked eye and does not require technical equipment [67].

There is a 0–4 score scale introduced by Barry and Woodford [72], based on which the skin blanching is classified (0: no blanching, 1: faint, 2: faint-moderate, 3: moderate-strong, and 4: strong-intense skin blanching). In chromametry, the reflection of white light is quantified with three parameters of red-green, yellow-blue, and light-dark. In digital image analysis, similarly, a 0.5 cm^2 skin site image at 300 dots per inch is captured and analyzed by the same three parameters [72]. A schematic representation of the vasoconstrictor assay is illustrated in Figure 10.21.

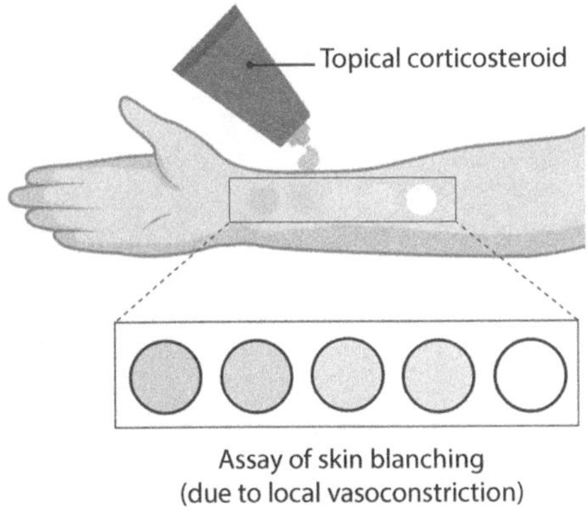

Figure 10.21 Vasoconstriction technique (created with BioRender.com).

10.7.5 Skin Absorption Kinetics

To be transdermally absorbed, a drug molecule must penetrate different layers of the skin and reach systemic blood circulation [6]. The stepwise process of percutaneous drug absorption is schematically illustrated in Figure 10.22.

Figure 10.23 shows a cross-sectional view of the percutaneous diffusion pathway for drug molecules. Based on Fick's law of diffusion, the transport

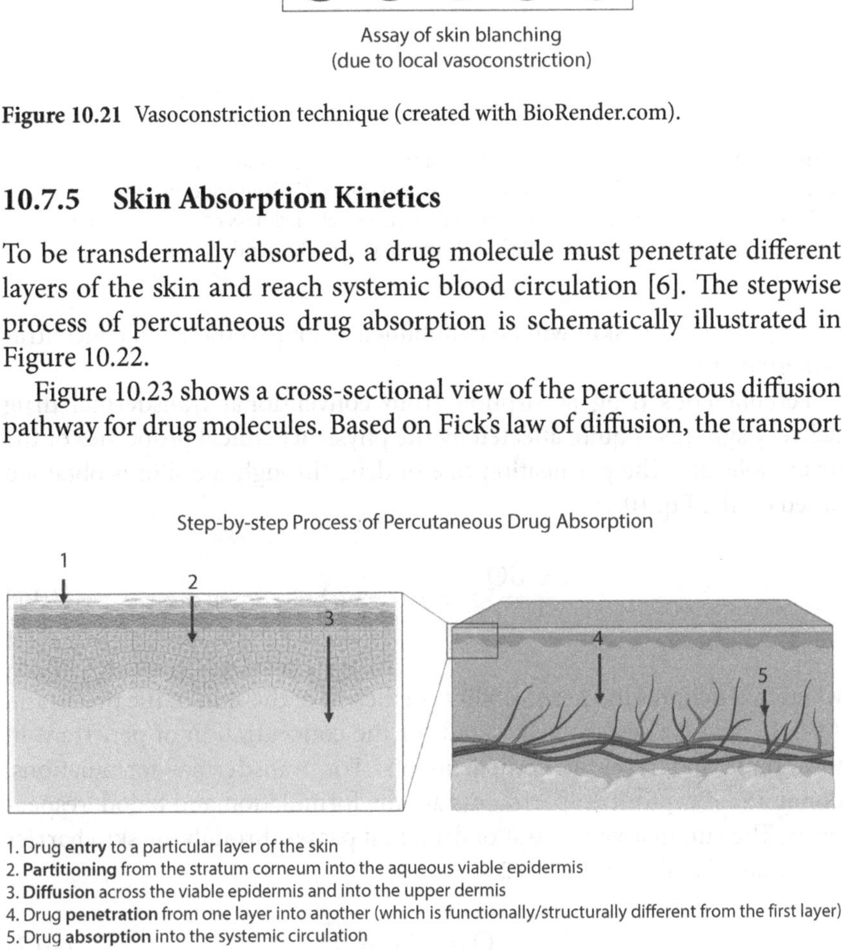

Figure 10.22 The stepwise process of percutaneous drug absorption (data obtained from Ref. [6]) (created with BioRender.com).

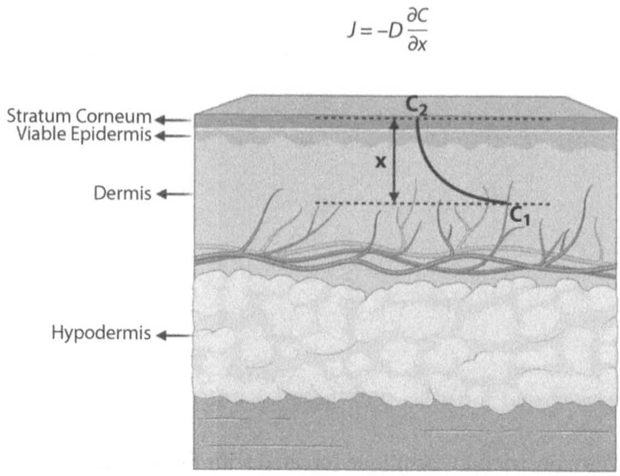

Figure 10.23 The flux across the skin from a transdermal formulation. The equation represents Fick's law of diffusion where J: molecular flux, C: drug concentration (C2 > C1), D: diffusion coefficient, x: the cross-sectional thickness of diffusion, and t: the diffusion time (Inspired by Ref. [73]) (created with BioRender.com).

of drugs across the skin will be maintained as long as there is a concentration gradient.

Percutaneous drug absorption from conventional transdermal drug delivery systems is quite affected by the physicochemical properties of the drug molecule. The permeation rate of drug through the skin is obtained based on the Eq. 10.1:

$$\frac{dQ}{dt} = P(C_d - C_r) \quad (10.1)$$

wherein P denotes the permeability coefficient of the skin to the penetrant. Also, C_d and C_r, respectively, stand for the concentration of penetrant in the donor and receptor compartments. For transdermal formulations, donor and receptor compartments are the formulation and blood, respectively. The cumulative amount of drug that passes through the skin barrier (Q) is calculated from Eq. 10.2:

$$Q = PAC_d t \quad (10.2)$$

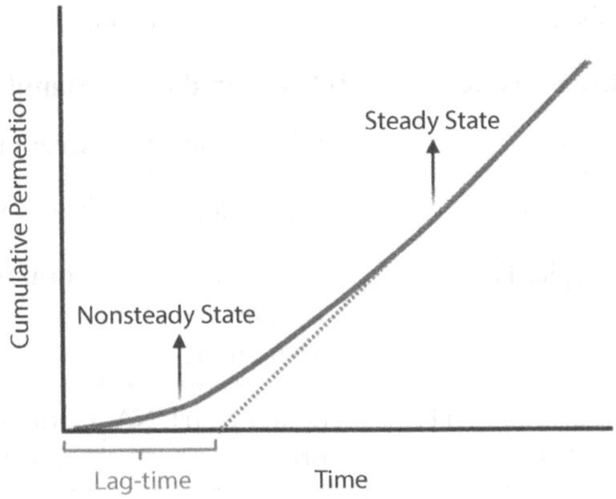

Figure 10.24 A typical plot obtained from skin absorption studies (lag-time is the time required to reach the steady state).

where A is the effective surface area at a given time (t). Accordingly, the permeability coefficient (*P*) can be obtained from the steady-state (linear zone) slope of the cumulative concentration of the penetrant-time curve (see Figure 10.24) [6]. To allow optimum passive diffusion through the skin, transdermal formulations are saturated with sufficient amounts of drug to provide maximum thermodynamic driving force, and thereby, maximum permeation [6].

10.8 Products in the Market

Skin formulations have been in the market for a long time as an indispensable component. A skin patch called Transderm Scōp® containing scopolamine was first introduced to the market in 1979 to treat motion sickness. From 1991 to 2004, transdermal contraceptives entered the market. From 2005 onwards, other skin formulations were developed, such as Zecuity® and Emsam®, which contain sumatriptan and selegiline, respectively [23]. In Table 10.1, some of the skin products that have been launched on the market or under clinical trials are presented.

Table 10.1 List of some of dermal and transdermal products.

Dosage form	Brand name	Active ingredient	Manufacturer
Microneedle	V-Go®	Basal-bolus insulin	Valeritas Inc., USA
	Qtrypta®	Zolmitriptan	Zosano pharma
	CorplexTM®	Donepezil & Corplex Memantine	Corium International Inc.USA
	MicroCor® PTH (1-34)	Teriparatide [rDNA origin]	Corium International Inc. USA
	JewelPUMPTM®	Insulin	Debiotech, Switzerland
	Intanza®	Split Influenza virus	Sanofi Pasteur
Metered Dose Transdermal Spray (MDTS)	Evamist®	Estradiol	KV Pharmaceuticals (from 2007 to 2014); Perrigo (from 2014)
	Recuvyra®	Fentanyl	Elanco Pvt. Ltd.
	Nicotine MDTS®	Nicotine	Acrux Pvt. Ltd. (Phase-I trial completed)
	Nestorone (NES) MDTS®	Nestorone	Acrux Pvt. Ltd. (Phase-I trial completed) partnered with Population Council and Fempharm Pvt. Ltd
	Ellavie®	Estradiol	Aspen Pharmacare and Vifor Pharma
	Lenzetto®	Estradiol hemihydrate	Gedeon Richter
	Axiron®	Testosterone	Acrux Pharma Pty Ltd.

(*Continued*)

Table 10.1 List of some of dermal and transdermal products. (*Continued*)

Dosage form	Brand name	Active ingredient	Manufacturer
Transdermal patch	Neupro®	Rotigotine	Schwarz Pharma
	Minivelle®	Estradiol	Noven Pharmaceuticals, Inc.
	Butrans®	Buprenorphine	Purdue Pharma L.P.
	Sancuso®	Granisetron	ProStrakan Group plc
	Twirla®	Ethinyl estradiol and levonorgestrel	Agile Therapeutics, Inc.
	Synera®	Lidocaine and tetracaine	ZARS, Inc.
	Zecuity®	Migraine	Teva Pharmaceuticals USA, Inc.
	Salonpas®	10% Methyl Salicylate & 3% 1-Menthol	Hisamitsu Pharmaceutical Co., Inc.
	Oxytrol®	Oxybutynin	Merck
	Exelon®	Rivastigmine	Novartis
	Combipatch®	Estradiol-norethindrone	Novartis
	Adlarity®	Donepezil	Corium
Dermal PowderJect	Zingo®	Lidocaine hydrochloride monohydrate	Anesiva, Inc.
Topical Gel and Hydrogel	Gelnique®	Oxybutynin chloride 10 % gel	Watson Pharmaceuticals, Inc.
	Divigel®	Estradiol	Upsher-Smith Laboratories, Inc.
	Astero®	Lidocaine	Gensco Pharma

(*Continued*)

Table 10.1 List of some of dermal and transdermal products. (*Continued*)

Dosage form	Brand name	Active ingredient	Manufacturer
	Collagen Hydrogel Mask®	Collagen and Sodium Hyaluronate	Skin Republic
	Flector® Patch	Diclofenac	IBSA Farmaceutici Italia
	Lidoderm®	lidocaine	Teikoku Pharma USA
	Voltadol®	Diclofenac	GlaxoSmithKline
Topical Emulsion	Estrasorb®	Estradiol	Novavax, Inc. Columbia, MD 21046
Topical Emulgel	Voltaren®	Diclofenac sodium	Novartis
Topical Cream	Opzelura®	Ruxolitinib	Incyte Corporation
Auto-Injector	EPIPEN® and EPIPEN JR®	Epinephrine	Viatris
	Tolts®	Ixekizumab	Eli Lilly
	Dupixent®	Dupilumab	Regeneron Pharmaceuticals, Inc.
	Cyltezo®	Adalimumab-adbm	Boehringer Ingelheim Pharmaceuticals, Inc.

References

1. Lai-Cheong, J.E. and McGrath, J.A., Structure and function of skin, hair and nails. *Medicine*, 45, 6, 347–51, 2017.
2. Chen, Y., Feng, X., Meng, S., Site-specific drug delivery in the skin for the localized treatment of skin diseases. *Expert Opin. Drug Deliv.*, 16, 8, 847–67, 2019.
3. Garg, U. and Jain, K., Dermal and transdermal drug delivery through vesicles and particles: Preparation and applications. *Adv. Pharm. Bull.*, 12, 1, 45, 2022.
4. Benson, H.A., Grice, J.E., Mohammed, Y., Namjoshi, S., Roberts, M.S., Topical and transdermal drug delivery: from simple potions to smart technologies. *Curr. Drug Deliv.*, 16, 5, 444–60, 2019.

5. Bonifant, H. and Holloway, S., A review of the effects of ageing on skin integrity and wound healing. *Br. J. Community Nurs.*, 24, Sup3, S28–33, 2019.
6. Zaid Alkilani, A., McCrudden, M.T., Donnelly, R.F., Transdermal drug delivery: Innovative pharmaceutical developments based on disruption of the barrier properties of the stratum corneum. *Pharmaceutics*, 7, 4, 438–70, 2015.
7. Moniz, T., Costa Lima, S.A., Reis, S., Human skin models: From healthy to disease-mimetic systems; characteristics and applications. *Br. J. Pharmacol.*, 177, 19, 4314–29, 2020.
8. Yousef, H., Alhajj, M., Sharma, S., *Anatomy, Skin (Integument), Epidermis*, StatPearls Publishing, Treasure Island, 2020, Ncbi nlm nih gov.
9. Nagy N.A., De Haas A.M., Geijtenbeek T.B., Van Ree R., Tas S.W., Van Kooyk Y., De Jong E.C., Therapeutic liposomal vaccines for dendritic cell activation or tolerance. *Front. Immunol.*, 12, 674048, 2021.
10. Gu, Y., Bian, Q., Zhou, Y., Huang, Q., Gao, J., Hair follicle-targeting drug delivery strategies for the management of hair follicle-associated disorders. *Asian J. Pharm. Sci.*, 17, 3, 333–52, 2022.
11. Ahmadi-Ashtiani, H.R., Bishe, P., Baldisserotto, A., Buso, P., Manfredini, S., Vertuani, S., Stem cells as a target for the delivery of active molecules to skin by topical administration. *Int. J. Mol. Sci.*, 21, 6, 2251, 2020.
12. Vestita, M., Tedeschi, P., Bonamonte, D., Anatomy and Physiology of the Skin, in: *Textbook of Plastic and Reconstructive Surgery*, M. Maruccia and G. Giudice (Eds.), pp. 3–13, Springer International Publishing, Cham, 2022, Available from: https://link.springer.com/10.1007/978-3-030-82335-1_1.
13. Javadzadeh, Y. and Bahari, L.A., Therapeutic nanostructures for dermal and transdermal drug delivery, in: *Nano-and Microscale Drug Delivery Systems*, Elsevier, pp. 131–46, 2017, Available from: https://www.sciencedirect.com/science/article/pii/B978032352727900008X.
14. N'Da, D.D., Prodrug strategies for enhancing the percutaneous absorption of drugs. *Molecules*, 19, 12, 20780–807, 2014.
15. Cheng, T., Tai, Z., Shen, M., Li, Y., Yu, J., Wang, J. *et al.*, Advance and challenges in the treatment of skin diseases with the transdermal drug delivery system. *Pharmaceutics*, 15, 8, 2165, 2023.
16. Singh, I. and Morris, A.P., Performance of transdermal therapeutic systems: Effects of biological factors. *Int. J. Pharm. Investig.*, 1, 1, 4, 2011.
17. Holmgaard, R. and Nielsen, J.B., Dermal absorption of pesticides: evaluation of variability and prevention, Danish Environmental Protection Agency Copenhagen, Denmark, 2009, Available from: http://www2.mst.dk/udgiv/publications/2009/978-87-7052-980-8/pdf/978-87-7052-981-5.pdf.
18. Singh, S., Dash, A.K., Agrawal, S., Semisolid Dosage Forms, in: *Pharmaceutics*, pp. 341–91, Elsevier, London, UK, 2024, Available from: https://www.sciencedirect.com/science/article/pii/B9780323997966000199.

19. Harding, C.R., The stratum corneum: structure and function in health and disease. *Dermatol. Ther.*, 17, s1, 6–15, 2004 Jan.
20. Konda, S., Meier-Davis, S.R., Cayme, B., Shudo, J., Maibach, H.I., Age-related percutaneous penetration part 1: skin factors. *Skin Ther. Lett.*, 17, 5, 1–5, 2012.
21. Moser, K., Kriwet, K., Kalia, Y.N., Guy, R.H., Enhanced skin permeation of a lipophilic drug using supersaturated formulations. *J. Controlled Release*, 73, 2–3, 245–53, 2001.
22. Cilurzo, F., Casiraghi, A., Selmin, F., Minghetti, P., Supersaturation as a tool for skin penetration enhancement. *Curr. Pharm. Des.*, 21, 20, 2733–44, 2015.
23. Ramadon, D., McCrudden, M.T.C., Courtenay, A.J., Donnelly, R.F., Enhancement strategies for transdermal drug delivery systems: current trends and applications. *Drug Deliv. Transl. Res.*, 12, 4, 758–91, 2022.
24. Nakamura, E., Iwase, H., Arima-Osonoi, H., Sakuragi, M., Effect of water content on stratum corneum penetration mechanism of W/O type micro-emulsions. *RSC Adv.*, 13, 26, 17742–9, 2023.
25. Tabatabaei, M.S., Kobarfard, F., Aboofazeli, R., Ramezanpour, S., Moghimi, H.R., Optimized cutaneous delivery of anti-wrinkle dipeptide KT via molecular modification: Preformulation, permeation, and the importance of conjugate chain length. *BioImpacts*, 2023, Available from: https://bi.tbzmed.ac.ir/Inpress/bi-27558.pdf.
26. Mortazavi, S., Kobarfard, F., Manafi, A., Maibach, H.I., Moghimi, H.R., Terpene Conjugation: A Novel Approach for Topical Peptide Delivery. *J. Cosmet. Sci.*, 72, 5, 547, 2021.
27. Wang, J., Li, M., Duan, L., Lin, Y., Cui, X., Yang, Y. *et al.*, Deep Eutectic Systems as Novel Vehicles for Assisting Drug Transdermal Delivery. *Pharmaceutics*, 14, 11, 2265, 2022.
28. Xu, Y., Zhao, M., Cao, J., Fang, T., Zhang, J., Zhen, Y. *et al.*, Applications and recent advances in transdermal drug delivery systems for the treatment of rheumatoid arthritis. *Acta Pharm. Sin. B.*, 29, 4417–4441, 2023, Available from: https://www.sciencedirect.com/science/article/pii/S2211383523001703.
29. Ita, K., Perspectives on transdermal electroporation. *Pharmaceutics*, 8, 1, 9, 2016.
30. Polat, B.E., Hart, D., Langer, R., Blankschtein, D., Ultrasound-mediated transdermal drug delivery: mechanisms, scope, and emerging trends. *J. Controlled Release*, 152, 3, 330–48, 2011.
31. Hsiao, C.Y., Yang, S.C., Alalaiwe, A., Fang, J.Y., Laser ablation and topical drug delivery: a review of recent advances. *Expert Opin. Drug Deliv.*, 16, 9, 937–52, 2019.
32. Sklar, L.R., Burnett, C.T., Waibel, J.S., Moy, R.L., Ozog, D.M., Laser assisted drug delivery: A review of an evolving technology. *Lasers Surg. Med.*, 46, 4, 249–62, 2014.

33. Schoppink, J. and Rivas, D.F., Jet injectors: Perspectives for small volume delivery with lasers. *Adv. Drug Deliv. Rev.*, 182, 114109, 2022.
34. Wang, R., Bian, Q., Xu, Y., Xu, D., Gao, J., Recent advances in mechanical force-assisted transdermal delivery of macromolecular drugs. *Int. J. Pharm.*, 602, 120598, 2021.
35. Akbarzadeh, A., Rezaei-Sadabady, R., Davaran, S., Joo, S.W., Zarghami, N., Hanifehpour, Y. *et al.*, Liposome: classification, preparation, and applications. *Nanoscale Res. Lett.*, 8, 1, 102, 2013.
36. Nsairat, H., Khater, D., Sayed, U., Odeh, F., Al Bawab, A., Alshaer, W., Liposomes: Structure, composition, types, and clinical applications. *Heliyon*, 8, e09394, 2022, Available from: https://www.cell.com/heliyon/pdf/S2405-8440(22)00682-X.pdf.
37. Lai, F., Caddeo, C., Manca, M.L., Manconi, M., Sinico, C., Fadda, A.M., What's new in the field of phospholipid vesicular nanocarriers for skin drug delivery. *Int. J. Pharm.*, 583, 119398, 2020.
38. Pilch, E. and Musia, W., Liposomes with an ethanol fraction as an application for drug delivery. *Int. J. Mol. Sci.*, 19, 12, 3806, 2018.
39. Sudhakar, K., Fuloria, S., Subramaniyan, V., Sathasivam, K.V., Azad, A.K., Swain, S.S. *et al.*, Ultraflexible liposome nanocargo as a dermal and transdermal drug delivery system. *Nanomaterials*, 11, 10, 2557, 2021.
40. Fernández-García, R., Lalatsa, A., Statts, L., Bolás-Fernández, F., Ballesteros, M.P., Serrano, D.R., Transferosomes as nanocarriers for drugs across the skin: Quality by design from lab to industrial scale. *Int. J. Pharm.*, 573, 118817, 2020.
41. Souto, E.B., Baldim, I., Oliveira, W.P., Rao, R., Yadav, N., Gama, F.M. *et al.*, SLN and NLC for topical, dermal, and transdermal drug delivery. *Expert Opin. Drug Deliv.*, 17, 3, 357–77, 2020.
42. Chaudhari, Y., Nanoparticles-A paradigm for topical drug delivery. *Chron. Young Sci.*, 3, 1, 82–82, 2012.
43. Shukla, T., Upmanyu, N., Agrawal, M., Saraf, S., Saraf, S., Alexander, A., Biomedical applications of microemulsion through dermal and transdermal route. *Biomed. Pharmacother.*, 108, 1477–94, 2018.
44. Souto, E.B., Cano, A., Martins-Gomes, C., Coutinho, T.E., Zielińska, A., Silva, A.M., Microemulsions and nanoemulsions in skin drug delivery. *Bioengineering*, 9, 4, 158, 2022.
45. Jeong, W.Y., Kwon, M., Choi, H.E., Kim, K.S., Recent advances in transdermal drug delivery systems: a review. *Biomater. Res.*, 25, 1, 24, 2021.
46. Ghasemiyeh, P. and Mohammadi-Samani, S., Potential of Nanoparticles as Permeation Enhancers and Targeted Delivery Options for Skin: Advantages and Disadvantages. *Drug Des. Dev. Ther.*, 14, 3271–89, 2020.
47. Ghasemiyeh, P. and Mohammadi-Samani, S., Hydrogels as drug delivery systems; pros and cons. *Trends Pharm. Sci.*, 5, 1, 7–24, 2019.

48. Mohamad, E.A., Rageh, M.M., Darwish, M.M., A sunscreen nanoparticles polymer based on prolonged period of protection. *J. Bioact. Compat. Polym.*, 37, 1, 17–27, 2022.
49. Alvarez-Román, R., Naik, A., Kalia, Y.N., Guy, R.H., Fessi, H., Skin penetration and distribution of polymeric nanoparticles. *J. Controlled Release*, 99, 1, 53–62, 2004.
50. Maiti, D., Tong, X., Mou, X., Yang, K., Carbon-based nanomaterials for biomedical applications: a recent study. *Front. Pharmacol.*, 9, 1401, 2019.
51. Kuche, K., Maheshwari, R., Tambe, V., Mak, K.K., Jogi, H., Raval, N. *et al.*, Carbon nanotubes (CNTs) based advanced dermal therapeutics: current trends and future potential. *Nanoscale*, 10, 19, 8911–37, 2018.
52. Salvioni, L., Morelli, L., Ochoa, E., Labra, M., Fiandra, L., Palugan, L. *et al.*, The emerging role of nanotechnology in skincare. *Adv. Colloid Interface Sci.*, 293, 102437, 2021.
53. He, S., Wu, L., Li, X., Sun, H., Xiong, T., Liu, J. *et al.*, Metal-organic frameworks for advanced drug delivery. *Acta Pharm. Sin. B*, 11, 8, 2362–95, 2021.
54. Barati, M., Hashemi, S., Sayed Tabatabaei, M., Zarei Chamgordani, N., Mortazavi, S.M., Moghimi, H.R., Protein-based microneedles for biomedical applications: A systematic review. *Biomed. Microdevices*, 26, 2, 19, 2024.
55. Jung, J.H. and Jin, S.G., Microneedle for transdermal drug delivery: current trends and fabrication. *J. Pharm. Investig.*, 51, 5, 503–17, 2021.
56. Khare, N. and Shende, P., Microneedle system: a modulated approach for penetration enhancement. *Drug Dev. Ind. Pharm.*, 47, 8, 1183–92, 2021.
57. Grammatikopoulou, M.G., Gkiouras, K., Dardiotis, E., Zafiriou, E., Tsigalou, C., Bogdanos, D.P., Peeking into the future: Transdermal patches for the delivery of micronutrient supplements. *Metab. Open*, 11, 100109, 2021.
58. Bird, D. and Ravindra, N.M., Transdermal drug delivery and patches—An overview. *Med. Devices Sens.*, 3, 6, e10069, 2020.
59. Wong, W.F., Ang, K.P., Sethi, G., Looi, C.Y., Recent Advancement of Medical Patch for Transdermal Drug Delivery. *Medicina*, 59, 4, 778, 2023.
60. Bakshi, A., Bajaj, A., Malhotra, G., Madan, M., Amrutiya, N., A novel metered dose transdermal spray formulation for oxybutynin. *Indian J. Pharm. Sci.*, 70, 6, 733, 2008.
61. Milutinov, J., Krstonošić, V., Ćirin, D., Pavlović, N., Emulgels: Promising Carrier Systems for Food Ingredients and Drugs. *Polymers*, 15, 10, 2302, 2023.
62. Shakeel, A., Lupi, F.R., Gabriele, D., Baldino, N., De Cindio, B., Bigels: A unique class of materials for drug delivery applications. *Soft Mater.*, 16, 2, 77–93, 2018.
63. Rapalli, V.K. and Singhvi, G., Dermato-pharmacokinetic: assessment tools for topically applied dosage forms. *Expert Opin. Drug Deliv.*, 18, 4, 423–6, 2021.

64. Choy, Y.B. and Prausnitz, M.R., The rule of five for non-oral routes of drug delivery: ophthalmic, inhalation and transdermal. *Pharm. Res.*, 28, 943–8, 2011.
65. Uchechi, O., Ogbonna, J.D., Attama, A.A., Nanoparticles for dermal and transdermal drug delivery, in: *Application of nanotechnology in drug delivery*, vol. 4, pp. 193–227, 2014.
66. Shah, V.P., Flynn, G.L., Yacobi, A., Maibach, H.I., Bon, C., Fleischer, N.M. *et al.*, Bioequivalence of topical dermatological dosage forms–methods of evaluation of bioequivalence. *Skin Pharmacol. Physiol.*, 11, 2, 117–24, 1998.
67. Pensado, A., Hattam, L., White, K.A.J., McGrogan, A., Bunge, A.L., Guy, R.H. *et al.*, Skin Pharmacokinetics of Transdermal Scopolamine: Measurements and Modeling. *Mol. Pharm.*, 18, 7, 2714–23, 2021.
68. Benfeldt, E., Hansen, S.H., Vølund, A., Menné, T., Shah, V.P., Bioequivalence of topical formulations in humans: evaluation by dermal microdialysis sampling and the dermatopharmacokinetic method. *J. Invest. Dermatol.*, 127, 1, 170–8, 2007.
69. Bodenlenz, M., Tiffner, K.I., Raml, R., Augustin, T., Dragatin, C., Birngruber, T. *et al.*, Open Flow Microperfusion as a Dermal Pharmacokinetic Approach to Evaluate Topical Bioequivalence. *Clin. Pharmacokinet.*, 56, 1, 91–8, 2017.
70. Patel, P., Schmieder, S., Krishnamurthy, K., Research techniques made simple: Drug delivery techniques, Part 2: Commonly used techniques to assess topical drug bioavailability. *J. Invest. Dermatol.*, 136, 5, e43–9, 2016.
71. Barry, B.W. and Woodford, R., Comparative bio-availability and activity of properietary topical corticosteroid preparations: vasoconstrictor assays on thirty-one ointments. *Br. J. Dermatol.*, 93, 5, 563–71, 1975.
72. Bronaugh, R.L. and Maibach, H.I., *Topical absorption of dermatological products*, vol. 21, CRC Press, eBook-PDF, 2002.
73. Subramony, J.A., Needle free parenteral drug delivery: Leveraging active transdermal technologies for pediatric use. *Int. J. Pharm.*, 455, 1–2, 14–8, 2013.
74. Hashemi S, Mortazavi SA, Moghimi HR, Darbasizadeh B. Development and evaluation of a novel methotrexate-loaded electrospun patch to alleviate psoriasis plaques. Drug Development and Industrial Pharmacy. 2022;48(8):355-66.

11

Ocular Novel Drug Delivery Systems

Payam Khazaeli[1]*, Mahsa Sayed Tabatabaei[2], Leila Sayed-Tabatabaei[3] and Mohammad Amin Raeisi Estabragh[1]

[1]Department of Pharmaceutics, Faculty of Pharmacy, Kerman University of Medical Sciences, Kerman, Iran

[2]HIDA Pharmaceutical Technologies Development Center, HIDA Pharmaceuticals, Tehran, Iran

[3]Department of Pharmacy, University of Copenhagen, Denmark

Abstract

The eye is a unique organ with different anatomical and physiological barriers and protective mechanisms against the entrance of foreign substances. So, ocular drug delivery faces serious challenges, namely low bioavailability. This is mainly due to the inability of drugs to pass through barriers and premature elimination of drugs from the intended site. Therefore, drug delivery to the eye remains one of the most challenging areas of drug delivery to be addressed by researchers. In recent decades, innovative approaches have emerged to conquer the unique challenges of the ocular environment. This chapter aims to, after a brief introduction on ocular drug delivery, provide a comprehensive overview of anatomical (e.g., corneal, conjunctival, scleral, retinal, and vitreous) and physiological (e.g., time, flow, efflux pumps, and enzymes) ocular barriers against drug permeation. In the following, different challenges against ophthalmic drug delivery as well as opportunities provided by the special structure of the eye will be reviewed. Also, general considerations that should be taken into account in ophthalmic formulations and novel ocular drug delivery systems will be briefly introduced. Subsequently, the strategies that can be used to overcome these challenges and take advantage of the opportunities (including ocular devices, hydrogels and *in-situ* gel systems, nanocarriers, viral vectors, absorption enhancement methods, and drug modification strategies) will be overviewed. Due to the complexity of the eye structure, the drug will take a tortuous journey after administration. This winding path will

**Corresponding author*: pkhazaeli@kmu.ac.ir

Mehrdad Hamidi and Mahsa Sayed Tabatabaei (eds.) Novel Drug Delivery Systems: Fundamentals and Applications, (393–418) © 2025 Scrivener Publishing LLC

be reviewed in the form of a conceptual pharmacokinetic map. Finally, some of the novel ocular drug delivery systems that have entered the market will be presented.

***Keywords*:** Eye, ophthalmic drug delivery, ocular drug delivery, ocular pharmacokinetics, ocular devices, nanocarriers

11.1 Introduction

The history of ocular drug administration dates back to ancient Egypt when it was tried to treat night blindness using animal extracts [1]. Still, this organ is in the spotlight as a target for drug administration. This is mainly because of the prevalence of eye-related disorders and diseases with a direct negative effect on patient's quality of life (e.g., cataracts, glaucoma, retinoblastoma, and fungal keratitis), which has become even more important with the increase of the elderly population in the world and the significant change in the patterns of eye use in recent decades [2–5].

The eye is a sophisticated and unique multi-compartment structure that provides several barriers against the permeation of xenobiotics, including drug molecules. The inherent ocular barriers include static (e.g., cornea, sclera, and retina) and dynamic (e.g., blood and lymphatic clearance, mucus layer, and nasolacrimal drainage) barriers. So, ophthalmic drug delivery remains one of the most complicated challenges to be overcome by researchers [6]. For effective delivery of ophthalmic drugs, the drug molecule must be able to remain in the precorneal region for a sufficient time (to create an appropriate drug level) and efficiently permeate the corneal barrier (to reach its target site) [3].

Although drug delivery through the eye is a straightforward approach, conventional ocular dosage forms (e.g., solutions, suspensions, etc.) suffer from inadequate bioavailability, mainly due to the early elimination of drugs and their inability to pass the ocular barriers. Semisolid dosage forms (e.g., ophthalmic ointments) may somewhat extend the maintenance time of the drug; however, they create new problems, such as blurred vision, which reduce patient acceptance.

Therefore, an upsurge of research has been conducted in recent decades to modify the topical formulations and conquer the ocular barriers, extend the residence time, and improve drug bioavailability. Among these, one can mention ocular drug delivery devices, hydrogels and thermo-gelling systems, nanocarriers, viral vectors, absorption enhancement approaches, and modified drugs (e.g., prodrugs and ion pairs). In parallel, thanks to the multiple compartmental structure of the eye, novel non-topical approaches have emerged to improve the retention time of drugs and therapeutic efficiency [5]. Still, some conventional methods (e.g., systemic drug administration to provide a sufficient concentration at the target site) are also available. Therefore, today, there are different routes of ocular administration, each with its advantages and limitations, including systemic, topical, periocular, and intraocular pathways [7]. Considering the topical route as a common, available, and non-invasive approach, in this chapter, different aspects of this route of ocular administration are mainly highlighted.

11.2 Anatomy and Physiology

Anatomically, the eye consists of three main layers. The outer layer is a connective tissue composed of cornea and sclera which are connected through the limbus. The middle layer, a vascular tissue, consists of the iris, ciliary bodies, and the choroid. The innermost layer is a neural tissue, called the retina, which is responsible for the capture of photons and transmitting them as electrical/chemical signals to the brain through the optic nerve pathway. Inside the eye globe, just behind the iris and between the two ciliary bodies, the lens is held in place by zonular fibers. Based on the location of the lens in the eye, the eyeball is commonly divided into two main segments (i.e., anterior and posterior). The front third of the eye is the anterior segment which includes the cornea, conjunctiva, iris, ciliary muscle, and lens. This chamber is filled with aqueous humor. The posterior segment is the back two-thirds of the eye, consisting of the sclera, choroid, and retina, and contains the vitreous humor [8, 9]. Different parts of the eye are shown in Figure 11.1.

In addition to its main physiological functions in visualizing objects, the eyes, as one of the organs exposed to direct contact with the external environment, create several barriers against the entry of foreign bodies. Obviously, drugs will not be an exception to this rule. Therefore, to reach its target site, the drug has to go through various obstacles. Although this is a challenge, the special structure of the eye with multiple compartments provides opportunities for drug administration through various routes. In the following, the main ocular barriers will be reviewed.

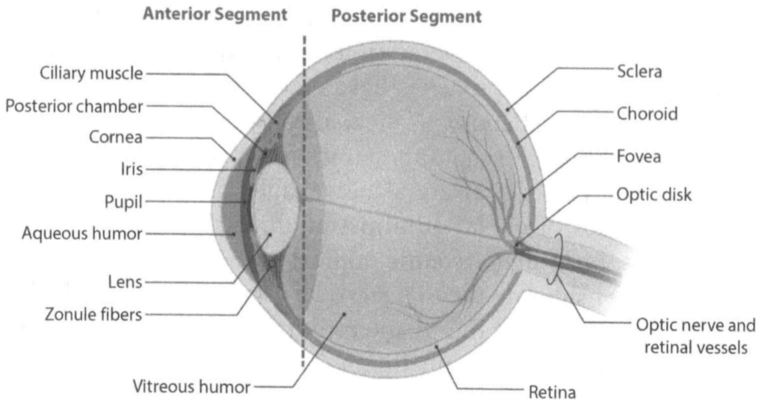

Figure 11.1 Anatomy of the eye (created with BioRender.com).

11.3 Barriers to Ocular Drug Delivery

Generally, as shown in Figure 11.2, there are two main categories of ocular drug delivery barriers that prevent drug access to its target site, including static and dynamic barriers, each with different subclasses [6, 10]. These barriers are explained below.

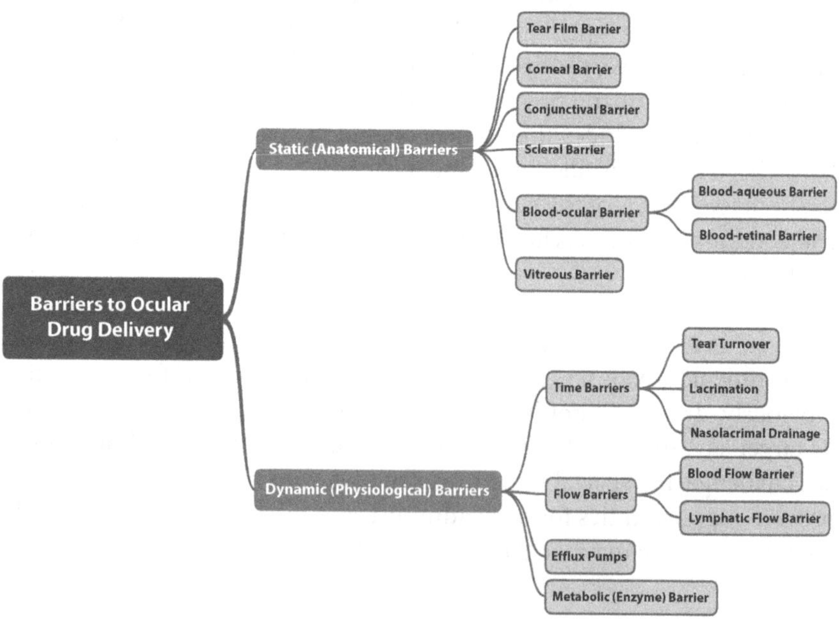

Figure 11.2 Classification of ocular drug delivery barriers.

11.3.1 Static Barriers

Static (anatomical or penetration) barriers refer to different parts of the eye that physically prevent drug penetration and include the tear film, cornea, conjunctiva, sclera, retina (blood-aqueous and blood–retinal barriers), and vitreous barriers. A schematic illustration of the main anatomical barriers against ocular drug delivery is presented in Figure 11.3.

11.3.1.1 Tear Film Barrier

The tear film, a thin transparent multilayered barrier with a thickness of 6–10 μm, is the outermost layer protecting the eye and, therefore, the first barrier against the entry of xenobiotics. As shown in Figure 11.3, this film is composed of three layers, from the outside in order:

(i) the lipid layer: a thin layer with an approximately 40 nm thickness as a barrier to hydrophilic drugs,

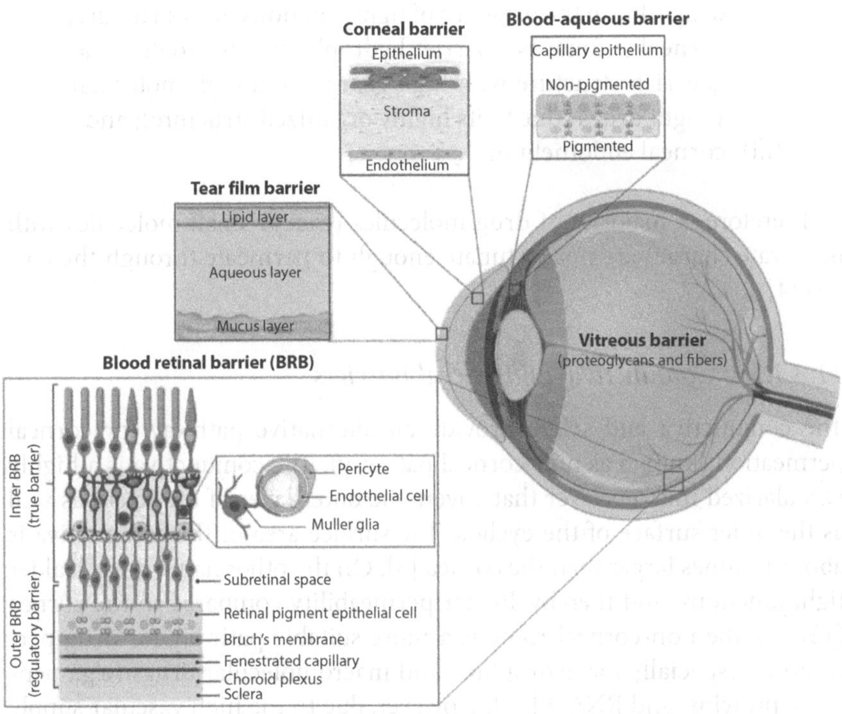

Figure 11.3 The main anatomical barriers to ocular drug delivery (created with BioRender.com).

(ii) the intermediate aqueous layer: the largest part of the tear film about 4 μm thick, as a barrier for lipophilic drugs, and
(iii) the mucus layer, which prevents foreign materials from permeation through electrostatic interactions.

In addition, there are several proteins in the tear (e.g., albumin) that can reduce the available drug to reach its target through non-specific interactions [5, 8, 11].

11.3.1.2 Corneal Barrier

The cornea, a collagen-rich transparent structure with an approximately 500-μm thickness, undertakes a major role as the primary barrier in preventing topical drug absorption. From the anterior to the posterior segment, there are three different layers (Figure 11.3):

(i) corneal epithelium; the first cellular defense against a variety of environmental factors, particularly for hydrophilic drugs due to the presence of tight junctions in its structure;
(ii) corneal stroma; as a barrier hydrophobic (due to its collagen-rich structure with high charge) and high molecular weight drugs (due to its highly organized structure); and
(iii) corneal endothelium.

Therefore, a majority of drug molecules (except small molecules with moderate charge) are not fortunate enough to permeate through the cornea [4].

11.3.1.3 Conjunctival and Scleral Barriers

The conjunctiva and sclera provide an alternative pathway for corneal permeation (known as non-corneal pathway). The conjunctiva is a highly vascularized mucous layer that covers the outer layer of the sclera as well as the inner surface of the eyelids. The surface area of the conjunctiva is about 17 times larger than the cornea [5]. On the other hand, it has leakier tight junctions, and thereby, higher permeability compared to the cornea [12]. So, the non-corneal route is a more suitable pathway for drug permeation, especially for hydrophilic and macromolecular drugs (e.g., peptides, proteins, and RNAs) [13]. However, due to the high vascular supply of the conjunctiva, it is more likely for drugs to be absorbed systemically than localized. In this case, a major share of the drug will be out of reach, which will require a higher dose of the drug, and possibly an increase in the dosing frequency, to compensate. However, it is not feasible since it can increase the risk of side effects and reduce the patient compliance. Then, the small remaining fraction of the drug goes through the trans-scleral route (i.e., from the sclera to the anterior segment) to reach the inner layers. In the non-corneal pathway, due to the larger surface area and higher permeability of the sclera compared to the cornea, the size of the drug molecules is a more important determinant than their lipophilicity. Also, the thickness of the sclera plays an important role in drug penetration [5].

11.3.1.4 Blood-Aqueous Barrier

Blood-ocular barriers encompass two different parts including blood-aqueous and blood-retinal barriers (Figure 11.3). The blood-aqueous barrier (BAB) is formed by the vascular endothelial layer of the iris and non-pigmented epithelial layer of the ciliary body. Both these endothelial and epithelial layers have several "leaky" tight junctions in their structure which provide a non-specific barrier. Diffusion of drugs across BAB is pressure-dependent and impacted by the osmotic pressure as well as the physicochemical characteristics of drugs. Generally, small and lipophilic drugs are better absorbed than those with hydrophilic nature or high molecular weights [5, 14, 15].

11.3.1.5 Blood-Retinal Barrier

The blood-retinal barrier (BRB), as the main posterior segment barrier against drugs to reach the intraocular space, is made up of two structurally different parts both with "non-leaky" tight junctions (Figure 11.3):

(i) the outer blood-retinal barrier (oBRB) consists of an epithelial monolayer of pigmented cells that regulates transportation between the fenestrated capillaries of the choriocapillaris and the retinal pigment epithelium; and
(ii) the inner blood-retinal barrier (iBRB), composed of the retinal endothelial cells that form retinal capillaries, with a regulating function for transport across the retinal capillaries.

The main mechanisms involved in the transcellular transport across this endothelium membrane are diffusion for lipophilic molecules with low molecular weights, and active (energy-dependent) transport for lipophilic drugs with higher molecular weights as well as hydrophilic ones [5, 15–17].

11.3.1.6 Vitreous Barrier

The vitreous chamber is an avascular connective tissue located in the posterior segment of the eye, between the lens and optic nerve, and filled with a clear gel-like structure of collagen, hyaluronic acid, and water, called vitreous humor [3]. Due to its unique structural properties, the vitreous is considered a barrier against the distribution of intravitreally injected nanoparticles to reach the retina [18].

11.3.2 Dynamic Barriers

Dynamic barriers are physiological factors that reduce the possibility of ocular drug absorption by removing the available drug or expelling it. These factors include time barriers, lymphatic clearance, blood flow (choroidal and conjunctival), and efflux pumps.

11.3.2.1 Time Barriers

One of the main barriers to ocular drug delivery is the inadequate retention time and early wash out of topically administered drug from the surface of the eye, in just 15 to 30 seconds, due to the rapid turnover of the tear, lacrimation, insufficient penetration into the mucin layer, or fast nasolacrimal drainage. Such a serious obstacle, alongside the corneal barrier, makes the bioavailability of topically administered drugs to the eye remarkably low – typically less than 5% and even lower in the case of macromolecules [4, 19].

11.3.2.2 Flow Barriers

Flow barriers are a general term to describe blood flow and lymphatic flow clearance from the choroid and conjunctiva. The choroid has a high clearance due to its incredibly high blood supply-to-surface area ratio as well as fenestrated choriocapillaries with a pore size of 70-80 nm which allows the small molecules to rapidly enter the blood circulation, and the macromolecules and even nanoparticles to extravasate. The systemic absorption of drugs is also high from the sub-conjunctival space which leads to limited bioavailability. Additionally, lymphatic flow is of importance in drug removal from the choroidal and sub-conjunctival spaces [20].

11.3.2.3 Efflux Pumps

Formerly, the low ocular bioavailability was mainly attributed to the presence of precorneal barriers (including tear film, tear clearance, lacrimation, and drainage [6]) and the inability of drugs to pass through. Recently, however, the importance of efflux pumps like P-glycoprotein (P-gp) and multidrug resistance-associated protein-2 (MRP2) has been highlighted in preventing drug absorption from the cornea [21].

11.3.2.4 Metabolic (Enzyme) Barrier

Considering the strategic location of the eye in direct contact with xenobiotics, the ocular tissue is equipped with a metabolic barrier containing several metabolizing enzymes like the CYP family, esterases, reductases, and peptidases. Although enzymatic degradation is considered a threat, it also provides an opportunity for the delivery of prodrugs [22].

11.4 Challenges and Opportunities

As stated so far, considering the inherent barriers of the eye, ocular drug delivery is too complicated, but at the same time, its special structure creates opportunities to improve drug delivery. A summary of the challenges and opportunities associated with ocular drug delivery is given in Figure 11.4.

11.5 Formulation Considerations

For optimal drug penetration to the eye, several physicochemical properties of drug molecules (e.g., logP, ionization state, molecular weight, etc.) should be carefully considered. In the cases where the drug lacks optimal features for penetration, different strategies (e.g., prodrugs and ion pairs) could be employed to modify its characteristics. For nanocarrier-containing formulations, different physicochemical properties of nanocarriers (e.g., size, surface modification, etc.) should be adjusted to attain maximum therapeutic efficiency. There are also some formulation-related factors (e.g., drug concentration, viscosity, and so on) that can potentially

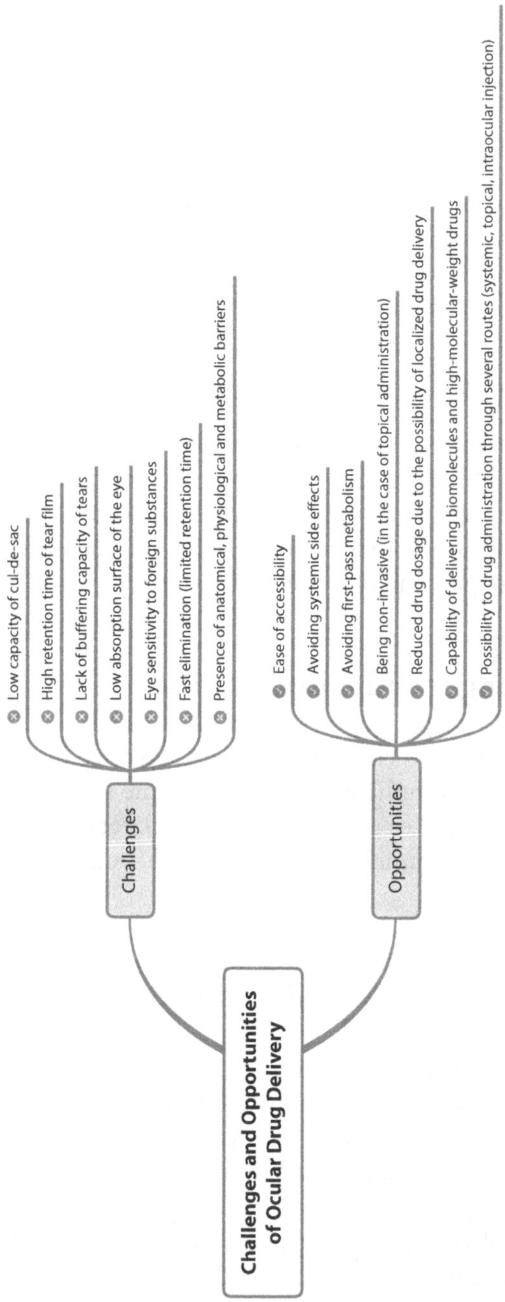

Figure 11.4 Challenges and opportunities of ocular drug delivery.

affect the treatment outcome. Here, the common considerations for the formulation of ophthalmic systems are briefly reviewed.

11.5.1 Physicochemical Properties of Drug Molecules

One of the most important factors determining the permeation of drugs through the cornea is the molecular weight (MW). Other factors influencing ophthalmic drug absorption are the charge of molecules and their lipophilicity/hydrophilicity ratio. In summary, drug molecules with MW < 500 Da are considered appropriate candidates for trascorneal permeation. Also, lipophilic compounds have a higher trans-corneal penetration than hydrophilic ones, and drugs with a positive charge are more fortunate to pass through. This is mainly due to the presence of the mucous layer with a negative charge on the epithelial surface. Positive molecules are attracted and show better penetration, while negative ones are repelled. It should be noted, however, that the positive charge is a double-edged sword. Since the drugs with a positive charge may interact with the proteoglycan matrix of the sclera with a negative charge and get trapped. Also, the negatively charged hyaluronic acid of the vitreous chamber may hinder the movement of drugs with a positive charge. Therefore, the optimal charge is determined depending on the drug delivery to which part of the eye is intended [23]. The conjunctiva allows the passage of compounds with higher MWs (about 5–10 kDa). It is easier for hydrophilic molecules to pass through the conjunctival path than the corneal route. But, in general, lipophilic molecules have a better permeation both in the corneal and conjunctival pathways. In the case of sclera, drug molecules with very high MWs (< 70 kDa) can permeate. In this pathway, hydrophilic drugs and those with a negative charge are preferred. In the vitreous, small molecules can easily diffuse. High molecular weight and hydrophilic drugs have a higher half-life in the vitreous and are commonly eliminated through the anterior pathway. In contrast, low molecular weight and lipophilic drugs can permeate the retina, so, they are generally removed from the posterior pathway [24].

Many drugs, inherently, lack optimal physicochemical properties to cross the cornea. As a powerful strategy, prodrugs can modify the hydrophilicity/lipophilicity, aqueous solubility, and stability of drugs, enhance their penetration and bioavailability, extend the residence time of the drug in the intended site, improve the patient compliance, and reduce the side effects of administered drug molecules [25]. Ion pairing is another simple but effective approach to improving the permeation of ionized drugs [26]. These two approaches will be explained more in the following sections.

11.5.2 Physicochemical Properties of Nanocarriers

Nanocarriers offer several advantages as ophthalmic drug delivery systems such as controlled drug release, improved trans-corneal permeation, providing high concentration of drug in the aqueous humor and cornea, and consequently, enhanced therapeutic effect of drugs. The physicochemical properties of the nanocarriers have a profound effect on the therapeutic efficiency [5, 7]. Generally, it has been shown that particles with smaller sizes have a faster penetration into the mucin layer of the tear film and a higher absorption into the aqueous humor. The optimal size of nanocarriers for corneal absorption is considered between 50 to 400 nm. To be easily absorbed from the pre-corneal pathway following topical administration, the optimal size of nanocarriers has been shown to be smaller than 200 nm. Hydrophilic nanocarriers (with about 20 to 80 nm in diameter) can diffuse through the pores of the sclera (with a size of about 30–350 nm) to the vitreous humor. Following intravitreal injection, nanocarriers with a smaller size than 350 nm have the potential to reach the retina, while the movement of the large particles is hindered by the vitreous humor. In addition to the particle size, the surface charge of nanocarriers can impact corneal absorption. The surface charge of the both conjunctiva and cornea is negative. So, cationic nanocarriers have a longer retention time in the ocular regions than those with a negative charge. However, as stated in the case of the surface charge of molecules, the positive charge of the nanocarriers can also have a negative effect on their permeation through the sclera matrix, probably due to immobilization by the electrostatic attraction [7]. Also, diffusion of anionic particles in the vitreous is easier than those with a positive charge [24].

11.5.3 Formulation-Related Factors

One of the most important factors to be adjusted in ophthalmic formulations is drug concentration. Generally, based on Fick's law of diffusion, the higher the drug concentration in the ocular solution, the higher the corneal permeation. However, there is an upper limit where the relationship between concentration and therapeutic effect reaches a plateau [23].

Drugs may be dissolved or dispersed in ophthalmic formulations. In cases where the drug is dispersed as solid particles (e.g., suspensions), the particle size undertakes a major role in drug permeation. This is particularly important for drugs with very low water solubility. In these cases, the ocular drug delivery is partly controlled by the dissolution of the drug. As a general rule, the smaller the particle size, the higher the surface area, and thereby, the higher the dissolution rate and permeation. However, it should be noted that smaller particles have also a faster clearance. On the other hand, the particle size has a maximum limit, beyond which, it may cause reflex lacrimation. In addition to the size, the shape of particles could also affect the fate of ocularly administered drugs; for instance, those with sharp corners may irritate the eye, induce reflex lacrimation, and subsequently lead to early clearance [23, 27].

It is also of crucial importance to control the pH of formulation since it can determine the ionization state of drugs, permeation and also, formulation stability. Generally, the corneal permeation is higher in the non-ionized form. In addition to the aforementioned factors, the isotonicity of ophthalmic formulation is a determining parameter. For instance, hypotonic solutions may lead to improved corneal permeation, possibly due to local edema. In contrast, hypertonic formulations lead to dehydration [28].

Common eye drops have a very short retention time on the surface of the eye (a few minutes at most). To improve ocular drug delivery, the contact time could be extended either through viscosity enhancement or mucoadhesion. Another strategy to help the drug retain longer (even to several days or weeks), is using drug delivery platforms with a relatively large size (e.g., inserts and contact lenses) [23, 27].

11.6 Classification

Significant advancements have recently transformed the landscape of ocular disorders treatment. These disorders are broadly categorized into anterior and posterior conditions. While many anterior eye disorders can be effectively treated by topical drug formulations, delivering conventional therapeutic doses to the posterior eye poses a considerable challenge. Consequently, a variety of drug delivery systems has emerged, aiming to transport drugs efficiently to either the anterior or posterior segments of the eyes [29]. Here, we focus on the exploration and application of ocular drug delivery devices, hydrogels and *in-situ* gel forming systems, nanocarriers, viral vectors, absorption enhancement methods, and prodrugs and ion pairs.

11.6.1 Ocular Drug Delivery Devices

11.6.1.1 Implants

Treating eye problems usually involves using eye drops or injections directly into the eye. Eye drops have been the mainstay for a long time, especially for conditions like glaucoma. But, making sure people use them as they should is a big challenge, especially for diseases without noticeable symptoms. Injecting medicine into the eye was a breakthrough for treating posterior eye diseases. However, invasiveness, low patient compliance, and accordingly not getting injections often enough, are big challenges in using injectable ocular treatments. The first implantable eye devices were reservoir-based devices which put into the eye through surgery, incision, and suturing to the sclera for long-term residence within the eye. The newer versions of ocular implants aimed to make the procedure less invasive, using an injector. These implants can be made of degradable or non-degradable materials and can be used for both small-molecule drugs and protein-based treatments. Ocular implants offer a minimally-invasive treatment, with a longer-lasting effect. While these devices have shown success, there's still room for novel improved treatments that could replace eye drops and injections [30].

11.6.1.2 Microneedles

Microneedles have emerged as a promising technology in ocular drug delivery, offering a novel and minimally invasive approach to administering therapeutic agents to the eye. The length of microneedles is often less than a millimeter. These microarrays are designed to penetrate the outer layers of the eye, reaching specific target areas with precision. One significant advantage of microneedles is their ability to overcome the natural barriers of the eye, such as the cornea, to deliver drugs more effectively. Unlike traditional methods like eye drops, which may have limited penetration, microneedles facilitate the direct delivery of medications into the targeted ocular tissues. This type of drug administration improves drug absorption, ensuring that the therapeutic concentration of medication reaches the desired site of action. The application of microneedles in ocular drug delivery holds great potential for various conditions, including posterior eye problems. This technology offers a more controlled and sustained release of therapeutic agents, reducing the need for frequent administrations and enhancing patient compliance. Additionally, microneedles can contribute to minimizing systemic side effects by delivering medications directly to the eye and bypassing the bloodstream. As the technology of microneedles in ocular drug delivery continues to evolve, they present a promising avenue for improving the precision and efficiency of ocular drug delivery, to ultimately enhance the treatment of various eye diseases [31, 32].

11.6.1.3 Microcannulation or Microcatheters

In the landscape of ocular drug delivery, microcannulation or microcatheters have emerged as innovative tools with the potential to revolutionize how therapeutic agents are administered to the eye. These miniature tubes, often on a scale of micrometers, are designed to navigate through the intricate structures of the eye, facilitating targeted drug delivery to specific ocular regions. The distinctive feature of microcannulation lies in its ability to precisely guide therapeutic agents to their intended destination within the eye. This targeted approach allows enhanced drug delivery to both anterior and posterior segments, surpassing some of the limitations associated with traditional methods like eye drops. By navigating through the eye's intricate pathways, microcannulas offer a means to overcome barriers and deliver medications directly to the desired treatment sites. The application of microcannulation in ocular drug delivery holds promise for improving treatment outcomes, particularly in conditions that require precise and localized therapy. This technology provides an avenue for controlled drug release, minimizing the need for frequent administrations and potentially improving patient adherence to treatment plans. Additionally, like microneedles, microcannulas offer benefits in terms of reducing systemic side effects, as drugs can be delivered directly to the eye, bypassing systemic circulation. Ongoing research endeavors are focused on refining the design and capabilities of microcannulation devices, exploring optimal materials, and fine-tuning drug formulations for improved safety and efficacy. As this technology advances, microcannulation stands as a compelling approach to enhance the precision and efficiency of ocular drug delivery, contributing to advancements in the treatment of various eye conditions [33, 34].

11.6.1.4 Drug-Eluting Lenses

Drug-eluting lenses represent a cutting-edge innovation in ocular drug delivery systems, combining the convenience of contact lenses with the targeted release of therapeutic agents for ocular disorders. These lenses are designed to release drugs directly onto the ocular surface, providing sustained and controlled drug delivery while overcoming the limitations of traditional eye drops. By integrating drug-releasing polymers or nanoparticles into the lens material, these devices offer a prolonged release of medications, improving patient compliance and reducing the frequency of administration. The drug-eluting lenses have shown promise in treating various ocular conditions, such as glaucoma, dry eye syndrome, and bacterial infections. This approach not only enhances therapeutic efficacy but also minimizes systemic side effects associated with conventional drug administration. Despite the promising advancements, challenges such as optimizing drug release kinetics, ensuring biocompatibility, and addressing individual variations in ocular physiology remain areas of ongoing research. The development of drug-eluting lenses signifies a remarkable stride towards personalized and patient-friendly ocular drug delivery systems [35].

11.6.1.5 Inserts

Ocular inserts refer to reservoir-type controlled-release dosage forms designed to remain in the cul-de-sac. They are generally composed of polymers in the form of solids or semisolids. They were first introduced to the market by Ocusert® for the delivery of pilocarpine. Ocular inserts allow enhanced bioavailability, improved local absorption, reduced dosing frequency, and high stability without the need for preservatives. Despite their advantages as long-acting systems, wide use of ocular inserts is hampered due to the sensation of a foreign body in the eye, the possibility of moving in the eye and interfering with vision, and difficulty in removing by the patients [36].

11.6.2 Hydrogels and *In Situ* Gels

Hydrogels and *in-situ* gels have emerged as promising platforms in the realm of ocular drug delivery, offering unique advantages for sustained and controlled release of therapeutic agents. Hydrogels, owing to their high water content and biocompatibility, mimic the natural environment of the eye, providing a comfortable and non-irritating application. These three-dimensional networks of hydrophilic polymers can encapsulate a variety of drugs, facilitating a gradual release over an extended period. *In-situ* gels, on the other hand, exhibit a liquid-to-gel transition in response to specific physiological stimuli, such as temperature, pH, or ions present in the tear fluid. This transition allows for easy instillation as a liquid that transforms into a gel upon contact with the ocular surface, enhancing drug retention and bioavailability. Both hydrogels and *in-situ* gels address the limitations associated with traditional eye drops, offering prolonged drug exposure, reduced dosing frequency, and improved patient compliance. These innovative delivery systems hold significant promise for treating ocular conditions ranging from infections to chronic disorders, showcasing their potential to revolutionize the landscape of ocular drug delivery [37, 38].

11.6.3 Nanocarriers

Nanocarriers have emerged as pivotal players in the dynamic field of ocular drug delivery, harnessing the unique properties of nanotechnology to address challenges associated with traditional therapeutic approaches. A diverse array of nanocarriers, such as liposomes, nanoemulsions, solid lipid nanoparticles (SLNs), nanostructured lipid carriers (NLCs), cubosomes, niosomes, nanosuspensions, micelles, dendrimers, polymeric nanoparticles, cyclodextrin complexes, exosomes, nanowafers, and nanofibers, offer a versatile platform for encapsulating and delivering a diverse range of drugs to targeted ocular tissues. These carriers, with their varied structures and compositions, allow for improved penetration and sustained release within the ocular structures, enhancing drug bioavailability while minimizing systemic exposure and potential side effects. The tunable characteristics of these nanocarriers, such as size, surface charge, and drug-loading capacity, enable customization for prolonged release kinetics and enhanced drug stability. Despite ongoing challenges, the application of this diverse array of nanocarriers in ocular drug delivery represents a transformative paradigm, offering the potential to revolutionize treatment strategies and elevate therapeutic outcomes for a multitude of ocular disorders [5].

11.6.4 Viral Vectors

Viral vectors have emerged as powerful tools in ocular drug delivery, harnessing the genetic engineering capabilities of viruses to transport therapeutic genes into target cells with remarkable precision. Adeno-associated viruses (AAVs) and lentiviruses are among the commonly employed viral vectors due to their ability to efficiently transduce ocular tissues. This innovative approach allows for the introduction of therapeutic genes directly into the cells of the eye, offering a transformative strategy for treating various ocular disorders at the genetic level. In gene therapy, viral vectors can be engineered to carry and deliver specific therapeutic genes, addressing the root causes of genetic-based eye diseases such as retinitis pigmentosa and Leber's congenital amaurosis. By integrating the therapeutic genes into the host cells, viral vectors have the potential to correct or replace malfunctioning genes, offering a unique and targeted approach for sustained therapeutic effects. Despite the promising advancements, challenges like immune response and vector-related toxicity are areas of active research. The application of viral vectors in ocular drug delivery represents a frontier in personalized medicine, paving the way for innovative treatments that could significantly improve the lives of individuals affected by genetic eye disorders [39].

11.6.5 Absorption Enhancement Methods

Absorption enhancement methods in ocular drug delivery play a crucial role in overcoming the formidable barriers presented by the ocular surface for effective drug delivery. These methods aim to improve the bioavailability and therapeutic efficacy of drugs by enhancing their penetration into ocular tissues. One common approach involves the use of penetration enhancers, which can temporarily alter the integrity of the corneal epithelium, facilitating the entry of drugs. Surfactants, bile salts, and cyclodextrins are examples of such enhancers that interact with the lipid layer of the corneal epithelium, promoting drug permeation. Physical methods, such as iontophoresis and ultrasound, have also been explored to enhance drug absorption. Iontophoresis involves the application of a low electrical current to drive drug molecules across the ocular tissues, while ultrasound can disrupt the corneal barrier temporarily, facilitating drug permeation. While these absorption enhancement methods hold promise for improving ocular drug delivery, it is essential to balance efficacy with safety and minimize potential side effects. Continued research in this field is crucial to refining these methods and expanding the repertoire of tools available for enhancing the delivery of therapeutic agents to the eye [40].

11.6.6 Prodrugs and Ion Pairs

Prodrugs and ion pairs represent innovative strategies in ocular drug delivery, offering tailored approaches to enhance drug solubility, stability, and bioavailability. Prodrugs are biologically inactive compounds that undergo enzymatic or chemical transformation in the body to release the active drug. In ocular drug delivery, prodrugs can be designed to improve the permeability of drugs across ocular barriers, prolong drug release, and reduce systemic side effects. By modifying the physicochemical properties of the drug molecule, prodrugs optimize its pharmacokinetics and enhance its therapeutic potential [41]. Ion pairs, on the other hand, involve the formation of charged complexes between the drug and an oppositely charged counterion. This technique can improve drug solubility and stability, crucial factors in enhancing ocular drug delivery. Ion pairing is often employed in combination with polymers or carriers to facilitate controlled release, especially in situations where sustained drug action is desirable. The resulting complexes can improve the residence time of the drug on the ocular surface, leading to enhanced absorption and therapeutic effects [42].

11.7 Pharmacokinetics

There are different routes for the administration of ocular drugs including systemic (oral and injection), intraocular (intravitreal, suprachoroidal, subretinal, intracameral, retrobulbar, sub-tenon and posterior juxta-scleral), and extraocular (subconunctival and topical) pathways. Among these, topical is the most common and straightforward strategy [36, 43]. So, here, we delve into the pharmacokinetics of topically administered ophthalmic formulations. As stated previously, the bioavailability of ocular drugs is generally less than 5%, mainly due to significant precorneal drug loss and inadequate corneal permeation [5]. In general, after applying a topical formulation to the eye, the drug molecule has three main pathways ahead:

(i) Pre-corneal pathway; which means drug elimination before reaching the cornea
(ii) Non-corneal pathway; drug absorption from the conjunctiva and sclera
(iii) Corneal pathway; that is drug permeation through the tear film and cornea into the anterior segment.

More details about the fate of the drug, if it enters any of these pathways, are given in the Figure 11.5 as a conceptual map.

11.8 Products in the Market

Global demand for ophthalmic drugs is projected to grow at a CAGR of 6.6% over a ten-year period from 2023 to 2033. It is predicted that the global ophthalmic drugs market will reach $72 billion by the end of 2033, from $38 billion in 2023 [44]. Such an incremental trend, emphasizes the importance of development in the field of ocular drug delivery, and particularly, the role of ocular novel drug delivery systems. In Table 11.1, some of the novel ocular products in the market, by their category is provided.

In conclusion, the diverse landscape of ocular drug delivery systems offers a rich tapestry of innovative approaches aimed at overcoming the unique challenges posed by the ocular environment. Each strategy presents a distinct avenue for optimizing therapeutic outcomes. These advancements underscore the ongoing commitment of researchers and practitioners to develop patient-friendly, effective, and targeted solutions for numerous ocular conditions. As the field continues to evolve, the potential for

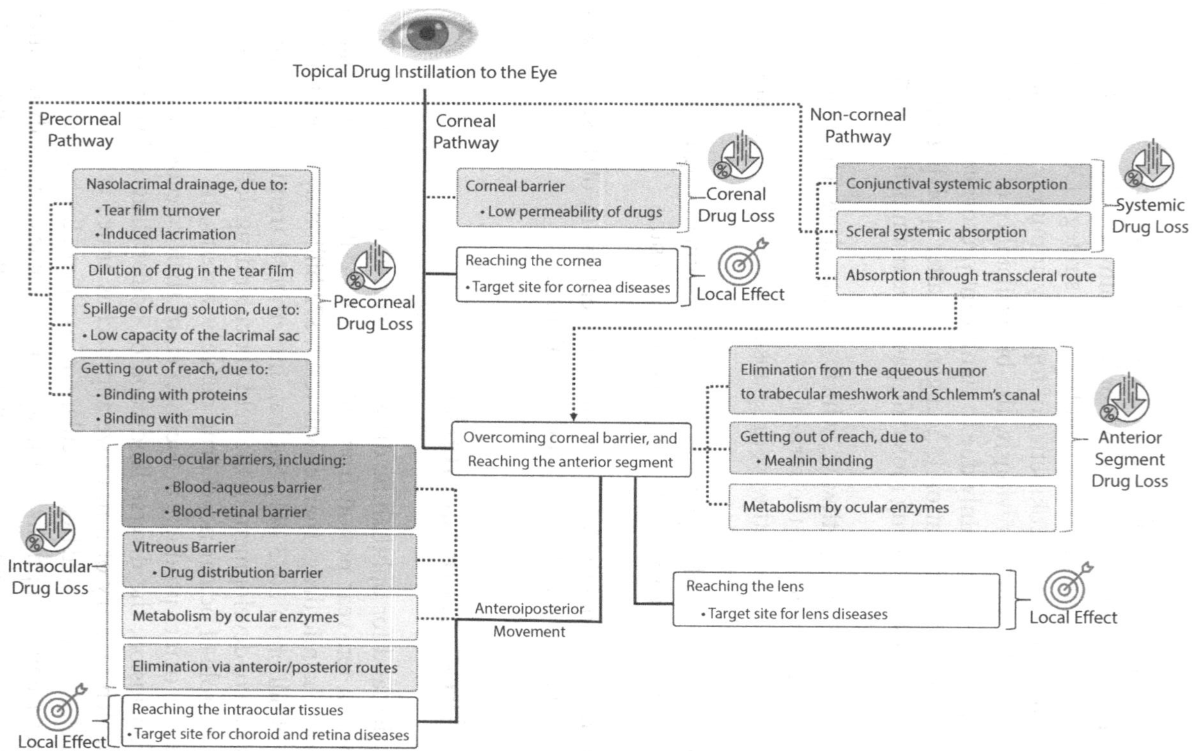

Figure 11.5 Ocular pharmacokinetics.

Table 11.1 List of some novel ocular products in the market.

Category	Product name	Active agent	Formulation type	Method of delivery
Ocular drug delivery devices	DURYSTA™ (bimatoprost implant)	Bimatoprost	Biodegradable implant	Intracameral injection
	Retisert® (fluocinolone acetonide implant)	Fluocinolone acetonide	Non-biodegradable implant	Intravitreal implantation
Hydrogels and *in-situ* gels	Systane® Ultra Hydration Preservative-Free Lubricant Eye Drops	Polyethylene glycol	Hydrogel eye drops	Topical application
	Timoptic-XE® (timolol gel-forming solution)	Timolol	Gel-forming solution	Topical application
Nanocarriers	Retisert® (fluocinolone acetonide implant)	Fluocinolone acetonide	Nanoparticles embedded in implant	Intravitreal implantation
	Visudyne® (verteporfin liposomal injection)	Verteporfin	Liposomal injection	Intravenous infusion followed by photodynamic therapy
	Restasis® (cyclosporine nanoemulsion)	Cyclosporine	Nanoemulsion	Topical application
	Cyclokat® (cyclosporine-loaded SLNs)	Cyclosporine	Solid Lipid Nanoparticles	Topical application
	Lipusu® (amphotericin B-loaded NLCs)	Amphotericin B	Nanostructured Lipid Carriers	Topical application

(Continued)

Table 11.1 List of some novel ocular products in the market. (*Continued*)

Category	Product name	Active agent	Formulation type	Method of delivery
	Cubicin® (daptomycin cubosomes)	Daptomycin	Cubosomes	Topical application
	Cloderm® (clotrimazole niosomes)	Clotrimazole	Niosomes	Topical application
	Liposic® (artificial tear bioliposomes)	N/A	Bioliposomes	Topical application
	Ilevro® (nepafenac nanosuspension)	Nepafenac	Nanosuspensions	Topical application
Viral vectors	Luxturna® (voretigene neparvovec-rzyl)	Voretigene neparvovec	Adeno-associated virus vector	Subretinal injection
Absorption enhancement methods	Prolensa® (bromfenac ophthalmic solution)	Bromfenac	Nanosuspension eye drops	Topical application
Prodrugs and ion pairs	Xalatan® (latanoprost ophthalmic solution)	Latanoprost	Latanoprostene bunod prodrug	Topical application

revolutionizing ocular drug delivery is both promising and transformative. Yet, the journey is not without challenges, and ongoing research efforts are crucial to refining these methods, ensuring safety, and addressing the intricacies of individual patient needs. The collective pursuit of innovation in ocular drug delivery reflects a commitment to improving the quality of life for individuals affected by ocular disorders, paving the way for a future where personalized and precise treatments redefine the landscape of ocular therapeutics.

References

1. Wolf G. A history of vitamin A and retinoids. The *FASEB Journal*. 10, 9:1102, 1996.-7.

2. Wang, L. and Zhang, H., Ocular barriers as a double-edged sword: preventing and facilitating drug delivery to the retina. *Drug Deliv. Transl. Res.*, 13, 2, 547–67, 2023. Available from: https://link.springer.com/10.1007/s13346-022-01231-5.
3. Ahmed, S., Amin, M.M., Sayed, S., Ocular Drug Delivery: a Comprehensive Review. *AAPS Pharm. Sci. Tech.*, 24, 2, 66, 2023. Available from: https://link.springer.com/10.1208/s12249-023-02516-9.
4. Kim, Y.C., Chiang, B., Wu, X., Prausnitz, M.R., Ocular delivery of macromolecules. *J. Control. Release*, 190, 172–81, 2014. Available from: https://www.sciencedirect.com/science/article/pii/S0168365914004520.
5. Li, S., Chen, L., Fu, Y., Nanotechnology-based ocular drug delivery systems: recent advances and future prospects. *J. Nanobiotechnol.*, 21, 1, 232, 2023. Available from: https://jnanobiotechnology.biomedcentral.com/articles/10.1186/s12951-023-01992-2.
6. Gaudana, R., Ananthula, H.K., Parenky, A., Mitra, A.K., Ocular Drug Delivery. *AAPS J.*, 12, 3, 348–60, 2010. Available from: http://link.springer.com/10.1208/s12248-010-9183-3.
7. Han, H., Li, S., Xu, M., Zhong, Y., Fan, W., Xu, J. *et al.*, Polymer- and lipid-based nanocarriers for ocular drug delivery: Current status and future perspectives. *Adv. Drug Deliv. Rev.*, 196, 114770. Available from: https://www.sciencedirect.com/science/article/pii/S0169409X23000856.
8. Agrahari, V., Mandal, A., Agrahari, V., Trinh, H.M., Joseph, M., Ray, A. *et al.*, A comprehensive insight on ocular pharmacokinetics. *Drug Deliv. Transl. Res.*, 6, 6, 735–54, 2016. Available from: https://www.ncbi.nlm.nih.gov/pmc/articles/PMC5319401/.
9. Cholkar, K., Dasari, S.R., Pal, D., Mitra, A.K., Eye: Anatomy, physiology and barriers to drug delivery, in: *Ocular transporters and receptors*, pp. 1–36, Elsevier, 2013. Available from: https://www.sciencedirect.com/science/article/pii/B9781907568862500010.
10. Kim, S.H., Lutz, R.J., Wang, N.S., Robinson, M.R., Transport barriers in transscleral drug delivery for retinal diseases. *Ophthalmic Res.*, 39, 5, 244–54, 2007. Available from: https://karger.com/ore/article-abstract/39/5/244/258008.
11. Xu, X., Li, G., Zuo, Y.Y., Biophysical properties of tear film lipid layer I. Surface tension and surface rheology. *Biophys. J.*, 121, 3, 439–50, 2022. Available from: https://www.cell.com/biophysj/pdf/S0006-3495(21)03957-6.pdf.
12. Ramsay, E., Del Amo, E.M., Toropainen, E., Tengvall-Unadike, U., Ranta, V.P., Urtti, A. *et al.*, Corneal and conjunctival drug permeability: Systematic comparison and pharmacokinetic impact in the eye. *Eur. J. Pharm. Sci.*, 119, 83–9, 2018. Available from: https://www.sciencedirect.com/science/article/pii/S0928098718301556.
13. Kwatra, D. and Mitra, A.K., Drug delivery in ocular diseases: Barriers and strategies. *World J. Pharmacol.*, 2, 4, 78–83, 2013. Available from: https://www.wjgnet.com/2220-3192/full/v2/i4/78.htm.

14. Chen, M.S., Hou, P.K., Tai, T.Y., Lin, B.J., Blood-ocular barriers. *Tzu Chi Med. J.*, 20, 1, 25–34, 2008. Available from: https://www.sciencedirect.com/science/article/pii/S101631900860004X.
15. Cunha-Vaz, J., The blood-ocular barriers. *Surv. Ophthalmol.*, 23, 5, 279–96, 1979. Available from: https://www.sciencedirect.com/science/article/pii/0039625779901589.
16. Díaz-Coránguez, M., Ramos, C., Antonetti, D.A., The inner blood-retinal barrier: Cellular basis and development. *Vision Res.*, 139, 123–37, 2017. Available from: https://www.sciencedirect.com/science/article/pii/S0042698917300986.
17. Schlosshauer, B., 24 Blood–Retina Barriers, in: *Handbook of Neurochemistry and Molecular Neurobiology: Neural Membranes and Transport.*, 2007.
18. Mains, J. and Wilson, C.G., The Vitreous Humor As a Barrier to Nanoparticle Distribution. *J. Ocul. Pharmacol. Ther.*, 29, 2, 143–50, 2013. Available from: http://www.liebertpub.com/doi/10.1089/jop.2012.0138.
19. Hsueh, P.Y., Edman, M.C., Sun, G., Shi, P., Xu, S., Lin, Y.A. *et al.*, Tear-mediated delivery of nanoparticles through transcytosis of the lacrimal gland. *J. Control. Release*, 208, 2–13, 2015. Available from: https://www.sciencedirect.com/science/article/pii/S0168365914008141.
20. Del Amo, E.M., Rimpelä, A.K., Heikkinen, E., Kari, O.K., Ramsay, E., Lajunen, T. *et al.*, Pharmacokinetic aspects of retinal drug delivery. *Prog. Retin. Eye Res.*, 57, 134–85, 2017. Available from: https://www.sciencedirect.com/science/article/pii/S1350946216300635.
21. Mitra, A.K., Role of Transporters in Ocular Drug Delivery System. *Pharm. Res.*, 26, 5, 1192–6, 2009. Available from: http://link.springer.com/10.1007/s11095-009-9862-x.
22. Duvvuri, S., Majumdar, S., Mitra, A.K., Role of metabolism in ocular drug delivery. *Curr. Drug Metab.*, 5, 6, 507–15, 2004. Available from: https://www.ingentaconnect.com/content/ben/cdm/2004/00000005/00000006/art00003.
23. Desai, M. and Kapadia, J., Factors Affecting Intra-ocular Bioavailability of Drugs. *Asian J. Pharm. Clin. Res.*, 15, 9, 15–24, 2022. Available from: https://www.academia.edu/download/101581061/26918.pdf.
24. Loescher, M., Seiz, C., Hurst, J., Schnichels, S., Topical drug delivery to the posterior segment of the eye. *Pharmaceutics.*, 14, 1, 134, 2022. Available from: https://www.mdpi.com/1999-4923/14/1/134.
25. Kour, J., Kumari, N., Sapra, B., Ocular prodrugs: Attributes and challenges. *Asian J. Pharm. Sci.*, 16, 2, 175–91, 2021. Available from: https://www.ncbi.nlm.nih.gov/pmc/articles/PMC8105420/.
26. Cristofoli, M., Kung, C.P., Hadgraft, J., Lane, M.E., Sil, B.C., Ion Pairs for Transdermal and Dermal Drug Delivery: A Review. *Pharmaceutics*, 13, 6, 909, 2021. Available from: https://www.ncbi.nlm.nih.gov/pmc/articles/PMC8234378/.
27. Subrizi, A., Del Amo, E.M., Korzhikov-Vlakh, V., Tennikova, T., Ruponen, M., Urtti, A., Design principles of ocular drug delivery systems: Importance

of drug payload, release rate, and material properties. *Drug Discov. Today*, 24, 8, 1446–57, 2019. Available from: https://www.sciencedirect.com/science/article/pii/S1359644618303994.
28. Dubashynskaya, N., Poshina, D., Raik, S., Urtti, A., Skorik, Y.A., Polysaccharides in ocular drug delivery. *Pharmaceutics.*, 12, 1, 22, 2019. Available from: https://www.mdpi.com/1999-4923/12/1/22.
29. Mostafa, M., Al Fatease, A., Alany, R.G., Abdelkader, H., Recent Advances of Ocular Drug Delivery Systems: Prominence of Ocular Implants for Chronic Eye Diseases. *Pharmaceutics.*, 15, 6, 1746, 2023. Available from: https://www.mdpi.com/1999-4923/15/6/1746.
30. Cao, Y., Samy, K.E., Bernards, D.A., Desai, T.A., Recent advances in intraocular sustained-release drug delivery devices. *Drug Discov. Today*, 24, 8, 1694–700, 2019. Available from: https://www.ncbi.nlm.nih.gov/pmc/articles/PMC6708500/.
31. Gadziński, P., Froelich, A., Wojtyłko, M., Białek, A., Krysztofiak, J., Osmałek, T., Microneedle-based ocular drug delivery systems – recent advances and challenges. *Beilstein J. Nanotechnol.*, 13, 1167–84, 2022. Available from: https://www.ncbi.nlm.nih.gov/pmc/articles/PMC9623140/.
32. Glover, K., Mishra, D., Gade, S., Vora, L.K., Wu, Y., Paredes, A.J. *et al.*, Microneedles for advanced ocular drug delivery. *Adv. Drug Deliv. Rev.*, 201, 115082, 2023. Available from: https://www.sciencedirect.com/science/article/pii/S0169409X23003976.
33. Kang-Mieler, J.J., Rudeen, K.M., Liu, W., Mieler, W.F., Advances in ocular drug delivery systems. *Eye.*, 34, 8, 1371–9, 2020. Available from: https://www.nature.com/articles/s41433-020-0809-0.
34. Kim, H.M. and Woo, S.J., Ocular Drug Delivery to the Retina: Current Innovations and Future Perspectives. *Pharmaceutics.*, 13, 1, 108, 2021. Available from: https://www.ncbi.nlm.nih.gov/pmc/articles/PMC7830424/.
35. Lovrec-Krstič, T., Orthaber, K., Maver, U., Sarenac, T., Review of Potential Drug-Eluting Contact Lens Technologies. *Mater. (Basel)*, 16, 10, 3653, 2023. Available from: https://www.ncbi.nlm.nih.gov/pmc/articles/PMC10222559/.
36. Bennett, L., Topical Versus Systemic Ocular Drug Delivery, in: *Ocular Drug Delivery: Advances, Challenges and Applications*, R.T. Addo (Ed.), pp. 53–74, Springer International Publishing, Cham, 2016. Available from: http://link.springer.com/10.1007/978-3-319-47691-9_5.
37. Fang, G., Yang, X., Wang, Q., Zhang, A., Tang, B., Hydrogels-based ophthalmic drug delivery systems for treatment of ocular diseases. *Mater. Sci. Eng. C*, 127, 112212, 2021. Available from: https://www.sciencedirect.com/science/article/pii/S0928493121003520.
38. Wu, Y., Liu, Y., Li, X., Kebebe, D., Zhang, B., Ren, J. *et al.*, Research progress of *in-situ* gelling ophthalmic drug delivery system. *Asian J. Pharm. Sci.*, 14, 1, 1–15, 2019. Available from: https://www.ncbi.nlm.nih.gov/pmc/articles/PMC7032175/.

39. Bordet, T. and Behar-Cohen, F., Ocular gene therapies in clinical practice: viral vectors and nonviral alternatives. *Drug Discov. Today*, 24, 8, 1685–93, 2019. Available from: https://www.sciencedirect.com/science/article/pii/S1359644618304720.
40. Moiseev, R.V., Morrison, P.W.J., Steele, F., Khutoryanskiy, V.V., Penetration Enhancers in Ocular Drug Delivery. *Pharmaceutics*, 11, 7, 321, 2019. Available from: https://www.ncbi.nlm.nih.gov/pmc/articles/PMC6681039/.
41. Barot, M., Bagui, M., Gokulgandhi, M.R., Mitra, A.K., Prodrug Strategies in Ocular Drug Delivery. *Med. Chem.*, 8, 4, 753–68, 2012. Available from: https://www.ncbi.nlm.nih.gov/pmc/articles/PMC4407659/.
42. Suresh, P.K. and Paul, S.D., Ion-paired drug delivery: an avenue for bioavailability improvement. *Sierra Leone J. Biomed. Res.*, 3, 2, 70–6, 2011. Available from: https://www.ajol.info/index.php/sljbr/article/view/71806.
43. Wu, K.Y., Tan, K., Akbar, D., Choulakian, M.Y., Tran, S.D., A new era in ocular therapeutics: advanced drug delivery systems for uveitis and neuro-ophthalmologic conditions. *Pharmaceutics*, 15, 7, 1952, 2023.
44. Ophthalmic Drugs Market Study by Anti-inflammatory Drugs, Anti-infective Drugs, Anti-glaucoma Drugs, Anti-allergy Drugs, and Anti-VEGF Drugs from 2023 to 2033. Available from: https://www.factmr.com/report/4799/ophthalmic-drugs-market.

12

Vaginal Novel Drug Delivery Systems

Yasaman Khajeamiri[1] and Solmaz Ghaffari[2]*

[1]Department of Pharmaceutics and Pharmaceutical Nanotechnology, School of Pharmacy, Shahid Beheshti University of Medical Sciences, Tehran, Iran
[2]Department of Pharmaceutics, Faculty of Pharmacy, Tehran Medical Sciences, Islamic Azad University, Tehran, Iran

Abstract
The vagina has been considered as an important route for drug delivery. Vaginal drug delivery offers benefits such as providing targeted systemic or localized treatment with lower potential side effects, as it bypasses the gastrointestinal (GI) tract and liver metabolism. Moreover, medications can be administered in a variety of dosage forms including solids, semisolids, and liquids. Despite numerous advantages of this route, vaginal drug delivery faces challenges due to the existence of various biological barriers including fluid flow, pH barrier, mucus barrier, enzymatic barrier, epithelial barrier, and immune barrier. Conventional vaginal dosage forms suffer from drawbacks such as the possibility of leakage, inadequate retention time (and hence, low bioavailability), and the need to frequent dosing. These all may lead to a lack of patient compliance. In recent years, to overcome the limitations of traditional vaginal drug delivery systems, various novel approaches (e.g., nanotechnology, thermo-gelation, and mucoadhesion) have been employed. In this chapter, we aim to investigate novel vaginal drug delivery systems by understanding the main concepts including vaginal anatomy and physiology, major barriers against vaginal drug delivery, necessary considerations in vaginal formulations, classification of novel vaginal drug delivery systems, vaginal pharmacokinetics, and some of the novel vaginal products available in the market.

Keywords: Novel drug delivery systems, vaginal drug delivery, vaginal barriers, vaginal dosage forms, vaginal pharmacokinetics

**Corresponding author*: ghafari.s@iaups.ac.ir

Mehrdad Hamidi and Mahsa Sayed Tabatabaei (eds.) Novel Drug Delivery Systems: Fundamentals and Applications, (419–448) © 2025 Scrivener Publishing LLC

12.1 Introduction

Vaginal drug delivery systems are considered as a fascinating area of pharmaceutical research and development, as they offer a targeted and efficient way to deliver drugs with improved bioavailability [1]. The vaginal route acts as a potential drug administration site for a non-invasive, controlled trans-mucosal local and systemic drug delivery of various therapeutic agents [2]. This route is particularly useful for delivering hormones, antifungal agents, contraceptives, and other medications for localized treatment [3].

Vaginal drug delivery allows localized drug delivery by allowing direct delivery of drugs to the site of action, providing targeted treatment with lower systemic exposure and potential side effects. This approach can also improve bioavailability, especially compared to oral administration, by bypassing the harsh environment of the GI tract and first-pass metabolism in the liver [1, 3–6]. Vaginal drug delivery systems can also offer convenient dosing options, making them suitable for self-administration [1, 3]. In addition, controlled-release formulations can be designed to provide sustained drug release over an extended period, reducing the frequency of dosing [3, 6].

Conventional vaginal drug delivery systems can be designed in various forms such as gels, creams, tablets, and suppositories [7]. However, the traditional vaginal drug delivery approach has several limitations that contribute to the exploration of novel delivery methods such as potential irritation, risk of infection, lack of sustained release, inconsistent drug absorption, lack of targeting, and low user acceptance [1, 3, 7]. Addressing these limitations through the development of novel vaginal drug delivery systems is an active area of research and innovation in pharmaceutical and biomedical sciences.

Novel approaches in vaginal drug delivery aim to address the limitations of traditional formulations and improve therapeutic outcomes for various conditions, including infections, reproductive health disorders, and gynecologic cancers. Researchers continue to explore new technologies and formulations to enhance the safety, efficacy, and patient acceptance of vaginal drug delivery systems [4]. Moreover, new research in the field of novel vaginal drug delivery not only seek to improve the local effect but also actively address the systemic delivery of biomolecules such as peptides, proteins, RNAs, and so on [2, 8, 9]. Novel intravaginal systems have been recently explored to deliver drugs into systemic circulation [2, 10]. In order to conquer the drawbacks associated with the conventional vaginal systems, various novel approaches have been introduced including [2, 4, 8–10] nanotechnology, mucoadhesive formulations, smart (responsive) drug delivery systems, vaginal devices, *in-situ* gelling systems, hydrogels, and nanocarrier-in-gel systems to allow controlled and extended drug release.

12.2 Anatomy and Physiology

The upper reproductive tract in females includes the fallopian tubes, ovaries, and uterus, while the lower reproductive tract includes the vagina, ectocervix, and vulva (external genitalia) (Figure 12.1a) [11]. The vagina is a muscular, tubular structure that extends from the vulva to the cervix of the uterus. It is approximately 3-4 inches in length. The inner layers of the vagina comprise inner mucosa, intermediate muscular tissue, and outer connective tissue (Figure 12.1b).

The mucosa of the vagina consists of multiple layers that include epithelial cells, connective tissue, and blood vessels. The inner layer covering the vagina consists of multi-layered squamous epithelium [11]. This mucosal lining provides a protective barrier against pathogens and foreign particles, helping to prevent infections. Moreover, the vaginal mucosa helps maintain an acidic pH, which is essential for the healthy microbial ecosystem of the vagina. Also, the vaginal structure has elastic properties that allow it to stretch during childbirth and sexual activity and then return to its original shape [11]. Understanding the physiology and anatomy of the vagina is essential for designing novel drug delivery systems. The main permeability barrier of the vaginal, squamous epithelium, is equipped with ceramides, cholesterol, and glucosyl-ceramides [12]. According to previous studies,

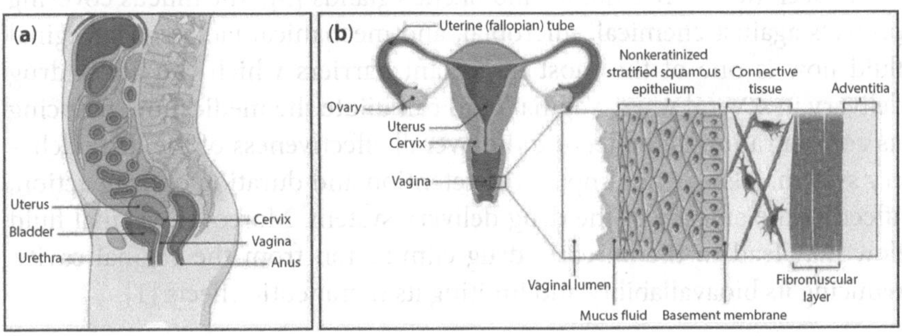

Figure 12.1 (a) Anatomy of vagina; (b) different layers of the vagina (created with BioRender.com).

the permeability of the vagina to water is almost 16 times higher than that of the skin [13]. The rate and extent of intravaginal drug absorption is dependent on the different physiological factors such as thickness changes of the epithelium layer, cyclic changes, hormonal changes, vaginal fluid volume, and change of vaginal pH [3]. As previously mentioned, one of the valuable features of vaginal drug delivery is that the active molecules absorbed from the vaginal route avoid the hepatic metabolism (first-pass effect). So, to attain a certain blood concentration, a lower dose in required than the oral route [1–3].

12.3 Barriers to Vaginal Drug Delivery

The vagina has been considered a vital route for drug delivery with a wide variety of dosage forms [1]. However, vaginal drug delivery remains a significant challenge due to various physiological and biological barriers in the vagina [1, 3, 14]. These barriers can contribute to poor vaginal drug delivery, impacting the absorption of drugs administered into the vaginal canal including fluid flow, pH barrier, mucus barrier, enzymatic barrier, epithelial barrier, and immune barrier [14]. Considering these barriers is essential for developing strategies to overcome these challenges and improve the effectiveness of vaginal drug delivery systems.

12.3.1 Fluid Flow

The outer epithelium of the vagina is commonly moisturized and covered by vaginal secretions. Vaginal fluid production varies with age so it is 50% less in postmenopausal women than in women of reproductive age [1]. This fluid is made up of a mixture of cervical mucus, endometrial and tubular fluids, secretion of Bartholin's and Skene's glands [3]. The mucus covering protects against chemical, microbial, and mechanical factors [1]. Vaginal fluid flow is one of the most important barriers which can affect drug delivery in several ways. Vaginal fluid can dilute the medication, reducing its concentration and affecting the overall effectiveness of the drug delivery system. Also, it can impact the retention and duration of drug action, affecting the efficacy of the drug delivery system. Moreover, vaginal fluid flow may lead to the an early drug elimination from the vaginal cavity, reducing its bioavailability and limiting its therapeutic effects [1].

12.3.2 pH Barrier

The vaginal pH barrier presents a challenge for drug delivery due to its dynamic nature. The pH of vagina is about 3.8 to 4.5, providing an acidic environment that can affect drug stability and solubility [1, 3]. This acidity can influence the performance of drug delivery systems and pose a barrier to achieving optimal drug concentrations [14, 15]. Moreover, pH-sensitive drugs may undergo degradation or precipitation, compromising their bioavailability and therapeutic effectiveness [16]. Overcoming the pH barrier is crucial for the successful development of vaginal drug delivery methods.

12.3.3 Mucus Barrier

In vaginal drug delivery, the mucus barrier plays a vital role in influencing the absorption, retention, and distribution of drugs within the vaginal canal. The mucus barrier consists of a complex network of glycoproteins, water, and other macromolecules, forming a viscous layer that overlays the underlying epithelial cells [17, 18]. The mucus barrier provides a gel-like matrix covering the vaginal epithelium that acts as a physical and molecular sieve, selectively allowing certain drugs to pass through while impeding the entry of others [1]. The composition of the mucin fibers within the mucus affects the permeability of the barrier, influencing the diffusion of drug particles. Moreover, the mucus barrier can trap drug particles, affecting their residence time and bioavailability. Additionally, the mucus layer can interact with drug molecules, potentially influencing their release kinetics and dispersion within the vaginal environment [19, 20].

It's important to note that the mucus barrier's properties and thickness can vary across the menstrual cycle and in different physiological states, impacting drug permeation and retention [20]. Understanding the dynamic nature of the mucus barrier is crucial for designing effective vaginal drug delivery systems.

12.3.4 Enzymatic Barrier

Vaginal fluid contains various enzymes including proteases, glucosidases, and phosphatases. These enzymes could interact with drug molecules and drug delivery systems, leading to potential metabolism and degradation of active substances. However, the genital tract has less enzyme activity than the GI tract leading to less degradation of drugs [3].

12.3.5 Epithelial Barrier

The vaginal epithelium serves as a protective layer that separates the vaginal lumen from underlying tissue. This layer actively participates in drug delivery by transporting drugs across the epithelial barrier [1]. The epithelial barrier significantly influences vaginal drug delivery by regulating the absorption and distribution of drugs administered into the vaginal canal. The thickness and composition of the epithelial barrier vary across different regions of the vaginal canal, impacting the penetration and distribution of drugs [21]. This barrier serves as a protective layer composed of tightly packed epithelial cells and mucus, forming a semi-permeable membrane. Its selective permeability allows certain substances, including drugs, to pass through while preventing the entry of harmful pathogens and toxins. When drugs are administered vaginally, they encounter the epithelial barrier, which serves as the first point of contact [1]. The barrier can selectively hinder or facilitate the passage of drug molecules based on their size, charge, and lipophilicity [22]. This selective permeability influences the rate and extent of drug absorption into the systemic circulation [3]. Moreover, this barrier can affect the bioavailability as some drug molecules may be absorbed directly into the systemic circulation by crossing the epithelial barrier, while others may undergo early metabolism [23].

12.3.6 Immune Barrier

Vaginal epithelial cells are the first line to defense against pathogens [1, 24]. Moreover, the presence of immune cells such as dendritic cells and macrophages in the vaginal tissue, which provide the immune response, can influence the drug delivery system [25, 26]. The interaction of immune cells with the drug delivery system affects drug absorption and distribution within vaginal mucosa by regulating the entry of foreign substances [1, 27].

12.4 Formulation Considerations

Effective formulation design and optimization can address the aforementioned hurdles, leading to enhanced drug absorption and therapeutic outcomes in vaginal drug delivery applications. Some key formulation considerations for efficient vaginal drug delivery are explained below.

12.4.1 Physicochemical Properties of Drug Molecules

The rate of diffusion of drug molecules into the mucosal tissue depends on the physicochemical properties of the drug molecules and the tissue properties. Optimizing the physicochemical properties of drug molecules including molecular weight (MW), size, lipophilicity, partition coefficient, and polarity is essential for successful vaginal drug delivery [28]. Generally, drug molecules with MW < 500 Da are considered appropriate candidates for penetration through the mucosal barrier. Moreover, the logP of medicinal molecules affects the process of absorption from vaginal mucosa. There are different ways to transfer molecules through the vaginal route, the main ones are transcellular and paracellular. Generally, lipophilic molecules are mostly absorbed through the transcellular route while hydrophilic therapeutics are mostly transported by paracellular diffusion [29]. Lipophilic drug molecules with low molecular weights are better candidates for vaginal absorption than larger substances or hydrophilic compounds [30]. In addition, in order to achieve proper water solubility and membrane permeability, it is important to consider the ionization of drug molecules at the vaginal pH [29, 30]. The pH of vaginal secretions affects the ionization state and determines drug penetration through the vaginal mucosa [29, 30].

Moreover, the choice of appropriate and compatible excipients in the formulation affects the stability, solubility, and permeability of the drug [31]. Novel excipients with a significant impact on the design of multipurpose vaginal pharmaceutical formulations have properties such as pH-responsiveness, swelling, and bioadhesive properties with low toxicity effects [31]. In this regard, researchers have suggested several biocompatible polymers, such as hydroxypropyl methylcellulose (HPMC), chitosan, methacrylic acid copolymers, polylactic co-glycolide (PLGA) and polyethylene oxide to promote mucosal adhesion and improve drug release [31].

12.4.2 Physicochemical Properties of Nanocarriers

Polymeric nanoparticles with at least 500 nm in diameter can be delivered through cervicovaginal mucus with a transport rate depending on the surface properties of nanoparticles. It has been shown in several studies that different nanoparticles with a diameter less than 300 nm are prone to penetrate the epithelium and even in animal studies, it has been shown that they are able to reach deeply into lamina propria. Decreasing size usually contributes to improving cellular uptake and diffusion rate of nanoparticles [32]. Surface chemistry has a critical impact on the diffusion behavior of nanoparticles. Nanoparticle mucin interactions and their long-lasting attachment may occur due to the existence of hydrogen bonding, physical entanglement, covalent bonding, electrostatic and hydrophobic interactions [32, 34]. Mucoadhesive properties of nanoparticles are highly beneficial by considering the natural self-cleaning mechanisms of the vagina, which cause poor retention and absorption of the drug in the mucosa [32].

Several studies have demonstrated that using mucoadhesive polymers in the production or surface modification of nanocarriers is associated with enhancing adhesive interactions, retention, and residence time of medication in the vaginal mucosa [35, 36]. Moreover, the use of positively charged nanoparticles can cause an increase in the bioadhesion of a vaginal formulation [32]. Electrostatic interactions between the mucin with negative charge and the positively charged nanoparticles increases binding to the mucin and as a result, resistance to vaginal clearance after application. In addition, the toxicity of nanosystems should be considered [29, 32]. The accumulation of nanocarriers or their components in the genital tract probably depends on biodegradability, the amount and frequency of administration, and penetration of nanoparticles into the tissue [32]. However, most of the retained materials are eliminated by the natural mucus turnover and discharge of the genital tract [37].

12.4.3 Formulation-Related Factors

There are several challenges in development of an appropriate formulation to overcome the vaginal barriers [14]. According to vaginal anatomical and physiological characteristics, the optimal vaginal formulation must have a series of properties including the ability to deliver hydrophilic or lipophilic compounds with low molecular weights to achieve optimal absorption rates, being partially soluble in water due to the aqueous nature of vaginal secretions, being easy to administer with minimal discomfort, and being bioadhesive to prevent removal from the vaginal canal [29, 38, 39]. Moreover, pH and osmolarity play crucial roles in the vaginal formulations. According to World Health Organization (WHO) recommendations, the optimal pH and osmolarity for intravaginal formulations are about 4.5 and lower than 1200 mOsm/kg (and ideally not more than 380 mOsm/kg), repectively [29, 40]. According to the findings of previous studies, hypotonic formulations (with < 260 mOsm/kg) provide more particle penetration through the mucus layers and a safer profile in comparison with hyperosmolar vaginal dosage forms [41–44]. Moreover, the size of the inner phase particles/droplet of formulation matters that the aim of vaginal drug delivery formulations can be classified into 3 groups based on this factor: i) 100–150 nm; aim at function within the vaginal mucus ii) 200–500 nm; aim at permeate through the vaginal epithelium iii) > 1000 nm; aim at function on the surface of vaginal mucus [29].

12.5 Classification

Vaginal administration has a historical background in the traditional medicine. In the past decades, intravaginal formulations have been developed for controlled drug delivery [45]. The most commonly used intravaginal dosage forms and devices are provided in Figure 12.2.

12.5.1 Solid Dosage Forms and Devices

12.5.1.1 Vaginal Tablets (Pills)

Vaginal pills or tablets include regular vaginal pills, effervescent pills, sustained-release pills, double-layer pills, and bioadhesive pills. In terms of shape, vaginal pills are usually oval or pear-shaped, which are easy to maintain in the vagina. These tablets can be inserted into the vagina with the aid of an applicator and should be easily dissolved or melted to release the drug in the medium. Sex hormone drugs can be prescribed simultaneously with anti-inflammatory and topically antimicrobial agents using vaginal tablets. Vaginal pills, like regular oral pills, may contain binders, disintegrants, and other excipients. Bioadhesive polymers are sometimes used in pill formulations to increase the residence time of the dosage form in the administration site. The release mechanism is according to drug diffusion *via* the swollen polymer and progressive erosion/dissolution of the hydrated matrix. The controlled-release profile of the vaginal tablets can be modified through the addition of soluble and insoluble fillers into the dosage form and alteration of their weight ratio [46, 47].

12.5.1.2 Vaginal Capsules

Both hard and soft capsules are employed for vaginal delivery. Hard capsules are generally prepared by filling powders, granules, pellets, mini-tablets, or nanocapsules into a hard gelatin shell. Clivon® Cap, as a commercially available vaginal hard gelatin capsule, contains hyaluronic acid, aloe vera and lactic acid. This product is administered as an adjuvant therapy for

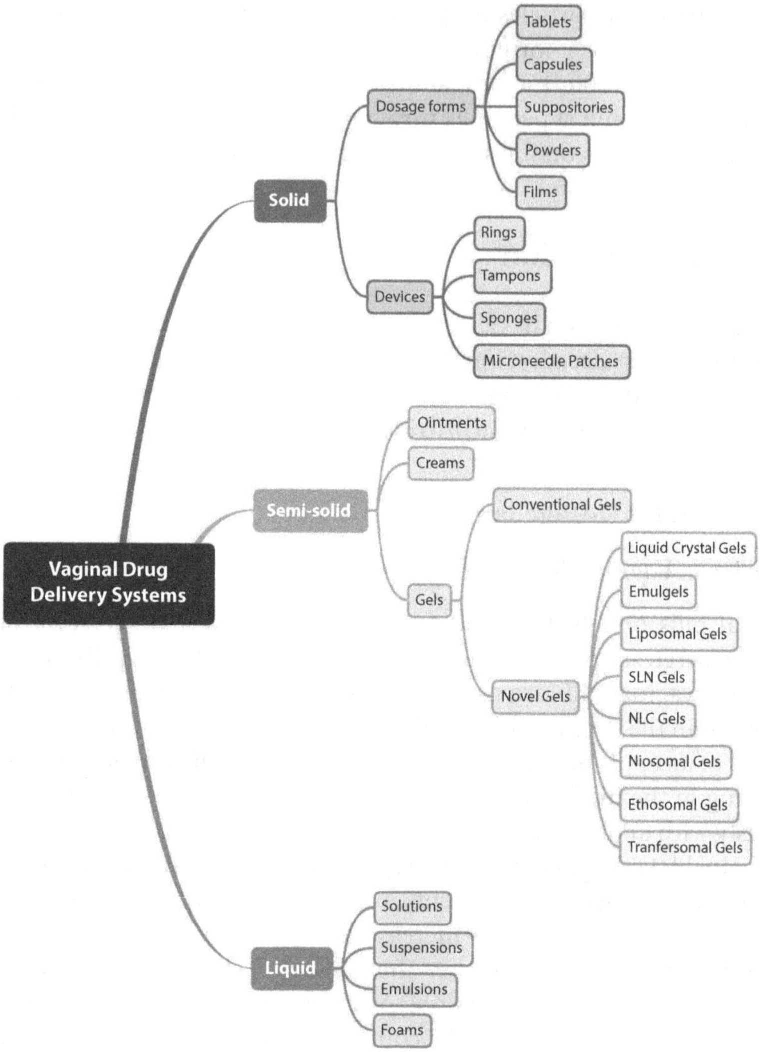

Figure 12.2 Classification of vaginal drug delivery systems.

vaginal dryness. Clivazol®, a combination medicine of an antibiotic and an antifungal, is an example of a vaginal soft gelatin capsule. Canesten® is also a vaginal soft gel capsule containing clotrimazole which is applied as a single-dose treatment of vaginal thrush [48].

12.5.1.3 *Vaginal Powders*

Powders were used by women for many years to maintain genital hygiene by absorbing moisture and unpleasant odors. Although powder use in the genital area has declined in recent decades, it remains a routine practice for some women [49]. Although there are not enough studies on the potential dangers of vaginal products, some studies provide insight into the health risks associated with lubricants, deodorant sprays, and vaginal powders that can cause bacterial vaginosis, urinary tract infections, intercourse and increase the risk of ovarian cancer in women [50, 51].

12.5.1.4 *Vaginal Suppositories*

A suppository is a solid dosage form used to deliver drug molecules by insertion into a body cavity, where it dissolves or melts to exert local or systemic effects. Some authors apply the terms pessaries and suppositories interchangeably. Different ingredients can be used to form the base of a vaginal suppository including cocoa butter or other similar lipid-based substituents, polyethylene glycol (PEG), and glycerinated gelatin. A wide range of vaginal medications is available in suppository form. Suppositories systems are currently the most commonly used dosage forms for the administration of drugs for cervical ripening before the childbirth and the local drug delivery in such situations [52, 53].

Vaginal suppositories containing antimicrobial agents (e.g., clindamycin, metronidazole, miconazole, and so forth) have been investigated to treat vaginal disorders like vaginosis or trichomoniasis and are available on the market [54, 55]. Also, products including ingredients for maintenance of the vaginal health are on the market like boric acid vaginal suppositories which support vaginal health by balancing the pH level [56]. Healthy microflora undertakes a major role in maintaining the vaginal health and preventing it from irritations. So, some vaginal suppositories are in the market from the probiotic category [57].

12.5.1.5 *Vaginal Films*

Vaginal films are flexible thin layers that are quickly dissolved or dispersed when exposed to vaginal fluids and turn into a mucoadhesive viscous solution or gel which adheres to the vaginal wall. Applicators are not needed to insert these dosage forms inside the vagina. Vaginal films have been studied for therapeutic (to deliver antimicrobial agents), cosmetic (to deodorize the vagina), and prophylactic actions (to deliver contraceptives). For example, the vaginal contraceptive film (VCF) which is commercially available in the US market contains 28% nonoxynol-9 spermicide in a polyvinyl alcohol base. VCF^{R} Dissolving Feminine Deodorant Film contains vanilla bean fragrance in a polyvinyl alcohol (PVA) base to eliminate unpleasant vagina odor. In the case of formulation of vaginal films, water-soluble mucoadhesive polymers (e.g., hyaluronic acid derivatives, cellulose derivatives, and carbopol) and plasticizers (e.g., PEG and glycerin) are among the main ingredients. The solvent casting technique is commonly employed to manufacture vaginal films [21]. Despite of many advantages of vaginal films, some limitations should be considered during their design and manufacturing. Due to the presence of hydrophilic substances in the formulation of vaginal films, special packaging is definitely required. Besides, the low capacity of drug loading of films is a main challenge for the delivery of high doses of medicine through films [58]. Designing new techniques tailored to the volume of vaginal fluids as well as achieving a comprehensive understanding of the processes of hydration and dispersion/dissolution of films in the vaginal area deserves further investigation [21].

12.5.1.6 Vaginal Rings

Vaginal rings refer to circular ring-formed devices which are designed for releasing the drug in a prolonged manner after insertion into the vagina. Therefore, they show a remarkable advantage in comparison with traditional formulations. Commercially available vaginal rings are made from thermoplastic polymers (e.g., Ornibel® and Nuvaring®) or silicone elastomers (e.g., Fertiring®).

Vaginal rings which are used for the delivery of contraceptive hormones are highly acceptable between users as they can release their medications very slowly and can be effective for a long time varying from 3 weeks to 1 year. Despite of diversity in sizes between rings, most of them have an approximate diameter of 58 mm with a cross section of 8.4 mm. The benefits of vaginal rings are being user-controlled, avoiding the need for daily intake of pills, and the possibility of the continuous delivery of low-dose steroids. In simple vaginal rings, the drug is simply dispersed in a polymeric matrix. The drug release rate from the surface of rings is higher than the inner parts, which may lead to the burst release. To overcome this drawback, different designs of ring have been proposed such as sandwich-type and reservoir-type. Sandwich-type devices include a small drug reservoir which is located between the non-medicated central core and outer band.

It is possible to have several cores within a single ring which allow delivery of different drug using one device. The core diameter and thickness of non-drug coating could be modified to achieve optimal drug release. The vaginal rings are commonly made up of polymers. The most commonly used polymers is poly (dimethylsiloxane) or other silicones, though other elastomeric polymers such as poly(ethyl vinyl acetate) and styrene-butadiene block copolymers have been recently investigated. Antimicrobials homogeneously dispersed in a polymeric vaginal ring could be released faster from the surface of the ring. Therefore, the main challenge associated with these systems is to find the optimal dose that produces the least amount of antimicrobial required to ensure antimicrobial activity. One strategy to minimize the initial drug concentration is the addition of extra rate-controlling drug-free elastomeric layers [59–64].

12.5.1.7 Vaginal Tampons

Vaginal tampons, fabricated of cellulose and/or cotton, are placed in the vagina for a limited time to absorb menstrual discharge and then removed [65]. To develop medicated tampons, drugs can be incorporated into the device. The first studies used tampons on the market that were impregnated with the drug through a simple immersion process. The FDA (Food and Drug Administration) has made medicated vaginal tampons available as an approved medical device. This dual-purpose tampon composed of a polymer delivery system (strips) that absorbs the menstrual fluid and, at the same time, releases lactic acid and citric acid, to preserve the acidic environment of vagina and prevent proliferation of potentially pathogenic bacteria [39, 91].

12.5.1.8 Vaginal Sponges

Vaginal sponges, as soft and round devices, fit inside the vaginal area and cover the cervix physically. The vinegar-soaked sponges were historically recommended as contraceptives. These devices have been criticized by some researchers because their use causes bacteria to colonize and multiply in the vagina, making women more susceptible to vaginitis and other reproductive disorders. Medicated sponges containing spermicides are still known and used as common birth control devices. These vaginal sponges can deliver drugs, absorb semen and block the cervical canal. They are inexpensive devices; and easy to insert and remove. However, these systems are less effective than other methods of contraception such as diaphragms. In the United States, vaginal sponges made of soft polyurethane foam (5.5 cm in diameter and 2.5 cm thick) and impregnated with 1 gram of nonoxynol-9 were introduced in 1983 as the authorized spermicide. The advantages of this formula were the possibility of using it for up to 24 hours without the need for additional use of spermicide. Due to non-compliance with FDA regulations in 1995, Allendale Pharmaceuticals purchased the sponge rights and re-launched its product in the US and Canada. Similar sponges are available outside the US containing benzalkonium chloride alone (Pharmatex, Innotech International) or a combination of nonoxynol-9 and sodium cholate (Protectaid, Pirri Pharma) [66–68].

12.5.1.9 Vaginal Microneedle Patches

Microneedle patches, an array of tiny needles located on a support base, are novel efficient devices to bypass physiological barriers without causing pain. Few studies have been published on the use of such systems in the vaginal area [69]. Although different types of microneedles have been fabricated, when it comes to vaginal drug delivery, dissolving microneedles are highlighted. A dissolving microneedle patch containing an antiretroviral drug has been introduced as a self-administered HIV pre-exposure prophylaxis [70]. In addition to vaginal immunization, treatment of vulvovaginal candidiasis using dissolving microneedle containing probiotic loaded multilayer microcapsules has also been investigated. However, vaginal delivery using microneedles is still in its infancy and further investigations are required in this respect [71].

12.5.2 Semisolid Dosage Forms

Semisolid dosage forms (including gels, creams, and less commonly, ointments) are among the most widely used vaginal formulations. Being easy to use and inexpensive, they are generally highly acceptable for patients. However, they also face several challenges such as low retention time and the possibility of leaking, which could negatively affect patient acceptance [45]. Therefore, recently, the role of novel semisolid formulations (e.g., mucoadhesive gels and thermo-gelling systems [72]) in the vaginal field has been highlighted. Here, some of these novel formulations are discussed.

12.5.2.1 Liquid Crystal Gels

Liquid crystals are mesophases between liquid and solid states with specific physicochemical and morphological characteristics. Liquid crystals are broadly categorized into lyotropic and thermotropic systems.

Thermotropic liquid crystals form upon the temperature change, while lyotropic ones are both dependent on the temperature and solvent. Different types of lyotropic liquid crystals (including lamellar, hexagonal, and cubic phases) have been recently in the spotlight as drug delivery systems, particularly in the vagina [73–77]. For instance, in a recent study, for sustained delivery of amphotericin B and miltefosine for local treatment of vaginal candidiasis, a drug-loaded microemulsion was prepared with the ability of swelling and phase transition into the hexagonal gel upon contact with vaginal fluids. This *in-situ* gel-forming system showed several advantages as a vaginal dosage form. First, it was a ready-to-use formulation with no need for further incorporation in a vehicle. Also, due to the higher viscosity of the formed gel, the retention time of formulation in the vagina could be improved, and hence, the dosing frequency may be reduced. Moreover, the possibility of leakage may decrease due to the *in situ* gelation [78]. More recently, a lamellar phase formulation with the ability of *in situ* swelling and forming the cubic liquid crystal was developed for controlled-release drug delivery. This formulation is composed of a precursor solution containing a model drug, sertaconazole nitrate, as a local vaginal treatment. Upon contact with the simulated vaginal fluid, the solution turned into the cubic phase with high stability, strength, and mucoadhesion [79]. In conclusion, such studies provide insights into the potential of *in situ* gel-forming mucoadhesive liquid crystals as promising sustained-release vaginal drug delivery systems.

12.5.2.2 Emulgel (Emulsion-in-Gel)

Emulgels, also known as gelled emulsions or creamed gels, are emulsions with a gelled continuous phase. So, they combine the advantages of both. They offer a dual-controlled drug release with a high loading capacity, especially for lipophilic compounds. Also, emulgels possess a transparent and pleasant texture (non-greasy, fresh cooling, washable, and easily spreadable) that improves patient acceptance [80]. To date, several microemulgel and nanoemulgel formulations have been studied for drug delivery to the vagina. When a polymer is added to a microemulsion, a microemulgel is formed with reduced mobility and higher stability [29]. Vaginal microemulgels have been investigated so far for the delivery of antifungal [81–83], microbiocide [84], and contraceptive [85] agents. Also, nanoemulgel (nanoemulsion-in-gel) formulations have garnered attention in recent years for the delivery of antimicrobials to treat vaginal infections [86–88].

12.5.2.3 Liposomal Gel

Liposomal gels (liposome-in-gel systems) are semisolid formulations in which liposomes are embedded in a hydrogel matrix. These systems combine the advantages of liposomes (e.g., incorporation of lipophilic drugs) and hydrogels (e.g., mucoadhesion and improved contact time). Also, this platform can overcome the limitations of both; such as the stability of liposomes against environmental situations and the low drug-loading capacity of gels [89, 90]. In recent years, liposomal gels have attracted significant attention as vaginal drug delivery systems. In this regard, chitosan as a natural hydrogel base has attracted much attention. This is because chitosan is a cationic polymer with the capability of dissolving and swelling in an acidic environment. So, in the normal acidic pH of the vagina, it could provide excellent mucoadhesive properties [91].

For instance, for localized treatment of vaginal inflammation and infection, natural polyphenols (resveratrol or epicatechin) with poor water solubility were loaded into liposomes. Then, drug-loaded liposomes were incorporated within a chitosan-based hydrogel with intrinsic antimicrobial properties for prolonged drug release [90]. In a following study, a similar platform (resveratrol-loaded liposomes-in-chitosan hydrogel) was developed for local therapy of vaginal infection caused by *Chlamydia trachomatis* [92]. Apart from conventional forms of liposomes, novel liposomal systems can also be incorporated in gels. For instance, Vanic *et al.* prepared deformable liposomes containing propylene glycol in a hydrogel base for treatment of vaginal microbial infections. These liposomes encapsulated metronidazole or clotrimazole for controlled drug release. Compared to conventional liposomes, deformable liposomes showed a faster penetration through the gel matrix. Also, they provided an efficient sustained antimicrobial delivery [93]. Altogether, liposomal hydrogels seem promising platforms for localized vaginal drug delivery.

12.5.2.4 Solid Lipid Nanoparticle (SLN) Gel

In addition to liposomes, other lipid nanocarriers for vaginal drug delivery have been also studied to be incorporated within a hydrogel matrix. Solid lipid nanoparticles (SLNs) are an alternative to liposomes with greater stability and high loading capacity. It has been shown that SLN-based gels could provide a prolonged vaginal mucoretention [14, 94].

Particularly, SLNs have been studied with the aim of delivering antifungal agents, with high lipophilicity to improve their therapeutic outcomes in the target site [95]. For instance, itraconazole-loaded SLNs were dispersed in a polymeric gel to provide a thermosensitive and mucoadhesive system. More recently, luliconazole-in-SLN-in-thermosensitive gel was developed for the treatment of candidal vulvovaginitis [96]. Apart from the delivery of antifungals, SLN gels have been used for other therapeutic purposes. For example, estradiol-containing SLNs coated with chitosan were investigated as a sustained-release mucoadhesive gel for intravaginal estrogen therapy [91]. In addition to hydrogel-based systems, SLNs have also been employed to form lipid-based creamy gels with no need for further gelling agents. Anjum and coworkers prepared dolutegravir sodium-loaded SLNs for prophylaxis of HIV. During a one-step manufacturing process, the lipid phase was added to the aqueous phase (under the temperature of 70 °C) to provide a creamy semisolid formulation. This SLN-based gel demonstrated a sustained release profile with high local deposition [97].

12.5.2.5 Nanostructured Lipid Carrier (NLC) Gel

Nanostructured lipid carriers (NLCs) are the second-generation lipid nanoparticles which composed of a mixture of solid and liquid lipids. Compared to the SLNs, NLCs have an improved drug loading capacity and minimized risk of medication leakage. NLC-containing hydrogel has been previously studied for local antifungal (miconazole) delivery to the oral mucosa [98]. As a promising platform, clotrimazole-containing NLC gel has been also explored for the treatment of fungal infections of the vagina. Drug-containing NLCs were then dispersed in a hydrogel matrix with thermogelling characteristics. The experiments showed that the activity of this formulation against *Candida Albicans* was 4 times higher than Fungizone® [99].

12.5.2.6 Niosomal Gels

Niosomes are the analogs of liposomes, composed of non-ionic surfactants with high stability and prolonged contact time. The importance of niosomes has been extensively demonstrated for dermal applications. However, their potential as vaginal drug delivery systems has been less noticed. In a study, Guo and coworkers proposed the formulation of niosomal gel for the long-term delivery of clotrimazole into the vagina for the treatment of candida vaginitis. Drug-loaded multilamellar niosomes were prepared and incorporated in a gel matrix to achieve a self-administrable dosage form with improved stability and viscosity. The results showed that this platform could be beneficial for reducing the dosing interval of clotrimazole [100].

12.5.2.7 Ethosomal Gels

Ethosomes are lipid-based vesicles containing a high amount of ethanol (as penetration enhancer) in their structure which promotes permeation through the biological barriers. Ethosomes, as transdermal drug delivery carriers, have been extensively studied. Recently, their potential for vaginal drug delivery has been explored. A pH-responsive ehtosomal gel was prepared for sustained delivery of metronidazole. *In vitro* evaluation provided clues about the capability of this system to create a high drug accumulation in the vagina with improved residence time, sustained release, and enhanced drug permeation. However, more studies are needed to investigate the *in vivo* capability of these systems in improving vaginal drug delivery [101].

12.5.2.8 Transfersomal Gels

Transfersomes, owing to the presence of edge activators in their structure, are considered ultradeformable version of liposomes. Thanks to their extreme flexibility, they can squeeze themselves into the membranes and improve penetration [102]. Additionally, transfersomes have a high loading capacity both for lipophilic and hydrophilic drugs. Also, they are useful for the delivery of small and macromolecule drugs due to their ability to pass the biological barriers [103]. The number of studies on the vaginal application of these promising carriers is still small. In one study, transfersomes have been investigated for vaginal delivery of nystatin to ameliorate vulvovaginal candidiasis [103]. Also, ketoconazole-containing transfersomal vaginal gel was prepared for the treatment of vulvovaginal candidiasis [102].

Taken together, these studies encourage researchers to further explore the potential of semisolid systems containing nanocarriers as promising candidates for future vaginal drug delivery.

12.5.3 Liquid Dosage Forms

Vaginal liquid dosage forms include solutions, suspensions, emulsions, and foams. Such vaginal products are commonly applied for irrigation and cleansing purposes. Since the residence time of liquid systems in the vagina is very short, they are not suitable for controlled drug release [104].

Vaginal foams have been tested to deliver drugs (mainly germicides or spermicides) into the vagina. VCF® vaginal contraceptive foam contains

12.5 % nonoxynol-9 and is inserted in the vagina one hour before intercourse using an applicator. The use of foams creates a good coverage of the vaginal mucosa with minimal leakage. Also, most foam bases are non-irritating, unlike some other conventional formulations which are reported to cause burning and itching. Therefore, improving their formula can be of great help in obtaining new vaginal drug delivery systems. Antimicrobials, local anesthetics, and hormones vaginal foams have the potential to gradually replace several existing dosage forms [105].

12.6 Pharmacokinetics

Understanding vaginal pharmacokinetics is important in the context of the development of efficient targeted vaginal drug delivery [44]. Pharmacokinetics of vaginal drug delivery refers to the process of absorption, distribution, metabolism, and elimination of drugs that are administered through the vaginal route. Drug absorption from vaginal systems includes two principal processes including dissolution in the vaginal lumen and permeation through the membrane. Thus, any physiological and formulation factor that affects drug dissolution and membrane penetration can potentially affect absorption from the vagina. Drugs administered vaginally may have different distribution patterns compared to other administration routes, due to the bypassed hepatic first-pass metabolism [106]. Distribution of a drug throughout the vaginal environment is a mass transfer process with the intervention of several types of active forces (such as compression by the channel walls, pressure gradient applied to the channel, and surface tension) that interact with passive diffusion to drive drug transport [44]. Moreover, drug administration through the vaginal route can result in sustained plasma concentrations. Although peak plasma concentration is higher and achieved sooner after oral and parenteral administration, detectable plasma concentration after vaginal administration has a longer lasting effect [107, 108].

Molecules undergoing vaginal absorption may be subject to metabolism by enzymes. Vaginal external cell layers and basal cells create the most enzyme activity [106]. By all of the vaginal enzymes, proteases are the main barrier against peptides and proteins vaginal delivery. It has been suggested that the vaginal epithelium has the most of the exopeptidases and endopeptidases [106]. It is difficult to predict the elimination process following vaginal administration which depends on drug levels in the systemic circulation [107, 108].

12.7 Products in the Market

There are several types of vaginal drug delivery systems available in the market. Examples of novel vaginal marketed products by their category are given in Table 12.1. These products offer several advantages including dosing convenience, controlled release, and localized or systemic treatment for various indications.

Table 12.1 List of some marketed vaginal novel drug delivery systems.

Category	**Product name**	**Active agent**	Formulation type
Solid dosage forms	Gynepro®	Metronidazole + Neomycin Sulphate + Polymyxin B + Nystatin	Vaginal suppository
	Canesten®	Clotrimazole	Vaginal tablet
	Vagifem®	Estradiol	Film-coated vaginal tablet
	Canesflor®	Probiotic (lactobacilli)	Vaginal capsule
	Colpotrophine®	Promestriene	Vaginal capsule
	IntraFresh®	Lactobacillus, garlic, rose oil	Vaginal pessary
	VCF®	Nanoxynol-9	Vaginal contraceptive film
	Annovera®	An estrogen (ethinyl estradiol) and a progestin (segesterone acetate)	Vaginal birth control ring
	NuvaRing®	Etonogestrel and ethinyl estradiol	Vaginal birth control ring
	TODAY®	Nonoxynol-9	Vaginal contraceptive sponge

(Continued)

Table 12.1 List of some marketed vaginal novel drug delivery systems. (*Continued*)

Category	**Product name**	**Active agent**	Formulation type
Semi-solid dosage forms	Gino-Canesten®	Clotrimazole	Vaginal cream
	Sertopic®, Dermofix®	Sertaconazole	Vaginal cream
	Gyno-pevaryl®	Econazole	Vaginal cream
	Lomexin®	Fenticonazole	Soft vaginal capsules
	Gino Travogen®	Isoconazol	Vaginal cream
	Dalacin V®	Clindamycin	Vaginal cream
	Ovestin®	Estriol	Vaginal cream
	Blissel®	Estriol	Vaginal gel
	Replens®	Vaginal moisturizer and lubricant	Vaginal gel
	Nanosilver gel®	Nanosilver	Silver nanoparticles in gel
	Candid-V®	Clotrimazole	Microemulsion-based vaginal gel
	VivaGel®	SPL7013, or astodrimer sodium	Dendrimer-based vaginal gel
Liquid dosage forms	Betadine®	Povidone-iodine	Vaginal solution
	Floragyn® Silver Vaginal Foam	Vaginal foam based on hyaluronic acid and wheat sprouts extract	Vaginal foam
	HyaluronV™	Anhydrous emollient base infused with hyaluronic acid (sodium hyaluronate)	Vaginal emulsion

References

1. Iqbal, Z. and Dilnawaz, F., Nanocarriers for vaginal drug delivery. *Recent Pat. Drug Deliv. Formul.*, 13, 1, 3–15, 1872-2113, 2019.
2. Gupta, S., Gabrani, R., Ali, J., Dang, S., Exploring novel approaches to vaginal drug delivery. *Recent Pat. Drug Deliv. Formul.*, 5, 2, 82–94, 1872-2113, 2011.
3. Sahoo, C.K., Nayak, P.K., Sarangi, D.K., Sahoo, T.K., Intra vaginal drug delivery system: an overview. *Am. J. Adv. Drug Deliv.*, 1, 43–55, 2013.
4. Mahant, S., Sharma, A.K., Gandhi, H., Wadhwa, R., Dua, K., Kapoor, D.N., Emerging trends and potential prospects in vaginal drug delivery. *Curr. Drug Deliv.*, 20, 6, 730–51, 1567-2018, 2023.
5. Srikrishna, S. and Cardozo, L., The vagina as a route for drug delivery: a review. *Int. Urogynecol. J.*, 24, 537–43, 0937-3462, 2013.
6. Hani, U., Bhat, R.S., Sisodiya, R., Shivakumar, H.G., Novel vaginal drug delivery systems: a review. *Curr. Drug Ther.*, 5, 2, 95–104, 1574-8855, 2010.
7. Ashok, V., Kumar, R.M., Murali, D., Chatterjee, A., A review on vaginal route as a systemic drug delivery. *Crit. Rev. Pharm. Sci.*, 1, 1, 8–9, 2012.
8. Garzon, S., Laganà, A.S., Barra, F., Casarin, J., Cromi, A., Raffaelli, R. *et al.*, Novel drug delivery methods for improving efficacy of endometriosis treatments. *Expert Opin. Drug Deliv.*, 18, 3, 355–67, 1742-5247, 2021.
9. Baxi, K., Sawarkar, S., Momin, M., Patel, V., Fernandes, T., Vaginal siRNA delivery: overview on novel delivery approaches. *Drug Deliv. Transl. Res.*, 10, 4, 962–74, 2190-393X, 2020.
10. Ensign, L.M., Cone, R., Hanes, J., Nanoparticle-based drug delivery to the vagina: a review. *J. Control. Release*, 190, 500–14, 0168-3659, 2014.
11. Cook, M.T. and Brown, M.B., Polymeric gels for intravaginal drug delivery. *J. Control. Release*, 270, 145–57, 0168-3659, 2018.
12. Patton, D.L., Thwin, S.S., Meier, A., Hooton, T.M., Stapleton, A.E., Eschenbach, D.A., Epithelial cell layer thickness and immune cell populations in the normal human vagina at different stages of the menstrual cycle. *Am. J. Obstet. Gynecol.*, 183, 4, 967–730002–9378, 2000.
13. van der Bijl, P., Thompson, I.O.C., Squier, C.A., Comparative permeability of human vaginal and buccal mucosa to water. *Eur. J. Oral. Sci.*, 105, 6, 571–5, 0909–8836, 1997.
14. Mirza, M.A., Panda, A.K., Asif, S., Verma, D., Talegaonkar, S., Manzoor, N. *et al.*, A vaginal drug delivery model. *Drug Deliv.*, 23, 8, 3123–34, 1071-7544, 2016.
15. Kim, S., Chen, Y., Ho, E.A., Liu, S., Reversibly pH-responsive polyurethane membranes for on-demand intravaginal drug delivery. *Acta Biomater.*, 47, 100–12, 1742-7061, 2017.
16. Yoo, J.-W., Giri, N., Lee, C.H., pH-sensitive Eudragit nanoparticles for mucosal drug delivery. *Int. J. Pharm.*, 403, 1-2, 262–7, 0378-5173, 2011.
17. Cone, R.A., Barrier properties of mucus. *Adv. Drug Deliv. Rev.*, 61, 2, 75–85, 0169-409X, 2009.

18. Boegh, M. and Nielsen, H.M., Mucus as a barrier to drug delivery–understanding and mimicking the barrier properties. *Basic Clin. Physiol. Pharmacol.*, 116, 3, 179–86, 1742-7835, 2015.
19. Ensign, L.M., Tang, B.C., Wang, Y.-Y., Tse, T.A., Hoen, T., Cone, R., Hanes, J., Mucus-penetrating nanoparticles for vaginal drug delivery protect against herpes simplex virus. *Sci. Transl. Med.*, 4, 138, 138ra79–ra79, 1946-6234, 2012.
20. Rossi, S., Vigani, B., Sandri, G., Bonferoni, M.C., Caramella, C.M., Ferrari, F., Recent advances in the mucus-interacting approach for vaginal drug delivery: From mucoadhesive to mucus-penetrating nanoparticles. *Expert Opin. Drug Deliv.*, 16, 8, 777–81, 1742-5247, 2019.
21. Machado, R.M., Palmeira-De-Oliveira, A., Martinez-De-Oliveira, J., Palmeira-De-Oliveira, R., Vaginal films for drug delivery. *J. Pharm. Sci.*, 102, 7, 2069–81, 0022-3549, 2013.
22. Mohammed, Y., Holmes, A., Kwok, P.C.L., Kumeria, T., Namjoshi, S., Imran, M. *et al.*, Advances and future perspectives in epithelial drug delivery. *Adv. Drug Deliv. Rev.*, 186, 114293, 0169–409X, 2022.
23. Johal, H.S., Garg, T., Rath, G., Goyal, A.K., Advanced topical drug delivery system for the management of vaginal candidiasis. *Drug Deliv.*, 23, 2, 550–63, 1071-7544, 2016.
24. Wira, C.R., Patel, M.V., Ghosh, M., Mukura, L., Fahey, J.V., Innate immunity in the human female reproductive tract: endocrine regulation of endogenous antimicrobial protection against HIV and other sexually transmitted infections. *Am. J. Reprod. Immunol.*, 65, 3, 196–211, 1046-7408, 2011.
25. Dezzutti, C.S., Brown, E.R., Moncla, B., Russo, J., Cost, M., Wang, L. *et al.*, Is wetter better? An evaluation of over-the-counter personal lubricants for safety and anti-HIV-1 activity. *PloS One*, 7, 11, e48328, 1932–6203, 2012.
26. Kumamoto, Y. and Iwasaki, A., Unique features of antiviral immune system of the vaginal mucosa. *Curr. Opin. Immunol.*, 24, 4, 411–6, 0952-7915, 2012.
27. Cauci, S., Vaginal immunity in bacterial vaginosis. *Curr. Infect. Dis. Rep.*, 6, 6, 450–6, 1523-3847, 2004.
28. Van Eyk, A.D., Van der Bijl, P., Moll, L.M., Physicochemical characteristics of molecules and their diffusion across human vaginal mucosa. *Eur. J. Inflamm.*, 6, 2, 65–71, 1721-727X, 2008.
29. Smoleński, M., Karolewicz, B., Gołkowska, A.M., Nartowski, K.P., Małolepsza-Jarmołowska, K., Emulsion-based multicompartment vaginal drug carriers: from nanoemulsions to nanoemulgels. *Int. J. Mol. Sci.*, 22, 12, 6455, 1422–0067, 2021.
30. Machado, R.M., Palmeira-de-Oliveira, A., Gaspar, C., Martinez-de-Oliveira, J., Palmeira-de-Oliveira, R., Studies and methodologies on vaginal drug permeation. *Adv. Drug Deliv. Rev.*, 92, 14–26, 0169-409X, 2015.
31. Dedeloudi, A., Siamidi, A., Pavlou, P., Vlachou, M., Recent advances in the excipients used in modified release vaginal formulations. *Materials*, 15, 1, 327, 1996–44, 2022.

32. das Neves, J., Nunes, R., Machado, A., Sarmento, B., Polymer-based nanocarriers for vaginal drug delivery. *Adv. Drug Deliv. Rev.*, 92, 53–70, 0169-409X, 2015.
33. Lai, S.K., Wang, Y.-Y., Cone, R., Wirtz, D., Hanes, J., Altering mucus rheology to "solidify" human mucus at the nanoscale. *PloS One*, 4, 1, e4294, 1932–6203, 2009.
34. Peppas, N.A. and Huang, Y., Nanoscale technology of mucoadhesive interactions. *Adv. Drug Deliv. Rev.*, 56, 11, 1675–87, 0169-409X, 2004.
35. Das Neves, J., Bahia, M.F., Amiji, M.M., Sarmento, B., Mucoadhesive nanomedicines: characterization and modulation of mucoadhesion at the nanoscale. *Expert Opin. Drug Deliv.*, 8, 8, 1085–104, 742-5247, 2011.
36. Sosnik, A., das Neves, J., Sarmento, B., Mucoadhesive polymers in the design of nano-drug delivery systems for administration by non-parenteral routes: A review. *Prog. Polym. Sci.*, 39, 12, 2030–75, 0079-6700, 2014.
37. Brown, J., Hooper, G., Kenyon, C.J., Haines, S., Burt, J., Humphries, J.M. *et al.*, Spreading and retention of vaginal formulations in post-menopausal women as assessed by gamma scintigraphy. *Pharm. Res.*, 14, 1073–8, 0724-8741, 1997.
38. Neves, J.D., de Oliveira, R.P., de Oliveira, A.P., Rodrigues, F., Sarmento, B., *Mucoadhesive Materials and Drug Delivery Systems*, Wiley Chichester, UK, 2014.
39. Bruning, E., Chen, Y., McCue, K.A., Rubino, J.R., Wilkinson, J.E., Brown, A.D.G., A 28 day clinical assessment of a lactic acid-containing antimicrobial intimate gel wash formulation on skin tolerance and impact on the vulvar microbiome. *Antibiotics*, 9, 2, 55, 2079–6382, 2020.
40. World Health O., *Use and procurement of additional lubricants for male and female condoms: WHO/UNFPA/FHI360: advisory note*, Department of Reproductive Health and Research, World Health Organization, Geneva, Switzerland, 2012.
41. Machado, R.M., Palmeira-de-Oliveira, A., Martinez-de-Oliveira, J., Palmeira-de-Oliveira, R., Vaginal semisolid products: Technological performance considering physiologic parameters. *Eur. J. Pharm. Sci.*, 109, 556–68, 0928-87, 2017.
42. Ensign, L.M., Hoen, T.E., Maisel, K., Cone, R.A., Hanes, J.S., Enhanced vaginal drug delivery through the use of hypotonic formulations that induce fluid uptake. *Biomaterials*, 34, 28, 6922–9, 0142-9612, 2013.
43. Lacey, C.J., Woodhall, S., Qi, Z., Sawant, S., Cowen, M., McCormack, S., Jiang, S., Unacceptable side-effects associated with a hyperosmolar vaginal microbicide in a phase 1 trial. *Int. J. STD AIDS*, 21, 10, 714–7, 0956-4624, 2010.
44. Katz, D.F., Yuan, A., Gao, Y., Vaginal drug distribution modeling. *Adv. Drug Deliv. Rev.*, 92, 2–13, 0169-409X, 2015.
45. Osmałek, T., Froelich, A., Jadach, B., Tatarek, A., Gadziński, P., Falana, A. *et al.*, Recent advances in polymer-based vaginal drug delivery systems. *Pharmaceutics*, 13, 6, 884, 1999–4923, 2021.

46. Gürsoy, A. and Bayhan, A., Testing of drug release from bioadhesive vaginal tablets. *Drug Dev. Ind. Pharm.*, 18, 2, 203–21, 0363-9045, 1992.
47. Genç, L., Oğuzlar, C., Güler, E., Studies on vaginal bioadhesive tablets of acyclovir. *Pharmazie*, 55, 4, 297–9, 0031-7144, 2000.
48. Zhang, L., De Salvo, R., Ehret, A., Young, K., Trapp, S., Vulvovaginal candidiasis: A real-world evidence study of the perceived benefits of Canesten®. *SAGE Open Med.*, 10, 20503121221085437, 2050–3121, 2022.
49. Gossett, D.R. and Del Carmen, M.G., Use of powder in the genital area and ovarian cancer risk: examining the evidence. *JAMA*, 323, 1, 29–31, 0098-7484, 2020.
50. Jenkins, A. and O'Doherty, K.C., "It was always just a sacrifice I was willing to make": Understanding Women's use of vaginal cleansing products in spite of adverse health effects. *SSM-Qual. Res. Health*, 2, 100133, 2667–3215, 2022.
51. Rosenblatt, K.A., Weiss, N.S., Cushing-Haugen, K.L., Wicklund, K.G., Rossing, M.A., Genital powder exposure and the risk of epithelial ovarian cancer. *Cancer Causes Control*, 22, 737–42, 0957-5243, 2011.
52. Choi, H.-G., Jung, J.-H., Ryu, J.-M., Yoon, S.-J., Oh, Y.-K., Kim, C.-K., Development of *in situ*-gelling and mucoadhesive acetaminophen liquid suppository. *Int. J. Pharm.*, 165, 1, 33–44, 0378-5173, 1998.
53. Kumar, A., Kolay, A., Havelikar, U., Modern Aspects of Suppositories: A Review. *Eur. J. Pharm. Med. Res.*, 3, 4, 23–9, 2795-8035, 2023.
54. Huang, W., Zhou, X., Huang, X., Peng, Y., Su, Y., Qiao, L. *et al.*, Clindamycin soluble suppository-production and its efficacy in treating bacterial vaginosis. *J. West China Univ. Med. Sci.= Huaxi Yike Daxue Xuebao*, 32, 3, 452–3, 8, 0257-7712, 2001.
55. Schwebke, J.R., Lensing, S.Y., Sobel, J., Intravaginal metronidazole/miconazole for the treatment of vaginal trichomoniasis. *Sex Trans. Dis.*, 40, 9, 710–4, 0148-5717, 2013.
56. Prutting, S.M. and Cerveny, J.D., Boric acid vaginal suppositories: a brief review. *Infect. Dis. Obstet. Gynecol.*, 6, 191–4, 1064-7449, 1998.
57. Verdenelli, M.C., Cecchini, C., Coman, M.M., Silvi, S., Orpianesi, C., Coata, G. *et al.*, Impact of probiotic SYNBIO® administered by vaginal suppositories in promoting vaginal health of apparently healthy women. *Curr. Microbiol.*, 73, 483–90, 0343-8651, 2016.
58. Bahrainian, S., Abbaspour, M., Kouchak, M., Moghadam, P.T., A review on fast dissolving systems: from tablets to nanofibers. *Jundishapur J. Nat. Pharm. Prod.*, 12, 2, 2228–7876, 2017.
59. Sharifzadeh, G., Hezaveh, H., Muhamad, I.I., Hashim, S., Khairuddin, N., Montmorillonite-based polyacrylamide hydrogel rings for controlled vaginal drug delivery. *Mater. Sci. Eng.: C*, 110, 110609, 0928–4931, 2020.
60. Ayton, R.A., Darting, G.M., Murkies, A.L., Farrell, E.A., Weisberg, E., Statistician, I.S., Fraser, I.S., A comparative study of safety and efficacy of continuous low dose oestradiol released from a vaginal ring compared with conjugated equine oestrogen vaginal cream in the treatment of postmenopausal

urogenital atrophy. *BJOG: Int. J. Obstet. Gynaecol.*, 103, 4, 351–8, 1470-0328, 1996.
61. Roumen, F., Contraceptive efficacy and tolerability with a novel combined contraceptive vaginal ring, NuvaRing. *Eur. J. Contracept. Reprod. Health Care*, 7, sup2, 19–24, 1362-5187, 2002.
62. Malcolm, K., Woolfson, D., Russell, J., Tallon, P., McAuley, L., Craig, D., Influence of silicone elastomer solubility and diffusivity on the *in vitro* release of drugs from intravaginal rings. *J. Control. Release*, 90, 2, 217–25, 0168-3659, 2003.
63. Roumen, F. and Dieben, T.O.M., Clinical acceptability of an ethylene-vinyl-acetate nonmedicated vaginal ring. *Contraception*, 59, 1, 59–62, 0010-7824, 1999.
64. Farage, M.A. and Maibach, H.I., *The vulva. Anatomy, physiology, and patology*, Informa Healthcase, USA, New York, 2006.
65. Chien, Y.W., Oppermann, J., Nicolova, B., Lambert, H.J., Medicated tampons: Intravaginal sustained administration of metronidazole and *in vitro-in vivo* relationships. *J. Pharm. Sci.*, 71, 7, 767–71, 0022-3549, 1982.
66. Smith, C.B., Noble, V., Bensch, R., Ahlin, P.A., Jacobson, J.A., Latham, R.H., Bacterial flora of the vagina during the menstrual cycle: findings in users of tampons, napkins, and sea sponges. *Ann. Intern. Med.*, 96, 6_Part_2, 948–51, 0003-4819, 1982.
67. Kafka, D. and Gold, R.B., Food and Drug Administration approves vaginal sponge. *Fam. Plan. Perspect.*, 15, 3, 146–8, 0014-7354, 1983.
68. Kuyoh, M.A., Toroitich-Ruto, C., Grimes, D.A., Schulz, K.F., Gallo, M.F., Sponge versus diaphragm for contraception: a Cochrane review. *Contraception*, 67, 1, 15–8, 0010-7824, 2003.
69. Barati, M., Hashemi, S., Sayed Tabatabaei, M., Zarei Chamgordani, N., Mortazavi, S.M., Moghimi, H.R., Protein-based microneedles for biomedical applications: A systematic review. *Biomed. Microdevices*, 26, 2, 1–22, 1572-8781, 2024.
70. Mc Crudden, M.T.C., Larrañeta, E., Clark, A., Jarrahian, C., Rein-Weston, A., Creelman, B. *et al.*, Design, formulation, and evaluation of novel dissolving microarray patches containing rilpivirine for intravaginal delivery. *Adv. Healthcare Mater.*, 8, 9, 1801510, 2192–2640, 2019.
71. Enggi, C.K., Sulistiawati, S., Stephanie, S., Tangdilintin, F., Achmad, A.A., Putri, R.A. *et al.*, Development of probiotic loaded multilayer microcapsules incorporated into dissolving microneedles for potential improvement treatment of vulvovaginal candidiasis: A proof of concept study. *J. Colloid Interface Sci.*, 648, 203–19, 0021-9797, 2023.
72. Caramella, C.M., Rossi, S., Ferrari, F., Bonferoni, M.C., Sandri, G., Mucoadhesive and thermogelling systems for vaginal drug delivery. *Adv. Drug Deliv. Rev.*, 92, 39–52, 0169-409X, 2015.

73. Blanco-Fernández, G., Blanco-Fernandez, B., Fernández-Ferreiro, A., Otero-Espinar, F.J., Lipidic lyotropic liquid crystals: Insights on biomedical applications. *Adv. Colloid Interface Sci.*, 313, 102867, 0001–8686, 2023.
74. de Araújo, P.R., Calixto, G.M.F., da Silva, I.C., de Paula Zago, L.H., Oshiro Junior, J.A., Pavan, F.R. *et al.*, Mucoadhesive *in situ* gelling liquid crystalline precursor system to improve the vaginal administration of drugs. *AAPS Pharm. Sci. Tech.*, 20, 1–12, 2019.
75. De Araújo, P.R., Calixto, G.M.F., Araújo, V.H.S., Sato, M.R., Rodero, C.F., Oshiro-Junior, J.A. *et al.*, *In vivo* study of hypericin-loaded poloxamer-based mucoadhesive *in situ* gelling liquid crystalline precursor system in a mice model of vulvovaginal candidiasis. *Med. Mycol.*, 59, 8, 821–7, 1369-3786, 2021.
76. Rodero, C.F., Fioramonti Calixto, G.M., Cristina dos Santos, K., Sato, M.R., dos Santos Ramos, M.A., Miró, M.S. *et al.*, Curcumin-loaded liquid crystalline systems for controlled drug release and improved treatment of vulvovaginal candidiasis. *Mol. Pharm.*, 15, 10, 4491–504, 1543-8384, 2018.
77. Pisano, S., Giustiniani, M., Francis, L., Gonzalez, D., Margarit, L., Sheldon, I.M. *et al.*, Liquid crystal delivery of ciprofloxacin to treat infections of the female reproductive tract. *Biomed. Microdevices*, 21, 1–12, 1387-2176, 2019.
78. de Bastiani, FWMDS, Spadari, C.D.C., De Matos, J.K.R., Salata, G.C., Lopes, L.B., Ishida, K., Nanocarriers provide sustained antifungal activity for amphotericin B and miltefosine in the topical treatment of murine vaginal candidiasis. *Front. Microbiol.*, 10, 493166, 1664–302X, 2020.
79. Tarsitano, M., Mancuso, A., Cristiano, M.C., Paolino, D., Fresta, M., *In situ* swelling formulation of glycerol-monooleate-derived lyotropic liquid crystals proposed for local vaginal application. *Molecules*, 27, 19, 6295, 1420–3049, 2022.
80. Milutinov, J., Krstonošić, V., Ćirin, D., Pavlović, N., Emulgels: Promising Carrier Systems for Food Ingredients and Drugs. *Polymers*, 15, 10, 2302, 073–4360, 2023.
81. Bachhav, Y.G. and Patravale, V.B., Microemulsion-based vaginal gel of clotrimazole: formulation, *in vitro* evaluation, and stability studies. *AAPS Pharm. Sci. Tech.*, 10, 476–81, 2009.
82. Bachhav, Y.G. and Patravale, V.B., Microemulsion based vaginal gel of fluconazole: formulation, *in vitro* and *in vivo* evaluation. *Int. J. Pharm.*, 365, 1-2, 175–9, 0378-5173, 2009.
83. Patel, A. and Patel, J., Mucoadhesive microemulsion based prolonged release vaginal gel for anti-fungal drug. *Am. J. Pharm. Tech. Res.*, 2, 4, 649–61, 2012.
84. Mirani, A., Kundaikar, H., Velhal, S., Patel, V., Bandivdekar, A., Degani, M., Patravale, V., Tetrahydrocurcumin-loaded vaginal nanomicrobicide for prophylaxis of HIV/AIDS: *In silico* study, formulation development, and *in vitro* evaluation. *Drug Deliv. Transl. Res.*, 9, 828–47, 2190-393X, 2019.

85. D'Cruz, O.J., Yiv, S.H., Waurzyniak, B., Uckun, F.M., Contraceptive efficacy and safety studies of a novel microemulsion-based lipophilic vaginal spermicide. *Fertil. Steril.*, 75, 1, 115–24, 0015-282, 2001.
86. Kaur, A., Gupta, S., Tyagi, A., Sharma, R.K., Ali, J., Gabrani, R., Dang, S., Development of nanoemulsion based gel loaded with phytoconstituents for the treatment of urinary tract infection and *in vivo* biodistribution studies. *Adv. Pharm. Bull.*, 7, 4, 611, 2017.
87. Khattab, A. and Ismail, S., Formulation and evaluation of oxiconazole nitrate mucoadhesive nanoemulsion based gel for treatment of fungal vaginal infection. *Int. J. Pharm. Pharm. Sci.*, 8, 33–40, 2016.
88. Srivastava, N., Patel, D.K., Rai, V.K., Pal, A., Yadav, N.P., Development of emulgel formulation for vaginal candidiasis: Pharmaceutical characterization, *in vitro* and *in vivo* evaluation. *J. Drug Deliv. Sci. Technol.*, 48, 490–8, 1773-2247, 2018.
89. Mou, Y., Zhang, P., Lai, W.-F., Zhang, D., Design and applications of liposome-in-gel as carriers for cancer therapy. *Drug Deliv.*, 29, 1, 3245–55, 1071-7544, 2022.
90. Jøraholmen, M.W., Basnet, P., Tostrup, M.J., Moueffaq, S., Škalko-Basnet, N., Localized therapy of vaginal infections and inflammation: Liposomes-in-hydrogel delivery system for polyphenols. *Pharmaceutics*, 11, 2, 53, 1999–4923, 2019.
91. Abou-Taleb, H.A., Fathalla, Z., Naguib, D.M., Fatease, A.A., Abdelkader, H., Chitosan/Solid-Lipid Nanoparticles Hybrid Gels for Vaginal Delivery of Estradiol for Management of Vaginal Menopausal Symptoms. *Pharmaceuticals*, 16, 9, 1284, 424–8247, 2023.
92. Jøraholmen, M.W., Johannessen, M., Gravningen, K., Puolakkainen, M., Acharya, G., Basnet, P., Škalko-Basnet, N., Liposomes-in-hydrogel delivery system enhances the potential of resveratrol in combating vaginal chlamydia infection. *Pharmaceutics*, 12, 12, 1203, 999–4923, 2020.
93. Vanić, Ž., Hurler, J., Ferderber, K., Golja Gašparović, P., Škalko-Basnet, N., Filipović-Grčić, J., Novel vaginal drug delivery system: deformable propylene glycol liposomes-in-hydrogel. *J. Liposome Res.*, 24, 1, 27–36, 2014.
94. Takalkar, D. and Desai, N., Nanolipid gel of an antimycotic drug for treating vulvovaginal candidiasis—development and evaluation. *AAPS Pharm. Sci. Tech.*, 19, 1297–307, 2018.
95. Almawash S. Solid lipid nanoparticles, an effective carrier for classical antifungal drugs. *Saudi Pharm J.*, 2023;31(7):1167-1180. doi:10.1016/j.jsps.2023.05.011
96. Firdaus, S., Hassan, N., Mirza, M.A., Ara, T., El-Serehy, H.A., Al-Misned, F.A., Iqbal, Z., FbD directed fabrication and investigation of luliconazole based SLN gel for the amelioration of candidal vulvovaginitis: a 2 T (thermosensitive & transvaginal) approach. *Saudi J. Biol. Sci.*, 28, 1, 317–26, 1319-562X, 2021.

97. Anjum, R. and Lakshmi, P.K., Dolutegravir sodium loaded solid lipid nanoparticles: A vaginal drug delivery system for pre-exposure prophylaxis of HIV. *J. Res. Pharm.*, 24, 4, 552–61, 2020.
98. Mendes, A.I., Silva, A.C., Catita, J.A.M., Cerqueira, F., Gabriel, C., Lopes, C.M., Miconazole-loaded nanostructured lipid carriers (NLC) for local delivery to the oral mucosa: improving antifungal activity. *Colloids Surf., B*, 111, 755–63, 0927-7765, 2013.
99. Ravani, L., Esposito, E., Bories, C., Lievin-Le Moal, V., Loiseau, P.M., Djabourov, M. *et al.*, Clotrimazole-loaded nanostructured lipid carrier hydrogels: Thermal analysis and *in vitro* studies. *Int. J. Pharm.*, 454, 2, 695–702, 0378-5173, 2013.
100. Ning, M., Guo, Y., Pan, H., Chen, X., Gu, Z., Preparation, *in vitro* and *in vivo* evaluation of liposomal/niosomal gel delivery systems for clotrimazole. *Drug Dev. Ind. Pharm.*, 31, 4-5, 375–83, 0363-9045, 2005.
101. Mbah, C., Builders, P., Nzekwe, I., Kunle, O., Adikwu, M., Attama, A., Formulation and *in vitro* evaluation of pH-responsive ethosomes for vaginal delivery of metronidazole. *J. Drug Deliv. Sci. Technol.*, 24, 6, 565–71, 1773-2247, 2014.
102. Singh, S., Verma, D., Mirza, M.A., Das, A.K., Anwer, M.K., Sultana, Y. *et al.*, Development and optimization of ketoconazole loaded nano-transfersomal gel for vaginal delivery using Box-Behnken design: *In vitro*, *ex vivo* characterization and antimicrobial evaluation. *J. Drug Deliv. Sci. Technol.*, 39, 95–103, 1773-2247, 2017.
103. Hady, M.A., Darwish, A.B., Abdel-Aziz, M.S., Sayed, O.M., Design of transfersomal nanocarriers of nystatin for combating vulvovaginal candidiasis; A different prospective. *Colloids Surf., B*, 211, 112304, 0927–7765, 2022.
104. Borges, S., Barbosa, J., Teixeira, P., Drug delivery systems for vaginal infections. *Front. Clin. Drug Res.*, 2, 233, 2016.
105. Bernstein, G.S., Conventional methods of contraception: Condom, diaphragm, and vaginal foam. *Clin. Obstet. Gynecol.*, 17, 1, 21–33, 0009-9201, 1974.
106. Hussain, A. and Ahsan, F., The vagina as a route for systemic drug delivery. *J. Control. Release*, 103, 2, 301–13, 0168-3659, 2005.
107. de Oliveira RP, de Oliveira JM, Caramella C. Special issue on vaginal drug delivery. *Adv. Drug Deliv. Rev.*, 2015;92:1. doi: 10.1016/j.addr.2015.09.006. PMID: 26452325.
108. Tang, O.S., Schweer, H., Seyberth, H.W., Lee, S.W.H., Ho, P.C., Pharmacokinetics of different routes of administration of misoprostol. *Hum. Reprod.*, 17, 2, 332–6, 1460-2350, 2002.

13

Future of Novel Drug Delivery Systems

Mehrdad Hamidi[1,2]*, Mehraneh Kermanian[2] and Masoumeh Kurd[2]

[1]*Pharmaceutical Nanotechnology Research Center (ZPNRC), Zanjan University of Medical Sciences, Zanjan, Iran*
[2]*HIDA Pharmaceutical Technologies Development Center, HIDA Pharmaceuticals, Tehran, Iran*

Abstract

NDDSs are among the fields of pharmaceutical science and technology with the highest progress rates, which roots back to the ever-growing nature of the relevant science and technologies as well as the high and ever-expanding market demand for these delivery systems. A dual-oriented approach has been followed in this chapter to highlight and describe the future trends of NDDSs. The first wing determining the major trends of NDDSs is the currently notable challenges of these entities, which mainly include drug delivery barriers, burst drug release, real-time monitoring issues, high inter-individual variations, high cost of NDDSs development and manufacturing, and low patient compliance. While being significant threats to the NDDS development roadmap, each challenging factor can be regarded as an opportunity for the development of better products overcoming one or more drawbacks. On the other hand, the second wing of driving factors is based on the opportunity created around the NDDSs development and presentation atmosphere, mainly including artificial intelligence (AI), new techniques, new visions in biology, new therapeutic agents/protocols/strategies, emerging technologies, and interdisciplinary development.

***Keywords*:** Novel drug delivery system, future trends of NDDSs, emerging technologies, challenge-based trends, opportunity-based trends

**Corresponding author*: hamidim@zums.ac.ir; hamidi@hidapharma.com

Mehrdad Hamidi and Mahsa Sayed Tabatabaei (eds.) Novel Drug Delivery Systems: Fundamentals and Applications, (449–466) © 2025 Scrivener Publishing LLC

13.1 Introduction

Novel drug delivery systems, so-called NDDSs, are a typical example of the emerging areas of pharmaceutical science and technology. Accordingly, it experiences remarkable changes every day with the advancements made in related science and technology and, on the other side, the ever-changing needs and expectations of the target market.

In this chapter, the future trends of the NDDSs area are reviewed and evaluated using a dual opportunity/challenge model. In other words, we have categorized the contents of this chapter into two classes of challenges and opportunities, with the majority of the new trendy events being organized as associated with overcoming the challenges or seizing the opportunities. Since the structure of this chapter is based on the identification of the main trends and opportunities shaping the future direction of the NDDSs fields, our intention was an 'all-inclusive' survey of the literature, both formal and informal resources, and even the non-related materials, to provide a comprehensive view to the readers, the chapter does not have a standard 'documentary' fashion which necessitates the referral of every and each point to the relevant resources at the end of the chapter. Here, instead, all the discussions are open take up, analysis, and conception with the aim to provide maximum insights to the researchers to shape their next prospective research projects. Accordingly, the readers are referred to the related literature to find more about each discussion.

13.2 Drug Delivery Challenges to be Overcome in the Future

13.2.1 Drug Delivery Barriers

A series of physiological and physicochemical factors within the host body act in concert against the fulfillment of the ideal purposes of the drug delivery system, which are collectively known as drug delivery barriers. Considering the exertion of temporal and/or spatial control on the drug release within the body as the ultimate target of the administration of an NDDS, different drug delivery barriers can be recognized based on the LADME system, namely liberation, absorption, distribution, metabolism, and excretion.

In terms of the nature of the barrier, four categories can be listed as chemical, physical, biochemical, and immunological barriers.

Chemical barriers can be mainly assigned to the pre-absorption stage, i.e., drug dissolution/release, pre-systemic and systemic chemical stability as well as drug binding to the biological factors, e.g., mucin in the GI tract. Two main strategies attempted to overcome these issues are using different drug carriers, e.g., nanocarriers, and drug conjugation to different compounds, e.g., polymers, to create new chemical entities.

Physical barriers are mainly related to the cell membranes as well as tissue partitioning. A part of these barriers are physically based but the main contribution is biochemical in nature, mainly imposed by the enzymes and membrane transporters. Again the two major approaches to overcome this kind of barrier are using drug carriers and drug conjugates. The main intention in all these efforts is to make the drug capable of permeating across the barrier.

Biochemical barriers, with no doubt, are the main role-playing ones in terms of both the effectiveness in prohibiting the achievement of the drug delivery objectives, and the wide distribution throughout the body. These kinds of barriers include the enzymes and transporters located in biological membranes and tissue 'dividers', intra- and extracellular enzymes in different body compartments, e.g., hepatic first-pass metabolism, and blood-tissue barriers surrounding the sanctuary sites, e.g., blood-brain-barrier. To overcome this kind of barrier, in addition to the two former classes of the carriers and conjugates, modulators of the enzymes as well as transporters, e.g., blockers, have been used with different NDDSs.

Immunological barriers are mainly related to the premature capturing of the drug delivery carriers by the host body's reticuloendothelial system, so-called RES, as well as the identification of the drug delivery system as foreign entities, thereby provoking the different immune system reactions. The main solution for this kind of barrier is trying to prepare delivery systems capable of being ignored, e.g., tiny nanocarriers, or, alternatively, identified as part of the host body, e.g., biocompatible implants.

As described, the variety of barriers acting against the drug delivery success, are, from another viewpoint, opportunities to present new more efficient systems with superior characters more suitable for our purposes from NDDSs.

13.2.2 Burst Release

Burst release is the uncontrolled and rapid release of a substantial quantity of the active ingredient in a formulation during the first few minutes or hours. While being known as a problem, burst release can also be regarded as an opportunity to provide an initial drug concentration to reach the therapeutic window to be, then, followed by the slow drug release in extended time intervals. The early surge of the drug is, however, typically undesired as it reduces the overall length of the drug's therapeutic impact. Additionally, an excessive release of the drug at once may lead to toxicity, which raises safety concerns. In addition to the health-related concerns, the burst release has economic implications as the 'waste' of a significant part of the drug dose, for example as a result of the extensive hepatic first-pass effect. The burst drug release is problematic basically with the NDDSs with large surface-to-volume ratios such as the generation one polymeric nanoparticles. Moreover, burst release occurs frequently for the hydrophilic small molecules as well as the large molecules such as proteins. Several methods have been attempted to overcome the burst release issue mainly including the development of multi-modal delivery systems, surface coatings, surface functionalization, cross-linking, molecular imprints, layer-by-layer (LbL) assembly, two-part gelatin capsule, and the use of a smart drug release system.

Smart in the context of drug delivery systems means the capability to release the cargo at a certain time and location in response to selected stimuli, whether internal or external.

Among the internal stimuli, biomarkers, enzymes, hormones, and other biochemicals can be mentioned, while external stimuli include a wide variety of environmental factors such as pH, temperature, ion concentrations, etc.

Sequential release delivery systems utilize various micro- or nanocarriers and have the potential to be highly effective in treating diseases like cancer. Their unique feature is the capacity to systematically release one or more drugs at specific intervals or locations.

Controlled release of medicine involves quickly reaching and maintaining a specific concentration in tissues or organs while reducing systemic adverse effects. Some controlled release systems allow customization of the medication release rate. Choosing the optimal carrier for pharmacological payloads and planning the route for the carrier to reach the target tissue in the body are key aspects of smart delivery systems. Controlled release systems, by definition, consistently dispense certain quantities of medications over a specified duration and/or to a specified location of the host body.

Sequential-release drug delivery systems, a type of controlled release systems, provide the ability to control the release of one or more active substances either temporally or spatially, resulting in improved therapeutic effectiveness. Sequential release systems involve carriers that can be modified with several components that react to various stimuli, allowing for the accurate discharge of multiple cargos at the intended location at different intervals.

Dual/multi-responsive drug delivery systems are formed by combining carriers that react to two or more distinct internal or external stimuli, thus enhancing the precision and effectiveness of controlled release. Combining multiple stimuli-responsive systems can address the limitations of individual systems, resulting in improved medication pharmacokinetics while supporting simultaneous release.

13.2.3 Real-Time Monitoring Issues

Real-time monitoring of drug delivery refers to the ability to track and monitor the delivery of drugs to a patient in real time. Novel drug delivery systems employ various mechanisms to achieve real-time monitoring. These include biosensors, microfluidics, nanoparticles, and wearable devices that can measure and transmit data about the drug's concentration, location, and effects in the body, as well as monitoring physiological parameters such as heart rate, blood pressure, and glucose levels.

Real-time monitoring of drug delivery offers several advantages over traditional methods of drug administration. First, it allows for personalized and optimized drug therapy, as healthcare providers can adjust the drug dosage and delivery rate based on the patient's individual response. Second, it can improve patient safety by detecting and preventing potential adverse effects or drug interactions. Third, it enables remote monitoring of patients, reducing the need for frequent hospital visits or clinic appointments.

However, real-time monitoring of drug delivery also presents challenges, including the need for reliable and accurate sensors, data security and privacy concerns, and the potential for device malfunctions or technical issues. Additionally, the cost of implementing and maintaining such systems can be a limiting factor.

For drug release monitoring, several multi-functional drug delivery systems have been developed from a range of building blocks, including organic molecules, polymeric nanoparticles, micelles, and inorganic nanoparticles. Various types of organic and inorganic nanomaterials, due to their distinct physicochemical features, have been used as imaging contrast agents, as well as their potential use as drug carriers for early disease detection and therapy monitoring.

In recent years, there has been a growing emphasis on the use of nanomaterials to develop non-invasive imaging modalities such as fluorescence imaging, magnetic resonance imaging (MRI), and surface-enhanced

Raman scattering (SERS) imaging for efficient drug release monitoring and delivery in both real-time and long-term fashions. Furthermore, multi-modal imaging has been developed to incorporate the advantages of each modality into a unified drug delivery system, offering comprehensive insight into the drug uptake process, pharmacokinetics, biodistribution, local distribution at the target site, drug release kinetics, and therapeutic efficacy. Such detailed spatiotemporal data would be so beneficial for the optimization of therapy with respect to the administration strategies, engineering of the drug delivery system, and drug release mechanisms. Overall, real-time monitoring using novel drug delivery systems represents a significant advancement in healthcare, enabling more effective, safer, and personalized drug therapies.

13.2.4 High Interindividual Variations

The incorporation of drug delivery systems with cutting-edge technologies such as using stimuli-responsive nanoparticles and developed electrochemical sensors has led to eminent targeted therapy and remarkable accuracy at the tissue and cellular levels with reduced side effects. Accordingly, nanotechnology-based NDDSs have the potential to revolutionize the therapy outcomes via different capabilities of these nanocarriers to act specifically at the nano-bio interfaces within the host body. Also, personalized dosage forms and wearable devices would enable customizing therapy according to individual needs, enhancing therapeutic efficacy and compliance. Companion diagnostics, often known as "theranostics," facilitate superior monitoring of the therapy response and specific modifications to the therapy plan.

13.2.4.1 Personalized Controlled-Release

Design and development of controlled-release drug delivery systems possess crucial importance in the field of therapeutics while tailoring the release of drugs and therapeutic agents at a desired rate, duration, and target site will lead to improved efficiency of the therapy as well as patient compliance.

There are several platforms and formulations for the fabrication of controlled-release systems, including various kinds of microspheres, nanoparticles, hydrogels, etc. These delivery systems are commonly designed to respond to a variety of stimuli, including pH, temperature, enzymes, and also external triggers, thus enabling individualized and targeted drug delivery.

These systems would be beneficial in synergistic cancer immunotherapy due to their intrinsic features of regulating the release profile of the bioactive agent as well as a precise recognition of the biological targets which, in turn, provide the basis for the individualized drug regimens in these patients.

13.2.4.2 Personalized System Design

While the development of the generation one NDDSs was one step forward towards the 'manipulated drug release' they still lack the capabilities needed by a real personalized delivery system. Accordingly, efforts have been continued to find more tunable and responsive candidates to have a 'dialogue' with the biology and to receive the data from the host body needed to implement a personalized system of drug delivery.

Continuous breakthroughs in this area are poised to open doors to personalized medicine along with personalized drug delivery systems which will consequently lead to creative and efficient drug delivery strategies and enhance treatments' efficacy to individual patient's needs. Insulin dosage management in diabetic patients is an unambiguous demonstration of the necessity for personalized drug delivery and therapy.

The aforementioned demands support the need for a new generation of pharmaceuticals that emphasize individualized and patient-centered care and treatment.

13.2.5 High Cost of Novel Systems

Extended/sustained-release drug delivery systems have been demonstrated to significantly enhance the efficacy of medications in clinical experiences as well as clinical trials. Natural, semi-synthetic, and synthetic polymers are efficient in developing extended-release drug delivery systems utilizing various polymeric materials. Polymer research in NDDSs primarily concentrates on synthetic polymers. Researchers have attempted several combinations of polymers and also polymers with other chemical classes such as lipid-based materials, inorganic compounds, and different surface-active agents, in the way to reach the intended drug release patterns. In parallel to the success stories with these innovations, different candidates or combinations have been shown to make the process more complex increasing the expenses of formulations.

Inexpensive natural polymers like chitosan, alginate, cellulose, and hyaluronic acid and their derivatives are ideal for creating drug delivery systems in terms of relative costs. These polymers can be readily derived from renewable sources such as prawn shells, algae, plants, and animal tissues, making them a sustainable choice for drug delivery applications.

Drug delivery systems using natural polymers require the drug to be integrated into the polymer matrix through physical entrapment or chemical conjugation. These devices can be engineered to provide the medication in a regulated fashion, thus maximizing therapeutic effectiveness while reducing adverse side effects. Natural polymers are advantageous for drug delivery applications because of their cost-effectiveness, sustainable properties, biocompatibility, biodegradability, and various manufacturing techniques available. These characteristics make them appropriate for many NDDS applications, such as oral, transdermal, injectable, and implanted systems.

13.2.6 Low Patient Compliance

The fact that NDDSs are new and emerging causes some kind of trust and certainty issues in the mindset of the general population. This factor, in fact, is a major social barrier against the widespread use of these novel therapeutic systems. In line with some social/cultural solutions, which are out of the scope of this chapter, some technical approaches can be regarded as part of the future strategies for the successful development of the NDDSs. Considering the priority, whenever practically possible, to the oral solid dosage forms, as mostly known and trusted pharmaceutical forms of the products, using the well-known excipients, e.g., polymers, in formulations, using generally familiar words and expressions in product artworks and brochures, and avoiding sophisticated devices and application methods for the NDDSs are among the hints which can change the threats to the opportunities for the popular acceptability, thus increased patient compliance and adherence to different NDDSs.

13.3 Drug Delivery Opportunities to be Seized in the Future

13.3.1 Artificial Intelligence

Artificial intelligence (AI) is a sophisticated tool that utilizes human-like understanding to offer quick solutions to intricate problems. Significant progress in AI technology and machine learning offers a revolutionary chance to improve medication development, formulation, and evaluation of pharmaceutical dosage forms. Researchers can use AI algorithms to scan vast biological data, such as genomics and proteomics, to find disease-related targets and forecast their interactions with possible medication candidates. This allows for a more effective and focused method for drug development, therefore enhancing the probability of successful drug approvals.

The combination of AI and large-scale data in the pharmaceutical area has resulted in the emergence of computational pharmaceutics, which seeks to improve drug delivery methods through the use of multiscale modeling techniques. Computational pharmaceutics uses AI algorithms and machine learning methods to examine extensive information and forecast medication performance. Researchers can assess different scenarios and enhance drug delivery systems by simulating drug formulation and delivery processes while eliminating the necessity for numerous trial-and-error studies. This expedites the drug development schedule, cuts down expenses, and enhances efficiency. Computational pharmaceutics focuses on modeling drug delivery systems across several dimensions, from molecular interactions to macroscopic behavior. AI algorithms may examine intricate connections among medicinal characteristics, formulation components, and physiological parameters to forecast drug activity at various levels algorithms which can help personalized medicine by analyzing real-world patient data, thus resulting in better treatment outcomes and more patient adherence. In brief, AI is a powerful tool with promising applications in drug discovery and drug development including different NDDSs form designs, process optimization, evaluation, and pharmacokinetic/pharmacodynamic (PK/PD) investigations.

Researchers can detect trends and even weak signals of the potential opportunities as well as risks provided by particular materials, combinations, constructs, and assemblies to be used in the preparation of NDDSs from different classes. When coordinated with the host organism biological data, e.g., enzymes and transporters, an inevitable source of data will be available which would result in the development of incredible new delivery systems with brilliant features enabling the formulator as well as clinical practitioner to touch different targets in therapy being un-touchable with the current level of knowledge in the field. This enables preemptive alterations and fine-tuning to reduce hazards and enhance therapeutic effectiveness. AI and computer modeling decrease the need for time-consuming and costly trial-and-error trials, hence reducing the likelihood of unexpected results.

13.3.2 Novel Techniques

13.3.2.1 Microfluidic

In recent years, there has been an increasing interest in microfluidic technology. It is grounded in the scientific field that focuses on creating microscopic devices with small channels and chambers and managing the movement of small amounts of fluids within these microstructures. The size of these channels and chambers typically ranges from tens to hundreds of micrometers. Microfluidics is a technique that combines various fields such as nanotechnology, biotechnology, biochemistry, physics, and engineering. These platforms offer numerous advantages, including the reduction of reagent and sample consumption to milliliters, the decrease of reaction time to seconds, and the facilitation of rapid diffusion as well as mass/heat transfer. Microfluidic systems provide accurate control and manipulation of very small quantities of liquid in a consistent and adjustable manner. Therefore, this technique has been used to create intricate drug carriers with accurate sizes and compositions, resulting in a predictable and adjustable release pattern.

Microfluidic systems can be used to deliver drugs and other bioactive agents in small amounts in a predetermined temporal and/or spatial manner. This feature makes it easier to administer medications with a short half-life or those that pose a risk of cytotoxicity systemically but are capable of providing effective concentrations whenever and/or wherever needed. Moreover, conventional delivery techniques like uncomfortable and risky injections can be improved using micro-technologies through the creation of microneedles or needle-free injection systems. Recent advancements in biosensing platforms and microfluidics integration have led to the development of a new class of drug delivery systems that can give medications as needed, via creating 'smart' devices. This technique allows for the creation of drug carriers that are uniform in size and have many functions, with adjustable physical and chemical characteristics to enhance the effectiveness of drug release, distribution, and elimination during the treatment process.

13.3.2.2 3D Printing

Three-dimensional (3D) printing is a fast and developing manufacturing method that creates 3D intricate items accurately by layering materials based on a computer-designed model. It is possible to build 3D items with exact and controlled architecture, geometry, and shape, thus ensuring great reproducibility. Recently, there has been a growing focus on utilizing this technology in the fields of medicine and pharmaceuticals.

3D printing technology is utilized to create various NDDSs with specific geometries tailored to achieve customized drug release profiles. This technology is used to produce NDDSs from different categories and dosage forms, mainly oral modified-release novel pharmaceuticals. It is recently attempted for BCS class II and class IV drugs, to enhance their solubility and bioavailability.

3D printing technology shows promise in creating solid oral dosage forms. 3D printing may create various sizes and intricate shapes with customized release properties to meet the need for personalized pharmaceuticals.

3D printing in drug formulation is an innovative method that allows production to be done near patients and provides real personalized therapy. Recent technological advancements and expanded research efforts can guarantee safer and more efficient treatment. Despite being in the early stages of development, this technology appears to be a groundbreaking tool that provides increased flexibility in drug manufacturing and is anticipated to elevate drug delivery systems to a new level in the near future. 3D printing has demonstrated its capability to create sophisticated medication delivery systems that provide various pharmaceuticals at varying rates of release.

Most 3D-printed biomaterials used in healthcare applications are still static and cannot adapt to dynamic changes in the body's internal environment and biological signals. Progress in dynamic materials, which may react to external stimuli for a specific duration, has expanded possibilities for developing future healthcare goods. Four-dimensional (4D) printing is a cutting-edge research area that utilizes intelligent materials in 3D printing to create objects that can alter their shape post-production in a controlled manner when subjected to specific external non-mechanical triggers such as moisture, electric/magnetic field, UV light, temperature, pH, and ion composition. 4D printing utilizes robust materials that induce physical bending or stretching to achieve intricate structures beyond the capabilities of traditional technologies. Objects created from intelligent materials using 3D printing are commonly regarded as smart gadgets since they are capable of altering their structure in a predetermined manner as time progresses. They are classified as 4D-printed systems since time, considered as the 4th dimension, plays a role in re-definitions of these architectures.

Various systems used in the biomedical area, like stents and scaffolds, and in drug delivery, are examined, focusing particularly on indwelling devices designed for retention in the urine bladder and stomach.

In conclusion, 4D printing will revolutionize the healthcare industry in the future. Ongoing global research aims to uncover novel applications of 4D printing in healthcare and drug delivery systems.

13.3.3 New Visions in Biology

Nowadays, biology is rather a system-based entity instead of being a collection of some static structures like cells, tissues, and organs each having its own anatomy and physiology. Today, biology is defined as an entirely dynamic, interrelated, and ever-changing complex system consisting of an enormous number of 'micro-systems' working in concert to fulfill plenty of functions and interactions. Since the majority of interactions and functions are still unknown, system biology tries, using a mathematical and computational approach, to model the complex aspects in a way to identify, recognize, and simplify the events.

NDDSs is a field of pharmaceutical sciences that would, undoubtedly, benefit from the new approach in biology. Complex events crucial to the NDDSs design and evaluation, like membrane transport, intracellular trafficking, and binding to biological macromolecules are among the inherently sophisticated phenomena thereby being good candidates for system biology re-definition. Consequently, new insights will be obtained from these evolutions which, in turn, will potentially lead to the design and preparation of safer, more efficacious/capable delivery systems with different types and levels of interactions with the biological structures.

13.3.4 New Therapeutic Agents, Protocols, and Strategies

13.3.4.1 Biopharmaceuticals

Biopharmaceuticals, also known as large-molecule drugs, are designed and produced through molecular biology techniques and cutting-edge biotechnological procedures from living cells or microorganisms, recombinant proteins, nucleic acids, and tissues. There are two main types of biopharmaceuticals: advanced innovator biologics and biosimilars, which, collectively, possess great potential to revolutionize the treatment of a wide range of diseases, including currently hard-to-treat diseases such as different types of cancers, diabetes, and neurodegenerative diseases, through novel bioactive agents like monoclonal antibodies, recombinant proteins, gene-based therapeutics, and cell therapy protocols. Since biopharmaceuticals can potentially be adjusted to an individual patient's genetic makeup and biomarkers profile, this makes them ideal candidates for establishing personalized medicine. However, there are some big challenges, including the high cost of development and manufacture, the complexity of production and storage, and accessibility and affordability issues. Another challenge ahead is the drug delivery of these compounds, which is due to physiological barriers such as their poor cell membrane permeability, low stability, and large size. Thereby, drug delivery systems for biopharmaceuticals compared with conventional drugs often need to offer higher specificity due to the complex nature of biologics, the complicated interactions of these classes of pharmaceuticals with biology, and the need for targeting specific cells or tissues. Biopharmaceuticals require specialized drug delivery systems due to their sensitivity to environmental conditions such as temperature and pH, which are critical to preserving stability and efficacy.

Using nanocarriers in the NDDSs context is considered to become a reliable strategy to improve their safety profile as well as therapeutic efficiency. In this way, integrating nanotechnology with biopharmaceuticals holds great potential for future drug delivery approaches. In this regard, recent developments and advancements have focused on engineering nanostructures for specialized delivery of biopharmaceuticals, which have been accomplished using a variety of strategies, such as physical fabrication, covalent molecular modification, encapsulation by biodegradable hydrogels based on natural and synthetic polymers, and nano-structuring processes.

Since intracellular targets are the main sites for establishing the therapeutic effects of biopharmaceuticals, it is crucial to develop drug delivery systems that can effectively transport biopharmaceuticals to their intended sites for promoting the necessary intracellular interactions. Endosomal escape has been a traditional strategy for intracellular delivery, typically called cytosolic delivery, of biopharmaceuticals; however, such a strategy is limited by the low release rate of the cargos in the cytosolic environment (less than 2%). Direct internalization is the other cytosolic delivery route that enables releasing 100% of the entrapped biopharmaceutical to the cytoplasm. This strategy involves the caveolae-mediated pathway, contact release, intermembrane transfer, membrane fusion, direct translocation, membrane disruption, and microinjection.

Nevertheless, safety concerns still need to be taken into account. Moreover, achieving the development of a scalable, flexible, and biocompatible nanocarrier remains to be considered as a significant issue.

13.3.4.2 Combinatory Drug Delivery Systems

Combinatory drug delivery systems, also known as combination therapies or co-delivery systems, aim to overcome the limitations of single-drug therapies and provide synergistic or complementary effects. These systems can be designed in various types, including fixed-dose combinations (FDC), in which two or more drugs are combined in a fixed ratio within a single dosage form, sequential delivery systems, in which the delivery pattern of drugs is sequential, and simultaneous delivery systems, where either a single dosage form or multiple dosage forms are delivered at the same time.

A great deal of research has been conducted on the use of such types of co-delivery systems for cancer therapy, mainly including liposomes, micelles, dendrimers, inorganic nanocarriers, hydrogels, and antibody-drug conjugates (ADCs). In the meantime, extensive studies have been carried out on liposomes in the development of co-delivery for anti-cancer drug delivery. However, a few commercial liposome-based formulations have gone to clinical trials and been able to complete phases one or two of clinical trials.

Moreover, combinatory drug delivery systems may be an emerging approach for overcoming cancer multidrug resistance (MDR) in the future, as they regulate the MDR pathway in tumor cells, protect unstable and easily decomposed drugs, and reduce side effects on normal tissues.

Nevertheless, there are significant limitations to developing and manufacturing these systems, including the time-consuming and expensive synthesis process on a large scale as well as careful consideration of biosafety and drug delivery efficiency.

13.3.4.3 Theranostics

Theranostics, as a rapidly growing field, combines therapy and diagnosis in a single platform and has great potential for personalized medicine and targeted therapy by enabling accurate disease detection, real-time monitoring, and customized drug administration. Actually, they are drug delivery systems incorporated by some kind of molecular detection, biosensor, or imaging agents. Theranostic drug delivery systems are expected to become more reliable due to improved imaging techniques as well as biomarkers and nanotechnology advancements. These platforms will allow clinicians to adapt therapies to individual patients, resulting in improved outcomes and lower costs for healthcare. A broad range of nanocarriers, including liposomes, cyclodextrin-based inclusion complexes, conjugates and molecular complexes, dendrimers, vesicles, micelles, core-shell particles, microbubbles, carbon nanotubes, metal-organic frameworks, mesoporous silica and so on can be applied in the preparation of a wide variety of theranostic drug delivery systems.

Theranostics has been applied to a wide range of diseases, including rheumatoid arthritis and inflammatory bowel disease (IBD), cancers, cardiovascular diseases, neurodegenerative diseases, and infections, to improve therapeutic outcomes by discovering related specific biomarkers or genetic signatures and profiles and employing advanced imaging techniques such as PET, SPECT, and MRI. Also, theranostic may be particularly efficient in bacterial infections, in such a way that by recognizing a specific bacteria metabolite, the antibacterial agents would be activated and a localized treatment would occur.

Toxicity and safety issues are the main limitations of theranostic drug delivery systems, hence more toxicology and biocompatibility studies on these systems should be pursued to have efficient systems in the future. Furthermore, an accurate combination of biomarkers and therapeutic agents in such delivery systems needs great attention to achieve favorable therapeutic efficacy.

13.3.4.4 Bio-Inspired Drug Delivery Systems

Bio-inspired drug delivery systems are innovative platforms with inspiration from nature and, thus, can mimic biological systems to deliver drugs to specific target sites in the body. These systems are designed to improve drug efficacy, reduce adverse effects, and enhance patient compliance. Improving therapeutic efficiency can be achieved by taking inspiration from natural nano-topographical materials and bio-interfaces, due to the fine-tuning of bio-functionalities like bio-adhesion and cellular uptake. Such systems can be applied to all types of drug carriers, such as thin films, patches, implants, stents, and micro- and nanocarriers, for the treatment of a wide range of diseases, including cancer, diabetes, infectious diseases, and regenerative medicine. These systems can be engineered to release drugs in response to specific stimuli, such as pH, temperature, or enzyme activity, thereby allowing for precise control over drug delivery.

The fabrication techniques for nano-topography engineering of drug carriers include both top-down and bottom-up methods such as lithography, polymer demixing and porogen leaching, electrospinning, chemical and electrochemical etching, surface roughening and particle deposition, and all other methods and platforms of particle fabrication (e.g., emulsion, crystallization, microfluidic, precipitation, self-assembly, hydrogelation, and micromoulding).

Recently, various types of bioinspired drug carriers have been developed, like virus-inspired spiky particles, which can enhance both cell targeting and uptake as well as subcellular trafficking, pollen- and gecko-inspired bio-adhesive particles, devices that can affect the retention of drug carriers and lead to local and sustained drug release, and insect-inspired antimicrobial nano-spikes, which can release drugs through bacterial membrane disruption and cell lysis.

Wet adhesion is one of nature's inspirational concepts. It has been also used as a basis for the design and manufacturing of drug delivery systems using nano-topography platforms. The main types of wet adhesion that are used in targeted drug delivery systems are chemical adhesives based on molecular interactions (like those found in muscles), mucus-covered nano-pillars and hierarchical pillars based on capillarity and Stefan adhesion (like those found in male diving beetles and tree frogs), suckers based on suction (like those found in octopuses), setae based on friction (like those found in may fly larvae), and hooks based on mechanical interlocking (like those found in teleosts).

The SOFUSA nano-topographical microneedle patch and the Nano+ (Lepu Medical) drug-eluting stent are the bioinspired systems that have been commercialized and evaluated in clinical studies.

The bioinspired adhesives provide promising prospects for the future of targeted drug delivery through oral and skin, including micro/nano-motors and drug delivery patches.

However, the main challenge in the nano-topographical fabrication of drug carriers is achieving a proper balance between scale-up and reproducibility, which is not currently attainable with the all aforementioned approaches.

13.3.5 Emerging Technologies

The impact of emerging technologies on the advancement of the NDDSs is self-evident. In fact, these technologies are the main driving forces for the progress of these systems. This influence is such that it is one of the two main factors pushing the NDDSs ahead, i.e., technology push and market pull. Among the most influencing mega-technologies on the progress of these delivery systems, nanotechnology, biotechnology, material technology, medical technologies, and information technology can be mentioned. It is obvious that any discovery and achievement in any of the fields described can potentially open new horizons beyond the existing boundaries of different branches of the NDDSs. The advancements in those technologies can potentially provide new knowledge, insight, vision, tools, and materials for preparation of the NDDSs with advantages in application, pharmacokinetic features, and, finally, the clinical outcome of the currently available NDDSs.

13.3.6 Interdisciplinary Development

NDDSs have presented indubitable benefits for drug administration in the last half-century. This field has been a convergence point for different scientific/technological disciplines with the ultimate goal of developing customized treatments for many diseases, including cancer and diabetes. An important advantage of interdisciplinary development is the possibility of generating novel and revolutionary results. By combining knowledge across several fields, researchers and practitioners can produce novel ideas, thoughts, and solutions that would not have been attainable within the confines of a single discipline.

This approach promotes ingenuity, challenges the limits of understanding, and cultivates novel modes of cognition. By integrating diverse disciplines such as pharmaceutics, nanotechnology, clinical medicine, biochemistry, physiology, anatomy, histology, chemical engineering, molecular biology, artificial intelligence, polymer science and engineering, tissue engineering, molecular genetics, material engineering, biomaterials, genetic engineering, biotechnology, pharmacology, robotics, chemistry, physics, material science, engineering, bioengineering, and biology, researchers have been able to create novel approaches for the delivery of medications by NDDSs thus offering better safety and efficacy for the active drug molecule. Shortly, the development of highly effective customized drug delivery systems is expected due to advancements in various interdisciplinary fields. These systems, indeed, utilize science and technological progress to exert some kind of control to intelligence on the time and/or place of drug release within the host body. These systems can be customized according to particular patient traits, leading to enhanced treatment results.

13.4 Concluding Remarks

A series of both challenge-based as well as opportunity-based trends for the evolution of NDDSs are reviewed in this chapter with the aim to draw the attention of researchers, policy-makers, the pharma industry, and the global technology ecosystem to these driving forces. The real spirit behind the ever-improving feature of NDDSs is two main mega-factors: i) the close interrelationship between the advancements made in pharmaceutical and related science and technology and the expansion of NDDS atmosphere, and ii) the high and ever-growing demand of the pharma market for the NDDSs with superior characteristics collectively focused on the two main determinants of safety and efficacy of the pharma products. From the first face, the emergence of new technologies and applications as well as the expanding vision and insight into the host body's biological features pushes the NDDSs ahead to achieve more favorable products and present them to clinical uses. The second face of the coin shows the expanding needs and expectations of the target market which, again, reflects somehow the progress made in medical sciences in terms of the availability of more and deeper knowledge and insights about diseases and treatments.

The world of NDDSs, with no doubt, will continue its fast-growing trend while being inspired and powered by the scientific and technological evolutions that present new concepts, techniques, tools, and materials to feed the promising journey of the NDDSs as one of the future engines of pharmaceutical innovation.

Index

Printed in the USA/Agawam, MA
September 16, 2026

910661.151